THE PHILOSOPHY OF ROBERT GROSSETESTE

Paten, chalice, episcopal ring, metal ring linking staff to head of crozier, found in Grosseteste's tomb in 1782

(Photograph by S. J. Harrop. Reproduced by kind permission of the Dean and Chapter of Lincoln.)

THE PHILOSOPHY OF
ROBERT GROSSETESTE

JAMES McEVOY

CLARENDON PRESS · OXFORD

Oxford University Press, Walton Street, Oxford OX2 6DP

London New York Toronto
Delhi Bombay Calcutta Madras Karachi
Petaling Jaya Singapore Hong Kong Tokyo
Nairobi Dar es Salaam Cape Town
Melbourne Auckland
and associated companies in
Beirut Berlin Ibadan Mexico City

Published in the United States by
Oxford University Press, New York

© James McEvoy 1982
First published 1982
Reprinted (New as Paperback) 1986

British Library Cataloguing in Publication Data
McEvoy, James
The philosophy of Robert Grosseteste.
1. Grosseteste, Robert
1. Title
180.0942   B765.G7
ISBN 0–19–824645–5
0–19–824939–X (pbk)

Library of Congress Cataloging in Publication Data
McEvoy, J. J.
The philosophy of Robert Grosseteste.
Bibliography: p.
Includes index.
1. Grosseteste, Robert, 1175?–1253.   I. Title
B765.G74M38   192   81–22438
ISBN 0–19–824645–5
ISBN 0–19–824939–X (pbk)

Printed in Great Britain
at the University Printing House, Oxford
by David Stanford
Printer to the University

Patri et doctori suo
Theodore Crowley
filio fideli Sancti Francisci
dedicavit opus discipulus.

Vicem amicitiae
vel unam me reddere
oportebat tempore . . .
(Planctus, *Peter Abelard*)

Preface

OFTEN it is the faces of bare acquaintances that are more easily called to mind, whereas of those whom we love, whom we have seen variously, in different lights, and over many years, no unambiguous image will form at will before the mind's eye. Later and deeper acquaintance with Robert Grosseteste so crosses and covers the unfocused knowledge of earlier years that I cannot now remember exactly when, where, and what I first learned of him. When I began to look more closely at him (with the predatory eye of the young thesis-hunter) it was the rather definite profile of the translator from the Greek that caught my attention; only gradually did I begin to be aware that he was much more besides. As my knowledge of him grew so did my admiration, even awe. Grosseteste's achievements were numerous and were spread over several domains of thought and activity, his contribution to any one of which would have ensured him a place in history. Sir Maurice Powicke hailed him as the greatest of Oxford's sons; and indeed he contributed more to the foundation of the intellectual tradition of Oxford than did any other of her early masters or chancellors. Franciscans everywhere know him as the founder of the school which, in the course of time, produced Roger Bacon, Duns Scotus, and William of Ockham. Historians of Scholasticism give him an honoured place in the nascent Aristotelean movement in the Europe of the early thirteenth century. Historians of science underline his intellectual leadership in the movement which culminated in the emergence of the early modern mathematical science of nature. General historians of the period admire in him the administrator of the largest diocese in medieval England, the leading personality in the contemporary episcopate of the country, and the energetic reformer of ecclesiastical abuses.

Grosseteste was a man so far outside the common run that his own contemporaries found it difficult to evaluate him and impossible to agree as to where precisely his greatness lay;

that he was great, no one, not even his opponents, could doubt. It is still true that he was one of those medieval figures of universal import who remain irrevocably particular: for the historian, he does not conform to type, either in the shape of his career or in his writings. He was a voluminous writer, yet the first half of his life appears to have been unproductive, so that had he died at fifty (a respectable age in those times) he would have been remembered only hazily, as the first known Chancellor of the University of Oxford. Only during the last forty years of a very long life did he become productive; but then, with the vigour and energy of a much younger man, he proved to be prolific in half a dozen genres. What is one to make of a man who, at sixty years of age, conscious though he was of declining powers of memory, took up the study of Greek and established an atelier for the production of translations, and who was learning Hebrew in his eighties? Late development is generally reckoned, after all, to have its limits.

It is more than sixty years since the last overall discussion of Grosseteste's philosophical ideas appeared (Ludwig Baur, 1917). In the meantime, historical studies bearing on the authenticity and chronology of the writings attributed to him, together with editions, and the re-examination of important areas of his thinking, have altered the older picture to the point where a new assessment of his philosophy appears at once possible and required, if we are to reach a better understanding, not only of the significance of his ideas, but also of the salient directions of the intellectual movement of his times. I hope to have provided in this volume some provocation to the deeper and fuller study of his writings.

I first took up the study of Grosseteste at the insistence of my teacher and predecessor at Queen's University, Belfast, Professor Emeritus Theodore Crowley, who is known for his work on his fellow-Franciscan, Roger Bacon (the reversed symmetry of master and disciple did not strike me at the time); but one does not thank a friend, one acknowledges him for what he is. It is my hope that the dedication of this book to him will give him much satisfaction.

My other acknowledgements go mostly in the direction of Belgium and Germany. I owe much to Professor Fernand Van

Steenberghen for the freedom of research which he encouraged, the deep interest he took in the matter and form of my doctoral thesis, and the sharpness of his critical eye; that has spared me many an embarrassment. Over the years I have run up a debt with Father Servus Gieben, that is as difficult to define as it is impossible to repay. I gratefully acknowledge the use which I have made of his unpublished doctoral work; but that is only the heading on my bill. Surely few who put so much of themselves into their work are so generous in sharing its fruits, even unpublished ones, with others. For all his undocumentable help, in particular, my thanks, although inadequate, are sincerely returned.

Of my professors at Louvain I have many happy memories of kindness and indulgence and I wish to thank Professor Suzanne Mansion, whose sudden death has deprived us of a remarkable Aristotelean scholar, for ready and practical help, given as director of the *Centre de Wulf-Mansion*; Professor Georges Van Riet, who presided at my *défense de thèse*; Professor Simone Van Riet, who found me ignorant of palaeography and did her best to enlighten me (*sed ars magistrae discipuli stultitia minuitur*); Professor Christian Wenin, whose benevolent kindness to foreign students was always in evidence, and Professor Herman Van Breda, in whom I (like so many others) lost a friend full of humour, inspiration and resource; he died within a few days of our leave-taking at Louvain. I am permanently indebted to my teachers at the *Institut Supérieur de Philosophie* and the *Institut d'Etudes Médiévales* there.

Two periods of scholarly hospitality in West Germany marked the progress and the termination of this work. For twelve months in 1972 and 1973 I studied as a *Stipendiat* of the *Deutscher Akademischer Austauschdienst* at the *Grabmann Institut* of the University of Munich, experiencing the kindness of Professor W. Dettloff and his team. The last additions to the volume have been made in the same happy circumstances, as the guest of the *Alexander Von Humboldt-Stiftung*, during sabbatical leave from Queen's University. To both the *DAAD* and the *Stiftung* my warmest thanks are returned. Queen's University and the Ministry of Education, N. Ireland helped to finance my years of postgraduate study at Louvain and awarded grants for the purchase of microfilms, for which I

am grateful. Oxford University Press were generous in allowing me space, their readers helpful and their representatives
unfailingly courteous. The British Academy made a useful
grant towards the cost of preparing the typescript.

Many deserve my thanks, who cannot all be named here.
My Bishop, Dr Philbin, never questioned the years I spent in
research but only encouraged me. I have lived for five years
with the Franciscans, in the Irish College at Louvain and at
St. Annastrasse in Munich; we have borne each others' burdens
and mine has been the lighter for it; but here also Grosseteste
created the precedent. The Librarian and staff of Queen's have
always been helpful. My pupil, Miss Fiona Lynch M.A., read
most of the chapters in typescript and made innumerable
stylistic improvements; my warmest thanks are due to her, as
also to Mrs Smyth, who did most of the typing, and numerous
friends and colleagues who helped with the proof-reading.

From all these thanks, and the many more which cannot
be expressed here, it is evident that no work is of one man
alone. An author's mistakes are the only things that belong to
no one but himself.

James McEvoy
Department of Scholastic Philosophy
Queen's University,
Belfast

Table of Contents

Abbreviations

Works of reference

Catholicisme	*Catholicisme Hier, Aujourd'hui, Demain* (Paris).
BGPM	*Beiträge zur Geschichte der Philosophie [und Theologie] des Mittelalters* (Münster i. W.).
Corp. Chr. C.C.	*Corpus Christianorum. Continuatio Mediaevalis* (Steenbrugge).
CSEL	*Corpus Scriptorum Ecclesiasticorum Latinorum* (Vienna-Leipzig).
DHGE	*Dictionnaire d'Histoire et de Géographie ecclésiastiques* (Paris).
DS	*Dictionnaire de Spiritualité* (Paris).
DTC	*Dictionnaire de Théologie Catholique* (Paris).
Pauly-Wissowa, *Realencyclopädie*	*Paulys Real-Enzyclopädie der Klassischen Altertumswissenschaft* (neue Bearbeitung von Georg Wissowa) (Stuttgart).
P.G.	Migne, *Patrologia Graeca* (Paris).
P.L.	Migne, *Patrologia Latina* (Paris).

Books and articles

Baur, *Die Philosophie* . . .: Ludwig Baur, *Die Philosophie des Robert Grosseteste, Bischofs von Lincoln (+1253),* (*BGPM* Bd. XVIII, H. 4-6) (Münster i. W., 1917), XVI + 298 pp.

Baur, *Die Werke*: Ludwig Baur (ed.), *Die philosophischen Werke des Robert Grosseteste, Bischofs von Lincoln.* Zum erstenmal vollständig in kritischer Ausgabe besorgt von Ludwig Baur (*BGPM* Bd. IX) (Münster i. W., 1912), XV + 183 + 778 pp.

Epistolae, ed. Luard: *Roberti Grosseteste Episcopi quondam Lincolniensis Epistolae*. Edited by Henry Richards Luard (Rerum Britannicarum Medii Aevi Scriptores) (London, 1861), CXXXIII + 467 pp. (reprinted New York, 1965).

Franceschini, *Roberto Grossatesta, vescovo di Lincoln*: E. Franceschini, *Roberto Grossatesta, vescovo di Lincoln, e le sue traduzioni latine*. (Atti del Reale Istituto Veneto di scienze, lettere ed arti. Anno accademico 1933-4. T. XCIII, parte seconda) (Venice, 1933), 138 pp.

Scholar and Bishop (ed. Callus): *Robert Grosseteste, Scholar and Bishop*. Essays in Commemoration of the Seventh Centenary of his Death. Edited by D. A. Callus. With an Introduction by Sir Maurice Powicke (Oxford, 1955), XXV + 263 pp.

Thomson, *The Writings*: S. H. Thomson, *The Writings of Robert Grosseteste, Bishop of Lincoln 1235-1253* (Cambridge, Mass., 1940), XV + 302 pp.

Periodicals*

Arch. Franc. Hist.	*Archivum Franciscanum Historicum* (Firenze).
Arch. Hist. Doctr. Litt. M.A.	*Archives d'histoire doctrinale et littéraire du moyen âge* (Paris).
Bodl. Libr. Rec.	*Bodleian Library Record* (Oxford).
Bull. J. Ryl. Libr.	*Bulletin of the John Rylands Library* (Manchester).
Coll. Franc.	*Collectanea Franciscana* (Roma).
Engl. Hist. Rev.	*English Historical Review* (London).
Ét. Franc.	*Études Franciscaines* (Paris).
Franc. Stud.	*Franciscan Studies* (St. Bonaventure, N.Y.).
Giorn. Metaf.	*Giornale di Metafisica* (Torino).
Harv. Theol. Rev.	*Harvard (The) Theological Review* (Cambridge).
Journ. Eccles. Hist.	*Journal (The) of Ecclesiastical History* (London).
Journ. Theol. Stud.	*Journal (The) of Theological Studies* (Oxford).

---

* Titles consisting of a single word have not been abbreviated, nor have those which have been quoted only once.

Med. Human.	*Medievalia et Humanistica* (Boulder, Colorado).
Med. Renaiss. Stud.	*Medieval and Renaissance Studies* (London).
Med. Stud.	*Mediaeval Studies* (Toronto).
Miscell. Med.	*Miscellanea Mediaevalia* (Berlin).
New Schol.	*New (The) Scholasticism* (Washington).
Phil. Jahrb.	*Philosophisches Jahrbuch* (Fulda).
Proc. Am. Cath. Phil. Ass.	*Proceedings of the American Catholic Philosophical Association* (Washington).
Proc. Brit. Acad.	*Proceedings of the British Academy* (London).
Rech. Théol. anc. méd.	*Recherches de Théologie ancienne et médiévale* (Louvain).
Rev. Néoscol. Phil.	*Revue néoscolastique de philosophie* (Louvain).
Riv. Crit. Stor. Fil.	*Rivista critica di storia della filosofia* (Milano).
Riv. Fil. Neo-scol.	*Rivista di filosofia neo-scolastica* (Milano).
Studi Franc.	*Studi Francescani* (Arezzo/Firenze).
Studi Med.	*Studi Medievali* (Torino).

PART ONE

THE MAN AND THE SCHOLAR

Chapter 1
A Portrait of Robert Grosseteste

ROBERT Grosseteste, or Greathead, Bishop of Lincoln, died in the night of the eighth to the ninth of October, 1253, leaving behind him a reputation for learning and sanctity which had spread in his lifetime, not through England alone, but throughout Europe. The Franciscan chronicler, Salimbene, regarded him as one of the greatest clerics in the world. Even Matthew Paris, the Benedictine chronicler of St. Albans, though not in all respects well disposed to Grosseteste, referred to him as a saint in the entry recording the bishop's death. Roger Bacon, his devoted admirer, never tired of praising him as the most illustrious scientist and translator of the Schools. Humbler admirers such as the Franciscan, Thomas of Eccleston, and the Friar Hubert who composed a verse tribute to the dead bishop, praised the great teacher, the distinguished and humane guide of souls, the dedicated pastoral bishop and the fearless reformer of the Church. The Schools, not of England merely but of all Christendom, knew him simply as *Lincolniensis* and held his authority in philosophical and theological matters in high regard in pre-Reformation days. The ordinary faithful of the Diocese of Lincoln revered his memory for upwards of a hundred years after his death and spread far and wide the stories of miracles wrought at his tomb in the cathedral and attributed to his intercession. Yet despite the persistence of his memory in later medieval times, his unique place in English Church life, and the fame of his voluminous writings, Grosseteste found no biographer until John de Schalby included him in his *Lives of the Bishops of Lincoln*, written around 1330. The result is that little is known of the shape of his life up to the year 1229, twenty-four years before his death at a very advanced age. Only after that year, when he was nominated Archdeacon of Leicester and was requested by the Franciscans to lecture at their newly-built school at Oxford, does the documentary evidence, in the form of correspondence and other dateable writings, journeys, official acts and public interventions in important affairs of the Church and the Realm of England, begin to abound, since for the last period of his life he was a public figure. However, the historian who sets out to provide even a sketchy biography of him as a prologue to an assessment of his character and the impact of his personality cannot be dispensed from attempting an outline of the general shape of his life, despite the host of minor problems that surround the earlier part of it.

We are not entirely without evidence upon which to hang the narrative of Grosseteste's life up to the year 1229. The first appearance of his name in literature is the mention by Gerald of Wales, in one of his letters, that Master Robert Grosseteste was attached to the *familia* or household of Bishop William de Vere of Hereford.[1] We know from this that Grosseteste was a Master (of Arts, presumably), some time before the death of the bishop in 1198. Gerald's encomium of the young clerk's abilities is generous: he praises his proficiency in the arts and his wide acquaintance with literature, and adds that he is able in the handling of various affairs, in the determination of causes and in securing bodily health. In plain language, he had some skill in canon law (at least, perhaps in civil also), and some knowledge of medicine.

The appearance of the name *Magister Robertus Grosteste* as a witness to a charter dating from the reign of Bishop Hugh of Lincoln (1186 to 1200), and written between 1186 and 1189-90, encourages the placing of Grosseteste's year of birth before 1170, since if he was a Master of Arts in 1189 at the latest, and must by then have been not younger than the early twenties, he cannot well have been born after around 1168.[2]

---

[1] Giraldus Cambrensis, *Opera*, ed. J.S. Brewer, vol. I (London, 1861), p. 249.

[2] Callus argues cogently for this conclusion: "Robert Grosseteste as Scholar", in D. Callus (ed.), *Robert Grosseteste, Scholar and Bishop* (Oxford, 1955), pp. 3-4. (References to this work will be given henceforth to Callus, *Scholar and Bishop*). My debt to this article of Fr Callus, and also to his study, "The Oxford career of Robert Grosseteste", *Oxoniensia* 10 (1945), 42-72, is heavy.

It is the (as yet unpublished) view of Sir Richard Southern and Mrs Cheney that the *Magister* who witnessed the charter in question was another Robert Grosseteste, otherwise unknown to us, and that the date of birth of the famous Grosseteste should be placed around 1175, just as the biographical tradition had it before Callus proposed to bring it back to 1168 or earlier. As I understand it, this view is meant to make better sense of the early career of Grosseteste by reducing in length the period concerning which we cannot do more than conjecture, and at the same time to render less miraculous-seeming his learning of Greek, the great volume of his writings stemming from the last twenty years of his life, and his

The identification of this unbeneficed young master with the master of the same name referred to by Gerald of Wales cannot be positively established, but the relative uncommonness of the patronymic gives it some plausibility; and we know in any case from one of Roger Bacon's allusions to the length of his life (*'propter longitudinem vitae suae'*), that he attained an advanced age. If the witness to Bishop Hugh's charter was indeed our Robert Grosseteste, then he must have been at least eighty-five years old at the time of his death in 1253, and already sixty-seven or more at the time of his elevation to the See of Lincoln.

Of the years between the death of Bishop de Vere and 1225 we can say nothing with certainty concerning the outer course of Grosseteste's life. In that year, on April 25, he was presented by the Bishop of Lincoln to the Church of Abbotsley. In the deed of presentation his name appears without the title of any other benefice, and he is given as deacon. The presentation of a deacon to a church benefice need not disturb us, for it was common. We do not know positively how Grosseteste had supported himself up to that date, or where he had passed his time; we must work to a large extent on general assumptions to fill out certain hints supplied by tradition.

Although Grosseteste was born into the poorest class of feudal society, according to the unanimity of the chroniclers, Matthew Paris tells us—and there is no reason on this occasion to disbelieve him—that he was 'educated in the schools from his earliest years'. He was born in Suffolk, but the tradition associating his family with Stradbroke is too late to be given

---

prodigious outpouring of energy as a bishop (1235-53). The hypothesis of the contemporaneous existence of two masters of the same name represents, of course, a possibility which cannot be ruled out altogether, especially since the name appears to have become an established patronymic by the end of the twelfth century—two instances of its occurrence have been authenticated (Callus, *Scholar and Bishop*, p. 2). However, until documentary evidence can be adduced which positively indicates the existence of a second Master Robert Grosseteste, the sole grounds for the hypothesis remain *a priori* ones. To me it seems scarcely more likely that a man of fifty-eight years (b. 1175) should have learned Greek than that someone already in his middle sixties (b. 1168 or shortly before) should have done so. In the famous Robert Grosseteste one is dealing with a very exceptional human being, to whom general considerations are *eo ipso* less applicable than they are to those closer to the mean. Moreover, if there was only one Robert Grosseteste, then the great age at which he died (eighty-five years at least) would encourage us to take more seriously than anyone has thus far done the praise of Gerald of Wales for the medical knowledge of the young master whom he met in the 1190s.

much credence. Of his family we know only that one of his
sisters, Ivette, became a nun, and that two of his kinsmen
wished to enter the diocese while he was bishop; his great
friend Adam Marsh wrote to him on their behalf.[3] Perhaps the
two Grossetestes who figured in his Rolls were these, or other
kinsmen of his. Thomas Gascoigne, who was Chancellor of
Oxford in the early fifteenth century and always a great admirer of his early predecessor in that office, asserts that
Grosseteste was a Master of Arts of Oxford, and we may accept
his unsupported word, in the lack of any contrary evidence.
Grosseteste's presence at Hereford is tantalizing, for the
Cathedral School of that city was celebrated in a surviving
poem of Simon du Fresne, addressed to Gerald of Wales around
1195 to 1197, as a flourishing centre for the study of the liberal arts, astrology and the natural sciences, theology, and
both laws. The association of Roger of Hereford and Alfred of
Sareshel with the school, and the possibility that Daniel of
Morley was active in the study of Arabic science there before
1187, tempt us to believe that Grosseteste, living in the same
city, made contact with the relatively advanced scientific
tradition which these masters represented, and continued to
further his studies while in the bishop's employ. It is worth
recalling in this connection that Gerald of Wales praised the
young master's knowledge of law, a subject which featured in
the curriculum of the School of Hereford, and of medicine;
now, it was through medical writings that Arabic science made
its earliest impact on the Latin twelfth century.

It is not unlikely that sometime after the death of William
de Vere on 24 December 1198 Grosseteste went to Oxford to
teach the arts. It is known that the nucleus of three faculties,
of theology, arts, and law, were there by around 1200, and
that the 'new' Aristotle was lectured on at Oxford before 1209.
The *suspendium clericorum* of 1209 to 1214 interrupted many
a career there, for while some masters continued to teach
during it, and were later penalized for their lack of solidarity
with the rest, prominent masters such as John Blund and
Edmund of Abingdon migrated to Paris to take up the study
of theology. The way to Paris was an easy one, and already
a traditional one for scholarly English clerks, and it is quite
possible that Grosseteste chose it. Bulaeus asserts that he

[3] Adam de Marisco, *Epistolae*, ed. J.S. Brewer (London, 1858), p. 137.

pursued and finished his theological studies at Paris, and this late testimony fits well enough with a number of much earlier hints to be acceptable.[4] When Grosseteste wrote to the Oxford Regents in theology as their Ordinary (after 1240) he showed familiarity with the course of theological studies at Paris, and moreover he recommended the same programme as a model for Oxford. The general character of his exegetical works is in harmony with the methods and outlook of what Powicke called the biblico-moral school of Parisian masters like Stephen Langton (unfortunately it has not proved possible as yet to identify Grosseteste's teacher at Paris), and is consistent with the hypothesis that he had studied at Paris. Furthermore, Grosseteste counted among his many friends a number of Frenchmen, including prominent figures like Thomas Gallus (*Vercellensis*), William of Auvergne (Bishop of Paris 1228–49), and William of Cerda, and these may have been the companions of his student years at Paris. Matthew Paris certainly believed that Grosseteste was known to the French clergy as well as to his countrymen, and seems to have assumed that he had lived in France. He puts into the mouth of Cardinal Gilles de Torres, the Archbishop of Toledo, a laudatory reference to the fame of Grosseteste's learning and devout life among the French and English clergy. The discourse which he attributes to Grosseteste on his death-bed is in many major respects suspect, but Matthew had no evident reason simply to phantazise the following element in it:

Caursini enim manifesti usurarii, quos sancti patres et doctores nostri, quos vidimus et audivimus, videlicet magister eximius in Francia praedicator, abbas quoque de Flay Cisterciensis ordinis, magister Jacobus de Viteri, Cantuariensis archiepiscopus Stephanus exulans, magister Robertus de Curcun, praedicando a partibus Franciae, quia antea hac peste Anglia non laborabat, ejecerunt, iste Papa [that is to say, Innocent IV] suscitavit et protegit suscitatos.[5]

Matthew is stating here as a fact something which a large number of scholars and clergy of his time could have denied had it been untrue, namely that the bishop was in a position to claim personal acquaintance with outstanding figures of the French Church of the previous generation: the Cistercian Abbot, Eustace of Flay, Robert Curzon, who was regent at Paris from 1204 until 1210–11, Stephen Langton *exulans*,

[4] Bulaeus, *Historia Universitatis Parisiensis* III, pp. 154, 260, 709.
[5] Matthew Paris, *Chronica Majora*, vol. V, p. 404.

who was absent from England from 1207 until July 1213 and was in Paris in 1211, and the illustrious preacher, Jacques de Vitry. Presumably Paris's phrase, *'patres et doctores nostri'*, was intended to carry its full meaning with regard to several of these famous names.

A final straw in the wind is Grosseteste's ignorance of the *libri naturales*, which he apparently only began to study in depth after *c*.1220. This would fit quite well with a Parisian course of study during the years after the first prohibition of the Aristotelean physical writings (by the same Curzon referred to above, acting as papal legate, in 1210), if we assume that conservative theological circles at Paris approved the ban on the public reading of these works in the faculty of arts, until they should be corrected, and gave but small encouragement to their private circulation. Only a dozen or more years after the ban did the more adventurous theologians give evidence in published writings of a more thorough study of the natural works; and only in the course of the decade 1220–30, when there is good evidence that Grosseteste was already back at Oxford, did he take up the study of Aristotle's physical treatises with an enthusiasm which was at no stage uncritical. Had he remained at Oxford during the secession he might have furthered his studies of Aristotle's natural philosophy, but it is hard to see how he could then have become Master of the Schools, since he would have lost the trust of his academic colleagues; and those precious contacts and friendships with outstanding French scholars and personalities might never have developed.

That Grosseteste was Master of the Schools at Oxford we know through the clear and trustworthy assertion of Bishop Oliver Sutton, his successor in the See of Lincoln (1280–99), whose testimony lends unequivocal support to the tradition that Grosseteste was chancellor, even the first chancellor of the university:

Bishop Robert, formerly Bishop of Lincoln, who occupied this office [i.e. the chancellorship] while he was regent in the same university, said when he was just being elevated to the episcopate that his immediate predecessor as Bishop of Lincoln did not permit the same Robert to be called chancellor, but Master of the Schools (*Magister Scholarum*).[6]

What are we to make of the assertion that Grosseteste was

[6] H.E. Salter, *Snappe's Formulary*, O.H.S. lxxx (Oxford, 1923), p. 52.

chancellor but was obliged by his ordinary to retain the title traditional for the head of the Oxford Schools? One conclusion is sufficiently evident: Grosseteste was a regent at the time, for only a regent master in theology would be nominated to the responsibility of overseeing all the faculties in the name and by the appointment of the *ordinarius*. Furthermore, the statement must refer to the university which was chartered by the Ordinance of the Cardinal Legate, Nicholas, Bishop of Tusculum, in 1214, the act which marks the end of the *suspendium* and the re-establishment of the guild of masters and scholars in the city. The reference made by this charter to 'the chancellor whom the Bishop of Lincoln shall set over the scholars'[7] makes it clear that the establishment of a chancellorship was envisaged in the Legatine Ordinance. Sutton's statement presupposes that the office of chancellor existed at the period of which Grosseteste was speaking, which then cannot be before 1214, but that some controversy or difficulty arose concerning the title of its holder. We can only guess at the reason for the bishop's decision. Callus surmises that the bishop's refusal to recognize the title 'shows assuredly a period of transition when the status of chancellor was not yet definitely settled, in all probability on the occasion of the first appointment to the new office.'[8] This is plausible. It is worth remembering that the chancellorship by its whole nature was a diocesan appointment, and the man on whom it was conferred attained *ipso facto* a status within the Diocese of Lincoln. Now if Grosseteste was considered to be the man best fitted to lead the new university, a regent newly returned from Paris, still the evidence is that he was not yet a priest but only at most in deacon's orders (as his presentation to a benefice in 1225 records); and priest's orders were normally conferred in direct association with a *cura animarum*. Bishop Hugh of Wells may have had his reasons for not wishing to appoint a deacon to an important, newly-established office within the diocesan structure, and may at the same time not have wished immediately to confer on him a benefice with cure of souls until the whole nature and the practical working of the new and exceptional post, as well as the full qualities of its occupant, were more fully assessible.

---

[7] The text of the ordinance is published in *Medieval Archives of the University of Oxford*, ed. Salter, I, p. 4.

[8] Callus, "Robert Grosseteste as Scholar", in *Scholar and Bishop*, p. 9.

What seems quite certain is that the title of chancellor was in use by 1221, for in that year Pope Honorius III addressed a letter to the Chancellor of Oxford. This may well have been Grosseteste himself, for no fixed term of office was assigned the chancellorship during the thirteenth century, and we do not know at what date he ceased to hold office, though it must have been before 1231, when Ralph of Maidstone was chancellor. The title does not appear on any of Grosseteste's letters. The earliest of these that survive dates from *c.*1226-28 and Grosseteste styles himself *Magister* (*Ep.* 1); letters 2 to 6 inclusive are headed 'Robert, Archdeacon of Leicester', an office he occupied from 1229 to 1231; and letter no. 8 is again superscribed *'Magister'*, after Grosseteste had resigned his archdeaconry.

We can be tolerably certain, then, that Grosseteste was the first head of the Oxford Schools after the creation of the university by the Legatine Ordinance of 1214; he was the first occupant of the new office of chancellor. His association with the university probably continued unbroken until 1235, when he was elevated to the See of Lincoln. He was at Oxford during the visit of Jordan of Saxony, St. Dominic's successor; he, the chancellor and other masters obtained royal pardon for some students imprisoned for forestry offences, in 1231; and together with the chancellor and Robert Bacon O.P. he was nominated to deal with some disciplinary matters, in 1234.[9]

We shall see in due course how Grosseteste's intellectual interests evolved during these years of hidden life, for his writings during this time give ample testimony to his mental and spiritual growth. Let us for the moment follow out the course of his official career.

Grosseteste was presented as deacon to the Church of Abbotsley by the Bishop of Lincoln on 25 April 1225. He must have been raised to priest's orders then or within a few years, for he was Archdeacon of Leicester from 1229 to 1231, with a prebend in Lincoln Cathedral which seems to have been the Church of St. Margaret of Leicester. This prebendary church he retained when in 1231, in the wake of a serious illness, he resigned his other benefices, judging in conscience

[9] Callus, "The Oxford career of Robert Grosseteste", in *Oxoniensia* 10 (1945), p. 53.

that he could not discharge all his responsibilities.[10]

The year 1229-30 was an important one in Grosseteste's life in more ways than one, for it was then that he was invited to become the first lector to the Friars Minor, who had just completed a school-house of fair dimensions, according to Eccleston.[11] Until that time, 'the student friars had gone daily to the schools of theology, however distant, barefoot, in bitter cold and deep mud', as the same chronicler relates. It was their Minister Provincial, Agnellus of Pisa, who approached the ageing master and former chancellor with the invitation to assume the direction of the new school. The fact that he took the unusual step of inviting a secular master, despite the presence of so many Franciscan masters in the English province and in the order, was, as A. G. Little has remarked,[12] a high tribute to Grosseteste's exceptional ability and to his position within the university. More than that, it was recognition of a spiritual kinship between Grosseteste and the friars, as we shall see.

Grosseteste lectured in that humble school for five or six years, until on 27 March 1235 he was elected Bishop of Lincoln. Matthew Paris records the death of Bishop Hugh on 7 February 1235, and continues:

After his burial, when the canons had to elect someone to succeed him, a division manifested itself among them; this one could not bear that so-and-so should be appointed, no more could that one suffer the other candidate, all from jealousy and resentment. At last, after many disputes among themselves, the canons voted unanimously for Master Robert, *cognomen* Grosseteste, to the surprise of all, even though it was being said that he was bound to the Franciscans.[13]

Matthew, of course, cannot be trusted when attributing motives, for he invariably assigned the worst, unless sometimes to fellow Benedictines; but I am rather inclined to accept the substance of his story, in part because its sequel can be verified (we know from Grosseteste's correspondence with St. Edmund, the Archbishop of Canterbury, that the monks of Canterbury

---

[10] For the date of Grosseteste's withdrawal from the cure of souls and an original and convincing interpretation of his action, see the excellent article of Fr L. E. Boyle O.P., "Robert Grosseteste and the Pastoral Care", in *Med. Renaiss. Stud.* 8 (1979), 3-6.

[11] Thomas of Eccleston, *De Adventu Fratrum Minorum in Angliam*, ed. Little, p. 48.

[12] Little, "The Franciscan School at Oxford in the thirteenth century", in *Arch. Franc. Hist.* 19 (1926), 803-74; p. 807.

[13] Matthew Paris, *Historia Anglorum*, ed. F. Madden, II (London, 1866), p. 376.

demanded that Grosseteste be consecrated there, but gave way
to his objections concerning the unnecessary expense involved
in that plan);[14] in part also because on grounds of age and
health Grosseteste surely did not appear a very promising candidate. Doubtless Matthew is right in implying that many expected him to join the friars: some of his colleagues at Lincoln,
observing his career with friendly interest, may have had good
reason to feel that he was in closer sympathy and more in
touch with the friars than with his own diocese. His election,
as a compromise candidate only, makes good sense.

Grosseteste was consecrated Bishop of Lincoln, the largest
diocese in England, at Reading, on 2 June 1235, by St. Edmund
of Abingdon, his old colleague at Oxford.

We have now reached the point in Grosseteste's biography,
and indeed have already passed it, where the documentary
evidence for his activities becomes extensive, and where whole
chapters could be devoted to studying his administration of
Lincoln Diocese, his relations with the crown, the papacy, the
dean and chapter of his Cathedral Church, the extensive official
correspondence that survives, his visitations of his diocese, and
his continuing intellectual labours. Fortunately, each of these
aspects of his achievement has been extensively studied by
outstanding scholars, so that the main lines, and even much
of the detail, of this part of his life have been clearly drawn.[15]
It remains true, however, as Sir Maurice Powicke remarked in
1953, on the seventh centenary of his death, that 'the definitive
life of him has still to be written and can only be written by
a very learned, versatile, and penetrating scholar indeed.'[16]
What I hope to offer in the remainder of this chapter is more
a series of impressions resulting from my researches into his
thought and his impact, both as a thinker and as a leader of
men, upon his age. In reading Grosseteste's works over a period
of years I have been deeply impressed by the mental ability,
the versatility, and the consistent growth of character to which
they attest. While I have learned from practically everyone
who has written on him in more recent times, and profited

[14] Grosseteste, *Epistolae*, ed. Luard, no. 12, pp. 54–6.

[15] The reader is referred to Fr Servus Gieben's complete bibliography of Grosseteste-studies published in *Coll. Franc.* 39 (1969), 362–418.

[16] Powicke, 'Introduction', in Callus (ed.) *Scholar and Bishop*, pp. xiii–xxv.
This is a slightly abridged form of a centenary lecture of Powicke's, first published
in *Bull. J. Ryl. Libr.* 35 (1953), 482–507.

especially from the impressions left on record by some who knew him, my aim differs in some measure from all of these, to whom I am never the less deeply indebted. It differs indeed almost as much from, say, Matthew Paris's, who set out in his own opinionated and perversely subjective way to chronicle and comment the events of his times, as it does from Powicke's erudite, wise, and moving appreciation of Grosseteste. Powicke conceived it as the historian's duty 'to make Grosseteste live so clearly that we can better understand the age in which he lived, thought and strove to make men what he believed they ought to be' (p. xiv). My concern is not with the broader canvas of the historian, but, as a portraitist, so to speak, or even a miniaturist, to depict the character of the man as known from his own writings, his friendships, his decision, his tastes, interests and enthusiasms, and from his ideals. Much of what I put into the portrait which follows the reader will find argued at different points in the succeeding chapters; here I shall not raise the dust of those discussions, but simply present the conclusions at which I have arrived.

Four medieval authors of very uneven fame wrote notably about Grosseteste. Roger Bacon venerated his memory and dinned in men's ears the praise of his intellectual gifts and achievements.[17] Matthew Paris, the monk of St. Albans who kept a large-scale and ambitious chronicle, made numerous entries concerning his actual or alleged doings and sayings while bishop, right up to the time of his death.[18] Thomas of Eccleston has left us precious materials of an anecdotal nature which illustrate Grosseteste's character and personality,[19] and these are supplemented by the recent, accidental discovery by Dr Hunt of a verse tribute composed by an otherwise-unknown friar, Hubert, who rises on several occasions above the level of the conventional lament for the irreplaceable defunct to paint a charming and attractive picture of Grosseteste's home life at the episcopal palace.[20] In what follows, I take my cue

[17] References occur in the following works of Bacon: *Opus Tertium*, ed Brewer, pp. 59, 74-5, 86, 91; *Op. Minus*, ed. Brewer, pp. 10, 33, 91, 434, 471 (etc.); *Op. Maius*, ed. Bridges, iii, i. 67; *Compendium Studii Philosophiae*, ed. Brewer, pp. 433-4, 471-2.

[18] Chiefly in his *Chronica Majora*, vols. III-V and *Historia Anglorum*, vols. II-III.

[19] Thomas of Eccleston, *De Adventu*, pp. 48, 91-4, 98-9. Thomas himself may have attended Grosseteste's lectures.

[20] Hunt, "Verses on the life of Grosseteste", in *Med. Human.*, N.S. 1 (1970), 241-51.

from these contemporaries of Grosseteste, and, criticizing their offerings where needs be, follow the lines which their very individual perceptions suggest.

Bacon's was a lone and oftentimes strident voice, which for better or worse went largely unheeded by his contemporaries. Though in his own times he spoke for the most part for no one but himself when he addressed the subjects nearest his heart, many historians have tended to see in him a privileged witness to Grosseteste's intellectual influence on the Friars Minor and the University of Oxford, since they assumed that Bacon was the old man's pupil and friend. In fact, Bacon was too young to have studied either arts or theology under Grosseteste, and while the rather personal warmth of admiration he quite evidently felt for Grosseteste suggests that he had known him, his tone throughout is one of somewhat breathless reverence for a hero-figure and a much older man.

If Bacon's numerous references to Grosseteste are assembled together they make up an unprecedented encomium of a notable translator and an original mathematical scientist and naturalist. He considers that Robert and Adam Marsh were 'perfect in divine and human wisdom'. He claims that, with some others (unnamed), they revived the ancient tradition of applying mathematical explanations in diverse fields to discover causes, and he lists the titles of a half-dozen of Grosseteste's surviving scientific works, attributing their merits to both men. Grosseteste, he tells us, completely neglected the books of Aristotle and their method and forged his own original scientific path through applying mathematics to experience. As a translator of patristic works from the Greek, he was, Bacon insists, vastly more competent than the celebrities of the age, Cremona, Michael Scotus, Alfredus, and Hermannus, because, though he began late to learn the language and his knowledge of it was not first rate (he required the assistance of helpers), his grasp of the material which he was rendering was unrivalled. Only Boethius excelled him in merit as a translator. Adam Marsh and Thomas Wallensis shared his linguistic studies.

Let us take things in order. What did Grosseteste achieve in science and theology to merit such enthusiastic praise from this testy and frustrated solitary?

If Roger Bacon was impressed by Grosseteste's independence

of Aristotle, the modern reader of his philosophical and scientific works is struck on the other hand by the remarkably large part they occupied in his attention. Grosseteste spent a considerable portion of his time over a period of at least ten years, while he was *regens* in theology, in reading Aristotle, commenting on him, and discovering his themes. He displayed a lively interest in Aristotle's scientific method and its applications. That was around the years 1220 to 1230. This had, admittedly, not always been the case. Much of his teaching as a master of arts (before 1209) must have been devoted to the logic of Aristotle, which by then figured rather heavily on syllabuses throughout Europe. But the early treatises of Grosseteste, which represent, it may be, only chance survivals of a much larger output, if they manifest some independent thinking on the generation of sound and the subjects of the *quadrivium*, do so, not due to neglect of an Aristotle read and spurned, but out of total ignorance of all but his logical works. Grosseteste's real debt at this stage of his career was twofold, to the giants of the Latin tradition, St. Augustine and Boethius, and to speculations deriving from Arabic science. Moreover, his developing interest in scientific questions, which he cultivated on the margin of a conservative theological taste, retained for almost a decade after 1209 a late twelfth-century atmosphere of the kind that was already outdated, for he was attracted in the main to astronomy and, almost inevitably, to astrological and alchemical lore. Such interests might have had some relevance to theology and liturgy, for as everyone knew the ecclesiastical calendar badly needed updating; but of themselves they would lead away from, rather than towards, sober Aristotelean science. And if the reading of the Doctor of the Soul, St. Augustine, and of his pseudonymous imitator (Alcher of Clairvaux?) stimulated a theologian to inquire into non-Christian psychology, once again a book by an Arabic writer like Avicenna or Algazel was simply easier to read and understand, in translation, than Aristotle's *De Anima*, and Grosseteste probably read Avicenna's book on the soul quite early on in his career.

For many years, while presumably he expounded the Bible in his school and held offices in the growing university, Grosseteste's main intellectual hobby was the study of the heavens. He had copies made of Arabic tables and of extracts from such Muslim astronomers as he came across. Probably before 1220 he began to take to the pen and write the first of a series of works in which he explained what he had learned of the heavenly globe (*De Sphaera*), and published his ideas, and revisions of those ideas, and revisions of those revisions, as to how the Christian year could be fitted once more into the cosmic rhythm ordained by the Creator as the pattern of man's works and days. All in all, he must have devoted many months during the decade 1220 to 1230 to the preparation of these works of computation. It was an interest, more, a reforming programme, that he never laid aside, for he published his last word on the subject as late as 1244, when many an onlooker must have felt that the Bishop of Lincoln should have had much better things to do. Grosseteste as he aged became sceptical of the philosophical accounts of the heavenly spheres, their number and kind of motion, and, borrowing a phrase from St. Ambrose, he once referred to them as 'no more substantial than spiders' webs':[21] but we can be certain that he did not regret the efforts he had personally expended on the study of the computational side of astronomy. Bacon did not allow his efforts to go unsung, nor the cause to be forgotten.

Of all Aristotle's scientific works, or what the thirteenth century referred to as *libri naturales*, the one which would least strike a modern reader as likely to inspire or enthuse him is probably the *Meteorologica*. We can still stretch the historical imagination sufficiently far to appreciate the grandeur of the speculation that lies behind the *De Caelo*, say, or the *Physics*, or the *Metaphysics*; we can admire the theoretical power of the *De Anima*, appreciate the originality of the little books on sleep, dreams, and the memory, and wonder at the observational finesse of the biological works. We can, moreover, dimly realize the revelatory effect which the reading of all of these produced on more ignorant ages. Yet it seems plain to me that Grosseteste was bowled over by his study of the book on meteorological phenomena, when he took it up shortly after 1220. The *Meteorologica* was to govern the whole trajectory of his scientific interests for years afterwards, as they moved from the heavens and the stars across the great cosmic watershed of the sublunary sphere into the atmosphere, finally

---

[21] St. Ambrose, *Hexaemeron* I, 7, *C.S.E.L.* xxxii, 6.

to reach the earth. No doubt the comprehensive scope of the work made a deep impression upon his mind, for in it Aristotle explained the causes of all phenomena that he considered to be of sublunary origin: shooting stars, comets, the Milky Way, the different forms of precipitation, the winds, seas, earthquakes, thunder and lightning, and the rainbow. A sidelong glance at the titles of Grosseteste's scientific treatises picks out a whole set devoted to such themes, and written probably between 1220 and 1233 or 1235: *On Comets, On the Ebb and Flow of the Tides, On Heat, On Colour,* and *On the Rainbow.* Other works of Aristotle he had indeed read or perused, at least in part, like the *De Caelo* and *De Generatione*; but I think we can say that the *Meteorologica* was the first he really came to understand, then to imitate, and finally, in his late treatises, to surpass in notable ways, thanks mostly to his reading of writers like Euclid and Al Kindi on optics, Alpetragius on the tides, and Avicenna on stones and minerals, but in part also— and a notable part for his historical influence on science— thanks to the originality of his own fertile mind.

Which brings us back to the point of Bacon's remark, underlining that originality by contrast with the vulgar academics who simply expounded the books of Aristotle *'et vias eorum'*, without much gain to science or to their own students. Grosseteste did not comment on the scientific works of Aristotle, he developed their themes in independent treatises which do not fit into the pattern of scholastic writings at all. That is something which strikes the modern student no less forcibly than it did Bacon, for it would appear that Grosseteste wrote primarily for himself, and only secondarily for a readership. He did not teach these scientific topics at his school, for there he did the work of a theologian, lecturing and disputing. Scientific writing was his literary hobby. Yet much of the originality of his scientific work derived in fact from his professional interest. There is nothing paradoxical about this; Grosseteste was simply too intelligent to compartmentalize his mind. As a theologian he had to explain the Bible, from creation to apocalypse, lecturing concurrently on a book of the Old Testament and one of the New; and the former notably lent itself in exegesis of the literal sense to some degree of scientific treatment. The two obvious surviving examples of this are the *De Luce* and the brief (and possibly fragmentary) commentary on Ecclesiasticus 40:1–5.[22] In the former, Grosseteste sets out in treatise form his own speculations on the creation of the first three days; in the latter we find an extended exegesis of the literal sense, as the Hebrew poet's exaltation of the sun is expounded, with all the scientific learning and ingenuity desirable. Grosseteste's *De Luce* is his philosophical masterpiece and shows him at his speculative best. Taking as his starting-point the light produced by the command of Yahweh, and amorphous matter, Grosseteste set himself to 'think God's thoughts after him' and to imagine how the omnipotent mathematician planned the process which gave as its result a world-machine that is plainly Aristotelean in its essential features. It is far beyond any 'Aristotelean' work of its kind; indeed it is beyond that categorization altogether, for it contains Grosseteste's one and only contribution to mathematics (in the discussion of relative infinities), and it attempts to convince the reader that the cosmogonic process was of a mathematical nature—that, in short, nature's regularity and economy derive from its mathematical construction.

Grosseteste's last scientific opuscules illustrate his belief that the knowledge of mathematics, by which he meant especially geometry, is the clue to all natural action and passion; material action derives from an energy whose inner workings are similar to the behaviour of rays of light and conform to the same geometric laws as to do these. Applying these bold and grandiose metaphysical beliefs, Grosseteste works out explanations of the variation of the world's climates (in terms of the different angles of incidence of the sun's rays), of heat, of colour, and of the rainbow. He was too bookish to be bothered experimenting, and he did not grasp either the method or the importance of experimental procedure in the way that Thierry of Freiberg did at the beginning of the following century. He relied on the various authorities in his extensive library for accounts of their experience, and he transmitted them or adapted them to his purpose. He himself discovered no new scientific law, and was not a great, nor even a very good mathematician. Yet his writings and his fame helped to establish at Oxford an interest in scientific and mathematical learning that

---

[22] *De Luce*, ed. Baur, *Werke*, pp. 51–9; McEvoy, "The sun as *res* and *signum*: Grosseteste's commentary on Ecclesiasticus ch. 43, vv. 1–5", in *Rech. Théol. anc. méd.* 41 (1974), 38–91 (this is an edition of the so-called *De Operationibus Solis*).

flourished in the fourteenth century, and which might well not have got off the ground without his inspiration; and his metaphysical ideas on light-energy, the fundamental unity of celestial and terrestrial space, and the central role of the sun in the universe's workings, were all anticipations of notions which fertilized the scientific speculation of more modern times.

I have suggested that without Grosseteste there might not have been a notable mathematical-scientific tradition at Oxford. In a parallel way it can, I think, be claimed that without the stimulus of his commentaries and translations the Aristotelean movement at Oxford would not have become what it did, or at least not so quickly. The *libri naturales* were not condemned at Oxford, as they were at Paris from 1210 onwards; this we may attribute in large measure to the influence of the chancellor-turned-bishop, whose continuing interest in the university, for which as its local ordinary he was responsible until his death in 1253, assured the harmonious and tranquil growth of studies there, and whose overt patronage of Aristotelean scholarship and learning guaranteed their sane and uninterrupted development until a much later period. When Roger Bacon arrived at Paris from Oxford in the early 1240s he found himself in the unusual position of introducing onto the syllabus of the arts faculty certain works of Aristotle whose official reading had been proscribed there, but which had been prescribed at Oxford. And it is clear that young masters like Adam of Buckfield were lecturing and commenting on the whole known corpus of Aristotle at Oxford a decade or more before the works in question were finally put upon the arts course at Paris in 1255, one of the crucial dates in the history of Aristotle's influence in the West. That it was Oxford and not Paris which took the lead in this respect is largely due to the favour which Aristotelean studies found with Grosseteste and to the generosity of spirit with which he listened to the wisdom of the Lycaeum. That his generosity did not blind him to the naturalism of Aristotle's thought was due to his alert critical sense.

It should cause us no surprise to find that Grosseteste spent considerable time in writing commentaries on Aristotle during his busiest academic years, around 1228 to 1232, for throughout the thirteenth century it was the theologians and not the more youthful masters of arts who turned out much of the superior sort of philosophical commentary. Grosseteste's scientific interests led him to the awareness that a new and exciting phase of intellectual change was at hand, and he naturally wished to play a part in it. He completed a full, literal commentary of Aristotle's *Posterior Analytics* (a work on which he may have lectured in the arts faculty many years before, shortly after Master Hugh brought it onto the syllabus at Oxford), and wrote a series of notes on the *Physics*, which he left, however, in an unfinished state. These were edited much later in the century and became well known at Oxford during and after the time of Duns Scotus. His commentary on the *Posterior Analytics*, Aristotle's great work on the principles of scientific knowledge, became a classic of scholastic literature, for its combination of secure grasp of Aristotelean principle with clarity of exposition were to ensure it constant use for three hundred years after its author's death. Grosseteste both understood and approved of Aristotle's basic thesis, that sense-knowledge is the foundation of all higher dimensions of knowing. Very typically, he set out in his commentary to place Aristotle's views within a Christian perspective, which of course shows up their partial character as truths. Yet it is plain that Grosseteste thought of himself as an Aristotelean, and regarded Aristotle as quite simply the Philosopher; he did not think of himself as a member of a rival philosophical sect, say, of Augustinians. He made evident his view that Aristotle's philosophy represents one level of truth and no more than that, and his commentary became something like a celebration of the superiority of revealed Christian truth over the *sapientia mundi*.

It may well be the case that part of the importance of Grosseteste's marginal notes on the *Physics* of Aristotle was purely personal, in that the struggle to understand that difficult work in translation decided him to learn Greek and read Greek writings in their original language. No doubt even more attractive to him than the prospect of being able to read Aristotle was the ambition to read the Old Testament in the Septuagint, and the New Testament, and to be thus liberated from dependence upon the Vulgate. Whatever his motives were, he applied himself to the study of Greek grammars and dictionaries at an age when most men are looking forward to their retirement. From around the year 1232 the evidence of this new, absorbing interest begins to appear in his biblical commentaries. Shortly after he had settled down in his new office at Lincoln the first of his translations from the Greek was published. Circumstances favoured him. It would appear that between his illness in 1231 and his elevation four years later Grosseteste had more leisure than he had enjoyed for years beforehand; and he was not a man to idle time away. Moreover, after becoming bishop he had the material resources to hand and could command the help he needed in terms of personnel for great projects like the rendering of the works of Pseudo-Dionysius and the Aristotelean ethical corpus. Master Nicholas Graecus appears from entries in the Rolls of the diocese to have been with him throughout his pontificate, and Matthew Paris tells us that he was *'natione et educatione grecus'*.[23] John of Basingstoke, who was Archdeacon of Leicester from 1235 until his death (in 1252—the year before Grosseteste's) had studied Greek at Athens; he drew Grosseteste's attention to the *Testaments of the Twelve Patriarchs*, which the Bishop ordered from Greece and translated with help of Nicholas.[24] No doubt there were others. Grosseteste repeatedly acknowledged the help given him by his *adiutores*, and both Bacon and Paris refer to them. But it was Robert himself who conceived the projects, directed the acquisition and copying of the MSS (which was done largely at St. Denis in Paris in the case of the Dionysian works), supervised the preliminary work, took responsibility for the translations, and added glosses or commentary. The products of his workshop, scholars agree, bear the stamp of his individuality, right down to the very details.

We may pause for a moment and wonder at the energy of this man of seventy or more, who, at a time when he was launching a programme of pastoral reform in his diocese, and when he was conscious of his own failing powers of memory, determined to devote the greater part of his leisure time to translating. He must often have brought clerks like Nicholas Graecus to his study after vespers, or with him to his country

---

[23] Russell, "The preferments and *Adiutores* of Robert Grosseteste", in *Harv. Theol. Rev.* 26 (1933), 161–72; p. 169, no. 45; Major, "The *Familia* of Robert Grosseteste", in Callus (ed.), *Scholar and Bishop*, pp. 216–41.

[24] Russell, art cit., pp. 168–9.

manors during holidays, and there settled down with dictionaries to render the current object of his interest. Indeed, one
of his own letters allows us a glimpse of just such employment
of a week's holiday. Addressed to the Abbot and community
of Bury St. Edmunds, the letter opens:

> While resting last week from the clamour of business, I was passing
> some of my time in reading, when on a certain day of the week I came
> across a writing on the life of monks which does that life full justice. I
> believed it would afford you great pleasure and advantage if I passed on
> to you what I have managed to understand of it, though not in a literal
> translation, since the expression is of a different cast to the Latin. Para-
> phrasing the words as well as I could, and adding a few explanations here
> and there, I had it written up, as you will see, and sent to you.[25]

Now a kindly interest in a friendly community is an excellent reason for a man of Grosseteste's disposition to devote a
day or two's leisure to such a congenial task; but what was it,
we may ask, that justified in his own eyes the enormous expen-
diture of time which his greater translating projects demanded?
I am convinced that Grosseteste's hobby was more than an
escape from his daily duties as ordinary and that he chose with
care what he felt scholars of his own times could most profit-
ably study. Part of the clue to his choice lies, I think, in his
increasing scholarly interest in the life of the early Church
during the Apostolic period, for Church reformers of all ages
are forced willy-nilly to go back for inspiration to the life of
Christ, first and foremost, but also to the life of the Church
he founded, that community which 'shared all its goods in
common' and 'had but one heart and soul'. For the reformer,
conscious as he is of the divisions and abuses within the Church
(or the monastery) of his own times, that vision represents a
hope of what may still be recovered, and of what must be
striven for amid the failings of the age. Grosseteste had lectured
and written on the Gospels and the commandments while in
the schools, but he felt himself drawn more and more in his
episcopal period to search out the essence of Christianity. He
was evidently fascinated, as Beryl Smalley has shown, with the
mentality of the Judaeo-Christians and the pastoral dilemma
posed to the Apostles and the bishops of the early Church by

---

[25] Letter no. LVII, ed. Luard, *Epistolae*, pp. 173–8. Thomson (*The Writings*,
p. 71) identifies the work in question as consisting of sections from the *Longer
Rules* of St. Basil, with a short extract from the *Ecclesiastical Hierarchy* of Pseudo-
Dionysius interpolated.

the failure of many Jewish converts to realize the newness of the life to which they had been introduced.[26] His *Expositio in Galatas* developed the struggle of Paul to convince the Church of Jerusalem, and even his fellow-Apostles, that the life of the Spirit is destined for the whole world's redemption, and the Jewish Law is all at once become an obstacle to true liberty. Just about the time of his elevation Grosseteste devoted a lengthy treatise to the themes of the *Cessation of the Law*, and the pivotal moment of history which marked the supervention of incarnate Grace upon the old world. It was in continuity with this theological interest that he turned to the translation of works which promised the Latin Church a new insight into Christian origins, and hence a new instrument of self-reformation.

It was certainly in this spirit that Grosseteste translated and commented on the four treatises of Pseudo-Dionysius, accepting that author's spurious claim to be the Athenian convert and intimate confidant of St. Paul. *The Testament of the Twelve Patriarchs*, a second-century Christian rewriting of a Jewish apocryphal work, he valued for the light it shed upon the Jewish expectations of the imminent Messianic age. His version secured an immediate and long-lasting success in Europe and made its translator celebrated. Evidently the world was avid for just this kind of discovery, for Grosseteste's translation of the *Letters of St. Ignatius the Martyr to the Blessed Virgin Mary* enjoyed a like triumph. The translation of the genuine Ignatian letters of the Middle Form (which is probably an authentic version by Grosseteste) represented a further offering of a precious early Christian work. Grosseteste would have been justified in feeling that these writings would illuminate the origins and the beliefs of the primitive Church and so help to further that spiritual purification of the contemporary Church which was the goal of his active ministry.

But why then did he divide his attention between these religious works and Aristotle's *Ethics* together with its pagan and Byzantine commentators? One can readily understand his undertaking the correction of the existing translation of John Damascene's works, for that was his first attempt at the craft, and the book itself was a deservedly famous survey of Greek theology. But Grosseteste had put his philosophical studies

[26] Smalley, *The Study of the Bible in the Middle Ages*, p. 343.

well behind him when he took the decision to resume them in this new, scholarly form as a translator of Aristotle. I consider that some sense can be made of his decision, though more so in the case of the *Nicomachean Ethics* than in that of the *De Caelo* and its Greek commentator.

By 1240 or so Grosseteste had got well into his stride as a translator and was unwilling to stop while the going was good. But there was much more than this, for as Bacon justly remarked, he stood out among the Latin translators as the only thinker of real power and originality among them, just as he was in reverse the only great scholastic thinker who was capable of translating from the Greek. It was quite simply his instinct for a great book that prompted him to replace the existing, unsatisfactory versions of the *Ethics*. He knew that the Aristotelean movement in the Schools was still on its upward swing; here then was his chance to win an honoured place in it. Moreover, although he felt bound once or twice on the way through to give the old pagan a belt of the crozier (for permitting exceptions to the moral law, for instance), Grosseteste no doubt felt that the themes of the book as a whole were not only not offensive, but actually quite edifying. They did not offer a direct challenge to any article of Christian teaching (as the *Physics* did to the doctrine of a temporal creation), but described the universal human search for happiness and its working out in virtuous habits, and recommended the contemplation of such divine and eternal realities as the human mind can attain knowledge of. Grosseteste welcomed this, and felt sure other theologians would do likewise. He had, no doubt, a special degree of admiration for the discussion of friendship in Books VIII and IX of the *Ethics*; for that was one of his life-themes (as is evident from his letters to his friends and the sections of commentary he wrote to accompany Book VIII), and Aristotle's treatment of it no doubt appeared to him to be full of sane, moral wisdom, and really the expression of rather edifying views. Grosseteste must have felt that a pagan wise and good enough to declare that a man should lay down his life if needs be for a friend should not prove too difficult to instruct and baptize, and that there was much to learn from him.

Even more surprising than Grosseteste's conversion to the study of Greek is the fact that he began to learn Hebrew, probably only a few years before his death. Nicholas Trivet, himself a Hebrew scholar, attests the persistence of a tradition at Oxford to this effect, declaring roundly that 'he drew a great deal of material from glosses of the Hebrews';[27] and Bacon also makes reference to his study of Hebrew. Grosseteste was the patron of a new Latin translation of the Psalter, which he ordered from an unknown translator and had written as an interlinear gloss of the Hebrew text, word for word so that he could follow the original. The Hebrew and Vulgate are written out in parallel columns in this MS, and the *Hebraica* of St. Jerome is put next to the Hebrew. The author of the prologue of this book, which came to be known as the *Superscriptio Lincolniensis*, took responsibility for its publication and explained its aim, but did not claim to be the translator. Miss Smalley attributes this prologue to Grosseteste: 'The freshly devotional tone of the prologue, the decision and the wide outlook, all remind one of Grosseteste, scholar, bishop and Saint.'[28] The original of the *Superscriptio* is lost, but two complete copies survive. It is a pity that we do not know who his collaborators were.

If I have concentrated thus far on Grosseteste's philosophical development and on his achievements as a translator, that is largely because these were the fields in which he attained greatest celebrity in the schools of his own times and later, and not because his production under these headings amounted to the whole of his output. A perusal of Thomson's catalogue suggests the bewildering variety of his literary heritage:[29] theological treatises, biblical commentaries, pastoral works on confession, conferences to clergy, many sermons, memoranda compiled by Grosseteste himself, addresses which he delivered in the presence of Pope Innocent at Lyons in 1250, popularizations of theology in Latin (*Templum Domini* was the most famous of these) and Anglo-Norman (the *Chasteau d'Amour* survives in innumerable copies), and of course those of his letters, mostly of an official nature, which were carefully collected some time after his death.

The truth is that Grosseteste had an unusual series of talents

[27] Trivet, *Annales*, p. 243.
[28] Smalley, "The Biblical Scholar", in Callus (ed.) *Scholar and Bishop*, p. 81.
[29] Thomson, *The Writings of Robert Grosseteste, Bishop of Lincoln, 1235–1253* (Cambridge, 1940), pp. 302.

which could not be confined within the straitened forms of the current scholastic literature. It is somewhat paradoxical that, though he was for a period of years in the van of the new philosophical thinking of his time, the Aristotelean movement, he was by then already too old to conform to the changing pattern of academic expression which younger men found to their taste. His life spans the latter part of the twelfth and the first half of the thirteenth century, and it is no wonder if some of his attitudes appeared quaint and rather archaic by his sixtieth or seventieth year; in many matters of note he thought the old ways best. One example of this is his instruction to the theological regents of Oxford to keep to the exposition of the Bible in their lectures and forget about the *Sentences* of Peter Lombard. Another is his refusal to adopt the newer, systematic theology which was to triumph already within his lifetime. He clearly held disputations on the literal sense of the Bible and on theological questions in his school (he himself refers to these once or twice, and so does Eccleston), but he resisted the scholastic trend towards the methods of *divisio textus* and the multiplication of *quaestiones*; he really only included questions in his writings when he chose to do so, and that was rarely. This is not to say, however, that his conservatism blinded him to progress or blunted his wits. He was probably the first Scholastic to take seriously the question of whether Christ's Incarnation was purely a consequence of the Fall, and he was certainly the first to answer it in the negative, albeit with some caution. He was one of the first to defend the doctrine of the Immaculate Conception, something on which he admitted he had found cause to change his mind. Both of these positions were to enjoy a considerable career at Oxford, in the Franciscan School especially. Grosseteste was original, too, in claiming that the chief head of theological dispute between the Latin and the Greek Church, the procession of the Holy Spirit from the Father 'and the Son' (*filioque*) was a mere difference of words. This is what he had to say in the brief note which he devoted to the controversial matter, and which Scotus reproduced in full in his *Ordinatio*, thereby giving it a well-deserved currency in the schools:

It may well be, however, that were this apparent contradiction to be discussed by two doctors, a Greek and a Latin, each of them a true lover of the truth and not of his own expression of it for its own sake, it would eventually be clear to both that the difference which opposes them is not so much real as verbal; the only alternative is that either the Greeks or we, the Latins, are nothing less than heretics. But who is so foolhardy as to accuse the author of this work, John Damascene, together with the Blessed Basil, Gregory the Theologian, Gregory of Nyssa, Cyril, and other such Greek Fathers, of heresy? And who on the other hand dares to make a heretic of Blessed Jerome, Augustine, Hilary, and other such Latin Doctors? It is probable, consequently, that the opposing statements quoted do not correspond to any real conflict between the Saints, for this reason, that what is said is said in a variety of ways; for example, in this context, 'of this person', and similarly 'from this or that person', or again 'by him'; and it may be that if this wide range of expressions were more subtly understood and analysed it would emerge clearly that the doctrine which finds opposing expressions is in fact the same.[30]

Modern scholars are agreed that Grosseteste on the whole preferred theological scholarship to dialectic. I would add to this, that as he grew older he wrote more and more for himself and cared less whether his writings would find a public. If he felt like spreading himself and covering the vellum on a matter which interested him (such as the celebrated disagreement between St. Augustine and St. Jerome), then he did so, regardless of whether others might find the whole thing disproportionate, not to say dull. He had little sense of proportion when he sat down to compose a treatise, or to write up his copious marginal notes on a given point—for he was an inveterate scribbler; however, he seems eventually to have written up the majority of his jottings and the half sheets or *cedulae* which he had stuck in his books, and to have edited them with care as *authentica*. The use over a long period of years of a system of indexing-symbols which he and Adam Marsh developed enabled him to retain confident command of his scholarly reading and to turn up related passages in different books almost at will, thus speeding up the process of composition considerably. The reader of his theological treatises soon learns that he can never predict what lies ahead; here is an author who is highly individual in his literary habits, and who is besides capable at the moment least expected of bringing the full warmth of his humanity to bear on the discussion of his problems. His blend of lively, imaginative curiosity and broad human experience is at the opposite remove from the dialectical habit of mind.

[30] Grosseteste appended this *Notula* to his own corrected version of Damascene's little work on the *Sanctus (De Hymno Trisagio)*. For its text and a comment see McEvoy, "Robert Grosseteste and the reunion of the Church", in *Coll. Franc.* 45 (1975), 39-84.

He shows himself aware that errors like the pagan belief in the eternity of the world, and popular superstitious practices, and mentalities like those of the Judaizers in the primitive Church, and schisms like that between Greeks and Latins, are resistant to abstract logic, because they are complex human situations and are surrounded by non-rational factors like habit, custom, and partisan loyalty. He attempted to understand such things and to explain them historically. It goes without saying that Grosseteste had no conception of historical explanation in the modern sense of the word, but he possessed the patience of a true scholar and an acute perception of detail, which, when combined with his practical concern, gave him a lively interest in human situations and a distaste for superficiality and oversimplification.

When after protracted disagreement and a stalemate the canons of Lincoln finally settled on a compromise candidate to succeed Bishop Hugh, they can have had little idea of what lay in store for themselves and for the diocese. Not for the first nor the last time in the history of the Church, a caretaker was elected to high office and proved in the event to be a memorable choice. The electors may have supposed that Grosseteste would be an able enough administrator and capable of supervising a host of delegated functions; they would have known that he had done the university no harm while chancellor, nor the diocese, during his three years as archdeacon. True, he was well on in years and had been forced once already by illness to limit his obligations and harbour his forces; but on the other hand, he was known as a learned and pious cleric, and to pass his last years as a bishop would crown his distinguished career quite suitably. The canons no doubt hoped that the next election to the see might be held in a better atmosphere, and were relieved that the immediate problem was solved and no heads broken.

It is easier to imagine how Grosseteste came to be offered the diocese than it is to say why he accepted it. My impression is that he was one of those rare people who never stop growing, and that he could not resist the challenge which God seemed to be throwing before him. When he looked back over the shape of his life and saw with the eyes of faith where its turnings and meetings had led him, he got the courage to place his trust in God and prepare for whatever might come. His life had given him full opportunity to develop his gifts, pursue his intellectual interests, and realize his professional ambitions, many years before 1235. Over a period of years he had been led through study, reflection, and prayer towards a vision of the Church and a clear view of his own responsibilities within it. We can assume that with Edmund Rich, Adam Marsh and other like-minded reformers at Oxford, such as Sewal, the future Archbishop of York, and Richard Wych, later Bishop of Chichester, he had canvassed projects for pastoral renewal. It had in time become plain to him that he must take priest's orders and assume the active care of souls, as a rector. Then within a few years the bishop had offered to promote him, just around the time when he himself had realized that the new mendicant orders, recently arrived in England and Oxford, were the ideal instrument of renewal in the Church, men dedicated to a life of poverty and preaching outside both the secular and the monastic tradition, lacking the comfortable security alike of cloister and of benefice; and that he himself was in a position to lend them his support. Should he accept promotion in the diocese, or work in whatever ways he could with the friars at Oxford? Typically, Grosseteste had resolved his dilemma by reorganizing his timetable and accepting both the bishop's offer and the invitation to teach the Franciscans theology. It had proved to be too much, and after a couple of years he had fallen ill. But even his illness had proved salutary; it had shown him a higher way, which was to renounce the archdeaconry and its emoluments and, accepting a degree of voluntary poverty, to concentrate on his work with the young Franciscans. He had not entered their order, but he shared their vision. Now a similar choice seemed to be placed before him once again, whether to remain at Oxford near the friars and try to carry through what he had begun, or to accept the see. Typically again, Grosseteste sought the best of both sides: he accepted the call and was consecrated, then introduced both Franciscan and Dominican orders into his diocese to be the agents of his programme.

Here is the beginning of Matthew Paris's entry covering the death of Grosseteste in 1253 and summarizing his achievements as bishop:

Thus in the night of St. Denis, near Buckden, St. Robert, Bishop of Lincoln, left his exile in this world; a plain confuter of the Lord Pope and of the King, a rebuker of prelates, a corrector of monks, an adviser of priests, a teacher of the clergy, a supporter of scholars, a preacher to the people, a persecutor of the incontinent, an unwearied student of every kind of Scriptures, a hammer and despiser of the Romans. At the table of bodily refreshment he was bountiful, liberal and courteous, pleasant and affable, while at the spiritual table he was devout, tearful and contrite. In the episcopal office he was sedulous, deserving of reverence and untiring.[31]

This tribute bears the usual stamp of Matthew's own prejudices against the king, the pope, and the Roman Curia. Despite his effort for once to be fair to Grosseteste's intentions, in calling him 'the corrector of monks', Matthew did not consider that the Benedictines required correction, and the attempt of a bishop to supply it made him smoulder with the sense of outrage which had led him in earlier entries to spread mischievous and quite scurrilous stories about Grosseteste's visitations of monasteries, and more so still of convents. His generous praise of the bishop's efforts to reform his diocese is the more sincere, as he resented some of those efforts and misunderstood others.

Grosseteste was one of a new breed of bishops, scholars who had been educated in the universities. The best type of twelfth-century bishop had been a monk, called from the cloister into the public arena; St. Hugh of Lincoln is a good example. The new currents of the thirteenth century, however, quickly threw up another kind of leader: an Edmund of Abingdon, a William of Auvergne, a Robert of Lincoln. Robert's is the most striking case of all, because he brought to the pastoral ministry a breadth of vision acquired over a lifetime's study of philosophy and theology. It is the interaction of the university-formed vision with the pastoral situation which lends to his reform of Lincoln diocese an interest that is altogether exceptional, in the issues that were raised if not in the results obtained. It is sufficient to read his letter to the Dean and Chapter of Lincoln, which is couched in biblical and Dionysian terms, in order to appreciate the degree to which his conception of the Church and of society was moulded by his studies and his attitude to learning.[32] No treatment of any aspect of

[31] Matthew Paris, *Historia Anglorum*, vol. III, pp. 146-7. Cf. *Chron. Maj.*, vol. V, p. 407.

[32] The letter entitled *'De Cura Pastorali'* (*Epistolae*, ed. Luard, no. 127, pp. 357-431).

his activity, whether as teacher of the Franciscans, as preacher and writer, or as reformer, can afford to juxtapose his philosophical and theological with his practical preoccupations without showing their interrelationship, at the risk of rending a seamless cloth. Grosseteste was everywhere consistent with himself.

To such an extent is Grosseteste's life a unity that no biography of him that was not in large measure an 'ideography' could succeed in grappling with his true greatness. Theory and practice cannot in his life be dissociated. He was not a professor first and then afterwards a reforming prelate, rather was his entire practical life in the public eye a commentary upon his intellectual positions. To quote Aristotle on kingship and tyranny to Henry III, or Dionysius on the nature of hierarchy to Innocent IV, was in him no affectation; long habit had accustomed him to cite the authorities who were in support of his position. The mark of the schoolman remained on him in all that he did, because the mental world he inhabited had been formed during most of a lifetime spent in the schools.

On the purely formal level, the level, that is, of study as a mental discipline, the scholastic life taught him the tidy habits of an ordered mind. The self-discipline which he acquired over many years bore fruit in his pastoral work, for idleness and slackness remained foreign to him. His sermons were prepared and written with care and attention, and his correspondence, though copious, was composed in a style that is far from negligent (Cicero was a firm favourite with him). The discipline with which he applied himself to the study of Greek and finally of Hebrew was the fruit of a life-time's training in method, which turned him in his old age into the largely unbeaten path of scholarship. Such painstaking philological scholarship as he exhibited in his commentaries on Pseudo-Dionysius was not to be surpassed until the Renaissance.

The discipline so characteristic of him made him deliberate in word and deed, so that even in anger he could choose the word he wanted. His thinking had a clarity and purpose which often disconcerted his contemporaries. He abhorred bad logic, and could no more temporize with it than he could suffer gladly the fools who were guilty of it. Nowhere is this more clearly seen than in his attitude to authority. Means serve ends, not the reverse. *Ergo*, authority serves its subjects, whose

salvation, temporal and eternal, is its *raison d'être*. The true
end to which all else in the universe is subordinated is the
salvation of souls; to this, all aspects of Church life quite logically stand as means. All abuse of these pastoral means stirred
to wrath, not only the Christian in him, but the Aristotelean,
for bad logic could not, he considered, be justified by good
intentions, should they be those of the Lord Pope himself. In
the schoolman's habit of reasoning from principle to conclusion is to be sought the origin of much of Grosseteste's reforming spirit. The dismay with which he regarded anything
that savoured of expediency was only another expression of it.

More than the gift of mental exactitude, his years in the
schools gave him a standpoint from which he could survey
the whole universe. The mental world in which he lived was
formed by reading of the Scriptures and by the life-long study
of the Fathers of the Church and the classics of ancient philosophy. The order of the cosmos was the chief message of their
wisdom. God has created all things in number, weight, and
measure. Creation is therefore pervaded by the divine law,
which operates through the stability of the natures of things;
these in turn are known to us through their acts. Natures are
participations in God's ideas, hence their symphony. Since
they do not share equally in the divine light which constitutes
them, they form a hierarchy or order of subordination, from
the highest angelic choir down to the lowest plant or mineral.
All acknowledge a common allegiance, which is a response to
their place within the created order. Man's place is special, for
in him alone do material and spiritual meet. The dignity and
responsibility of man are the expression of his place in the
scale of perfection constituted by creation. Responsibility for
those beneath one is the fundamental law of hierarchy; as the
angels hand on to the lower ranks the illuminations received
from their superiors, so must a bishop or priest feed his flock
on the fruit of his own contemplation, on the lights he has
received. That is the law of all hierarchy, expressed at the
level of pastoral concern. The aged bishop who enunciated it
as a guide for the conscience of his priests was the same man
who may have sat up the previous night translating Dionysius,
disciple of Paul, or reading pagan Aristotle on the last end.

One aspect of Grosseteste's episcopate which strikingly
confirms the over-all impression of consistency and principle

is his relations with the papacy. His attitude to the papacy has been greatly misunderstood, especially since the Reformation. That Grosseteste was right in attacking the general medieval attitude to benefices as useful sources of income for papal or royal servants, no-one has ever seriously denied. That he was right when, speaking before the pope and cardinals at Lyons in 1250, he pointed to the Roman Curia as the fountainhead of the ensuing pastoral evils, is probably equally true; he justified his denunciation on that occasion by the claim that only the curia could, as the centralized authority, clear away the abuses in question, and that instead of doing so it itself 'by its dispensations, provisions and collations, appoints these bad pastors, and so leads patrons to fill benefices on carnal and worldly motives.' But there is a decided difference between energetic criticism and the outright refusal to obey with which Grosseteste met a papal mandate a few months before his death. In the month of January 1253 he learned that Pope Innocent IV had conferred a Lincoln canonry with the right to the first vacant prebend upon his young nephew, Frederico da Lavagna. The letter which informed Grosseteste of this decision emanated from the two executors whom the pope had appointed to effect the provision, namely Master Innocent, the papal scriptor then resident in Yorkshire, and the Archdeacon of Canterbury, Stephen de Montival. Only the confirmation of the original provision survives, by way of quotation in Grosseteste's deservedly famous reply to the two provisors, who, according to the Burton annalist, forwarded the text of the reply to the pope. It was an unusual experience for a medieval pope to read—and it cannot be doubted that Innocent IV did read—the following words from the conclusion of the bishop's letter:

Nor can someone who is subject and faithful in unsullied and sincere obedience to the Holy See and not cut off from the body of Christ and the same Holy See through schism, obey mandates or precepts or any other instructions of this kind, wherever they emanate from, even should they come from the highest order of angels, but he has no choice save to speak against the whole business with all his might and to rebel. Therefore, reverend Sirs, on account of the duty of obedience and fidelity . . . I do not obey, I resist, I rebel.[33]

The apparent contradiction between belief in the *plenitudo*

[33] Letter no. 128 (*Epistolae*, ed. Luard, pp. 432-7); see n. 37 and Appendix A.

potestatis and refusal to obey a papal command has disconcerted many historians and been explained away by the remainder. It vanishes, once we become fully aware of the nature of Grosseteste's reply to the pope and the canonical grounds which could be invoked by a subject against his spiritual superior. Behind all that Grosseteste wrote concerning the *plenitudo potestatis* there lie certain assumptions to which only the long series of medieval commentaries on the Decretals, discussing the nature, range and limits of papal authority, provide the key. Professor Tierney, who has studied Grosseteste's famous letter in the context of the canonical discussions of the times, has pointed to the existence of an elaborate jurisprudence concerning the validity of papal rescripts, including the kind which the Bishop of Lincoln chose to disobey.[34] Only when the *non obstante* clause was invoked and when the pope stated that he was acting *ex certa scientia* was the mandate to be considered an incontestable expression of the plenitude of power. That was precisely the sort of letter which Grosseteste received in 1253. Yet he rejected it. In attempting to make sense of the rejection we must bear in mind that papal authority was never considered to be arbitrary or irresponsible. In the case of heresy a pope could be deposed, and the commentators on the *Decretum* considered him to be bound also by other traditions of the Church: 'Certain canons cannot be abrogated even by the pope, namely those concerning articles of faith or the general state of the Church'.[35] The problem was often raised, whether obedience must be accorded to an unjust command of the pope; and in reply a distinction was forged by canonists between the person of the pope and the Apostolic See, since it was not assumed that every papal command was supported by divine authority.

A special interest attaches in this matter to the canonical works of Innocent IV, Grosseteste's adversary. In an article of his commentary Innocent proposes the question as to what a subject must do if he receives an unjust command from the pope. His reply is that he should obey, 'unless the command contains heresy, for then it would be sinful; or unless there arose from the unjust order a strong presumption that the

---

[34] Tierney, "Limits to obedience in the thirteenth century", in *Contraception, Authority and Dissent*, ed. C. Curran, p. 78.

[35] Quoted from the *Summa* of Huguccio by Tierney, p. 85.

state of the Church would be upset, or perhaps even other evils would come about; for then obedience implies sin'.[36] Now we have here a statement of Grosseteste's case which in essentials parallels his own letter rejecting the papal command. The irony of the whole affair is that Grosseteste is justified in his refusal of obedience out of the very mouth of his adversary. Grosseteste's case, as he himself stated it, was that the pope's policy on provisions was disrupting the purity and peace of the Church, wherefore to obey the mandate would be for him a mortal sin, because it would involve doing an injury to the pastoral ministry of the Church.

Grosseteste was, of course, on the firmest of ground in proclaiming that the mandates of the highest office in the Church must respect the teaching of Christ and the Apostles and must furthermore be in accord with the holiness of the authority from which they came; to 'defraud and cheat' the pastoral ministry of the Church would be to subvert its foundation. We can infer from the language in which Grosseteste expressed his rejection of 'the things contained in this letter' that it was his own episcopal care of souls which he felt to have been 'defrauded and cheated'.[37] The complete failure of the Curia to consult the lawful patron of the benefice before providing represented the subversion of his own frequently-expressed principle that a bishop is personally responsible for every soul in his diocese. The telescoping of the traditional legal steps (the letter of request to the patron, the formal admonition and order to provide, and the executory letter) into one peremptory missive displayed disregard for the usual process of consultation. The repeated invocation of the *non obstante* clause audaciously overrode the special privilege granted by Pope Gregory IX in 1239, which had guaranteed Grosseteste that he might disregard any future papal provision unless explicit mention were made of that privilege. The whole added up in Grosseteste's eyes to a grievously sinful abuse of a power which the Holy See possessed for the building up of the Church, but was applying instead to its destruction:

These provisions do not tend to edification, but to manifest destruction;

---

[36] Ibid., p. 97.

[37] I am indebted to Boyle for his recent edition and novel discussion of letter 128 in *Med. Renaiss. Stud.* 8 (1979), 3-51. Quotations here and immediately following are made from Boyle's text of the letter (pp. 40-44).

for flesh and blood, which will not possess the Kingdom of God, have revealed them, and not the Father of Our Lord Jesus Christ, who is in heaven.

There is, then, no new proposition in Grosseteste's letter of the kind that some historians thought to have discovered there, only common doctrine. There is no contradiction with his previous attitudes, because he remained fully faithful to them. There is no evolution in his beliefs, but rather knowledge and application of a moral and canonical principle universally accepted in his age. The decision he reached was not ordinary, nor can it have been easy, and it was liable to be misconstrued by less well-informed minds; yet, as Tierney says, 'Grosseteste was never more Catholic than when he refused to injure the Church at the behest of the Pope' (p. 100).

This view of the vexed question can claim support, oddly enough, in the reaction which Grosseteste's letter appears to have brought from the pope, and also in a statement of Matthew Paris which, if not properly understood, might appear mistaken at best, and at worst, mischievous and misleading.

In November 1253, a month after Grosseteste's death, Pope Innocent issued *motu proprio* the bull entitled *Postquam Regimini*, freeing all patrons, clerical and lay, from any restriction upon the exercise of their patronage.[38] It would seem from this that the pope understood the *gravamen* laid against him by the English bishop, admitted his fault, and tried to undo the harm he had done against the Church. He did not regard Grosseteste's act as being an example of disobedience to due authority, but rather as a deserved rebuke; and he had, it appears, the magnanimity to pay heed to the warning given him.

In detailing the supposed content of Grosseteste's deathbed conversation with his physician, John of St. Giles, Matthew Paris relates that he denounced the pope as a heretic in the terms of the Decretals, because of his attitude to provisions, in that he held and acted upon an opinion chosen by human judgement and contrary to the Gospel.[39] While not accepting

[38] The Burton annalist states that the bull was written in consequence of Grosseteste's letter; see Stevenson, *Robert Grosseteste Bishop of Lincoln*, p. 317.

[39] The last words of the reported discourse are: 'Item dicit Decretalis, quod super tali vitio, videlicet haeresi, potest et debet Papa accusari'. *Chron. Maj.*, vol. V, p. 402. A.L. Smith rightly remarks, *Church and State in the Middle Ages*, (Oxford, 1913), p. 115, that in these so-called last discourses of Grosseteste all those things and persons are attacked who were the objects of Matthew's perennial animosity. As he wisely points out, 'The modern historian is often faced by the

that his report of the discourse is accurate, I think one can none the less find in this reference an echo, distorted certainly, but still audible, of the true position. Matthew's words betray a knowledge of the famous incident and an awareness that Grosseteste had refused obedience to the pope in the name of the Church's order. His story, then, represents the somewhat confused awareness that in certain cases, of which heresy is the most evident one, the duty of the subject to submit to the bearer of supreme authority is replaced by an obligation to disobey a command which is sinful.

Even such a sincere admirer of Grosseteste as Powicke was inclined to feel that he was ultimately too severe and unbending to win more than the respect of his subjects, and to wonder whether he really came to know his priests and people as the ideal pastor should. This reservation is worth pondering, for it affects one's final judgement of the effectiveness of the man in his exercise of office, and indeed one's opinion of his humanity.

It is my impression that, while Grosseteste certainly preached to his people, as all the contemporary sources attest, his more continuous preoccupation was with the level of education of his clergy, and he regarded himself as the pastor of the priests first and foremost. How well he understood them it is difficult to say. His advice to those who were ill-educated was certainly compassionate enough: to learn off the following Sunday's Gospel story during the week and recount it at Mass, to borrow from another priest some notes on the *Pater Noster* and the Creed, and to learn the lives of the saints. His own homilies to the clergy, on the other hand, make no concessions to the ignorant, but are learned, demanding, and lengthy. The sermon on the resurrection which he preached to his clergy on an Easter Sunday morning, for instance, may have taken up to an hour to deliver, and it is hard to believe that there were many there, aside from the university men, who were capable of gaining much from the doctrinally rich but erudite and philosophical homily.[40] Perhaps, though, the majority of

demoralising alternative, whether he will be critical, cautious and dull; or will accept Matthew Paris and make a good story' (p. 170).

[40] McEvoy, "Robert Grosseteste's Theory of Human Nature. With the Text of His Conference, *Ecclesia Sancta Celebrat*", in *Rech. Théol. anc. méd.* 47 (1980), pp. 131-87.

the clergy were accustomed to make certain allowances and were prepared to admire on the mouth of their bishop those periods and cadences of sacred eloquence which they themselves neither expected to follow nor desired to emulate.

It may be that no man could have resolved upon doing what Grosseteste undertook to do without having some iron in his soul. In a limited social order, everything a prominent and powerful man does is subject to criticism, and a leader who believes in the justice of his cause cannot be faulted for doing what he feels he must, while knowing all the time that tongues will wag, and wags give tongue. Grosseteste was no more invulnerable than any other man, and was in some ways less so than many. It must be remembered that he had risen from the poorest of people to become by force of his natural endowments the leader and spokesman of the English hierarchy. Those who might normally have passed over his origins in tactful silence could not resist casting up his base birth when he chose, as they saw it, to stir the pot; and we know that the canons of Lincoln were not slow to do so. It must have hurt. It helped to make him over-careful to remain above all suspicion of nepotism. He avoided promoting members of his own family, and his closest friend, Adam Marsh, felt he had to press the legitimacy of the claims of two of Grosseteste's kinsfolk who were studying at Oxford. As bishop he was pressed by old friends and former colleagues like John Blund to favour their nephews, and he no doubt had to steel himself against affection and sometimes even regard it as the enemy of the right. His recourse was to procedure, in the shape of an examination for all candidates. He was never allowed to forget that one who challenged vested interests, hereditary rights, and precedent, would incur the enmity of all those who were at home within the comfortable, settled world of established ecclesiastical custom, and that he could not afford to make any exceptions. A certain severity is perhaps inevitable, if the alternative would destroy a worthwhile programme of action. The strain of isolation and unpopularity took its toll as the years passed, and added weight to the unavoidable burden of responsibility, bringing him depression. In the year 1250, after his visit to the Papal Curia at Lyons, he felt so discouraged that he considered resigning his see, feeling that he had failed completely in his task. He was dissuaded from that step only by the intervention of his close friends. There are some signs that he was sensitive, perhaps over-sensitive, to criticism and gossip about himself. We may be certain, however, that there was plenty of both going about, and it is as well to bear in mind that those who feel persecuted do not necessarily suffer from a complex; presumably in some cases they simply are persecuted.[41]

If we wish to get a clear picture of the man who was Bishop of Lincoln then we must leave aside the official acts of his episcopacy, the pages of Matthew Paris, and even in a measure his own writings, and consult instead those who knew him best, loved him, and left something on record. It was undoubtedly with the friars that Grosseteste felt most at home among men. With them he worked hardest and relaxed most completely, and by them he was remembered. Some of his students, like the young Master Adam of Oxford, whose death on his way to the missions Grosseteste deeply mourned, and a number of his closest friends, had joined the mendicants. And he made more friends, among the Franciscans particularly, through his close association with them at Oxford. These friends, men like William of Nottingham, Agnellus's successor as Provincial, and Peter of Tewkesbury, admired him and stood by him; they supported his campaigns, and their friendship prevented his becoming completely isolated in office. Several members of the Marsh family served him loyally and devotedly; Robert, a member of his *familia* and (from 1248) his Archdeacon of Oxford, who accompanied him to Lyons in 1250; his kinsmen William, the bailiff at Buckden, and Thomas; but above all Adam, with whom he had spent (so Bacon tells us) thirty years in the pursuit of common studies, and whom he regarded as 'his truthful friend and counsellor, one who looks upon the truth and not on mere appearances; who rests upon a firm and solid foundation and not on a hollow and fragile reed'.

Adam loved him and admired in him the gifts of leadership he himself did not possess. Adam had a genius for friendship, as his correspondence shows, and regarding Grosseteste he exercised it in all the ways the friend of a lifetime does, in continual concern for his health as he aged, in warning him

[41] Matthew Paris is the only source for the story that an attempt was made on Grosseteste's life two years after his elevation, but this may have been nothing more than a rumour.

that he was overdoing it and trying to do too many things at once, in recommending physicians to him, and in exercising the friend's privilege and duty of forthright speech to save him from the severe side of a temperament Adam knew better than did any other, as he showed when he counselled him to aim at being loved rather than feared. Adam professed gratitude to the older man for his kindness to him '*a juvenilibus annis*'. Although our knowledge of their relationship is lopsided (sixty-two letters from Adam to Grosseteste survive, as against only two letters from Grosseteste to Adam[42]), it is plain that their friendship overcame the inequality of its beginnings. 'Nothing in human affairs', he wrote to Adam (who had expressed the fear that his frequent letters might be a burden to answer), 'save only your holy conversation when we are together, causes me such joy as the consolation I get from your letters'. Grosseteste found in Adam the other half of his soul. Adam, he believed, would always understand what he did and the reasons for it, even when the world thought him a fool. When he resigned his benefices in 1231 he confessed to Adam that up unto receiving his letter he had found no understanding of his actions from a single soul, but had had to endure nothing but obloquy and contempt, even from his very kindred. Many times as a bishop it was to be the confidence that one at least would understand him that would make his lonely command bearable. Though he had many friends, and even in letters addressed half-a-dozen of them with the familiar '*tu*', it was Adam who was and remained '*dilectissimus sibi in Christo*', and he '*suus Robertus*', sending him in his greeting, 'health, and himself'.

When Grosseteste's friendship with the friars is mentioned people think in the first place of the Franciscans and forget too easily his friendship with Jordan of Saxony, the successor of St. Dominic, John of St. Giles, whom he requested to have by his side at Lincoln (and who attended him in his last illness), and Robert Bacon, his old colleague at Oxford University. It is certain that he had Dominican priests living at the palace as well as Franciscans. The Brother Hubert who wrote a lament for him shortly after his death was probably one of the former (one detects a slight emphasis on the Order of Preachers at one

---

[42] Grosseteste, *Epistolae*, nos. 9 and 20. Adam is mentioned in nos. 99, 114, and 120.

point in his verses), and he certainly knew what Grosseteste's life at Lincoln was like during his episcopacy. Hubert probably penned these lines in praise of *'sanctus Robertus'* during the vacancy following his death. He was prepared to hear no criticism of a man whom he revered already as a saint.

I give here some of the passages in which Hubert, who was no Dante, rises above what was conventional and recurrent in the medieval lament for the passing of the great.

The sun of the land has set, the vessel of light has left the world, the earth pales as its sun fades. / Weep, Lincoln, at this setting, and grieve as the buried sun yields to darkness. / The holy army of brothers may grieve at the loss of its father, the flock will bewail the death of its shepherd. / Weep, Britain, widowed by the loss of your great protector, weep, for the radiance of your sun has gone down to rest / . . .
Suffolk rejoices to have given birth to such a man, and is glad that it was given so great a son. / England has rejoiced at the gift of so great a protector, for he was the wise counsellor of his fatherland. / Lincoln exulted while he ruled it; under him the city grew and its honour flourished. / . . . Girded with the sword of zeal, he launched himself upon the Church's enemies; he was heedless of injuries, threats, and death itself. / The king's majesty did not affright him, nor the loss of goods: he chose the divine part, heedless of threats and losses. / He suffered much for the sake of the flock, and struggled with all his might to defend his sheep from the savagery of the wolves. / . . .
He would not grant the care of the sheep to boys, but appointed men of good judgement and life as shepherds / . . .
It was the wish of this man never to be inactive in times of leisure, not an hour of his time passed unfilled. / Whether at prayer or in contemplation, reading or teaching, he was always about something of value. / He was assiduous in the sacred studies, at the same time the liberal arts occupied his time. / This holy man fostered the sons of noble families, whom he had with him, and educated them in sanctity.[43] / He taught them the rudiments of Greek and Latin; this was his pastime and relaxation when he was free. / . . .
The grace of hospitality waxed strong while he lived and he closed his door on no one. / His table was open to the famous and the unknown both, with a sufficient abundance of food and drink. / He was not one to build treasure-houses on this earth, he scattered widely, the generous giver of generous gifts. / He made himself the father and protector of the poor, as members of God he had them beside him while eating. / Often he knelt washing tenderly the feet of the needy, kissing them and weeping over them. / At the disposal of the suffering and sick, the poor and the needy, in these members he studied the love of the Head. / . . . In the villages he gave his money generously to the sick wherever they were

---

[43] It is known that his friend, Simon de Montfort, sent his two sons to the palace in fosterage.

lying. / He ordered the ill to be fed from his own table, when he found out the different places of their confinement./
Two orders mourn the loss of a father: the Preachers of the Word, and the Minors. / He was the father and protector of them both, and won their undying gratitude. / He loved to have friars by him, loved to see them come; / the greater their numbers and the more frequent their talks with him, the more pleased he was. / He determined to have the friars as partners in his task, to lighten his burden of work. / With the tenderness of a mother for her offspring he favoured, loved, protected, cherished and valued them. / He could not live without friars, in his *familia* and outside; he kept them right through, and his devotion to them was lasting. / Who now will be the protector of his wards, the friars, to whom shall they fly, who will be their support? / Their stronghold is gone now with Robert's death, their helper is powerless, their spokesman is silenced / . . .
Might there only succeed to him one who is aflame for the honour of Christ, one concerned to set to rights the Lord's house, / one incapable of yielding to entreaty, one who does not have a price, / one who cannot be bent by pressure or bribery, a man who will destroy every wicked thing before it harms / . . . He bore himself admirably before the people, with no levity in act, word, or deed. / There was great edification in his countenance, the whole populace had him for a mirror. / None was more devout in Church than he, he was ever at prayer, weeping and beating his breast. / Who can describe him at the altar? Who can number his tears and sighs? / Borne in rapture beyond himself, raised in soul above himself, he seemed often unconscious of himself. / Grave in Church, he was invariably agreeable in the hall; in his chamber he was cheerful, and gentle as a lamb.[44]

It is strange that Hubert makes no mention of Grosseteste's love of music, which became legendary in medieval England. From his own writings we learn that, like Boethius, he looked on music as a therapy for the mentally ill. A passage in his *De Cessatione Legalium* (on the harmony that prevails among the Fathers) is a strong indication that he was *au fait* with the nascent contrapuntal style of his time.[45] The following verse

[44] Hubert's poem is 200 lines long; I have translated less than half of it. It concludes with the following fragment of a sequence, which, Dr Hunt surmised, Hubert may have composed in anticipation of Grosseteste's canonization:

O Roberte presulum
　　Decus et flos cleri,
Sanctitatis speculum
　　Et amator veri,
Causam piam exulum
　　Dignare tueri,
Reis placans oculum
　　Iudicis severi.

[45] This passage is printed in an appendix to my article, "Robert Grosseteste and the reunion of the Church", pp. 83–4.

translation by Robert Mannyng of William de Wadington's *Manuel des Péchés* celebrates the bishop's love of harping and singing:

> Y shall you tell as I have herd
> Of the bysshop seynt Roberd;
> His toname is Grosteste,
> Of Lyncolne, so seyth the geste.
> He lovede moche to here the harpe,
> For mans witte it makyth sharpe;
> Next hys chamber, besyde his study,
> Hys harpers chamber was fast the by.
> Many tymes, by nightes and dayes,
> He hadd solace of notes and layes.
> One askede hem the resun why
> He hadde delyte in mynstrelsy:
> He answerde hym on thys manere
> Why he helde the harpe so dere:
> 'The virtu of the harpe, thurgh style and ryght
> Wyll destrye the fendys myght;
> And to the cros by gode skeyl
> Ys the harpe lykened weyl'.[46]

The warmest and most personal illustrations we have of Grosseteste's friendship with the friars is found in Thomas of Eccleston's account of the coming of the Friars Minor to England, which was written around 1250 by a man who, though not himself a scholar, in all probability remembered the former lector. Eccleston first tells how Grosseteste came to the friars' school, and in a much later chapter he records a medley of anecdotes about the bishop which were stored in the collective memory of the English province, and which bring us closer to his warm, wise, and humorous personality than does any other source.[47]

Friar Peter of Teukesbury was worthily favoured by the special love of the Bishop of Lincoln, and more than once did he hear from him many secrets of his wisdom. For he remarked once to him, that unless the friars cultivated study and applied themselves studiously to the law of God, we would certainly end up in the same state as other religious, whom we see, alas! walking in the darkness of ignorance.

Eccleston with all his simplicity understood the nature of Grosseteste's central objection to harmful provisions far more

---

[46] The full passage is reproduced by Stevenson (p. 334) from Furnivall's 1862 edition of *Handlyng Synne*, by Robert Mannyng de Brunne.

[47] For details of references to Grosseteste in the *De Adventu*, see n. 19 above.

accurately than those who would make him out an English
nationalist:

> Again, he said to Friar John de Dya to look out for six or seven suitable
> clerics from his parts [France] whom he might appoint to offices in his
> church. Even if they did not know English they could preach by their
> example. It is evident from this that it was not for their ignorance of the
> English language that he refused the Pope's candidates and the nephews
> of cardinals, but because they were only out for temporal ends. Whence
> when an advocate in the curia said to him, 'The canons (*canones*) want
> this', his answer was, 'Exactly, the canines (*canes*) want this'.

The story which immediately follows upon the last indicates
that Grosseteste shared the medieval genius for the eloquent
gesture:

> He got up and went to confession in English on bended knees before the
> boys presented to him by the cardinals, and beat his breast, and wept
> and shouted; and they retired in confusion.

One can picture this scene, as the old bishop, finding that
for once words failed him, acted the part of a distraught peni-
tent before the uncomprehending eyes of the young Italian
candidates for benefices, no doubt to the great amusement of
his secretaries. Accounts of an incident like this must have
spread like wildfire round the town and through the neigh-
bouring parishes.

Grosseteste was particular in matters of money, when there
was any likelihood of usurious practice:

> Moreover, when the chamberlain of the Lord Pope asked him for a thou-
> sand pounds due on the occasion of his visit to the Curia, wishing him
> to raise them from merchants, he replied that he did not want to give
> them the occasion of sinning mortally; but if he came safely back to
> England he would deposit the sum at the Temple in London; otherwise
> he would never get a halfpenny.

Grosseteste's wit was not forgotten either.

> Moreover, he said to a friar preacher, 'Three things are necessary for
> temporal well-being: food, sleep and jest'. Again, he enjoined upon a
> certain friar who was melancholic to drink a cup full of the best wine
> for his penance; and when he had drained it, albeit with much aversion,
> he told him, 'Dear brother, if you had that penance a few times you
> would then have a better conscience'.

Grosseteste had a sharp eye for anything which resembled

a bribe, and Eccleston admired him for this, as he did St. Edmund:

And the same Brother, Friar Peter, told how, when Lord Robert the Bishop of Lincoln was in great need of horses at the time of his elevation, his seneschal came to him, found him seated at his books, and announced to him that two white monks had come to present him with two lovely palfreys. And when he pressed him to receive them and pointed out that they were exempt,[48] he would not agree, nor rise from his place, but said, 'If I took them, they would drag me by their tails into hell'.

There is also an amusing and very human story of the bishop in a huff:

The Lord Robert Grosseteste, the Bishop of Lincoln, was once so badly vexed that the Minister [Provincial] did not permit a certain friar whom he had once had to remain in his guest house, that he refused to talk to any friar, even to his confessor. And on that occasion Brother Peter said to him that if he were to give all his goods to the friars but did not give them the affection of his heart, the friars would not care for them. And the bishop began to weep and said, 'In truth it is you [the friars] who are at fault, in that you cause me too much pain, for I cannot not love you, even if I show you the face I do.' Yet the friars ate at his own table beside him, and still he would not talk to them.

Grosseteste was ready upon occasion with practical advice:

He remarked to him [Peter] that houses (*loca*) above water are not healthy unless placed well up. Again, he said that it pleased him greatly when he saw the sleeves of the friars patched. Again, he said that pure pepper was better than salted ginger.[49]

Students were the same then as now, but Grosseteste had learned to live with that:

Again, he said that he rejoiced when his students were not paying attention to his lecture, though he had prepared it very carefully; because, of course, he would have no occasion for vainglory, and also would not lose any of his merit.

Eccleston's last anecdote concerning Robert suggests that he understood the mind of St. Francis very well indeed:

The aforementioned friar, William [of Nottingham] once said that when the Lord Bishop of Lincoln of holy memory, at the time when he was

[48] 'Exempt', i.e. from episcopal jurisdiction, as Cistercian monks. The implication of the seneschal's remark is that the gifts were being offered with no strings attached, the monks not being Grosseteste's subjects. Grosseteste was not naive enough to behave as though a bishop's influence ended at the limits of his actual jurisdiction.

[49] As a preservative of food, presumably.

the official lector to the Friars Minor at Oxford, was preaching on poverty at a chapter of the friars, and placed the mendicant state at the next rung on the ladder of poverty to the embrace of heavenly things, he yet remarked to him in private that there was a still higher grade, namely, to live from one's own work; whence he said that the Béguines have the most perfect and holy religious order, because they live by their own work and do not burden the world with exactions.[50]

There can be no doubt that Grosseteste was well acquainted with the personality and ideals of St. Francis, the circumstance of the Franciscans' foundation, and the nature of their vocation. There would have been no better source of information on all these matters than the saintly Agnellus of Pisa, the Minister of the English Province, who is said to have received the habit from the hands of St. Francis himself, and whose trust in Grosseteste is evident in the invitation he gave him to lecture in the Oxford School of the friars. What he accepted in becoming their first lector was nothing less than the responsibility for forming their youth for the ministry of preaching the Gospel and living the evangelical life. His aim as a teacher in the minorite school was to prepare itinerant preachers who would instruct the masses. In that shabby, ill-equipped and crowded school room, before the fervour of the first converts, daring speculation and school logic would have been out of place, even had Grosseteste favoured them; but his exposition of the Bible gave them all the riches their rags could carry. It was an exceptional combination of energies and talents which was cloistered at Oxford in those years, before being released upon the country at large; but the most remarkable of them all was the ageing teacher, who impressed his own convictions deeply upon his hearers, and who himself drew strength from their warm and youthful response.

R. W. Chambers in his classic life of St. Thomas More puts his hero into an English tradition that begins with Langland and reaches forward to Burke, all three men being, as he says, 'among the greatest of our reforming conservatives'.[51] The tradition of reforming conservatism in England was, however, older than Langland, for it began with Robert Grosseteste;

[50] 'De Scala Paupertatis' is the title given to a sermon preached by Grosseteste before a chapter of the Friars Minor. Extracts from it were published by Little in his Paris 1909 edition of Eccleston's De Adventu. There seems to be some doubt as to whether this is the actual sermon preached on the occasion referred to in William's anecdote.

[51] Chambers, Thomas More (London, 1935), p. 346.

and indeed the resemblance that links him with the 'English Socrates' is profound. Like More, Grosseteste had a strong faith in the society in which he lived. It was the responsibility acutely felt for it by both men which prompted and lent vigour to their criticism. They were Europeans, bound by strong ties of loyalty to the pope as head of the Church, yet tragically aware of his human frailty. They were men who understood loyalty and obedience to their superiors in the Church and in the realm, but who acknowledged the higher loyalty owed to their *daimon*. Each was remembered for a high moment which marked the end of his life, and which appeared to later generations to epitomize the integrity so fundamental to the character of both. More chose his own death in refusing obedience to the king, though it is at least as true to say that it was forced upon him. Grosseteste too chose to disobey, though his last letter is charged with outraged grief and bitterness that such a terrible step should be his only option in good conscience; all his instincts were in revulsion against it. If they were Utopians in the reforms they envisaged, and if they expected therefore too much from human nature, the disillusionment they suffered in realizing the fragility of all humanly-based order yet reflected the nobility of their ideals. Both probably underestimated the extent to which lesser men are governed in their actions by expediency; they were themselves totally free from it, innocents who failed to understand its attractions. They set their own standards by the counsels of perfection, and were dismayed when others failed to keep the commandments. It is in their integrity and consistency that More and Grosseteste are most nearly alike; they had clean hands and awkward consciences, qualities the more dangerous for their possessors, as they are an inevitable reproach to their antagonists.

Grosseteste died in his manor at Buckden, Huntingdonshire, on the night of 8 October 1253, after several months of illness. He was buried in his Cathedral, in the grave where Adam Marsh was to join him a few years later. Leland recorded in his *Itinerary* that Grosseteste 'lyeth in the hygheste southe isle with a goodly tumbe of marble and an image of brasse over it'.[52] Cromwell's troopers made short work of tomb and effigy after the capture of the city in 1644, and nothing now remains

---

[52] See "The tomb of Robert Grosseteste with an account of its opening in 1782", by Hill, in Callus (ed.) *Scholar and Bishop*, pp. 246-50.

of it. Grosseteste was venerated as a saint immediately after his death, and for several years afterwards the annals refer to miracles which were attributed to his intercession. Three major attempts were made by his successors at Lincoln to have his cause introduced at Rome for canonization, the last, in 1307, having the support of King Edward II. All failed. English historians have been inclined to attribute the failure to curial resentment against Grosseteste's criticisms of abuses, and there is probably some truth in this view. As it happens, the requirements for canonization were being made more stringent at the time, and something more impressive than a purely local cult was being required. If Grosseteste had been a regular rather than a secular, his chances of canonization might have stood higher, for he would at least have had a biographer, and some clients at court. David Knowles records the judgement of Fr Daniel Callus on this question: 'I once asked him, rather maliciously, why Grosseteste had never attained the canonization for which he was so often recommended. "He was all right, but he was not a saint. That was it", replied Fr Daniel'.

The anonymous medieval author of the following sequence had no such reservations:

> Alme pater, doctor morum,
> Forma factus subditorum
> Vita, verbo opere, /
>
> Informator sui gregis,
> Verbo doces, actu regis,
> Vita das proficere. /
>
> Carne purus,
> Carni durus,
> Carnem domas aspere, /
>
> Iam solutus,
> Carne tutus,
> Celo gaudens vivere. /
>
> Quo, Roberte,
> Pater per te
> Possimus pertingere.[53]

---

[53] The text of this sequence, and also of verses entitled *'Planctus Roberti'*, are published by Reichl, *Religiöse Dichtung im englischen Hochmittelalter* (Munich, 1973), p. 482.

PART TWO

THE ANGELIC LIGHT

Chapter 1
The Influence of Augustine

A study of medieval philosophy which undertakes to follow
an author through the grades of created being is under no obligation to apologize for including a chapter on angelic being;
on the contrary, to avoid discussion of 'the first created light'
would be indefensible. The historian must break the hold of
later developments upon his thought, and attempt to return
within the cognitive horizon of an age whose philosophical
sources were filled with references to separate substances
(Aristotle), lesser gods (Plato), and intelligences (the Arabs)—
in short, to a region of being which later developments have
consigned to the theologian, and more specifically still to the
biblical exegete, on the assumption that even he can say little
enough about it. Every student of medieval literature and ideas
must win for his imagination the freedom to exercise itself
within a universe only half visible to the eyes of flesh.[1] He
must submit to the vast readjustment involved in seeing man,
not as the creature standing at the top of the evolutionary stair
the foot of which is lost in a biological underworld, but as
the figure visible at the bottom of Jacob's ladder, the summit
of which is invisible with light, the traffic of which intense.
Jacob's struggle with the angel gave him possession of an alien
land; the historian can achieve a new space of sympathy with
the Middle Ages only by repeating it. The territory newly
gained, however, alters the whole map, since the regional on-
tology of the separate intelligences pursued by the Scholastics
did more than add an essential element to their model of the
universe: it affected every other element in the plan by the
harmony it gave to the whole and the multiform interaction
and analogy it acknowledged among the angels, the visible
cosmos, and man. For Aquinas, just as much as for his theological predecessors, the beauty, harmony, and motion of the

[1] As C. S. Lewis has no trouble in showing at numerous points in his fine little
book, *The Discarded Image* (Cambridge, 1964).

world are due, under God, to the heavenly substances. The consistent effort made by the Scholastics to develop the concept of a pure spirit gave them a unique point of reference and of contrast for their anthropology, one which some seventeenth-century metaphysicians did not fail in their time to value and preserve. These factors made angelology a *Grenzgebiet* between philosophy and theology, and the historian of philosophy must take the delimitation between nature and grace as he finds it in the texts he studies; its very unclarity can be adequately recorded only within a respect for the unitary doctrinal whole which was its circumambient up to and during the thirteenth century.

Grosseteste inquired into the nature of the angels in many of his works, returning ever again to the same themes and problems. Consistent with our interest in the development of his thought, we shall place the main witnesses to his teaching in proper chronological order (*De Intelligentiis, Hexaemeron, Commentarius in Caelestem Hierarchiam*), and note at each new stage the new problematic of his enquiry and the new sources on which his reflection was based.[2]

I   DE INTELLIGENTIIS

This little treatise forms the second part of a long letter composed by Grosseteste in answer to an appeal addressed to him by Master Adam Rufus; the first part of the letter is often referred to as *De Forma Prima Omnium*, or *De Unica Forma Omnium*.[3] Rufus had asked Grosseteste's opinion about the proposition, '*Deus est prima forma et forma omnium*', and had added a supplementary question concerning the intelligences: '*utrum sint distinctae loco, an in quolibet loco simul?*'. The identification of Adam Rufus, who is referred to in two later epistles of Grosseteste (nos. 2, 38), with the young Master

---

[2] I have profited from Fr. Gieben's pages on the angelic nature in his unpublished thesis, thus far the only study made in this area.

[3] The letter has been edited twice: by Luard (*Epistolae*, no. 1, pp. 1-17), and Baur (*Die Werke, De Unica Forma*, pp. 106-11,*De Intelligentiis*, pp. 112-19). The two texts differ little, and none of the variants effects the doctrine. I give references to Baur's text, adopting, however, the following corrections from Luard: *existimabar* for *existimabam* (Baur, p. 106.18-19); *iuvando* for *iurando* (p. 117.16); *in mente artificis artificii* for *in mente artificii* (p. 109.23).

This letter is one of the best-authenticated treatises of Grosseteste, presumably because the letter-heading was retained in its copying. The division into two parts and the titles given to them are not authentic.

Adam of Oxford (or Exeter), the story of whose entry into the Franciscans and early death in Southern Italy is told by Eccleston, strongly suggests that *Letter no. 1*, from which *De Intelligentiis* is excerpted, was composed before 1231 or 1232, after which time Adam would have been addressed as *'Frater'*. There are good reasons for thinking that the letter cannot have been written much earlier than 1228, and that it comes from the period of Grosseteste's greatest interest and pro- ductivity in scientific matters, when he commented on the *Posterior Analytics*, jotted down the bulk of his notes on the *Physics*, and composed the best examples of his scientific works.[4]

The second problem posed to Grosseteste by Adam was: are the intelligences (or angels) in distinct places, or in any place at once? (p. 112.3-4) Grosseteste's approach towards an answer appears somewhat tortuous, for he discusses at length the ubiquity of God in terms of transcendence and immanence (contenting himself largely with a series of quo- tations from Augustine: pp. 112.4-114.7), and then treats equally extensively, but more originally, of the soul's omni- presence in the body (pp. 114.8-115.23), before feeling him- self ready at last to discuss the relationship of the angels to matter (pp. 115.24-119.26).

Just as the soul has neither *situs* nor *locus*, but is everywhere in the body united to it, so an angel which assumes a body for the sake of some ministry is present in it without site or place. Grosseteste's key-phrase, *sine situ praesens, et sine loco ubique totus*, is repeated here from the earlier part of his dis- cussion; he found it applied to God in a passage from the *De Trinitate*,[5] where Augustine forged it in order to express God's ubiquity and his transcendence of the categories.[6] The phrase becomes Grosseteste's guideline and is applied with the necess- ary adaptations to the soul and the intelligences. Augustine's list of categories has, however, vibrated the Aristotelean chords of Grosseteste's memory, so that he feels stimulated to invoke a more scientific conception of *situs* and *locus*; his definition

---

[4] McEvoy, "Der Brief des Robert Grosseteste an Magister Adam Rufus (Adam von Oxford, OFM): ein Datierungsversuch", in *Franziskanische Studien* 63 (1981) 221-6.

[5] St. Augustine, *De Trinitate* V, i, n. 2.

[6] *De Intell.*, p. 112.4-9. The categories named by Augustine are *situs, locus, qualitas, quantitas, habitus,* and *tempus*; he adds *indigentia* and *mutatio* for good measure, and according to the exigencies of his rhetoric.

of the latter is fully Aristotelean,[7] and his use of the former reveals the mathematician's imagination at work.[8]

The relationship of the angel to the body it assumes can be thought of on the model of one of the soul's functions with regard to the body. If we abstract from the soul considered as the perfection of the body united to it, there remains its other capacity, that of moving the body and undergoing its movements; if this were the only function it had, it would still be everywhere in the body, lacking site and place. This aspect of the soul's reality provides a model for the angel's relationship to the body it assumes: it is its mover and governor, though it is neither situated nor located. As angels in themselves have no place, so too they are not circumscribed within the body they move, nor diffused spatially within it as one material thing in another (such as light in the atmosphere); nor are they at a measurable distance from it or in proximity to it.

How do angels move the bodies which they assume when they minister? Grosseteste admits the difficulty of the question, but again recurs to the soul's activity as an analogy for the angel's. The soul moves the greater members through the agency of nerves and muscles, which it in turn moves by the bodily spirits; these it affects directly and immediately by a natural or voluntary desire which, like the soul itself, is immaterial, and affects the material thing whose subtlety most assimilates it to the spiritual: bodily spirit, or light. The angel produces motion in the same fashion, very likely: by affecting the subtlest element of the body through its immaterial desire (p. 116.21-30). Grosseteste is not satisfied that this explanation will work; after all, what enables the soul to act as mover of the body is its union with it; it is on account of this connection that the soul's immaterial and non-local motion can produce proportionate but material and local change in the body. The angel, however, has no union with the body such as would enable it to produce effects upon it. This, Grosseteste confesses, is a genuine puzzle; yet we must accept, it appears that the angel does somehow move the body it assumes by affecting it immaterially, much as the soul does its partner;

---

[7] Ibid., p. 115.13-15: '. . . sine loco ubique tota . . . i.e. sine superficiei ambientis ipsam circumscriptione.'

[8] Ibid., p. 115.8-10: 'Si enim esset [anima] situalis, posset a puncto extra situm ipsius sumpto duci linea ad ipsam et mensurari et determinari certis mensuris spatium inter ipsam et punctum signatum.'

though we cannot explain exactly how this is possible in the case of a pure spirit. We arrive, therefore, at defining the angel's presence in a body as a mode of moving and controlling it for the purposes of a ministry.[9]

An angel is sometimes said to be in a certain place even though it is not in a body, as when it exercises a sort of prefecture, furthering, helping, defending, and ruling the things that are in that place; its relation to them is then definable in terms of acting upon them and presiding over them. It can be said to be circumscribed in that place, though not of course in the Aristotelean sense of 'being contained in it'; here the description, *'sine situ praesens, sine loco ubique totus'*, applies once more. Angels can also be said to move from place to place, according as they change their prefecture, because, unlike God, they are not everywhere at once; but once again, a physical or mathematical sense cannot be given to their movement.

A spiritual being can rightly be said to be in a place when that spirit is passive to the physical motions of the bodies located there; as the fallen angels and the souls of the lost are said to be in hell. Indeed, evil spirits and lost souls, while they are in bodies, can suffer pain with regard to the motions of those bodies, but they cannot be said to suffer *by* their motions, for while the more noble can act upon the less noble, the reverse is never the case.[10] The bodily action is no more than a necessary occasion of the soul's change, which it does not causally produce, any more than the motion of the mirror causes the motion of the ray of light which falls upon it, though it is indeed the necessary occasion of its change of direction (p. 119.24-26).

The reading of this little treatise invites several reflections. The first is that, if Grosseteste has summarized Rufus's question with accuracy, then it would appear that he has answered it only indirectly. The question was, whether the angels differ in place, or are together in any place. The answer given is to the effect that the relationship of an angel to a material element is individual but not permanent; one angel can replace another

---

[9] 'Hoc est itaque angelum esse in assumpto corpore; comparationem moventis et regentis illud in usum alicuius ministerii ad ipsum habere.' (ibid., p. 117.9-11).

[10] p. 119.5. Grosseteste supports himself with the authority of Augustine, *De Musica* VI, 5, n. 9.

as term of the relationship, but there will not be a plurality of angels together in the same place or body.

Grosseteste felt the need to put the question into a wider horizon, so that what he wrote is in fact a little treatise on the spiritual world: God, the human soul, and the angel. Already evidenced in this option is the strong Augustinian trend of thought which will flavour the whole treatise. For Augustine, the convert to Christianity from materialism through Neo-Platonism, the most fundamental metaphysical difference is that which divides being into material and spiritual; this is the perspective which we find throughout the *De Intelligentiis*, and which dominates its method from end to end. A quotation from Augustine rebukes the sensible imagination for thinking the divine and spiritual in sensuous terms,[11] and Grosseteste reiterates this theme by contrasting the spiritual world with the material: space, place, bulk, distance, and passivity, are excluded from the discourse on the spiritual. The spiritual world is defined by its immateriality, its transcendence of matter, therefore, but also by its immanence within matter; transcendence and immanence are co-implicative, because the greater the power of the spiritual being, the more extensive will be its action upon material being. It is in order to situate the angel's relationship to the material world that Grosseteste feels compelled to discuss God and the soul first. God alone of all spirits is everywhere in the universe at once (pp. 112 f.), but his image, finite spirit united to a body, is everywhere in the body at once. The angel is neither omnipotent nor yet so finite as the soul, rather it is capable of a wider relationship with the material world than is the human spirit. The angel can move this body or that, and be present here or there as guardian, just as God requires it; whereas the soul, by its union with the body, is limited to producing its effects upon a single material thing. Never the less, the reality of spirit's dominion over matter is instanced in it too. Grosseteste's position is neatly summarized in a single proposition: 'In the place [over which it presides] the angel is present, though not situated. While not in place, it is yet to be found whole everywhere in

[11] 'In eo tamen ipso, quod dicitur Deus ubique diffusus, carnali resistendum est cogitationi, et mens a corporis sensibus avocanda, ne quasi spatiosa magnitudine opinemur Deum per cuncta diffundi, sicut humus aut aer aut lux ista diffunditur.' Augustine, *Epistola 187*, 11, *P.L.* 33, col. 836.

that place, just as the soul is present in the body without being situated, and, though not in place, is yet whole in each part of it; and God is present in the entire universe, not situated nor in place, but wholly everywhere.'[12] A single formula thus covers the relationship of all spiritual being to matter, though the relationship is differentiated into various degrees of transcendence and immanence.

Having located the spirituality of the angel in general terms between God and the soul, Grosseteste handles all particular problems associated with it by a method of contrast and analogy which takes the soul's union with the body as the primary, experiential instance of the matter-spirit relationship. The angel's moving a body is thought of on the model of one function of the soul, that of moving and governing the body. In the case of the soul this function is rendered possible by the union which makes it the perfection of the body (pp. 115.24-117.11). The angel must act on the body it assumes in the same immaterial way as does the human spirit: its will affects the subtlest and least material energy of the body, light or bodily spirit, which in turn moves the nerves and muscles. As the soul's virtue is everywhere in the body, so that the soul itself can be said to be everywhere in it, the angel too is located (though not circumscribed) in each member where motion originates. To the extent that the angel's relationship to a body differs from that of the soul to its partner, the *nexus* between it as a pure spirit and the body it uses remains impenetrable to the mind. In other words, Grosseteste has no metaphysical categories which apply to the angels as an autonomous region of spiritual being. The other problems, as regards the pure spirit's relationship to the region it protects (pp. 117.12-118.2), the suffering of the fallen angels (pp. 118.3-26), and the impossibility of direct action by the body upon the spirit (pp. 118.27-119.26), are all resolved by a similar application of analogy.

The problem of the relationship between the intelligences and the material order should in principle have given Grosseteste a free hand to discuss such questions as might occur to a Christian reader of Aristotle's *Physics*: is the motion of the

[12] 'Est itque in tali loco angelus sine situ praesens, et sine loco ubique in illo loco totus, sicut anima in corpore sine situ praesens et sine loco ubique tota, et Deus in universo sine situ praesens et sine loco ubique totus.' (p. 117.23-6).

heavens a function which falls under the presidency over matter given by God to the angels? How is this motion effectuated? What kind of causality does it represent? How many are the movers? Grosseteste's silence is as absolute as though he had never read a physical treatise. The reason is evident: the interest which the *De Intelligentiis* expresses is not remotely cosmological, but exclusively theological. All of the problems discussed derive directly from the Christian tradition, and concentrate upon the angel as messenger and minister. The philosophical categories and the technical appearance of the treatise cannot distract attention from its exclusively theological purpose, which is indeed explicit: in the opening lines of his treatment of the angelic nature, Grosseteste refers to the ministry of the angels who appeared in visible form to the patriarchs. This is the origin of the problem he proposes (p. 115.27-31). Grosseteste's problematic remains that of the Western Fathers. Angels are described in the Bible as being associated with God in his work of providence and salvation; they are his messengers to man, the guardians of man and of the earth deputed by God to intervene in history. The biblical perspective remained dominant in this tradition, and precluded any exercise of metaphysical curiosity concerning the angels as a separate region of being. This factor separated the Western Christian tradition from later Neo-Platonism, which tended in most of its forms to be a philosophy of the intelligences emanating from the One, and therefore a metaphysical angelology. The latter preoccupation had to await the quickening of Grosseteste's interest in Pseudo-Dionysius.

II ANGELIC KNOWLEDGE

Already in our first encounter with Grosseteste's theory of angelic nature it has become apparent how profoundly both his problematic and his solution derived from a Latin tradition whose crest-lines had been established by St. Augustine. The exploration of the problem of angelic knowledge in Grosseteste's writings between c.1228 and 1237, and above all in two major works, the *Commentary on the Posterior Analytics* and the *Hexaemeron*, provides a strong confirmation of the first impression. Only after 1239 or so, when the idea of hierarchy—not in angelic nature alone, but in creation universally—asserted strong claims over Grosseteste's thought, can we speak of a real modification of his perspectives, exemplified in the simultaneous attraction and tension which were the vectors of his assimilation of the *Hierarchies* of the Pseudo-Dionysius.

St. Augustine's personal contribution to angelology lay above all in his introduction of a new perspective, the originality of which allowed him to choose out a few central principles from the extremely fluid tradition he inherited, and thereby to determine the shape of western speculation in this area until the high period of Scholasticism. Augustine established the problem of angelic knowledge at the centre of his enquiry, and gave it a force that did not leave even the theory of creation itself untouched. Having to reconcile the temporal Genesis-narrative with the Wisdom-tradition of simultaneous creation,[13] Augustine made the six days of creation into noumenal developments within angelic cognition. Morning, noon, and evening of these days became essential moments in the angels' knowledge, and the latter was made a principle of differentiation within the whole work of creation, to distinguish an order of natural priority within the temporal simultaneity of universal production. Thus noonday light denotes the knowledge of things seen in the Word before their creation, evening light attains them as realized in the Word, and the dawn light sees the things in themselves and relates them to the Word, their origin. The angel's evening and morning knowledge of itself is the first day, of the firmament the second day, and so on up till the sixth day, the day of man.[14] The entire process of creation is represented by Augustine as a granting of light and illumination. The Word is eternal, spiritual light, physical creation is corporeal light or form, and the illumination of intelligence by the Word perfects the luminous angelic nature by direct impression as well as by reflection from the lower creation.[15]

An instruction to his priests on the imitation of angelic righteousness illustrates the extent to which Augustinian exemplarism underlay and enfolded Grosseteste's definition of angelic life.[16] Although the firmly-expressed belief in the

---

[13] 'Qui vivit in aeternum creavit omnia simul', Wisdom 18:1. See further Tavard in *Handbuch der Dogmengeschichte* Bd. II, Fasz. 2b, 'Die Engel', p. 47.

[14] *De Civitate Dei* XI, 7.

[15] 'Lux dicuntur angeli quia illuminati sunt', ibid, XI, 11.

[16] *'Sacerdotes tui induantur iustitiam'*, edited (with many corruptions) by Brown, *Fasciculus . . .*, p. 302. Delivered perhaps as early as 1229-31, when

spirituality of the angels represents an advance on Augustine's doctrine,[17] which had reflected the common patristic ambivalence concerning their freedom from matter, the view of angelic knowledge emerging from Grosseteste's study conforms fully to the teaching of his chosen master, as does the language in which it is expressed. The angels are defined as purely unembodied and intellectual substances, the energies of whose intelligence are inflexibly fixed upon the Trinity. The Word is their book of life and mirror of eternity, in which they see the power, wisdom, and goodness of God; He is their rule of life, by whose guidance and illustration the weight of their love is regulated—a favourite Augustinian emphasis of Grosseteste.[18] Their contemplation and activity are a single expression of a simplicity of spiritual life to which our fragility can aspire but may never in this life attain.

It was, however, for his *Hexaemeron*, the supreme achievement of his theological career, that Grosseteste reserved his fullest and most thorough study of the 'created spiritual light' and exemplarism. When read with fuller understanding, he points out, the creation-narrative does not omit the angels from its account, for in the production of light on the first day we find an expression of their generation from the first light, an expression whose congruency could not be improved upon,[19] for pure intelligence cannot be better described than

Grosseteste was Archdeacon of Leicester, this conference exhorts priests to imitate the supernal powers in contemplation and action. It derives its inspiration from St. Anselm, *Dialogus de Veritate*, c. 12. It is noteworthy that this form of 'individual' exemplarism passed over into the corporate imitation of the celestial by the ecclesiastical orders, during and after Grosseteste's study of Pseudo-Dionysius.

[17] Though this belief remained undeveloped right up until the Golden Age of Scholasticism, Grosseteste seems at no stage to have hesitated in subscribing to it; see for example the *Comm. in Anal. Post.*, fol. 4d: 'Dicitur secundo per se quod per causam materialem non est, et sic dicuntur intelligentie per se entes vel per se stantes.' Cf. fol. 4c; *De Finitate Motus et Temporis*, ed. Baur, *Die Werke*, p. 105; *De Ordine Emanandi . . .*, *ibid.* p. 150.

[18] 'Sunt itaque substantiae incorporeae intellectivae, irrepercussa acie intelligentiae ineffabilem pulchritudinem trinitatis contemplantes, ipsaque sapientia et verbum Patris est eis liber vitae et speculum aeternitatis in quo intelligibiliter legunt et limpide conspiciunt immensam patris potentiam, quam revereantur, universalem filii sapientiam, qua illustrantur, suavissimam spiritus bonitatem, quam amplexantur, ibidemque conspiciunt quid ad deum superiorem illis, quid ad sibi pares, quid ad inferiores illis sit agendum. In illum a quo ad haec videnda illustrantur, superfervido amore rapiuntur, et in eius visione et dilectione suavissima delectantur, omnes operationes suas secundum regulas quas in libro vitae legunt, regulariter dirigunt . . .' (Brown, *Fasciculus*, p. 302).

[19] *Hexaemeron*, Oxford *MS Bodleian lat. th. c. 17*, fol. 203a: 'Nec putet aliquis

in terms of light, as an image of the eternally active and generative light which God is. The identification of intellectual being with light is posited frequently by Grosseteste,[20] and always makes reference to the image-bearing nature of created spirits, those reflections of the first light. The light created on the first day contains a double level of reference in the symbolic language of the Bible, for it refers globally to the material light which presided as active created form over the work of the first three days, and to the angelic nature.[21] This double reference of light is a consequence of its transcending the dichotomy of spirit and matter, since as a synonym of 'form' it denotes the active and self-expressive character of being as such.[22]

The association of spiritual with material light leads naturally to the nexus of questions concerning the temporal duration of the creation-process and the order of priority and posteriority in the production of things. In the *De Cessatione Legalium* Grosseteste unhesitatingly affirms the simultaneity of all creation on the basis of Wisdom 18:1, but allots to the angels a priority of nature over the light of the material heavens.[23] In the *Hexaemeron* he discusses the problem at length, evidently preoccupied with exploring the reasons underlying Augustine's hesitancy in the matter.[24] Basil and Jerome, he points out, asserted the simultaneity of creation, whereas Augustine wavered, being at first tempted by the view that the creation-process involved temporal succession, then later on suggesting that the things of heaven and earth were

angelorum creationem inter opera horum sex dierum esse omissam, quorum creatio et consummatio nusquam congruentius in hiis sex diebus exprimitur quam per lucis conditionem.' A marginal note in Grosseteste's hand reads: 'quod non est hic omissa angelorum condicio'. Cf. St. Augustine, *De Civitate Dei* XI, 7.

[20] *Comm. in Anal. Post.*, fol. 8c.

[21] *Hexaemeron*, fol. 203a: 'Per lucem igitur conditam intelligitur primo sensu lux invisibilis primos tres dies temporaliter peragens, et secundo natura angelica in dei contemplationem conversa, tercio quoque intelligi potest quod lucis condicio sit informis materie usque formationem deducto. Omnis namque forma quedam lux est et manifestatio materie quam informat, ut ait Paulus, omne quod manifestatur lux est.'

[22] Ibid., fol. 200a: 'Omnis enim forma aliquod genus lucis est, quia omnis forma manifestativa est; et forte ista tria, inane, vacuum, et tenebrosum, correspondent forme, nature, et accioni.'

[23] *De Cess. Leg.*, MS cit., fol. 169c: 'Angeli possunt dici ante solem creati et conditi prioritate nature, sed non prioritate temporis, quia "qui vivit in aeternum creavit omnia simul".'

[24] *MS cit.*, fols. 201d–202d.

made in a single moment, but that, while the former were perfect from the beginning, the latter were made in seminal reasons which unfolded in successive phases to produce life. The firmest point in Augustine's reflections, Grosseteste notes, was his belief that the revolution of the six days in the angelic mind was simultaneous; there the principle of the angels' transcendence of time, being applied, brought clarity into an involved issue. Grosseteste's determination of the question hopes to recover the essence of Augustine's intention, and succeeds. Neither in the progress from morning to evening denoting a day, nor in the entirety of the seven days, are we to understand a natural temporal succession. The phases differentiating the single divine act of creation are moments occurring in the cognition of the angels. They are called 'days', only because in them there is a revolution analogous to the morning and evening of a temporal day. As moments they are distinct; as occurrences in a purely spiritual intellect they are simultaneous. Their distinction consists in an order, one may even say a natural order, since it is based upon the eternal ideas; its principle is priority and posteriority of nature, i.e. of the differential participation of a graduated creation in being and goodness. The best illustration of this play of order and coincidence is to be found in the action of physical light, which, passing through a transparent medium at infinite speed, lights all its contents simultaneously, but those nearer the origin of light more proximately.[25]

Grosseteste describes in detail how the revolution of what are called 'days' occurs in the angel's mind.[26] By the 'light of

[25] Ibid., fols. 202^d–203^c: 'Huiusmodi igitur circulatio a mane usque ad mane in cognicione angelica, dies [est *adds Grosseteste in margin*] naturalis. Sed dies iste temporali caret successione, et septem dies hic commemorati non temporaliter sibi succedunt, sed in cognicione angelica simul sunt, ibi itaque simul sunt dies et nox, et vespera et mane. [fol. 203^a] Habent tamen ibidem prius et posterius secundum naturam, quemadmodum solis splendor subito pertransit et simul tempore illustrat loca soli viciniora [et remociora *adds Grosseteste in margin*], cum tamen prius natura illustret loca proximiora.'

[26] Ibid., fol. 202^d: 'Item vespera et mane aliter intelliguntur, prima namque lux ut dictum est secundum Augustinum est angelica natura ad deum conversa, et conversione que ad deum est deiformis effecta, in qua deiformitate ipsa est quasi lux et dies post tenebras negacionis existentie sue, et post tenebras privacionis in se naturaliter precedentis [hanc lucem sue deiformitatis *adds Grosseteste in margin*], que erant quasi tenebrae super faciem abyssi. In hac vero luce et die cognovit creatorum et seipsam in ratione sua creatrice in mente divina. Huius itaque prime diei vespera est post lucem dicte cognicionis velud obscurior cognicio sue proprie nature in se, qua cognoscit quod ipsa non est hoc quod deus. Cum vero post hanc

the first day', we understand the angelic nature as turned to the first light and made deiform in the image of its splendour; as it were, the light of day succeeding to the darkness of non-being. The angel's bright vision of eternal light includes the sight of its own creative reason in the divine mind, by contrast to which the angel's reflexive knowledge of itself, a connate property of spiritual existence, is a less clear knowledge of itself, since it poses a distinction between its existence in the eternal reason, and in itself. This self-consciousness is ves-pertinal, a mere remnant of the noonday splendour, but to night there succeeds again morning, when the spirit is borne back by contemplation of the Word's presence within itself to the praise of the light that originated it. The end of the first day is the beginning of the second, for the angel contemplates the eternal reason of the firmament, and in the dimmer even-ing light perceives the firmament as created and posed in its finitude. Morning knowledge, ending the second day by the praise of the Creator as reflected in the heavens he has made, is simultaneously the beginning of the third day; and so for the remaining days up until the sixth. The principle guiding the process within the angelic mind is, of course, exemplaristic, and Grosseteste formulates it expressly and for its own sake.

A brief summary of the creation-process occurring much later in the *Hexaemeron* divides the production of things into three moments: their production in the eternal reasons, in created causes, and in form and species.[27] The *glissement* be-tween cognition and cosmology, a characteristic as we have seen of the Augustinian heritage, could not be clearer. What is more, Grosseteste finds it to be part of the meaning of the biblical text: when, in the narrative of events from the second to the sixth day God said '*fiat*', we are to understand the reason of the creature in the eternal Word; '*et sic est factum*', refers to the understanding of the thing imprinted in the angelic intellect, while the repetition of '*quod fecit deus*' signifies the coming-to-be of the creature in reality. When we turn back

obscuriorem cognicionem sui in se, refert se ad laudandam ipsam lucem quod deus est, cuius contemplatione formatur, et percipit in ipsa luce firmamentum creandum, sit mane finiens velud primum diem naturalem et velud inchoans secundum diem.'

[27] Ibid., fol. 236^d: 'Habent enim res triplicem fiendi modum, primus est quo fiunt in eternis rationibus increatis, ubi omnia sunt vita, sicut scriptum est quod "in ipso vita erat". Alter modus quo res fiunt in causis creatis. Tercius est modus quo res perficiuntur in forma et specie.'

only a few years in time to the *Commentary on the Posterior
Analytics*, the same three phases, with the moment of angelic
intellection interposed, are found stated, albeit in terms that
appear so abrupt that interpreters of Grosseteste's thought
have felt it incumbent upon them to explain what looks like
a dangerously pantheistic conception of emanation. I quote
the texts in question:

Cognitiones enim rerum subsequentium, quae cognitiones sunt in ipsa
mente intelligentiae, sunt formae exemplares et etiam rationes causales
creatae rerum posterius fiendarum; mediante enim ministerio intelligentiarum, virtute causae primae, processerunt in esse species corporales.[28]

Similiter intelligentiae recipientes irradiationem a lumine primo, in ipso
lumine primo vident omnes res scibiles universales et singulares, et etiam
in reflexione ipsius intelligentiae supra se cognoscit ipsas res scibiles
universales et singulares, et etiam in reflexione ipsius intelligentiae supra
se cognoscit ipsas res quae sunt post ipsam, per hoc quod ipsa est causa
earum.

Read in the knowledge of a strong Avicennian influence on
the commentary as a whole,[29] the two passages seem to teach
that the separate intelligences are mediators in the work of creation.[30] Pierre Duhem could find no other sense in the words,
and was forced to assume that Grosseteste, whom no one could
accuse of pantheism, was here simply reporting the beliefs of
pagan philosophers. I am reluctant to accept this explanation,
which is too benign and aprioristic to be offered as an interpretation of what are in fact highly personal passages, both of
which beyond all reasonable doubt express views actually held
by their author. Gieben, pointing out very correctly that
Grosseteste explicitly rejects emanation in the *Hexaemeron*,
offers an intrinsically attractive solution to this embarrassing
problem.[31] Respecting the integrity of the text as an expression
of Grosseteste's own views, he makes the connection between
angelic knowledge and the cosmological/exegetical problem
of simultaneous creation, suggesting that the *processio rerum
in propria forma* is creation only in a derivative sense, and that

[28] *Comm. in Anal. Post.*, fol. 8ᶜ. The second passage is at fol. 17ᵃ.

[29] Note in this respect the strong predilection for the more philosophical word
'intelligence', over against traditional terms.

[30] Avicenna taught that all knowledge of lower creatures is present in the
Intelligences, and that material things emanate from these; see Verbeke, "Introduction", in *Avicenna Latinus. Liber de Anima seu Sextus de Naturalibus*. Édition
critique . . . par S. Van Riet (Louvain, 1968), p. 57*.

[31] In his unpublished thesis.

it is in this phase that the angels mediate between the one and the many. The merit of this solution to our problem depends upon the degree to which Grosseteste admitted creatures as co-operators in the act of creation. Here we are lucky enough to have fairly complete information.

On three occasions in the *Hexaemeron* Grosseteste denounces and confounds the error which holds that God employed intermediaries in creating. On one occasion it is Plato's belief that comes under attack: the Demiurge was only an artificer who employed demigods created by himself to form mortal bodies; to himself he reserved the creation of souls.[32] Another time Grosseteste refutes 'the lies of some Jews' (as indeed the Fathers had done before him), to the effect that God addressed the angels saying, 'Let us make man in our image'.[33] The third and most important passage would fit Arabic Neo-Platonism very well, though if it really is aimed at the Arabs it is kind to at least some of their doctrines, restricting as it does the work of the intermediaries to the production of material bodies, just as in the case of Plato. Many philosophers, Grosseteste recalls, are of the opinion that God created the angel directly, and that the angel then created and formed material things through the interior word which God spoke in and through it. However, the authoritative commentators on the Bible—Augustine, Jerome, and Basil—reprove this opinion on the grounds that God achieved the work of the six days by his coeternal Word alone, without the assistance of any created power.[34] As with all decisions of divine wisdom,

---

[32] *Hexaemeron*, fol. 223[b]: 'Alii vero dicebant ipsum [i.e. God] formare substancias primas celestes et committere eis formacionem rerum inferioris mundi. Unde Plato inducit summum deum deos minores per se facientem, factis autem a se diis iniungentem formandorum corporum curam mortalium.' Grosseteste's defence of the creation of things by God alone must have been inspired, in a general fashion at least, by St. Augustine, *De Civ. Dei* XII, 27.

[33] See Muckle, "The *Hexaemeron* of Robert Grosseteste . . .", in *Med. Stud.* 6 (1944), 151-74: 'Et hoc exemplo superatus Iudaeus recurrens ad aliud mendacium, viz. quod locutus sit Deus angelis dicens: "Faciamus hominem ad imaginem nostram." Item confutatur quia non potest creatoris et creature una esse imago.' For his closely-reasoned argument which follows, see p. 159.

[34] 'Multi namque philosophi opinantur deum per se creasse angelum, et angelum creasse et formasse corpora, quod non faceret angelus nisi verbo intellectuali quod in illo et per illum loqueretur deus, sed auctores sacre pagine expositores hanc sententiam habent reprobatam, asserentes quod deus solo verbo sibi coaeterno et nullius creaturae ministerio fecerit opera sex dierum in mundi principio.' Grosseteste cites as his authorities St. Augustine, *De Gen. ad Litt.*, Bks. VIII and IX, Jerome, and Basil. He makes it clear that the Word of God is God the Son: 'Patet igitur auctoritate

there is a good reason for this, for it is an unambiguous manifestation of the divine omnipotence, and it provides a firm ground for our belief that God can create *ex nihilo*, and that he alone can do so. Grosseteste adds as an afterthought that, the divine power being thus proclaimed, it was open to the divine goodness to make use of creaturely ministers in the government of the world (fol. 201^{b-c}).

Such assertions are too emphatic and too unambiguous to be consistent with the hypothesis that the angels assisted *'in principio, in operibus sex dierum'*, in however attenuated a sense. It is possible, of course, that Grosseteste changed his mind during the years intervening between the composition of the two works; but before resorting to hypotheses, it is worth attempting first to find in the two passages of the commentary a sense congruent with Grosseteste's mature ideas on creation.

The context of the first passage quoted from the commentary is an attempt by Grosseteste to illustrate the truth of the Aristotelean dictum that all demonstration is of incorruptibles.[35] How can universals, he asks, be incorruptible, when singulars are not, and the universals are, after all, found in singular instances? Grosseteste appeals instinctively to Augustinian exemplarism for a wider view than Aristotle could achieve by unilluminated reason. The first light is the first cause, and the knowledge of things that are to be made, which is in the first cause from eternity, is the set of formal, exemplary, creative causes. What Plato called ideas, therefore, are principles both of being and of knowledge.[36] Now a pure intellect, says Grosseteste, with typically Augustinian indiscrimination between the angels and the blessed, can know creatures in the first light. In the created light of intelligence there is knowledge of the creatures that come after it, and this knowledge is communicable to man. Such *descriptiones*, or ideas, are exemplary forms and even created causal reasons of

quod solo eterno verbo quod deus loquitur eternaliter per suam substantiam, fecit omnia in principio, non usus alicuius creature ministerio.'

[35] For a more extensive analysis of this whole very significant question, see Part IV, ch. 5 (pp. 327–9).

[36] *Comm. in Anal. Post.*, fol. 8^c: 'Cogniciones enim rerum causandarum, que fuerunt in causa prima eternaliter, sunt rationes rerum causandarum et cause formales exemplares; et ipse sunt creatrices, et hec sunt quas vocavit Plato ydeas et mundum archetypum; et hee sunt secundum ipsum genera et species, et principia tam essendi quam cognoscendi.'

material things that are later to be made, for material species came to be through the mediating ministry of the intelligences.

This passage presents no insuperable difficulty when the entire context of discussion is borne in mind. The question is not of creation itself but of the illumination of created minds to know true universal concepts, and the accent falls upon the ideas as principles of knowledge. In the divine mind, of course, the principles of knowledge are identically the principles of reality for created things; this identification is not, however, extended in full degree to the angels. The universal ideas present in the created intelligence are none the less exemplary forms of things, because of course they are, like created spirit itself, an image of what is in the first light. They can even be called 'created causal reasons', because the angels are active co-operators with the divine causality. It is not stated by Grosseteste that the angels co-operate in the foundational act of creation; the text need imply no more causal action on their part than is required for their co-operation in the divine government and providence of the universe, the same activity, in fact, as is envisaged in the *Hexaemeron*, and was universally admitted by thirteenth-century Scholastics. Of course, the highly metaphysical language of the passage owes a debt to Avicenna, a fact which obscures the very real *parenté* between this thought and that of Augustine, from which it truly derives. The passage taken as a whole, therefore, does not suggest that Grosseteste dallied with the *dator formarum* theory,[37] nor that anything in it is out of harmony with either traditional Christian ideas on the act of creation, or Grosseteste's own unambiguous reaffirmation of these in his *Hexaemeron*.

The second text from the *Commentary on the Posterior Analytics* can be interpreted along the same lines, and without any strain.

Although in these years Grosseteste used other sources on the angels such as Jerome, Anselm, Gregory the Great, and Basil,[38] and sometimes revealed an interest in minor problems concerning their nature and activity,[39] there is no mistaking

---

[37] Mentioned by him on only one occasion that I know of, but clearly not adopted; see *Comm. in Phys.*, ed. Dales, p. 16.26: 'Cum enim aliquid fit, aut fit ex nichilo—et hoc concedunt qui ponunt datorem formarum . . .'

[38] See, for example, *Hexaemeron*, fol. 204ᵈ, 206ᶜ.

[39] As, for instance, his placing of angelic being relative to eternity and time in *De Ordine Emanandi*, ed. Baur, *Die Werke*, pp. 148-50.

the predominantly Augustinian inspiration of his thought. When he completed the *Hexaemeron* he was around seventy years of age; one might have expected that his ideas would not suffer any very drastic revision or extension. He still had before him, however, a long struggle with the works of Pseudo-Dionysius, the representative of a very different (though, of course not unrelated) Neo-Platonism from that of St. Augustine. The resulting encounter was to bring an entirely new dimension to his thought.

Chapter 2
The *Commentary on the Celestial Hierarchy*

I THE AIM OF GROSSETESTE'S COMMENTARY

GROSSETESTE'S work of translating and commenting upon Pseudo-Dionysius's four major writings was accomplished in the protracted Indian summer of his life. Although in his seventies, he displayed none of the usual signs of mental ageing; indeed, his mind remained alert and even adventurous. His reputation as a Greek scholar depends principally upon these most finished products of his learning, equipped as they are throughout with a systematic and thorough commentary, surpassing in scope and realization the notes with which he accompanied most of his other versions from the Greek.

Scholars have reached a number of firm conclusions about some general features of Grosseteste's Dionysian commentaries. They were finished in 1243, and must have been begun around four years earlier, after the completion of Grosseteste's revised translation of Damascene,[1] at the earliest in 1238.[2] The cross-references among them make clear the order of their composition: first came the translations and commentaries on the *Celestial* and *Ecclesiastical Hierarchies*, then those on the *Divine Names* and the *Mystical Theology*. Grosseteste almost certainly rendered the *Scholia*,[3] but he did not translate the epistles of Pseudo-Dionysius, even though he possessed their Greek text. The evidence for the authenticity of his commented translation is clear beyond dispute. No one doubts

---

[1] Callus, "The Date of Grosseteste's Translations and Commentaries on Pseudo-Dionysius and the *Nicomachean Ethics*", in *Rech. Théol. anc. méd.* 14 (1947), p. 196.

[2] McQuade, *Robert Grosseteste's Commentary on the Celestial Hierarchy of Pseudo-Dionysius the Areopagite.* An edition, translation, and introduction of his text and commentary (unpublished thesis presented for the degree of Doctor of Philosophy, Queen's University, Belfast, 1961), p. 9. I gladly avail myself of this opportunity of thanking Dr. McQuade for placing his learning and its fruits at my disposal, when I first began the study of Grosseteste. The memory of his kindness remains a vivid one.

[3] See our Appendix A, no. 4, "*Scholia Maximi Confessoris*".

any more than his knowledge of Greek was thorough,[4] and it is highly improbable that his reliance on *adiutores* weakens his exclusive claim to real authorship of translations and commentaries alike,[5] commentaries for which he took full personal responsibility.[6]

With so much already assured by scholarly enquiry it might be thought that the use of the four commentaries for doctrinal history would be relatively unproblematic. If they are authentic, is not their content then Grosseteste's own? His stated purpose as translator and annotator of Dionysius implies something different, however: he aimed at reaching the intention of his author and discovering his full meaning. Did he then regard his work with scholarly detachment, as a piece of exegesis rather than advocacy of his author's ideas? One of the qualities of Grosseteste's commentaries which has impressed modern scholars is the reticence of their author, who writes with an uncommon degree of self-effacement. In Gieben's judgement, for example, 'Grosseteste a grand soin de ne pas dépasser les limites de son rôle de commentateur et d'interprète. On le voit très rarement advancer de sa propre autorité et se servir du texte de Denys pour développer des idées personnelles . . .'.[7]

The problem of the commentator's responsibility for the ideas represented in his work is not, of course, limited to Grosseteste, for it extends to the whole genre of medieval commentaries. It is aggravated by the protestations of many expositors, particularly those of Aristotle, that they mean only

[4] A very ample discussion of the evidence for this claim can be found in Callus's study, "Robert Grosseteste as Scholar", in *Scholar and Bishop*, pp. 37–44.

[5] Cf. Franceschini, *Roberto Grossatesta, vescovo di Lincoln*, p. 18; Callus, art. cit.; McQuade, op. cit., pp. 65–70.

[6] Not only in the prologues and epilogues to the various works, as Franceschini and McQuade have brought out, but implicitly, in the first-person comments scattered throughout the commentaries, of which the following may serve as an illustration: 'Quidam autem transtulerunt "agelarchiam" in "angelorum principatum", quasi esset dictio composita ab "angelo" et "principatu"; sed "aggelos", quod est "angelus", scribitur per duo, "agelarchia" autem per unicum solum *g*, nec occurrit mihi nomen "angeli" in graeco scriptum per unicum *g*; forte tamen qui plus perscrutati sunt id viderunt.' (*Comm. in Hier. Cael.*, ed. McQuade, p. 43; cf. pp. 58, 130, 273; and further McEvoy, *Robert Grosseteste on the Celestial Hierarchy of Pseudo-Dionysius.* An edition and translation of his commentary, chs. 10 to 15 (thesis presented for the degree of Master of Arts to the Queen's University, Belfast, 1967), p. 18.

[7] "Denys l'Aréopagite en occident (xiiie s.): Robert Grosseteste", in *DS*, fasc. XIX, col. 341.

to explain the doctrines of the Peripatetics, not to defend nor to adopt them as their own opinions.[8] Few modern historians of Scholasticism would, however, doubt that scarcely any medieval author abstained from developing the thought he expounded in a personal manner, with degrees of originality depending upon his own genius or talent. Were it not so indeed, it would be difficult to claim any philosophical originality for the Middle Ages, for the vast majority of surviving works issuing from the arts faculties or from theologians with philosophical interests, take the form of commentaries or questions on Greek texts; the commentary was the fundamental literary form of the entire cultural renaissance, amounting almost to a definition of Scholasticism,[9] and the course in the arts faculties was, if anything, more tied to the exposition of authorities than were the theological courses to the Scriptures. If, however, the medieval teacher attempted to press ever closer to the *intentio auctoris*, he was guided in his exploration not by an historical interest in the reconstruction of an ancient system of thought, but by the ambition of discovering the truth of the reality with which he dealt. His research was conducted under the aegis of truth, not erudition. The whole aim of his hermeneutical method was to discover his author as a witness to the nature of things, rather than as the spokesman of an ancient and foreign culture—a method which it is difficult for a modern to appreciate, coming as he does long after methodological abstraction has posed a disjunction between exegesis and philosophical exploration. While the modern exegete has no need to disavow the tenets of his author, since he is not assumed in any case to share what he expounds, a medieval commentator was taken to accept implicitly all that he did not explicitly reject. Thus it was no contradiction for a medieval expositor to dissociate himself from his author as a preliminary

---

[8] Cf. the famous text of St. Albert, *Comm. in Librum de Animalibus* (quoted by Chenu, *Introduction à l'étude de saint Thomas d'Aquin*, p. 176 n.): 'Expletum est totum opus naturalium, in quo sic moderamen tenui, quod dicta Peripateticorum, prout melius potui, exposui; nec aliquis in eo potest deprehendere quid ego ipse sentiam in philosophia naturali . . .' That this literary tradition survived at least until Ockham is clear from the *Prologus in Expositionem super VIII Libros Physicorum*: 'Eapropter opiniones recitandas mihi nullus ascribat, cum non, quid iuxta veritatem catholicam sentiam, sed quid istum Philosophum approbasse vel secundum sua principia, ut mihi videtur, approbare debuisse putem, referre proponam.' (*Ockham, Philosophical Writings*. A selection edited and translated by Philotheus Boehner O.F.M., pp. 2-3).   [9] Chenu, op. cit., pp. 176-7.

safeguard against likely attack, and yet to favour that author's principles with the truth that he considered to be their due, albeit by way of extending and enlarging his views, even to the point of neglecting the conclusions of authority where they did not seem to agree with the truth as the exegete saw it. If respect for authority was a fundamental aspect of the scholastic situation, finding its appropriate expression in a concern to attain the literal sense of an ancient writer, it was not felt by the Scholastics themselves to exclude a sharp awareness of the author's limitations, and even errors, which were, however, corrected where possible according to a principle of sympathy: *benigne interpretari*.

If this were the whole story we could be dispensed from further investigation; unfortunately, in Grosseteste's case the matter is not so simply resolved. Each commentator was in greater or lesser degree individual, and addressed himself to his task with more or less flair, though always within the overarching cultural framework sketched above. Therefore in each case, the degree of a medieval commentator's submission to the common practice of his age must be estimated with all possible exactitude. Now it must be acknowledged that Grosseteste was not only individual but original in his manner of commentary, and so was in some measure an exception to the generalizations expressing the medieval hermeneutical situation. A brief examination of his intention and practice as a commentator is therefore indispensable.[10]

Grosseteste's difference from other medieval commentators appears by comparison with two of the most important medieval commentators on Pseudo-Dionysius, Hugh of St. Victor and St. Albert. The discovery of a double interpretation of a text from the *Celestial Hierarchy* has enabled H. Weisweiler to conclude that Hugh commented much of that work twice, on the second occasion amplifying the original exposition of the text by the insertion of long personal discourses on subjects arising from his own reflection.[11] These passages leave

[10] What follows here is of necessity limited to the Dionysian commentaries of Grosseteste, and in particular to his *Comm. in Hier. Cael.* We shall not refer here to the Aristotelean commentaries, but it must be remarked in passing that a comparison between the two groups would reveal an enormous and exciting evolution in hermeneutical method. No such study has yet been attempted.

[11] Weisweiler, "Die Dionysiuskommentare *In Coelestem Hierarchiam* des Skotus Eriugena u. Hugo V. St. Viktor", in *Rech. Théol. anc. méd.* 19 (1952), pp. 42-5.

exegesis far behind, and offer the reader Hugh's immediate, original reflections on fundamental themes such as unity, participation, and imitation. Despite Hugh's conscientious use of Scottus and the lost Latin commentator on Dionysius, his interest lay not principally in the words of the text before him, but in the realities discussed, and the result is a Dionysius systematized and partially overlain by the thought of St. Augustine.

With St. Albert the scholastic commentary appears for the first time and at full development. It is not strictly exegesis any more that is in question, but a school treatise on the angels, the divine names, etc., whose method of procedure bears a strong resemblance to Albert's *Sentences*.[12] The division of the text, *quaestiones*, and attention to conceptual clarification, constitute together the unmistakeable atmosphere of the school, and liberate the creative teacher from the servitude of exegetical techniques to follow his own interest and pursue his true *métier*: the search for the truth of things.

On the opening page of his *Commentary on the Celestial Hierarchy*, Grosseteste defends his right to add '*aliquid modicum et vile*' of his own to the famous and learned expositions of Dionysius's words.[13] Beneath the conventional modesty which refuses to compare his candle-like contribution to the great torches of his predecessors, we are justified, I think, in detecting a claim to originality such as he rarely makes elsewhere.[14] He never again refers to the commentators,[15] for in truth they did not offer help to achieve his rather singular goal: a humble search for the meaning of the author's words.

It is due above all to his systematic application of philological method that Grosseteste's aim, to recover the *mens*

---

[12] See Völker, *Kontemplation u. Ekstase bei Pseudo-Dionysius Areopagita*, p. 241.

[13] See ed. McQuade, pp. 1-2: 'Si autem, praeter eorum verba et praeter verborum eorundem expositiones profundas et praeclaras ante factas, nos de penuria nostri ingenii aliquid modicum et vile ad declarationis augmentum afferamus, non credimus ab aliquo id indigne ferri debere; inter ferentes namque magna luminaria sunt plerumque non repulsi sed admissi benigne etiam qui parva conferunt, maxime cum manus eorum non potest invenire maiora . . .'

[14] The only other example of such a claim that I can discover occurs at the beginning of his discussion of the reasons for the Incarnation, in the treatise, *De Cessatione Legalium*.

[15] He had studied Hugh of St. Victor's commentary, which is listed in his *Concordance of Scripture and the Fathers*, Lyons, MS *Bibl. Municipale 414*, fol. 23; perhaps too he had read those of Scottus and Sarracenus.

auctoris, detaches itself from the claim of the typical scholastic commentator, to the effect that the *intentio auctoris*—a cliché of the age—is the object of his labour. The preparation of the text, rendering, and commentary were governed by a degree of self-critical orderliness that make his undertaking unique and remarkable.

II GROSSETESTE'S METHOD AS TRANSLATOR

AND COMMENTATOR

We are fortunate in possessing the Greek MS from which Grosseteste translated the four Dionysian works: *MS Canonici Gr. 97* of the Bodleian Library.[16] Besides the text of Dionysius's whole corpus, including the ten letters, this MS contains an ample dossier of Greek material relative to the Pseudo-Areopagite: the prologue of Maximus the Confessor and the epigrams and chapter-headings to each work; the Eusebian extracts from Polycrates, Clement of Alexandria and Philo; a vocabulary-index of Dionysius; the *Encomium* of Michael Syncellus; Methodius's account of St. Dionysius's martyrdom, and the *Scholia* of Maximus and John of Scythopolis to the four chief works. The whole MS is written with meticulous care in Greek, but by a Latin scribe. The text was copied from the present *Bibliothèque Nationale MS gr. 933*, a tenth-century codex brought from Constantinople by William the Physician, later Abbot of St. Denis, in 1167. Someone, probably Grosseteste himself, went through the text of the Dionysian corpus and placed a dot in the margin by each line where a correction or a variant reading was required. The main body of variants was taken from the present *B.N. gr. 437*, the gift of Emperor Michael II to the Frankish court in 827. Since both Greek MSS were preserved as relics at St. Denis, Grosseteste and his helpers presumably did the work of copying and collating there.[17] At least one other, as yet unidentified, Greek MS supplied its quota of variants and the text of the Eusebian extracts, which is not found in either Paris manuscript. Marginal corrections, variants and subject-captions derive from three hands, one of them Grosseteste's own, and he also added the two notes drawing attention to the subject-matter, a few expansions of

---

[16] See further Barbour, "A Manuscript of Ps.-Dionysius Areopagita copied for Robert Grosseteste", in *Bodl. Libr. Rec.* 6 (1958), 401–16.

[17] Probably just before his elevation to the See of Lincoln.

abbreviated portions of text or scholia, and the Greek number-
ing of the paragraphs in each chapter of the four main works.
The identity of his helpers remains unclear, but they must be
counted among the *adiutores* whose assistance he warmly
acknowledged.

Grosseteste's more proximate preparations for his translation included a close study, based on all four treatises, of
Dionysius's vocabulary. The evidence for this conclusion is
clear. In his versions, more than in those of his predecessors,
uniformity of translation was valued, and each Greek word
was assigned a Latin equivalent from which Grosseteste rarely
departed. He hoped thereby to rule out the fluidity in the
Latin versions which in his opinion interfered with the strict
rendering of the Greek. It is very likely that he constructed a
table of equivalences, although before long he would have
had it memorized. A further stage of preparation involved a
thorough comparative study of the three existing translations,
those of Hilduin, Scottus and Sarracenus, as a further means
of establishing an accurate Greek text. This procedure, if it
did not bring new variant readings to light, often helped to
explain corruptions existing in the Latin translations.

Perhaps the finest example of this kind of detective-work
occurs in the commentary on the thirteenth chapter of the
Celestial Hierarchy.[18] One translation, Grosseteste remarks,
has '*ipsa aqua nonne et in omnes proportionaliter pervenit*',
where his own version gives '*donatum et in omnes analogice
pertransiens divinum lumen*'. The Latin version follows a
corruption in the Greek exemplar on which it was based. The
Greek, he explains, reads *tou doroumenou*, governed by a
Greek verb which is one of a restricted class taking the genitive
case. The scribe joined *t* and *o* to make the article, *to*, and the
u he attached to the initial *d* of the following word to give
the Greek *udor*, Latin *aqua*. The remainder of the second word
then was fragmented into *ou men ou*, construed *non quidem
non*, or *nonne*. Faced with this corruption, the translator had
to deprive the Greek verb of its attendant genitive object,
which object in turn he verbalized, and having elided the awkward reference to divine light, ended up with the phrase quoted
above.

It was thus that a mistake introduced by Eriugena, followed

18 See ed. McEvoy, pp. 65-6.

unsuspectingly by the anonymous commentator and by Hugh of St. Victor,[19] and corrected by Sarracenus (who used a new Greek MS), was finally run to earth, understood and explained in a way that could not be improved upon.

The stages of Grosseteste's working-method have been illumed by Callus's study of two manuscripts of his Damascene version, *British Museum MS Royal 5.D.x.* and *Pembroke College MS 20*.[20] The first of these contains Burgundio's translation in two columns. The margins are heavily annotated by notes headed *ex greco* and bearing etymological explanations, source-references or extracts translated from Greek in order to illustrate the words of the text. The divergences between Burgundio's reconstructed Greek MS and Grosseteste's own are recorded likewise in the margin, and the central space between the columns is reserved for his indexing-symbols, which relate the topics discussed to his *Topical Concordance of the Bible and Fathers* and the rest of his library. The second MS represents a more advanced stage of work, and may have been written under Grosseteste's direct supervision. Its text is Grosseteste's revised version, the *correcta translatio*, and its margins are copiously annotated to provide a critical apparatus to which philological, historical, and theological explanations are added. To this we may add Oxford, *All Souls MS 84*, containing the version of the *Nicomachean Ethics* and its Greek commentaries. The text of Aristotle is divided into brief sections, each followed by the co-ordinate passages of a Greek commentator. Interlinear notes are added, diversifying and expanding upon the wording of the text; mostly they are introduced by '*i.e.*', or '*sc.*' (*scilicet*). Symbols, affixed to important words in Aristotle's text, refer to marginal notes, some of great length, providing a battery of information on Greek words and pronunciation and on the history of concepts like 'metaphysics',[21] 'the good',[22] 'idea' (fol. 21[d]), 'art' (25[c]), '*finis ultimus*' (26[b]), 'the sufficiency of the good' (27[c]), the doctrine of good in the Stoic, Pythagorean, and other Schools (31[b]), the Olympic Games (32[a]), enjoyment and love (32[b]), and the angels (33[d]). He expounds the meaning

---

[19] Weisweiler, art. cit., p. 41.

[20] Callus, "Robert Grosseteste as Scholar", in *Scholar and Bishop*, pp. 47 ff.

[21] Oxford *MS All Souls Coll. 84*, fol. 11[b], recalls how the name 'metaphysics' came to attach itself to first philosophy.

[22] Ibid., fol. 21[b], recording Aristotle's opposition to Plato and his god.

of Aristotle's frequently terse text, summarizes the drift of his arguments, corrects his pagan laxity, and gives index-type references to the content and progress of discussion at the beginning of sections and at the foot of most pages. It is our belief that all this material stems from Grosseteste, and that it witnesses to the penultimate stage of composition of a full commentary. It only needed the interlinear and marginal glosses to be written up for the work to assume the appearance of his Dionysian commentaries, which were doubtless produced in the same methodical way.

Grosseteste's Latin version of the Dionysian corpus did not begin *ex nihilo*. He declares that he has based his efforts on the achievements of the previous translators (Hilduin, Eriugena, and Sarracenus), choosing from them the words which seemed most suitable to express the *mens auctoris*, and supplying his own only where the previous efforts were found wanting.[23] In general he seems to follow Sarracen, whose version, we may conjecture, he took as the basis of comparison; it was in the process of replacing the *translatio vetus* of Eriugena in the thirteenth-century schools. Often, however, he prefers a suggestion of Eriugena where he thinks it a more literal equivalent of the Greek, as where he puts *unifica* for ἑνοποιός as against Sarracenus's paraphrase, *'unum faciens'*. Where exactitude was at stake rather than simply self-consistency, he introduced a new word, as when he uses *extensive* for the Greek ἀνατατικῶς to replace Eriugena's *restituens* and Sarracenus's *suscitative*, or translates νοῦς by *intellectus* (invariably) instead of Sarracenus's uniform usage, *mens*, and the fluctuation of Hilduin and Eriugena between *mens* and *animus*.[24]

Grosseteste's avowed aim as an interpreter of Greek was to render the *mens auctoris et venustas sermonis* into Latin by means of a scrupulously exact word for word translation. Franceschini and McQuade have followed the working-out of this aim through the many undoubted exaggerations and bizarre consequences to which it led him.[25] Literalness of translation was, of course, the agreed ideal of the age, dictated

[23] *Comm. in Hier. Cael.*, ed. McQuade, p. 1: 'Eorum qui hunc librum de graeco in latinum transtulerunt, nunc huius, nunc illius verba, raro autem admodum nostra in textu ponimus, sicut nobis visum est convenientius ad mentem auctoris declarandam.'
[24] These examples come from McQuade's edition, pp. 70-1.
[25] Franceschini, pp. 74-84; McQuade, Introduction, pp. 70-4.

by its respect for the authorities, and adhered to strictly lest the meaning of their words be altered; but the scrupulosity with which Grosseteste pursued that elusive ideal seems to have derived from a presupposition which he never (so far as I know) made explicit, but which emerges in sharp relief from a close commerce with his versions: he was persuaded that each and every element of language has a semantic value—that there is no particle of a word, nor any detail of syntax, however small, that lacks a meaning and fails to register a demand for a corresponding element in the translation. It was from this underlying conviction, and a consciousness of the great responsibility it imposes upon the serious translator, that what we may term the extravagances of his translating-method derived: the effort sometimes made to render the Greek article;[26] the use of the preposition *'ut'* governing the infinitive to express purpose, in the manner of the Greek;[27] the retention of the Greek double negative, despite what he admitted to be its redundancy in Latin;[28] the construction of *'decet'* and *'sequitur'* with the dative case on the model of the corresponding Greek verbs;[29] the invention of a factitive form of the neuter adjective to represent the neuter adjective coupled with the Greek definite article, signifying an abstract property;[30] the coining of Latin composite words in order to transliterate Greek composites[31]—these are only some of the lengths to

[26] ἡ ἐκ τοῦ πάντων ἐπέκεινα = ex eo qui ultra omnia (Franceschini, pp. 94-5).

[27] *De Eccles. Hier.*, ch. 7: 'Hoc divinis nostris ducibus [in] intellectum veniens, approbaverunt suscipi infantes secundum hunc sacrum modum, *ut* et naturales adducti pueri parentes *tradere* puerum alicui edoctorum bono divina pedagogo et de cetero ab ipso puerum *perficere* ut a divino patre et salutis sacre anadocho.' (Franceschini, p. 78).

[28] *Comm. in De Div. Nom.* ch. 4: '. . . *non* omnino malum *neque* non ens, id est non omnino malum *et* non ens: una enim negatio nobis est superflua, quod non est in greco.' (Franceschini, p. 81).

[29] *Comm. in Hier. Cael.*, ch. 9: 'Nec miretur quis si hiis verbis "decet" et "sequitur" subiungimus plerumque dativos casus, quia sic semper fit in greco et sine multa absurditate potest hoc fieri in latino.' (McQuade, p. 436; Franceschini, p. 81).

[30] τὸ ἀνωφερές = sursumlativum.

[31] *Comm. in De Div. Nom.* ch. 1: 'Sunt autem hee dictiones hic divisim dicte: "spiritu mote", "Deo decenter", "procaciter enitentes", "bene possibiliter", "alis elevatos", in greco singule dictiones composite. Oportet autem huiusmodi dictiones grecas compositas, cum non habent correspondentes compositiones in latino nec equipollentes eis dictiones simplices, transferre in dictiones simplices idem significantes cum dictionibus grecis ex quibus fiunt grece compositiones. Cum vero inveniuntur in latino compositones grecis correspondentes et equipollentes, convenientissime transferuntur in huiusmodi compositas. Quando etiam possunt fingi

which he was prepared to go in finding or inventing equivalences.[32] Grosseteste had resort to these and similar—often ingenious—stratagems, without apparently feeling any very notable degree of compunction regarding their consequences for the Latinity of the resulting versions[33] (though he clearly was sensitive to criticism on this point).[34] Grosseteste, indeed, compositiones non multum absone latinitati, convenienter etiam in huiusmodi fit translatio, ut si dicatur composite "bonidecenter", "sacredecenter", "boniformiter", "omnisciens", "omnibonus" et huiusmodi. Mens enim auctoris et venustas sui sermonis per huiusmodi compositiones, ut existimo, magis est lucida. Existimo etiam quod diligenter consideranti que dictiones in latino divisim posite habent sibi correspondentes in greco compositas et fingenti, ut potest, ex simplicibus latinis correspondentes compositiones, licet latinitati absonas, sicut nos fecimus in principio expositionis angelice ierarchie, patebit dilucidius et mens auctoris et venustas sermonis. Ideo non incongruum nobis videtur si breviter tangamus que divisim posite dictiones in latino, habent in greco equipollentes sibi, quo ad sensum, dictiones compositas.' (Quoted by Franceschini, p. 77). This passage seems to attest the influence of Burgundio of Pisa's preface to his translation of Chrysostom's *Homilies on the Gospel of St. John*, more clearly than any other that I can find. Burgundio's preface is printed by Haskins, *Studies in the History of Medieval Science*, p. 151, n. 36.

[32] Only two Greek usages seem to have defied all his efforts in this direction, namely, the perfect active and aorist active participles, which he sensibly represents by the present participle in Latin, and the gender of certain Greek nouns which differ from that of their Latin equivalents. On the first see *Comm. in Hier. Cael.* ch. 8: '. . . philanthropia inquam, "sic disciplinans et sic tradens" hoc est "postquam disciplinavit et tradidit"; sunt enim haec duo participia in graeco praeteriti temporis.' (McQuade, p. 416; quoted by Franceschini, p. 96).

[33] '. . . praeeligentes minus latine loqui, ut mentem auctoris manifestiorem faciant, non erit, ut existimo, inutile', Grosseteste had remarked in the prologue to the *Comm. in Hier. Cael.* (cf. n. 35), and the option is reaffirmed in numerous places in the Dionysian Commentaries.

[34] Cf. the undisguised irony directed in the prologue to the *Comm. in Hier. Cael.* (n. 42 below) against over-willing critics of his translations, who yet themselves know no Greek; and compare with ibid., ch. 10: 'Unde et constructio ut existimamus est ordinanda sicut nos ordinavimus, et sensus accipiendus quem expressimus, licet aliqui linguae graecae ignari aliter forte senserint.' (McEvoy, p. 18). Two passages occurring at the very end of the commentaries on the *Celestial Hierarchy* and the *Divine Names* allow us to summarize the complaints made against him, which were chiefly two: obscurity and deformity in the Latin version, and prolixity in the exposition. Grosseteste admits his inadequacy ('*insufficientia mea*') to render Dionysius with limpidity and *éclat*, but issues at the same time a sharply-worded challenge to his critic, to get to work and do better than he himself has managed to do. The phrases '*si autem habet ille*', '*ille vel quis alius*', seem, as Callus suggests, to be aimed against one critic above all ("Robert Grosseteste as Scholar", p. 59). The *Comm. in Hier. Cael.* ends as follows:
'Si autem sermonis commensurationem censetur quis nos excessisse, et nichilominus inornatius et obscurius dixisse, fatemur quia verum est. Non enim se extendit nostra potentia ad brevem et delucidam et ornatam tam profundorum explanationem. Si autem habet ille, vel quis alius, breviora, meliora, et dilucidiora et ornaciora dicere, dico secundum quod aliquis incomparabiliter maior me dixit ante me, quod "erit ipse utique o domine et conditor crucifixe propter ineffabilem philanthropiam

says as much himself, when in the prologue to his version and commentary he writes as follows:

It must be realized that Greek has a large number of composite words for which Latin has no equivalents, so that Latin translators have no choice but to resort to paraphrase, an expedient which cannot render the author's intention adequately and fully, precisely as the single Greek composite word does. Hence I consider that translators render service to the fuller comprehension of their text, if they choose on occasion to represent such compounds by manufacturing corresponding ones in Latin, as far as may be, even if that means sacrificing Latinity in some measure in order to bring out the author's meaning more clearly.[35]

'The result', concludes McQuade, 'has little meaning, and no style; for this, Grosseteste was inclined to blame the limitations of the Latin language.'[36]

There is, however, another perspective than that of Franceschini and McQuade, justifiable as their judgement may frequently be, namely, that Grosseteste's exacting scholarship had, pro nobis, et amicorum omnino optimus, et doctor mihi venerabilissimus, et mea quidem sint igni cibus, quae autem illius, in animabus bonum semper amantibus et deiformissimis." Falsa autem et temeraria si quis nos dixisse invenerit, supplicamus ut fraterna caritate falsa rescindat, temeraria dilucidet, et nostrae compatiens ignorantiae et temeritati, oratione sua ab eo qui est veritas et humilitas nobis impetret indulgentiam.' (*Comm. in Hier. Cael., in fine*, ed. McEvoy, pp. 239–40). The concluding line of the *Comm. in De Div. Nom.* has even more fire: 'Nec quicquam in hiis invide vel malitiose carpat, vel derisorie subsannet, sed caritative meae compatiatur insufficientiae.' (Callus, art. cit. p. 59).

[35] 'Sciendum etiam quod graeci habent plurimas dictionum compositiones, quibus correspondentes compositiones non habent latini; unde necesse habent translatores ad hoc ut latine loquantur plerumque pro una huiusmodi dictione composita, plures dictiones latinas ponere, quae autem non ad omnem proprietatem vel sensus plenitudinem possunt exprimere mentem auctoris, sicut exprimit eam unica dictio graeci idiomatis composita. Quapropter si ad explanationem pleniorem quandoque ponant expositores ipsas dictiones graecas compositas confingentes quantum possunt eis correspondentes compositiones, praeeligentes minus latine loqui ut mentem auctoris manifestiorem faciant, non erit, ut existimo, inutile.' (ed. McQuade, p. 15*).

[36] Grosseteste was not the only medieval translator to draw this conclusion; here it was John Sarracenus who preceded and possibly influenced him; cf. the preface to his version of the *Celestial Hierarchy* (quoted in Franceschini, p. 76 n.): 'Nam apud grecos quedam compositiones inveniuntur quibus eleganter et proprie res significantur; apud latinos autem eedem res duabus aut pluribus dictionibus ineleganter et improprie et quandoque insufficienter designantur. Ad commendationem etaim alicuius persone vel alterius rei pulcre apud grecos articuli repetuntur, et per eosdem articulos multe orationes ad invicem connectuntur. Taceo de insigni constructione participiorum et infinitorum coniunctorum; huiusmodi autem elegantiae apud latinos nequeunt inveniri . . . Sepe enim ubi duas vel tres latinas dictiones pro una greca posui, eas quasi unam coniunxi, non quod unam dictionem ex hiis esse vellem, sed ut planior fieret intellectus, et quantum elegantie ex inopia latine locutionis iste perderet tractatus appareret.'

both as presupposition and as result, the willingness to see in language more than the merely external vestment of a thought which could somehow transcend the conditions of its verbal incarnation, and which could, in consequence of its indifference to its clothing, exchange one robe for another, remaining in its essential condition naked under all garments.[37] His experience as a translator seeking ideally exact verbal correspondences between Greek and Latin, and often being forced into the admission that none such existed, led him to acknowledge the differences between the two languages, differences which he made no effort to conceal, rather the opposite: he was consistently at pains not to disregard nor minimize such differences, nor to undervalue the effort needed by a Latin to appreciate the quality of otherness found in the object of his study. Instead of contenting himself with a compromise solution[38] and producing translations that were in the fullest sense Latin works, attaining therefore in their own right a high degree of clarity and even a certain grace, Grosseteste opted for a type of version which not only closely followed the Greek, but actually allowed the Greek text to appear through it.[39] In his eyes, the peculiarities of the Latin language had to be subordinated to and even sacrificed in favour of those of the

[37] Grosseteste reveals in his translations an acute awareness that the literal method of rendering *de verbo in verbum* cannot be understood as implying that each Greek word does in fact stand in a relation of direct, one-to-one correspondence to a Latin one, which it would suffice to seek in order to find. Precisely the lack of such direct correspondences between the two languages is exemplified at the level of lexicography in many permutations, which are normally exhibited in his commentary on the text. For some relevant examples from his translations, see McEvoy, *Coll. Franc.* 45 (1975), 81-2.

[38] As Sarracenus was eventually forced to do in translating Pseudo-Dionysius. Having tried to adhere to the strictest principles of literalness in his version of the *Celestial Hierarchy*, he was forced to modify his approach in rendering the *Ecclesiastical Hierarchy*, in favour of rendering the sense, sometimes at the expense of the nuances of meaning: 'Eundem autem transferendi modum in hac ierarchia quam in celesti observavi excepto hoc quod latinas dictiones multas pro una greca positas non ubique coniunxi, et alicubi sensum pocius quam verba sum secutus.' (*Praefatio*, transcribed in Franceschini, p. 76 n.). Cf. also Aquinas's sensible remark concerning the latitude permitted a translator, the result perhaps of his conversations with William of Moerbeke: 'Unde ad officium boni translatoris pertinet ut . . . servet sententiam, mutet autem modum loquendi secundum proprietatem linguae in quam transfert' (from the *Prooemium* to the *Contra Errores Graecorum*).

[39] Cf. Franceschini, p. 77: 'È il metodo della versione letterale, verbum de verbo, portato alle conseguenze estreme: per cui le traduzioni non sono che rivestimenti latini di brani che rimangono greci, aiuto preziosissimo per la ricostruzione degli originali, ma di nessuno o scarsissimo valore letterario e spesso di significato oscuro.'

original Greek; Latin required to be as it were emptied out and to take upon itself the form of a servant, if it were to be the faithful vehicle of the Greek in all its irreducible difference and foreignness. To accuse Grosseteste's translations of obscurity and deformity in their use of language is to say but half the truth; for he accompanied at least his most developed and finished versions (those of Pseudo-Dionysius's works) with commentaries, or *expositiones*, which were conceived and executed as the necessary complement to the study of his versions.[40] Each section of text as it was translated, with all its rebarbative crudities and its undoubted dissonance to the Latin ear, was taken up particle for particle in the ensuing exposition, to be woven into a more fluent discourse than the limitations of strict translation could admit of, a discourse which clarified, ordered, magnified, and diversified the wording of the text of his author. Taken together, these translations and commentaries are, *pace* Franceschini and McQuade, an admirable and not impracticable solution to the difficult and thorny problem of method which faced all medieval translators of Greek works.[41]

The concluding paragraph of Grosseteste's own prologue to the translation and commentary is a fine statement of the

[40] The same is true in some measure of practically all of Grosseteste's translations. He was ill-content to put a bare text in the hands of other scholars, and, where possible, he accompanied it with a version of one or more Greek commentators, and eventually also with his own *notulae*. Thus he translated the *Scholia* of Maximus (or rather, Pseudo-Maximus) to the four Dionysian works; the commentaries of four Greek writers on the *Nicomachean Ethics*, with his own extensive philological and historical notes; he rendered the commentary of Simplicius with Bk. II of the *De Caelo et Mundo* (and, presumably, also with Bk. I of the same work, which is not extant); he annotated the *De Orthodoxa Fide* of Damascene, and indeed elements of a philological commentary can be clearly distinguished in the text of his version of the *Dialectica*, though the modern editor has not picked them out (cf. *St. John Damascene, Dialectica. Version of Robert Grosseteste*, edited by Colligan, esp. pp. 13, 15, 20, 22, 24, 31, 50). Finally, *marginalia* of Grosseteste occur also in his version of the Ignatian Letters; see Callus, art. cit., pp. 34, 48–51, 63–65.

[41] I have nothing to add to the well-chosen words of Fr H.-F. Dondaine, O.P.; speaking of Grosseteste's Dionysian works he remarks that, 'Les qualités techniques de cette version auraient dû lui valoir une place de choix dans le corpus du xiiie siècle. Mais nous aurions tort d'en juger d'après nos exigences à nous; les théologiens du xiiie ne demandaient pas à leur *Corpus areopagiticum* ce que nous lui demandons aujourd'hui. Pour eux, la théologie est toute autre chose qu'une histoire des doctrines; et ils étaient généralement plus désireux d'assimiler et comprehendre ce qu'ils présentaient de vrai et d'éternel dans leurs *auctoritates*, que de rejoindre précisément çela que tel *auctor* avait prononcé et même pensé.' (*Le corpus dionysien de l'Université de Paris au XIIIe siècle*, pp. 116–17).

principle which joins version and exposition in an indissoluble unity:

It must also be realized that in a Latin translation, and in particular in one which seeks to interpret word for word, to the extent of the translator's ability, it is inevitable that a number of expressions will be highly ambiguous, and capable of many interpretations which are not supported by the original Greek. When therefore someone, who either does not have the Greek text to hand or does not know the language, is placed in the presence of such ambiguities in the course of expounding this book, he will inevitably be for the most part unable to tell what the author intended by such expressions: things will elude him, which someone with a moderate or even a slight knowledge of Greek could not miss. Wherefore, should people who know nothing of Greek manage, now and again, to comment with exactitude on such ambiguous texts, or even to handle them with greater subtlety than others who have studied Greek, they still cannot pride themselves on having given a truer account of the mind of the author than those who have some knowledge of the language; unless perhaps they feel themselves in a position to boast of their ability to arrive by guesswork and conjecture at more far-fetched explanations of ambiguities.[42]

Grosseteste's commentarial approach proves almost as systematic as his execution of the version, with the result that a number of constant features recur in his exposition of the text. In what follows, I give a synopsis of these predominant aspects, not all of which, however, are to be found in all of the sections into which his work is divided; they constitute rather something like a battery of resources upon which he may draw according to the requirements of the text, as these appear to him.

Grosseteste links up each section of text thematically to the foregoing; he assumes brief lemmata of the text consecutively into a continuous commentary; discusses variant readings (on about thirty occasions in the commentary); defends his translation, explains its difference from earlier ones, and if he has

---

[42] 'Sciendum quoque quod in translatione latina, et maxime quae sit de verbo ad verbum in quantum occurrat transferenti facultas, necesse est pluries esse multa ambigue et multipliciter dicta, quae in graeco idiomate non possunt esse multiplicia. Unde libri huius expositor non habens aut nesciens eum graece scriptum, cum occurrunt huiusmodi ambigua, quae fuerit in his mens auctoris plerumque necesse est ut ignoret, quae scientem mediocriter vel tenuiter linguam graecam, latere non potest. Quapropter etsi ignari graecae linguae in huiusmodi ambiguis exponendis aliquando dicant vera et subtiliora eiusdem linguae sciolis, non possunt tamen linguam illam non ignorantibus se praeferre velut mentem auctoris verius manifestantes, licet forte praeferre se possint tamquam subtiliora divinare et conicere in ambiguis valentes.' (ed. McQuade, pp. 15*-16*).

made a reluctant concession to these, offers an absolutely literal rendering; explains the individual words of the Greek text, displaying a fondness for etymological considerations; elucidates strange words, relates technical terms to Dionysius's central ideas and discusses their function and relative strictness of usage; clarifies ambiguous words in the version by referring to the Greek text, which he may adjudge more precise or more fluid in meaning than the Latin equivalent; sketches a constellation of Latin words, all of which partly cover the meaning of a single Greek one;[43] links up the words and phrases, with much attention to the articulation of whole sentences (syntax, rules of agreement, and punctuation being here his chief guides); where two interpretations of a sentence are possible, he introduces the second with '*vel ordina sic*' or some such formula, and states his preference; he explains Greek constructions when necessary; fills out the quotations from Scripture given in Dionysius's text, referring quite frequently to the LXX version; relates the progress of the argument or discussion in hand to the discourse as a whole, keeping all four Dionysian works in mind; deals with apparent internal contradictions and other related objections, arising as often as not from Dionysius's rather involuted style of thought; lists the Greek composite words employed in the passage; and often ends by commending the thought to the acceptance and appreciation of his reader.

It can be seen that almost every one of these recurrent expository features is focused upon the words of the Greek text and dedicated to the recovery of every nuance of meaning present in it. Words, words, words and their differences: the reader is never allowed to forget that he is studying a text emanating from a foreign language, with all its irreducible specificity of syntax and semantics; he may not relax as though he were in his native culture. Grosseteste protests at least once against the 'facile Latin' (*littera latina facilis*) of a previous

---

[43] An example of this device could be chosen from almost any page of the commentary; the following is an instance: 'Adicit autem partibus eius [i.e. of the definition of hierarchy], hoc adiectivum "sacrum" pro quo in graeco ponitur "hieron", quod ut graeci dicunt significat sacrum seu sanctum, seu pretiosum et honorabile, seu incoinquinatum, seu admirabile, seu magnum, seu purum et mundum. Unde diversi translatores in diversa hoc transtulerunt. Opinamur autem quod "hieron", pro quo nos ubique fere ponimus "sacrum", significat directionem in deum tanquam in finem ultimum et optimum . . .' (McQuade, p. 113).

translation. The efforts made to reinforce the sense of the foreignness of the text at the primary linguistic level are unrelenting, with the result that much of the commentary has the quality of a magnified translation; the actual version presented is revealed as a tentative and unsatisfactory approximation to its Greek base. In an effort at total recovery of the alien medium, the greater part of the commentary remains in the closest proximity to the linguistic atmosphere of the translation, to a degree not paralleled in the other medieval commentaries on Dionysius.[44]

Yet the immediate context of exposition is not allowed blind domination over the reader. The language of the entire Dionysian corpus, and beyond that all that he knows of Greek, informs the exegesis of each part. Grosseteste is always awake to fluctuations of usage, and he can point out that a word like θεοφάνεια is not in fact given a completely uniform sense by Dionysius. A flexible and perceptive orderliness informs his attention to language: he seeks the meaning of words through examination of their etymology and usage, not through the scholastic game of distinction and clarification. Fine nuances of thought used without explanation by Dionysius are recovered. A good example of this is the distinction of νοητός and νοερός as attributes of the angels, a Neoplatonic nuance which Dionysius had borrowed from Proclus;[45] *intelligibile* and *intellectuale*, remarks the commentator, refer to two aspects of the same immaterial substance, which is called 'intelligible' in so far as it is or can be known, and 'intellectual' in virtue of its capacity to know.[46] A few pages later, he attempts to recover the original Greek sense of ἔρως, an activity analogically ascribed to the angels in Dionysius's work.[47]

---

[44] Eriugena anticipated him notably in this respect.

[45] See Roques, *L'Univers dionysien*, p. 156.

[46] 'Idem enim subiecto est intelligibile et intellectuale, sed intelligibile dicitur in quantum intelligitur seu potens est intelligi, intellectuale vero in quantum intelligit vel intelligere potens est.' (*Comm. in Hier. Cael.*, ed. McQuade, p. 81). No one who lacked direct knowledge of Proclus could have got the distinction any clearer.

[47] 'Pro nomine autem amoris hic positi habetur in graeco "eroos", quod nomen apud eos multipliciter dicitur. Est enim eroos vehemens appetitus venereorum, ut Epicurus dicit, vel augmentum amoris benefactionis propter apparentem pulchritudinem, ut Stoici, vel recordatio visae pulchritudinis, ut Plato; vel eroos est concupiscentia amicitiae vel famulatus dei in templorum adornationem et bonorum, ut Andronicus. Est enim eroos, ut dicunt, primum elementum et maximum multipliciter visae historiae entium, et amor laboris, et solertia. Iste autem eroos est

'*Eros*', we are informed, has many meanings in the Greek: the forceful desire for physical love, as in Epicurus, or the beauty-inspired growth of benevolent love, as in Stoic usage, or the memory of beauty once seen, as in Plato, or the desire for the friendship and service of God together with its ritual and moral manifestation, as in Andronicus. *Eros* is referred to in myth as the primary element and the greatest of things. Or it can signify love of labour and skill. It is the most prudent seeker after all good. The connotations of vehemence and dynamism are an integral part of the Dionysian *eros*, concludes Grosseteste, though these, he adds, are more explicitly invoked in the *Divine Names* than in the *Celestial Hierarchy*. The effect of Grosseteste's researches is that '*amor*', as used in his commentaries, comes to don some of the richness of the Greek term it represents.[48]

We have just seen one example of the somewhat rare use of sources in the commentary, for Grosseteste derived his information on the history of the concept of 'eros' from the *Lexicon* so-called of Suidas (seventy-one articles of which he translated into Latin), supplementing it with a definition quoted from his own version of the treatise *On Passions* by Pseudo-Andronicus.[49] It is a measure of his concentration on Dionysius that in the vast work that is his *Commentary on the Celestial Hierarchy*, only three authors are actually named. The reader is referred on two occasions to St. John Damascene for a comment on the *Trisagion*, to be found among the works which Grosseteste declares he has recently trans-

omnium bonorum prudentissimus investigator. Significat quoque eroos amorem vehementem et in termino, de quo in libro de divinis nominibus plenius tractat. Talis autem est amor angelicus in deum, et haec amoris vehementia per vehementiam quae est in concupiscentia brutorum potest insinuari.' (ibid., pp. 86–7; printed by Franceschini, p. 115).

[48] ἀγάπη he renders by the more gentle term 'dilectio'.

[49] On Grosseteste's translation of Suidas (Suda) see Thomson, *The Writings*, p. 63; on his further use of Suidas in late works, Franceschini, pp. 65, 115. The original text of the Suda article on ἔρως, with source-references, is to be found in *Suidae Lexicon*, ed. A. Adler (Lipsiae, 1931), pars IIª, p. 417, n.3070. Grosseteste's Latin translation of Pseudo-Andronicus, from which he quotes directly in this passage of the *Comm. in Hier. Cael.*, has been edited by L. Tropia in *Aevum* 26 (1952), 97–112. In his introductory comments, Tropia prints side by side the article from Suidas, the extract from the *Comm. in Hier. Cael.*, and the definition from *De Passionibus*. The full text of the latter is printed in the same study, p. 109.58–61. For some further information on the version of the *De Passionibus* and its modern editions, see Appendix A.

lated;[50] and on a further occasion it is recalled that Damascene remained undecided as to whether the angels are all of the same species or not.[51] Augustine is named on three occasions, and his definitions of order and love are recalled.[52] The only source much referred to is the *'Graecus qui ponit notas in graeco quas nos in margine scribimus'*[53] (Maximus Confessor and John of Scythopolis); on about ten occasions, Grosseteste makes mention of him (or them), and now and then the discreet presence of these Scholiasts is traceable in the commentary itself.[54] Grosseteste may borrow the explanation of a word,[55] adopt a variant reading suggested,[56] discover a confirmation of his translation,[57] or simply refer the reader to supplementary information or biblical references contained in the scholia;[58] but my overall impression is that his achievement would have been but little diminished had he lacked the

[50] 'De hac autem hymnologia, "Sanctus, sanctus, sanctus", etc., sufficienter tractat Johannes Damascenus in epistola quadam quam de hac scripsit ad Iordanem Archimandritam, quam, quia nuper in latinum transtulimus, ad illam epistolam qui de hac hymnologia plenius scire satagit recurrat.' (ed. McQuade, p. 369; cf. ch. 13, ed. McEvoy, p. 94).

[51] Grosseteste quotes the text briefly and exactly from what he calls the *'Sententiae'* of Damascene; see Buytaert's edition, *De Fide Orthodoxa*, p. 72.

[52] Defending his own decision sometimes to employ *'ipsum'* to translate the Greek article, Grosseteste comments: 'Sic enim fecit saepe, significantiae cause, beatus Augustinus, ut in titulis psalmorum pro "too David" ponens "ipsi David".' (*Comm. in Hier. Cael.*, ed. McQuade, p. 8). St. Augustine's classic definition of *'ordo'* as 'parium dispariumque rerum sua cuique tribuens loca dispositio', is quoted, and attributed (p. 113); and Augustine is appealed to by name one last time for a definition of love: 'Ad declarationem autem intentionis huius nominis "amor", dicit Augustinus quod appetitus rei cognoscendae dum inhiat, fit amor rei cognitae dum tenet, ex quo dicto potest elici generalior amoris ratio, scilicet quod appetitus rei habendae dum inhiat, fit amor rei habitae dum tenet seu habet.' (p. 275).

[53] See ed. McQuade, p. 424; see further his introduction to the edition, pp. 26–29 (summarized in our Appendix A under 'Translations from the Greek', 4).

[54] For example, on p. 256 of McQuade's edition, where the identification of St. Paul as the direct source of Pseudo-Dionysius's mystical inspiration is taken over from the Scholiast.

[55] *'Transitiva'*, developing a bare hint in a Greek scholion (ed. McQuade, p. 93; cf. p. 76).

[56] 'Secundum notam autem sumptam ex graeco, aliqui libri habent "sic propriam existimandum esse ierarchiam uniformem ipsis secundum omne", pro eo quod nostra exemplaria habent, "similem existimandum esse et secundum omne uniformem ierarchiam", quemadmodum nos transtulimus . . .' (ed. McQuade, p. 312).

[57] '. . . sed graecus qui ponit notas in graeco, quas nos in margine scribimus, exponit hic *apokrinesthai* per *choorizesthai*, quod est separari seu dividi', ibid., p. 424.

[58] Ibid., p. 100: 'A quibus autem sacris theologis [i.e. sacred writers, authors

scholia, for on no occasion does their presence appear to have been decisive (as distinct from helpful, which they frequently were) for the interpretation of his author. His use of sources can only be described as occasional and sparse, not systematic. Had he, however, had in his possession more Greek reference-books, there is no doubt that he would have used them in the same way as he did Suda and Pseudo-Andronicus, that is to say, philologically, to bear upon the elucidation of the text.

In commenting on the text, Grosseteste manages to keep before his reader the work of Pseudo-Dionysius in its wholeness, at the level of ideas no less than that of language. When, for example, one fundamental aspect of the notion of hierarchy, the triadic division of the angels, risks obscuring another, the taxiarchic law subordinating each of the nine choirs to its superior, Grosseteste will bring out the general principles at stake by referring to the whole sense of Dionysius's thought, and will subsume the difficult passage into the wider context of interpretation, much as a modern interpreter like Roques does.[59] Material and ideas stemming from the other three books of Dionysius are worked into the *Commentary on the Celestial Hierarchy* to throw light on the meaning of the work. A compendious little treatise on the mode of signification of the divine names is introduced into the commentary on Chapter Two, at the point where the question of the symbolic knowledge of spiritual being is raised by Dionysius;[60] the reader is thus helped to make the contrast between our knowledge of angels and of infinite being. Again, in introducing Dionysius's definition of hierarchy, Grosseteste points out that it applies univocally to the orders of angels and to the Church on earth; the notion of hierarchy is the *fil conducteur* running between the two related works of Dionysius.[61] In numerous passages, Grosseteste draws parallels between the two hierarchies, entirely in the sense of Dionysius himself.

of biblical books] laudetur ipsa thearchia per praedicta in tribus generibus exempla, satis patet ex glosa marginali ex graeco sumpta.' (see pp. 110-11).

[59] Cf. Roques, *L'Univers dionysien*, p. 145.   [60] See ed. McQuade, pp. 58-66.

[61] See, e.g., McQuade, p. 127: 'Lex igitur teletarchica et thearchica est, ut quicquid hierarchice in teletis agitur fine revocationis et reductionis in deum agatur, nec usurpet agens quod supra se est, nec omittat quod sibi conveniens est, nec tradat alii quod est supra suam receptibilitatem, nec deneget ei quod suae congruit susceptibilitati.' This basic law is applied, in what immediately follows, to the angels (briefly), and to the visible Church, with examples from the offices of bishop, priest and deacon.

The conscientious philological investigation that marks Grosseteste's translation and commentary has impressed itself on all modern readers, and has led Père Dondaine in particular to comment that the Grosseteste corpus of *Dionysiaca*, as we may call it, rivalled the entire Parisian corpus in completion and excelled it in technique of execution and fidelity to the original. If Grosseteste's monumental work failed even to threaten the dominant position of the Parisian collection, then that was principally due to the fact that the theologians of that age, being more intent upon assimilating what they felt to be of eternal value in their authorities than upon investigating the history of doctrines, preferred the easier and more westernized body of comment which was compiled around Sarracenus's version.[62]

III CORRECTIONS AND EXTENSIONS OF THE *AUCTORITAS*

If the qualities of Grosseteste's exegetical work thus far examined were to amount to the whole of his contribution they would mark him out already as the Erasmus of the Middle Ages; but their interest would belong to the province of philology and not in any significant degree to the history of philosophy. Somewhat paradoxically, it is only in the measure that a medieval commentary on an ancient author failed to sustain a rigorous hermeneutical approach and passed over into the assimilation of ideas (and they all did so without fail), that it becomes of possible direct interest to the history of ideas; in other words, only when a commentary fails to achieve its task, as we conceive of it today, does it become a useful source-document for the development of philosophy or theology. Now it is beyond doubt that Grosseteste's effort to attain to the mind of his author was paralleled by a profound interest in the author's themes and by the desire to benefit himself and his reader by entering into living dialogue with the ideas of the Christian past and assimilating them into his own developing thought, and thereby into the whole movement

---

[62] Dondaine, *Le corpus dionysien de l'Université de Paris au XIIIe siècle*, pp. 33, 116. Grosseteste's version took time to establish itself at Paris; Peter Olivi seems to have been the first prominent Scholastic to have used it. It was studied extensively in England, however: Roger Bacon shows some familiarity with it, and Thomas of York, William de la Mare, Roger Marston, William of Alnwick, Tyssington and Wyclif valued it, as did Denys the Carthusian and Vinzent von Aggsbach in the German territories (fifteenth century).

of ideas of which his generation was witness. Scholars are
agreed that ideas and reactions personal to Grosseteste are
present in his Dionysian commentaries, but they have to admit
that the interlacing of the two minds is often so intricate that
it is difficult to separate them.[63] Some success in this venture
has been reported; Gamba and Völker have uncovered some
of Grosseteste's personal imports into the *Commentary on the
Mystical Theology*,[64] and Ruello has shown fundamental doctrinal differences as between Grosseteste's and Albert's commentary on the *De Divinis Nominibus*. McQuade, on the other
hand, conceived the doctrinal part of his work as an introduction to the Dionysian thought-forms, and his remarks on
Grosseteste's contribution, though helpful, are unsystematic.
What is needed, and what I hope here to supply, is a systematic
survey of all the main doctrines of Grosseteste to be found in
the entire *Commentary on the Celestial Hierarchy*, and a judgement as to how they affect his exposition. My aim is to arrive
at a synoptic view of the character of his doctrinal infiltrations,
something that will make it possible to trace the guiding presence of his general preoccupations within the commentary and
to arrive at guidelines regulating the use of the commentary as
a document witnessing to Grosseteste's personal thought and
its development.

In listing the constant features of Grosseteste's exposition, I
included what I called 'advocacy' (for want of a better term),
intending it to cover a whole series of personal reactions on the
part of Grosseteste to his author. Almost every section of his
commentary carries some remark by way of recommending the
author's words and praising their depth, or, it might be associating Dionysius with St. Paul, defending his ideas, or underlining the value of a particular thought, drawing a moral for
the Christian life, or illustrating some aspect of the text by adding an example, often drawn from natural light. Grosseteste

[63] Thus McQuade, p. 84: 'It is very difficult to know when the exposition of
Dionysius's thought ceases and the interpretation in terms of Grosseteste's own
thought begins.' Ruello, "La *Divinorum Nominum Reseratio* selon Robert Grosseteste et Albert le Grand", in *Arch. Hist. Doctr. Litt. M.A.* 34 (1959), p. 99: 'Robert
moule sa pensée sur celle de Denys après l'avoir traduit. Il est parfois difficile
d'extraire de ce fleuve son apport personnel.'

[64] Gamba, "Roberto Grossatesta traduttore e commentatore del *De Mystica
Theologia* dello Pseudo-Dionigi Areopagita", in *Aevum* 18 (1944), pp. 125 ff.;
Völker, *Kontemplation und Ekstase bei Pseudo-Dionysius Areopagita*, pp. 237–41.

had no doubt that Dionysius's standing was immediately sub-apostolic. He regarded him with the respect due to a converted philosopher of great learning and the confidant of St. Paul, and considered his work to be substantially a transcription in philosophical terms of the latter's mystical visions. The few hesitations concerning the authenticity of the *Corpus Areopagiticum* which were raised and answered by Maximus left Grosseteste unaffected. Even Dionysius's style and mode of argument he exempted from all criticism, defending him against the charge of *superfluitas sermonis* by referring to the subtle and difficult objects of his investigation, and the necessity of converging upon the same problem from multiple viewpoints in order to illuminate it adequately.[65] 'Far from meriting accusations of superfluity', concludes Grosseteste, 'this author deserves the highest praise for the perfection of his teaching'.

Sympathy was the *loi du genre* of the medieval commentary on an authority, and there can be no doubt that Grosseteste's sympathy extended itself beyond the style to embrace the system of ideas developed in the four works of the Pseudo-Areopagite. In particular, he accepted the validity of the whole scheme of ideas contained in the *Celestial Hierarchy*; he was convinced, for instance, that the angelology of Dionysius could be harmonized with the biblical data concerning the angels, and he followed him with the utmost sympathy even into the details of his attempts to show that all angels can be called *virtutes*, that the human hierarch can be called an angel, and that the law of taxiarchy was not broken when a seraphim was sent to cleanse the lips of the prophet Isaias with a burning coal.[66]

Placed before a scheme of ideas emanating from a writer whom he had good reason to acknowledge as being of high

---

[65] The image he employs is a truly lovely one: 'Sed sicut si res corporea, parva et obscura, et fere invisibilis, inter plurimas alias sit occultata, prudens ipsius quaesitor accendit non unicam sed plures simul lucernas, evertens omnia donec clarescat ei res quaesita; sic auctor iste, quia res quas investigat subtilissimae sunt, et ad videndum difficillimae, eandem rem sub diversis et pluribus enarrat formis et rationibus, ut quae sub declaratione unius rationis non est manifesta, sub multiplicatis declarationibus per rationes multas clarescat et manifestetur. Exigit etiam completa unius et eiusdem rei doctrina, ut omnes ipsius rei formae et rationes de ipsa secundum ordinem ostendantur. Est igitur auctor iste non de superfluitate redarguendus, sed de perfectione doctrinae extollendus.' (ed. McQuade, pp. 366-7).

[66] *Comm. in Hier. Cael.*, ed. McEvoy, pp. 24-5, 31, and 44-8.

authority, the medieval commentator inevitably allied himself
with the author, entered into dialogue with him regardless of
the gulf of time between them, and joined with him in solidarity in the common quest of the human reason for the eternal verities. This dimension of Grosseteste's commentary
bulks surprisingly small in the *Commentary on the Celestial
Hierarchy*, but proves fertile in ideas and suggestions. I have
counted eight excurses, ranging in length up to thirteen quarto
pages, in the work;[67] similarly, in the *Commentary on the
Mystical Theology* there are two personal passages,[68] and in
the first chapter of that on the *Divine Names*, three, of some
length.[69] In every case, these passages venture beyond the pure
exposition of the text and consider the problem-areas for their
own sake. Here Grosseteste permits himself a latitude sufficient
to allow entry to his own ideas, many of which prove upon
examination to be rooted in the Latin tradition, essentially
that of St. Augustine. The freedom he enjoys in such passages
is apparent from the wide vocabulary employed (for here he
readily leaves behind him the verbal atmosphere of his author's
text), and from the accustomed fluency of his style, familiar
to readers of his letters and treatises. The censor which governs
all aspects of his purely expository writing is lifted, and to the
extent that this is so we can claim to share his personal thought
and hear his own voice speaking in these passages.

These excurses provide us with an amount of material that
can be utilized directly and without any major reserve for the
study of Grosseteste's thought. They do even more than that,
however, for they are never islands in his commentaries, but
something more like peninsulas, and their climate is partly
determinant of the conditions on the contiguous mainland of
commentary and exposition. Their close study provides vitally

[67] These are the following: (ed. McQuade), on naming God (pp. 55-6); on
passion in brutes and man (pp. 81-5); on the participation of all things in God
(pp. 174-7); on the vision of God (pp. 194-8); a question, whether the angels are
all of the same species (the longest excursus, pp. 235-48); whether all orders of
the same hierarchy are equal (pp. 261-4); on angelic love (pp. 277-84); (ed.
McEvoy), how God is universal (pp. 81-3).

[68] See Gamba's edition, *Il commento di Roberto Grossatesta al 'De Mystica
Theologia' del Ps.-Dionigi Areopagita*, pp. 34-5, on Moses's ascent of the mountain;
39-40, on the contemplative ascent.

[69] See Ruello's edition, published in *Arch. Hist. Doctr. Litt. M.A.* 34 (1959),
pp. 139-40, that no created power can comprehend God; pp. 149-51, the unity
and trinity of God; pp. 165-7, whether there is an order of priority among the
divine names.

important clues to the interpretation of the commentaries in which they are inserted, and when the corrections and reservations they record are re-applied to Grosseteste's exposition of Dionysius taken as a whole, the result proves to be illuminating.

The most important and the best-known of these passages is Grosseteste's long defence of the beatific vision.[70] Grosseteste had identified Dionysius as one of the Greek patristic sources who made the object of the heavenly vision a created theophany emanating from the divine essence, and not the very face of God himself, as in the Western tradition and in the writings of St. Paul. Grosseteste neither attacked Dionysius nor interpreted him benignly, by soliciting his statements into harmony with the tradition of Augustine and Gregory; but he expounded the received Western doctrine and drew out its consequences for human nature with precision and vigour, and he defended the substantial identity of mystic and beatific visions. It was Grosseteste's last and his best word on the subject. I only remark here that Grosseteste, in some measure like all the Latin commentators of the period, was detectably more at ease when expounding the *Mystical Theology,* which for all its intrinsic difficulties was free from the system of created mediations everywhere present in the *Celestial Hierarchy,*[71] for the latter at times brought into peril the traditional Western principle maintaining the intimacy of the human soul with God. Grosseteste's rejection of any intermediary in the vision of God, writ large as it was in his central discussions, had repercussions throughout the commentaries: without explicitly recurring to it at every turn, he none the less repeated with quiet insistence in commenting the *Celestial Hierarchy,* the *Divine Names,* and the *Mystical Theology,* that the soul will see God '*sicuti est, sine medio*' in the next life, and that the mystical experience is at its summit '*ipsa non velata visio, sine symbolo et parabola*'.[72]

[70] In his comment on ch. 4 of the *Cel. Hier.* The passage has been edited by Dondaine, "L'objet et le '*medium*' de la vision béatifique . . .", in *Rech. Théol. anc. méd.* 19 (1952), 60–130. (See further analysis in Part IV, ch. 2).

[71] Albert the Great brought out this contrast between the two works quite explicitly in his commentary on ch. 1 of the *Mystical Theology*; see Völker, op. cit., p. 242.

[72] *Comm. Myst. Theol.,* ed. cit. pp. 23.29, 24.12; cf. 33;27, 35: '. . . interior

A consequence of some importance for the nature of the angels follows from this passage: the angel too, like the blessed, must enjoy the unmediated vision of God, unaffected by the hierarchization of the Dionysian thought-system. Having to interpret the Greek word ἄγαλμα, used by Dionysius to qualify the members of hierarchy universally as images of God, Grosseteste retains the Greek word in his version, but in his translation comments that all rational creatures, and hence both angel and man, are made to God's image and likeness and reformed in his image by grace.[73] Only man, and not the angel, is referred to as the image of God in Scripture, but Grosseteste thinks this can be accounted for without strain.[74] Now the angel's natural operation as image is *intelligentia*,[75] which is likewise man's highest operation, being no less than the natural capacity to see God face to face. This faculty we shall exercise fully in heaven, when, illuminated by the divine light, we shall imitate the angels in love and be their equals.[76] Grosseteste rejoins the scriptural doctrine—admitted of course by Dionysius —of man as eschatologically ἰσάγγελος.[77] Only one conclusion is possible, namely, that Grosseteste overrode the Dionysian doctrine (which he never the less faithfully reflected in his translation and explained without distortion) that only the first heavenly hierarchy sees God without intermediary, and that the lower ones receive the illuminations μέσως.

The extent to which Grosseteste admitted subordination within the hierarchies can only be surmised, for nowhere in his *Commentary on the Celestial Hierarchy* did he attempt to explain exactly how members of each angelic hierarchy can be at once in possession of the beatific vision and yet purified,

homo, quaerens videre deum incircumvelate et vere . . .'; 37.8: union as 'illustratio radii', 'locutio et sermocinatio'. See further *Comm. in De Div. Nom.*, ed. Ruello, ch. 1, p. 136, and especially p. 153: '. . . tunc autem, cum erimus docibiles dei [i.e. in heaven], sine velaminibus omnia nude conspiciemus.'

[73] 'Intelligimus autem hic omnem rationalem creaturam ad imaginem dei conditam, licet de homine specialiter dicatur quod conditus sit ad imaginem et similitudinem dei, et licet a subtilius perscrutantibus assignetur ratio quare homo solus et non ita expresse angelus in scriptura dicatur ad imaginem dei conditus.' (*Comm. in Hier. Cael.*, ed. McQuade, pp. 123–4).

[74] Grosseteste would wish to explain it in microcosmic terms; cf. Part IV, ch. 6, iii a.

[75] *Comm. Myst. Theol.*, ed. Gamba, p. 62.25: '. . . angeli vero, qui est intellectus, operatio naturalis est intelligentia . . .'

[76] *Comm. in De Div. Nom.*, ch. 1, ed. Ruello, p. 155, § 56.

[77] For the teaching of Pseudo-Dionysius on this point, see Roques, op. cit., p. 163 and n. 4.

illuminated, and perfected by the order above it (save, of course, in the case of the highest hierarchy). He could have harmonized these two theses in either or both of two ways. The revelation of the divine essence and of the divine eternal plan (constituting together the beatific vision) is made to each member of the angelic hierarchy, but the angel's knowledge and love of God are dynamic and growing, not static and completed qualities, since their object is infinite. The higher *analogia* of the superior hierarchies would place them continually in advance of their inferiors and put them in the position of exemplars and interpreters leading and guiding the lower by essence and activity to ever-increasing knowledge and love of God. The second possibility would be as follows. The receptive analogy of each hierarchy, and of each member of that hierarchy, is specific to it, and hence a relationship of natural priority and posteriority exists in their activity of contemplation; wherefore all could enjoy the direct vision, but some primarily, others only derivatively. The degree of actual subordination admitted in this explanation is evidently very reduced.

The reorganization which Grosseteste thus introduced into the Dionysian system is not inconsiderable. Mediation can never again imply the same strictly subordinating consequences, and all his references to it in his commentary must be read in the light of this fundamental corrective.

Another example of correction, albeit this time of a minor and relatively inexplicit nature, is found in Grosseteste's reflections on what were referred to in Pseudo-Dionysian terminology as 'like' and 'unlike symbols' of divine and heavenly reality. 'Like formations' are so called, Grosseteste reminds us, because they are removed in varying degrees from the grossly material symbols of divine and angelic nature sometimes employed by the Scriptures. The suggestion of Dionysius was that light and life exemplify the so-called 'like' symbols,[78] and Grosseteste extends the exemplification sympathetically, by remarking that heavenly light and animal life or entelechy approach the nature of the spiritual by their degree of distance from crude matter.[79] Now, in comparing the value of like and

[78] *De Hier. Cael.*, ch. 2.
[79] *Comm. in Hier. Cael.*, ed. McQuade, p. 41: '. . . hoc est, formis seu plasmationibus signare et manifestare ipsa incorporea figurationibus familiaribus, et,

unlike symbols for the discussion of spiritual reality, Dionysius unhesitatingly awarded the palm to the 'unlike', whose very materiality and crudity are sufficient to counteract superstition. The popular imagination is in less danger of believing that angels actually are horses, for instance, than that they are luminous, heavenly bodies; wherefore it follows that unlike formations are the more suitable manifestations of the immaterial.

Commenting on the remark, Grosseteste reveals a typical concern for the faith of the *multitudo*, and a willingness to defend the liceity of the material symbols used in the Bible to represent the angels.[80] His concluding exposition of the thought of Dionysius is a model of honesty in the interpretation of his author.[81] Only the last words become personal, as a diffractive element enters in. Dionysius had grounded his respect for unlike likenesses in the scriptural dictum to the effect that all things are good. Grosseteste extends the remark: everything that participates in the good is, according to its degree of sharing, a vestige and a distant imitation of the Absolute Good.[82] All lower creatures can be made the object of a discourse symbolizing the spiritual world of God and angel. It is the Augustinian word '*vestigium*' used here that alerts us to a personal development in the exposition. Although, of

quantum possibile est, hoc fieri cognatis ad ipsa incorporea figurationibus, dico, sumptis ex his quae apud nos sunt pretiosissima, quae sunt superpositae nobis substantiae, scilicet caelum, luminaria et sidera, et quae supra mundum hunc inferiorem ex quatuor elementis constantem sunt corporalia, quae non omnino et simpliciter sunt immaterialia, sed quodammodo, comparatione videlicet eorum qui in hoc inferiori mundo, propter materiae grossitiem et fluxibilitatem sunt alterabilia, generabilia et corruptibilia.' See p. 57: 'Lumen enim, etsi sit corporeum, ut solis et lunae et stellarum et ignis, inter corporea tamen est subtilissimum et maxime immateriale, et eorum respectu nominatum immateriale. Vita etiam animalis, licet sit corporis organici entelechia, in se tamen spiritus est.'

[80] 'Et est modus iste similis formationis propinquior ut faciat multitudinem errare circa divina, quam sit modus ille qui fit per formationes dissimiles, ut in proximo consequenter ostendet.' (ibid., p. 65).

[81] The summary begins, 'Manifestatio igitur divinorum per dissimiles formationes nos magis reducit ad comprehendendum divina esse intelligibilia omne corporeum excedentia, quam manifestatio eorundem per formationes similes, et ideo convenientior est etiam modus manifestationis per dissimilia quam per similia symbola.' (ed. McQuade, pp. 71-2).

[82] 'Participans autem bono in quantum huiusmodi vestigium aliquod et quaedam licet longinqua imitatio est per se boni; et ideo ipsum per se bonum quod est deus, et deiformes celestes substantias eius bonitate proxime participantes, omnis inferior creatura in quantum bona et vestigium ferens primae et ei proximae bonitatis, convenienter figurat.' (ibid.).

course, the concept of participation is in no way alien to the thought-world of Pseudo-Dionysius, Grosseteste's introduction of it at this point adumbrates his transition from a Dionysian to a distinctly Augustinian atmosphere of thought. The perspective thus introduced will clearly accord favour to higher vestiges (the so-called 'like symbols') and especially to the *imago*, over unlike likenesses, and in so doing will effectively reverse the privilege accorded by Dionysius to the latter over against the former.

When we turn to Grosseteste's fully personal contribution, one of the most notable excurses in the entire commentary, our initial intuition is verified: the vestige is present in force. The Platonists' image of the footprint left in the dust, so favoured by St. Augustine and so frequently repeated by Grosseteste himself, is given early mention. Lower traces are distinguished from what Grosseteste terms *vestigia formatiora divinitatis*, or traces which manifest a more express likeness or imitation of the Creator:[83] *ratio, intellectus, lux et vita*—all of them 'like' symbols. God, we are assured, is more appropriately praised when higher and nobler properties are denied of him, than when lower negations drawn from the sphere of generation and corruption are posited, for the former highlight his supereminence over all of creation.[84] But if all attribution of creaturely qualities to God is false, so likewise is the application of material predicates to the angels. In all of this, Grosseteste writes no single word in support of any supposedly greater aptitude on the part of lower material symbols and predicates to express immaterial being; indeed the whole tenor of his remarks runs directly counter to any such suggestion. It must never the less once more be admitted, in fairness to Grosseteste as commentator, that his actual exposition of the text of Dionysius is sympathetic and accurate; he does not

---

[83] 'Quae igitur sunt velut formatiora divinitatis vestigia expressius similitudinem seu imitationem ipsius retinentia, quemadmodum sunt in incorporeis ratio, intellectus et substantia, et in corporeis seu corpori colligatis, lux et vita, dicuntur symbola similia et formationes et similitudines et manifestationes similes; quae vero sunt velut informiora vestigia et minus expresse et obscurius imitationem ipsius retinentia, dicuntur symbola dissimilia . . .' (ibid., p. 59).

[84] 'Quanto itaque superior, subtilior, excellentior et incorruptibilior natura, evidentia suae possibilitatis quantum est de se corrumpi, et evidentia suae defectionis a per se necesse esse, clamat se non esse deum, tanto ipsa abnegatione plus laudatur et honorificatur deus, quia tanto excelsior et maior ipsa abnegatione praedicatur.' (ibid., pp. 61–2).

traduce, but faithfully represents his author's intention. Only in the free treatise does he allow himself to take distance from the authority.[85]

Besides these two corrections to actual ideas of Dionysius, the one of fundamental, the other of purely marginal importance, neither involving a direct hermeneutical infidelity, Grosseteste serves his reading-public on one notable occasion by expressing a far-reaching reservation concerning ideas of his authority. In his comment on the sixth chapter he draws attention to something that is indeed a genuine tension in the thought of Dionysius, namely, the necessity to decide whether the taxiarchic law of subordination applies not only to the interrelationship of the three hierarchies, but within each one to the three orders constituting it.[86] Dionysius, he tells us, seems to affirm that all three orders of the first hierarchy are equal, each being illuminated 'immediately and primarily' by God. Does the vertical subordination, implied by Dionysius in the foregoing chapters as holding true within each hierarchy, now reduce itself to the equal participation of all orders of a given hierarchy? One way of resolving the difficulty would be to maintain that the three orders can be equal in some respects to each other, but unequal in others; but this approach the commentator finds to be facile and unhelpful, for it turns up no evidence from authority, nor any basis of reason helping to substantiate the distinction and to determine in which respects the orders are equal or unequal to each other. Grosseteste brings forward another hypothesis. The three orders compose a hierarchy, in that they participate in unity and are co-ordained to one goal; yet the unity of a hierarchy as a whole is perfectly reconcilable with inequality among its constituent members, as can be seen from the example of the ecclesiastical hierarchy. Some properties possessed by wholes are not possessed by their parts, as not all attributes of a battle-line are transferable to the foot-soldiers, archers, or cavalry alone; nor are properties of the Apostolic College as such attributable simply to each Apostle. Thus Dionysius could here be attributing immediacy

[85] See further on this topic Part IV, ch. 5 (pp. 354–68).

[86] '. . . hic autem videtur auctor asserere quod parificantur ad invicem singuli tres ordines singularum trium hierarchiarum. Videtur enim affirmare quod primi tres ordines complentes primam hierarchiam in hoc parificentur, quod quilibet primo et immediate praeoperatrices suscipit a thearchia illuminationes . . .'(ed. McQuade, p. 261).

to the first hierarchy collectively, but not extending it distributively to the cherubim and thrones. On the other hand, however, Grosseteste claims that Dionysius's later, more detailed explanation of each order certainly implies an order of dignity within each hierarchy.[87] Having alerted his reader sufficiently to the difficulty and drawn his attention to the wider context to which the awkward passage must be referred for interpretation, Grosseteste adds a personal remark of considerable significance. Interpretations of the text aside, whether in reality the three orders of a hierarchy share all properties equally or none, or some equally and other unequally, and if so which, he finds himself unable to say with any certitude; a surprising remark at first sight, because in a previous chapter he had devoted a notable personal discussion to this very matter, and had come to the conclusion that all nine choirs (without reference to the triple hierarchical division) share in the same properties, but in unequal participative grades. Recalling his treatment there, he now warns his reader that any reference he has already made or may still make in the course of his work to this *'tam profunda materia'* is not intended by him as anything more than a conjecture about probabilities in an area which does not admit of certitude.[88] The admonition is revealing, in that it poses a clear distinction between exegesis (which has its own problematic and approach) and 'the truth of things'. It is indicative of the distance which the commentator has managed to interpose between himself and his author, and without which good interpretation cannot be achieved. It testifies to Grosseteste's consciousness that the doctrine of Dionysius goes beyond what has been revealed in the Scriptures (though without suggesting that it is irreconcilable with the biblical data), and it implies that Grosseteste as commentator is passing on the sound, pious, and orthodox doctrine of a saint of the early Church, without, however, thinking himself able

---

[87] 'Et forte ideo non observavit hic ordinem in nominando cuiuslibet hierarchiae ordines tres ipsam complentes, quem in nominando ipsos observaturus est inferius, ordine nominationis insinuans ordinem dignitatis, ut per hoc innueret se non intendere hic attribuere sua attributa singulis divisim, sed hierarchiae ex his complete coniunctim.' (ibid., p. 263).

[88] 'Qualiter autem praetacta se habeant secundum veritatem et certitudinem . . . non est nostrae modicitatis determinare. Si quid enim de tam profunda materia supra diximus, vel infra dicturi sumus, non secundum certitudinem affirmando, sed probabiliter coniectando a nobis dictum intelligat, qui hoc legere dignabitur.' (ibid., p. 264).

to vouch for its truth in all matters of detail. This is not the only index of a certain distance he takes from Dionysius's thought. In a free and personal discussion of angelic species, Grosseteste assigns to the archangels the function of announcing '*maiora et principaliora*', and to the angels that of announcing simply, even though knowing very well that Dionysius's system demands that only angels communicate directly with the human hierarchy.[89] Later, when writing closer to the text of Dionysius, he allows for the same possibility, but points out that Dionysius can still save his system without being unfaithful to Holy Scripture.[90]

The greater part of Grosseteste's personal reactions to Pseudo-Dionysius consists of neither corrections nor reservations (for he was evidently in broad agreement with his authority), but extensions and developments of his ideas. Such prolongations of the thought of Dionysius are very numerous; they occur in some way on almost every page of commentary, but their multiplicity and variety make them little susceptible of systematic classification. Many of them emerge only after the closest commerce with the commentaries, and some are additions so well grafted onto Dionysius's own ideas that only an extensive comparison between the commentaries and Grosseteste's personal works, and even beyond that, between the latter and his favourite sources, alerts the reader to their real nature. Familiarity with Grosseteste's ideas and interests reveals their presence, one after another, in the Dionysian commentaries, some clearly (the doctrines of light, exemplarism, and the powers of the soul, for instance), some elusively, by way of adumbration or reminiscence. It is above all the changing atmosphere of his language that advertises the transitions from strict commentary to development of the doctrines, and reminds us that we are commercing in the commentary with

[89] 'Octavus vero [ordo] archangeli, nonus autem et ultimus angeli: hi, ut praetactum est, ab annuntiando maiora et principaliora, hi vero a simpliciter annuntiando, quia angelus nuntius dicitur.' (ibid., p. 247).

[90] '. . . ordo archangelicus prophetans et subenarrans ipsi ordini angelico et per ipsum nobis, vel aliquando forte sine medio enarrans nobis sicut videtur Gabriel archangelus, licet in evangelio angelus solum nominetur, annuntiasse beatae virgini incarnationem filii dei in ea fiendam. Sed de hac annuntiatione, utrum facta fuerit immediate per archangelum, an mediate per ipsum et immediate per aliquem simpliciter dictum angelum, non est a nobis temere definiendum.' (ibid., p. 444).

two creative minds at once. The commentator's function was not to transmit doctrines by sheer transparency or by simply reflecting them back; he made himself rather into a refractive medium for the Dionysian ideas. Unlike the angelic *agalmata* of the Areopagite, those perfect mirrors that give back without distortion the illuminations they have received, the human commentator has had to struggle with a text, the meaning of which has been rendered opaque by a cultural distance he can only very partially traverse. He has had to assimilate the text in terms of his own pre-understanding, and make the effort at transmitting the resultant comprehension of the work within a linguistic medium and a cultural milieu which he shares with his reading-public, but not with his author.

The need to present Grosseteste's contribution in a synthetic way makes the historian very conscious of his subjective intervention. For the purpose of clear exposition, I have chosen, in what follows, to divide up the material according to the sources of the leading ideas which Grosseteste invoked to aid him in the understanding and development of the Dionysian system, for it is apparent that each of the Augustinian tradition, Aristotle, Avicenna and Pseudo-Augustine's *De Spiritu et Anima*, all of them major inspirers of Grosseteste's mature thought, has left a traceable influence on his commentary.

We have already referred to Grosseteste's use of St. Augustine's vestige-motif. To this idea and its associations (imitation, image, and likeness) Grosseteste frequently returns in the course of commenting on the Pseudo-Areopagite. His use of them is flexible, for they are a prominent element in his own thought. Sometimes *vestigium* is placed in relationship to the philosophy of participation;[91] on other occasions the emphasis is allowed to fall upon the limitation of the imprint, and the illustration of the *pes in pulvere* is invoked;[92] then again, the Augustinian theme finds itself in alliance with the negative theology of Dionysius. In his commentary on the first chapter of the *Divine Names*,[93] Grosseteste points out that just as the imprint indicates the length and breadth of the foot without revealing its height and other qualities, so the *vestigia*, whether

[91] See n. 82.

[92] '. . . sicut figura pedis imprimentis vestigium manifestatur per vestigium.' (ed. McQuade, p. 59).

[93] The original text can be read in *P.G.* 4, 190C: God is invisible and incomprehensible.

singly or in their totality, fail to represent the totality of God as he is in himself, transcendent and simple.[94] In this case Grosseteste reads into his text the Augustinian associations of *vestigium*.

The strongest association of 'vestige' for an Augustinian is with 'image', and we have seen that Grosseteste applies this concept to both angel and man in his commentary on the *Celestial Hierarchy*, once again reading into a Dionysian text some typically Augustinian ideas: *imago naturalis, reformata per gratiam, imago maxima veri solis iustitiae.*[95] Furthermore, as in the vestige of light Grosseteste found a trinitarian symbol (*lumen, splendor, calor*), so too in the *imago reformata* of the rational creature hierarchized by grace into the order of love, knowledge, and operation, he finds a possible reference on Dionysius's part to the Trinity. This extension of an Augustinian thought-mechanism finds no support in the text commented.[96] Similarly, in the commentary on the *Divine Names*, Chapter One, a whole series of triplicities found in the created world and the human spirit is examined as the basis of our knowledge of God, in a passage marked by the closest verbal associations with the *Hexaemeron*, Part VIII and *Dictum 60*. On the basis of these features, as likewise of the whole treatment of the second chapter of the *Divine Names* (the Trinity), it becomes quite clear that Grosseteste has no perception of the different conjugation of unity and trinity in Dionysius and Augustine.[97] However, the passage mentioned above (*Divine Names*, Chapter One) is another of those long, free excurses which do not bear directly upon the exposition of the text, but do of course establish an atmosphere affecting the commentary as a whole.

[94] 'Etsi enim res omnis secundum sui totalitatem sit creatoris vestigium, non tamen possibile est ut ipsa sit vestigium totalitatis ipsius qui creavit; sicut vestigium pedis in pulvere, imitatio et enarratio quaedam est longitudinis et latitudinis et figurationis subterioris superficiei ipsius pedis et per consequens simpliciter pedis, sed non ipsius profunditatis . . .' (ed. Ruello, 141, § 22).

[95] '. . . rationalis creatura existens imago dei reformata est imago veri solis iustitiae et intelligentiae, non solum magna valde, sed et maxima.' (ed. McQuade, p. 124).

[96] 'Et forte dicta trinitas in rationali creatura refertur ad trinitatem quae deus est, amor viz ad spiritum sanctum, scientia ad filium, operatio ad patrem . . .' (ibid., p. 116).

[97] For more light on this, see Koch, "Augustinischer und Dionysischer Neuplatonismus und das Mittelalter", in Beierwaltes (ed.), *Platonismus in der Philosophie des Mittelalters*, pp. 317-42.

Perhaps the clearest example of a direct Augustinian importation into a Dionysian text is to be found in Grosseteste's discussion of the nature of the seraphim. The name of the highest angelic order means 'burning', and Pseudo-Dionysius interprets the symbol of fire as referring to their eternal movement around the divine: fire rises of its very nature, and preserves in the heavens an eternal circular movement. Neither in Chapter Seven nor later in Chapter Fifteen (on the symbols of fire applied to the angels in Scripture), does Dionysius extend the symbolism of fire and warmth to include love. Grosseteste evidently chafed under the restraint of his authority, for no sooner does he mention the symbol of fire in his commentary than he sets about extending it by associating it with scriptural usage and with ideas borrowed from St. Augustine.[98] The fire with which the Scriptures are filled is spiritual love, the fire that entered the prophet (Jeremiah 20:9) and that the Lord came to send on earth (Luke 12:49). Grosseteste has in mind the 'holy love of God's beauty', with which (according to Augustine) the angels are on fire,[99] and he is certain that Dionysius intended the enumeration of the properties of fire to be transferred to the attributes of angelic love.[100] Dionysius is in fact quite innocent of the intention thus uncritically attributed to him. Grosseteste launches with enthusiasm into a personal treatment of the nature of love; one of the most attractive episodes in his entire writings, it is charged with the language and thought characteristic of the Augustinian tradition. Though St. Augustine is quoted only once (for a defence of love), the warmly personal quality of Grosseteste's writing derives its inspiration from him, and is further vivified by the century-old analysis of love initiated by Latin Augustinian theologians under the influence of the Victorines and the Cistercians. Grosseteste's treatise barely touches the Greek text. It offers several definitions of different aspects of love and is interested both in the method of defining it and in the reason why the reality of love should be so *'multiplex'*, in how

---

[98] 'Dicitur quoque quandoque amor ipsa res amata, et maxime cum vehementer amatur. Est itaque primus amor sicut naturale pondus vel naturalis levitas rei ponderosae vel levis.' (ed. McQuade, p. 274).

[99] See *De Civitate Dei*, IX, 22.

[100] 'Volens igitur auctor amoris quo fervent seraphim proprietates manifestare sicut ea significavit per ignem, sic eorum deiformes habitus significat per ignis proprietates.' (ed. McQuade, p. 274).

angelic love differs from human love, in whether general goods and truths are loved before their particular exemplifications, and in how even human love can be described as being unlimited in growth, because it has a transcendent object and appetite: *'amat omnia scibilia scire'*.[101] The treatise concludes with a reference to the joy that follows true love, and to the very Augustinian distinction of angelic love, after the model of angelic knowledge, into matutinal and vespertine.[102] All this is admittedly offered as Grosseteste's own poor contribution to the understanding of the Pauline revelation to Dionysius; but it is attached—however loosely for the most part—to a text which offers it in reality no point of insertion.[103]

Once the reader of Grosseteste's commentary has found the theme of love written large, so to speak, in this passage, he is alerted to its presence throughout the book. Grosseteste adds love to the definition of hierarchy, offering it explicitly as an extension of Dionysius's reflections; he thinks his author would have approved, and no doubt in this instance he is quite right. But would Dionysius have recognized that definition of *amor* as a *'pondus et collocatio rationalis essentiae'* which gives each of its objects its due place—an idea replete with reminiscences of the Augustinian *amor ordinatus*?[104] The goal of hierarchy is identified by Dionysius as being union; but Grosseteste feels

---

[101] 'Praeterea, si scibilia sunt infinita saltem ex infinitate specierum, numerorum, et figurarum et hiis per se accidentium, amat autem amor intellectualis naturaliter omnia scibilia scire, primo et maxime in rationibus eorum eternis in mente divina, et consequenter in seipsis, ac per hoc amat omnia scibilia, inquantum scibilia sunt, numquid amor huiusmodi terminum sortitur magnitudinis?' (ibid., p. 279).

[102] 'Vel potest dici quod sicut cognitio dei in se est velud mane et cognitio ipsius in creaturis est velud vespere, sic amor dei in se est elevatior et amor ipsius in creaturis est depressior.' (ibid., p. 282).

[103] 'Ad hos deiformes habitus tam excelsos, et a nostra infirmitate longinquos et visui nostro absconditos exponendos, omnino minores sumus. Verisimile enim est hos esse de arcanis verbis quae Paulus in raptu suo vidit, quae licet non liceret homini loqui, id est Dionysio, huic tamen quasi specialissimo discipulo, et ex discipulatu ei concorporato et quasi caelestium concivi locutus est, quae iste ut existimo non communiter omnibus sed puritate intelligentiae et divinis illuminationibus et gratia privilegiatis, ac per hoc velud super hominem elevatis, decrevit communicare. De quorum numero licet me non agnoscam, sed omnino ut praedixi ad tantum opus minorem, confisus tamen de salvatoris gratia, qui facit lucem de tenebris splendescere, quod ipse dederit pro modulo meo communicare temptabo.' (ibid., p. 273).

[104] 'Et quia amor est pondus et collocatio rationalis essentiae, non solum ipsum amantem, sed et amata ab ipso quantum est in se singula et simul omnia in sibi competentibus locis statuens, existimamus quod in nomine ordinis voluit auctor amorem comprehendi . . .' (ibid., p. 115).

impelled to specify the urge to unitive love in ways derived
from Augustinian psychology, with all its refined attention to
the inner experience: *'amor sincerus melius se amat, magis
quam se imitari desiderat, quantum sibi est possibile, et sic
ipsum ducem habere'.*[105] Dionysius's angels are pure, with-
drawn from the region of dispersion; Grosseteste must show
reason why interior dispersion and disharmony are the inevitable penalties paid when the essentially unitive nature of
love is not properly ordered: *'Amor namque amantem cum
amato commiscet . . .'*[106] What for Dionysius is a cosmic event
(love, illuminative knowledge, the triad of purification, illumination, and perfection), is for Grosseteste the rising of the
rational creature in its individuality towards its *beatitudo*,
located essentially in wisdom.

An Augustinian idea which is present elsewhere in the commentary is the notion of sacred order, and Grosseteste reverts to
the *De Civitate Dei* for its definition: that arrangement of equal
and unequal things which gives each its due place.[107] Seen from
another perspective, order is the God-given power to govern
all one's subjects. Grosseteste finds that the Dionysian notion
of hierarchic structure, which, together with sacred knowledge,
operation, and love, constitutes hierarchy *simpliciter*, is implicitly present already in these two ideas, and draws it forth
with satisfied ease. He understands and does not distort it, so
that the rapprochement of Augustine and Dionysius has no
serious, negative hermeneutical consequences. However, having to explain the term *kosmos* in the *Mystical Theology* ('God
is not "order"'), Grosseteste reads the full Augustinian sense of
ordo into the Greek text, and seeks by this means to establish
harmonious connections with the negations that follow ('God
is not the great nor the small, is not equality, nor likeness nor

[105] 'Sincere love loves something better than itself and longs to model itself
upon that, so far as it can, rather than on itself, and so to have it as its guide.'
(ibid., p. 122).
[106] 'Amor namque amantem cum amato commiscet et ita ea unit . . . Unde
cum se vilius et sibi dissimile amat, aut quidquam amat aliter et dissimiliter quam
deberet, impuritatem contrahit, sicut aurum commixtum argento fit aurum impu-
rum ex ipsa commixtione cum viliori se et sibi dissimili. Si vero non amat quod
amandum vel sicut amandum, cum quiescere non possit ab amando, necessario
convertitur ad amandum inordinate . . .' (ibid., pp. 151-2).
[107] 'Ordinatio autem seu ordo est parium dispariumque rerum sua cuique
tribuens loca dispositio, ut dicit beatus Augustinus.' (ibid., p. 113; cf. *De Civitate
Dei* XIX, 13).

unlikeness'), on the plea that each of these terms is pre-contained in the concept of order.[108] No doubt he thought he had found a valid key to the jumble of negations in the text of Dionysius. The appeal to familiar Augustinian themes in order to introduce logic and intelligibility into the deliberately opaque text of the *Mystical Theology* is characteristic of Grosseteste's interpretation. A final westernization of the Neo-platonic hierarchy: Grosseteste leaves no doubt that its whole essence comes from grace and not from nature; he thus extends the distinction far beyond anything that Pseudo-Dionysius would have recognized.[109] For the latter, hierarchy is part of the structure of things, a cosmic element, a transcendental category of spiritual being.[110] For the disciple of Augustine and the child of Scholasticism, the locus of its origin must be drawn with exactitude in terms of the developing categories of Latin theology. This again is a mark of Grosseteste's inter-pretation, to distinguish, whenever he felt it appropriate, between nature and grace, a preoccupation which does not correspond to the less differentiated and very Eastern theology of divinization.

If Grosseteste summons the Augustinian notion of love to his aid in interpretating Dionysius, it is only to be expected that Augustine's ideas on knowledge will likewise be invoked. Knowledge is the second note of the Dionysian hierarchy, and it is subject in all its development to the law of taxiarchy, according to which all illumination comes from above and is received and handed on by each hierarchy according to its proper analogy. Grosseteste, it is true, lends this doctrine faithful and extensive attention, but many of the minor re-ferences to illumination that we find scattered throughout his Dionysian commentaries betray a strong contamination of Western ideas derived principally from Augustine. Thus Grosseteste extends the concept of illumination to cover both practical and speculative powers and to comprehend all knowledge attained in art and science, when directed to its

---

[108] Gamba, *Il commento . . .*, p. 64.15-33.

[109] 'Sequitur autem ex praemissis quod hierarchia non est aliquid de naturalibus bonis datis ex ipsa naturali conditione, sed quod ipsa est naturali conditioni gra-tuito apposita, quod naturale est pulchrificans, decorans et adornans sacre in deum videlicet directive; ex naturali namque conditione habetur solum esse, hierarchia autem confert bene et perfecte esse.' (ed. McQuade, p. 137).

[110] Roques, *L'Univers dionysien*, pp. 81 ff.

last end.[111] When he writes that God, who is light, shines immediately upon every truth and every intellect in order to reveal all truth,[112] or that the first light causes the being of light in things illuminated as well as in the act of sight,[113] there can be no doubt left that we are fully in the world of Augustinian Neoplatonism, with its emphasis on the immediate presence of God to the soul in all its acts, rather than in the world of Dionysius, who prefers to highlight the hierarchical and subordinating reception of illumination and to restrict the latter to sacred and salvific knowledge. Grosseteste makes no distinction between the two tendencies, namely, the psychology of interior experience and the theology of cosmic and structural illumination. The same interest in psychology is verified in his reference to the morning and evening knowledge of the angels,[114] a distinction resulting from a purely Augustian preoccupation, as well as in his frequent references to the illumination of the will.[115] Yet it must be admitted that the unity of intellect and will expressed in the pair *affectus-aspectus*, and likewise most of the light-illustrations of psychological doctrine in which Grosseteste's commentary abounds, are not after all unsympathetic to the doctrine of Dionysius, which they prolong rather than distort; they flavour the text without changing its diet of ideas. When, however, '*illustratio radii divini*' is referred to as being an integral element in the mystical union, itself conceived of as a realization of *sapientia*, the reader becomes aware of an interference with the genuine Dionysian forms of thought.[116]

[111] 'Illuminari autem est per virtutes et scientias, vel simpliciter per virtutes, quia sub nomine virtutis generaliter dictae ad activas et speculativas, comprehenditur omnis cognitio artis et scientiae ad suum finem optimum directa.' (ed. McQuade, p. 141; cf. p. 334, and *Comm. in De Divinis Nominibus*, ch. 1, ed. Ruello, p. 164–76).

[112] 'In hoc autem quod illuminativus est ignis gerit typum divinitatis, quae . . . radians indistanter super omnem veritatem et omnem oculum intellectualem, oculo intellectuali veritatem manifestat.' (*Comm. in Hier. Cael.*, ed. McEvoy, p. 145).

[113] '. . . causat enim [i.e. God, as *luminis substantia*] et esse ipsius luminis in illuminatis, et videre verum et bonum per lumen susceptum.' (ibid., p. 68).

[114] See n. 101 above.

[115] Grosseteste refers to the double significance of the illuminations, 'tam intelligentiae oculum illustrantes, quam amoris affectum simplificantes'. (ed. McQuade, p. 182).

[116] 'Et haec huius radii illustratio et illustrationis susceptio mistica est theologia, quia secretissima Dei et cum Deo locutio et sermocinatio.' (*Comm. Myst. Theol.*, ed. Gamba, p. 38.6–8). 'Maxime autem et excellentissime dicitur homo, qui vivit secundum supremam virtutum speculativarum, quae est sapientia, id est ipsius

Illumination is an expression of a special presence of God in creation; the doctrine of exemplarism as developed on Augustinian lines by Grosseteste attempts to characterize God's general presence in each thing he has made: he is *forma seu essentia omnium*. This favourite doctrine of Grosseteste appears on a number of occasions in his commentary on the *Celestial Hierarchy*, always in connection with the philosophy of participation. God is the most abstract form sustaining all things in being;[117] he is the being of things that simply exist, the power of living things, the light of knowledge, the wisdom of the wise, the essence of all things.[118] The illustration of the jar giving form to the water it contains is a recurrent one, and is repeated in the commentary from Grosseteste's *De Veritate*. The most curious occurrence of the doctrine that God is the form of all things involves an elaborate attempt on Grosseteste's part to show how Dionysius can legitimately call him 'universal'.[119] Three explanations are viable, he suggests. A logical universal is many in its many predications, whereas what is predicated of the three persons is wholly one and the same, and therefore universal in a unique way. Secondly, God is called universal by the Platonists, as subsisting before all things and giving subsistence to them through the universal ideas in his mind. Finally, God can be called universal from the universality of his causal action; as principle of all existence he pre-contains all things, and as the cause that gives them subsistence he forms them from matter.[120]

If Grosseteste's Augustinian stamp is heavily present in his

divinitatis cognitio et . . . ille est summe homo, qui palam, non per enigmata vel figuras, sed immediate, contemplatur divinitatem.' (*Comm. in Hier. Cael.*, ed. McQuade, p. 197).

[117] '. . . ipsa [i.e. *divina beatitudo*] est essentia et forma abstractissima proprie magis nominibus abstractivis quam concretivis nominata.' (*Comm. in Hier. Cael.*, p. 145).

[118] 'Quapropter ipsa deitas, quae est essentia et esse super omnem essentiam et omne esse, et vivifica virtus super omnem vitam, et sensifica virtus super omnem sensum, et per se perfecta et prae omnibus perfecta, sapientia super omnem rationem et intellectum, verius et perfectius dicitur esse entium et vivifica virtus viventium et sensifica sentientium et sapientia sapientium, quam dictae virtutes . . .' (ibid., p. 175). Perhaps Grosseteste was aware of Scottus Eriugena's relevant teaching concerning God as *forma et essentia omnium*, but this has never been proven.

[119] This attempt is based on his own translation: he renders the Greek phrase ὡς καθόλου (ch. 13) as '*universale*', and not '*universaliter*' (as is mistakenly printed in Chevallier's *Dionysiaca*), with Scottus Eriugena.

[120] 'Potest etiam divinum dici universale ab universalitate causandi, secundum quam se habet ad omnia ut superexistens eis, et superhabens et praehabens ea in

commentaries, other interests of his accompany it. Though the scholastic element in his exposition is kept to a minimum (by comparison with that of St. Albert, for example), discreet indices of an Aristotelean influence can be detected. There is only one case of what is probably a direct borrowing: discussing the operations of the Dominions, Grosseteste opposes justice to tyranny, and Aristotle's *Nicomachean Ethics* is the source of his remarks on the latter.[121] Apart from this instance, Aristotle's presence in the text is of a very diffuse kind and does not interfere with the interpretation. A few inflections of language reveal Grosseteste's Aristotelean reading, such as 'the speculative intellect', or 'experimental knowledge', or the definition of the animal soul as *entelechia corporis organici*.[122] On one occasion the *intellectus agens* is mentioned twice, but the content of the notion is patently non-Aristotelean.[123] The only real influences of Aristotle which I can discern are much more diffuse and less easily detectable than those mentioned.

In the first place, Grosseteste brings to his reading of Dionysius's text a differentiated outlook on knowledge which contrasts with the exclusively sacral and hieratic interests of his author. Having to comment, for instance, on the conclusion that the angel's knowledge is intuitive and not analytic or *resolutoria*, Grosseteste is led to expatiate on the conditions of our knowledge of God. In a broad sense, he remarks, the method we must use is '*resolutoria*', consisting in analysing the various kinds of teaching and senses we find in the Scriptures, syllogizing and arguing, dividing concepts and classes, and in brief, using every means given us of showing the less known by means of the more known. The angels are spared a great deal indeed![124]

seipso superintelligibiliter et supersubstantialiter, producens ea ex non ente in esse . . .' (*Comm. in Hier. Cael.*, ed. McEvoy, p. 82).

[121] Aristotle, *Nicomachean Ethics* VIII, 12, 1160a, 34–43.

[122] 'Vita etiam animalis, licet sit corporis organici entelechia, in se tamen spiritus est.' (*Comm. in Hier. Cael.*, ed. McQuade, p. 57).

[123] See Part IV, ch. 5 (pp. 346–51).

[124] 'Et est huius varietatis sacrae scientia resolutiva, non solum quia plerumque syllogistica vel alio genere argumenti argumentativa in quo indigetur arte resolutoria, sed et quia plerumque unius linguae per aliam interpretativa, plerumque omni genere divisionis divisiva, et plerumque a propria significatione vocis in rem primo significatam, et a re primo significata in secundo significatum, et sic forte gradatim in tertium et quartum significatum progressiva, et si quo alio modo sit ignoti per notius manifestativa. Omnis enim huiusmodi manifestationis modus . . . potest communiter dici resolutorius.' (*Comm. in Hier. Cael.*, ed. McQuade, p. 331).

Again, Grosseteste has a much more exact theory of how the physical cosmos functions than had Dionysius, and in interpreting the scriptural symbolism of fire (which he does at great length in his commentary on the fifteenth chapter of the *Celestial Hierarchy*), he explains the physical properties of fire with an interest and a facility that derive from his scientific studies. Thus a single phrase of Dionysius, to the effect that 'fire moves other things', occasions a paragraph of explanation: no bodily thing is moved unless by the agency of a principle that is either of elemental or of celestial and unmixed fire, i.e. the heavenly bodies. The principles of gravitational motion to a natural place are internal ones, but the heavenly bodies are the first moving cause of the nature (i.e., the principle of rest and motion) that moves things from within; wherefore fire is a symbol of divinity.[125] In language and thought, such glosses owe much to Aristotle himself and almost everything to the new scientific trend. Their introduction does not harm the exegesis of the Dionysian symbols, rather it grounds them in reality;[126] but it does paint a huge backdrop for them, and it underlines once again the ambiguity of the medieval exegetes's intention, namely, to recover an alien and historical thought, while at the same time updating and developing it as knowledge of the nature of things. In general, however, Grosseteste was discreet in using Aristotelean elements in the 'interpretation' (in the full medieval sense) of Dionysius, and his patiently-acquired knowledge of the latter forbade him the over-enthusiastic concordism of the young Aquinas, who announced on one occasion that Dionysius follows Aristotle in 'just about everything'.[127] None the less, it is plausible to find a psychological and thematic continuity between Grosseteste's Aristotelean

[125] 'Nihil enim corporeum movetur cuius motus principium non sit aut ignis elementum, aut ignes caelestes, hoc est corpora caelestia. Licet enim quae naturaliter feruntur sursum vel deorsum in seipsis dicantur habere principium motus, primum tamen motivum naturae interioris moventis est, ut existimo, virtus caelestium corporum. Secundum itaque hanc proprietatem suam gerit ignis vestigium ipsius de quo scribitur, "stabilis manens, dat cuncta moveri."' (*Comm. in Hier. Cael.*, ed. McEvoy, 149–50).

[126] I argue for this conclusion in Part IV, ch. 5 (pp. 354–68).

[127] 'Dionysius fere ubique sequitur Aristotelem, ut patet diligenter inspicienti libros eius. (*In II Sent.*, d. 14, q. 1, a. 2). Aquinas later abandoned this mistaken perspective, when he came to write his commentary on the *Divine Names* and realized that Dionysius 'in pluribus fuit sectator sententiae platonicae . . .', (*De Malo*, q. 16, a. 1). See further, Chenu, *Towards Understanding Saint Thomas*, p. 227 and n.

studies (especially his *Commentary on the Posterior Analytics*) and his Dionysian undertaking. The differentiated Aristotelean methodology mediated by the logical works raised the problem concerning the basis of discourse on God and the invisible universe; Dionysius's theory of symbolic knowledge and of God-talk seemed to the Scholastics generally to offer the elements at least of an acceptable logic of theological discourse, one which was capable of withstanding an Aristotelean-inspired critique.

If the physical universe is present as a backdrop to the *Commentary on the Celestial Hierarchy*, the microcosm could almost be said to share the stage with the choirs of angels who constitute the main programme of the work. The commentator's interest in human nature is, however, significantly greater than that of the author, with the result that the former constantly creates opportunities of situating man as an intermediary being between angelic and animal natures—an Augustinian perspective once again.[128] Just as Grosseteste's notion of physical knowledge is far more differentiated and nuanced than that of Dionysius, so also are his anthropological ideas, which turn out to be drawn largely from Avicenna's *De Anima*, the work which was unanimously regarded by contemporary writers as the best introduction to the Aristotelean science of the soul. When they are collected together, Grosseteste's essentially occasional glosses comprise a series of contrasts between man and angel which consistently develop the knowledge of both, in their similarity and difference; and it is true to say that a good portion of his mature anthropology could be reconstructed from the Dionysian commentaries alone.[129] A fundamental community in being of all rational or intellectual natures is posited by the commentator as underlying the differences between the two hierarchies, and he explicitly asserts that both angels and men are made in the

[128] A large part of Grosseteste's commentary on Chapter fifteen is given over to this conception of man's place among creatures; the following statement is typical of the entire perspective: 'Superat enim homo omnia post caelestes substantias duobus, scil. virtute intellectus et superdominatione secundum haec tria, scil. secundum rationalem scientiam, et secundum inservile animae natura, et secundum insuperabile ipsius natura.' (*Comm. in Hier. Cael.*, ed. McEvoy, p. 165).

[129] In practice, however, the reader would require much prior contact with these doctrines in their macroscopic form in the other works of Grosseteste, before being sufficiently alerted to their presence in the commentaries, which is for the most part relatively unobtrusive.

image of God. As rational and free beings, reformed into symmetrical hierarchies by the operation of divine grace, all intellectual beings are destined for the same direct vision of God through the mediation of Christ. The rational soul differs from the angel only by its natural capacity and desire for union with a body—an Avicennian emphasis.[130] The angel's being is simple, but the powers of the single rational soul in man are multiple; they are specified, again with the help of Avicenna, as being rational, animal, and vegetable.[131] Further enumerations of the faculties of the soul (the external and internal senses) conform to Grosseteste's assimilation of Avicenna's psychology.[132] Here and there in his commentaries Grosseteste draws upon his psychological sources (Aristotle, Avicenna, Algazel, and Pseudo-Andronicus) as occasion demands it for the most diverse psychological data: a phenomenology of passion which adverts to its *materialis dispositio*;[133] the training of responses to produce reflex action in animals;[134] the observation that some senses are weaker in man than in the animal.[135]

The psychological element in Grosseteste's commentary is derived also from Pseudo-Augustine (Alcher of Clairvaux?), the last of the authorities notably influencing his exegesis. The division of the specifically intellectual powers which issued from him underlies much of the language and thought of the Dionysian commentaries, just as, more generally, it sustains a

---

[130] 'Ad haec, rationalis anima non videtur differre ab aliqua caelesti substantia alia specifica differentia quam potentia et appetitu naturali unitionis cum corpore organico in personalem unitatem.' (*Comm. in Hier. Cael.*, ed. McQuade, p. 241).

[131] *Comm. Myst. Theol.*, ed. Gamba, p. 65.15–17.

[132] *Comm. in De Div. Nom.*, ed. Ruello, pp. 137–8.

[133] 'Est enim propassio materialis, quia animae motio quae passio est habet aliquam materialem dispositionem secum semper comitantem, velut ira quae formaliter describitur appetitus repunitionis habet secum ut materialem dispositionem fervorem seu ebullitionem sanguinis circa cor, et timor qui est fuga animae sanguinis recursum ad cor, et consimili modo in caeteris passionibus est aliqua motio in corde et sanguine ad imitationem motionis animae.' (*Comm. in Hier. Cael.*, ed. McQuade, p. 84).

[134] 'Est etiam in brutis quandoque passio non ex natura sola sed ex consuetudine, ut quod bos et asinus ad propria praesepia, in quibus pasci consueverunt, moventur, et animalia disciplinata, ut simiae ex disciplina saltantes et canes ex disciplina venantes, ad apprehensa signa actuum quibus assuefacta sunt statim moventur ad actiones.' (ibid.)

[135] 'Habet quoque homo respectu irrationalium animalium minimum secundum sensum, quia sunt plura de illis quae homine acutius vident, clarius audiunt, melius olfaciunt, et sapores discernunt, et subtilius tangunt . . .' (*Comm. in Hier. Cael.*, ed. McEvoy, p. 162).

great deal indeed of Grosseteste's later psychology. When our commentator divides the operations of the soul into *imaginatio, opinio, ratio actu ratiocinans*, and the *ratio* further still into *intellectus* and *intelligentia*, the influence becomes unmistakable. Moreover, his own development of this distinction is apparent when the *intellectus* of the Pseudo-Dionysius is subdivided in the commentary into *virtus intellectualis*, the equivalent of the Aristotelean νοῦς which intuits the first principles of knowledge, *virtus intellectiva*, which knows spiritual being, and *scientia rationalis*, which is acquired by reasoning.[136] The intelligence, which is the faculty of direct vision in angel and man alike, assumes a very major role throughout the commentaries, defining as it does the ultimate capacity of rational natures and representing the fundamental possibility of the mystical and beatific union.[137] Needless to say, it has no corresponding element in the thought of Dionysius, but represents a pure importation into the commentary. In addition to his psychology, Alcher also provided Grosseteste with the doctrine of the theophany of the glorified Christ to the blessed,[138] with which the commentator overlaid the similar but less detailed teaching of Dionysius found in the first chapter of the *Divine Names*.[139]

IV EMPATHY AND ITS LIMITATIONS

How do we estimate the strengths and weaknesses, the relative success or failure, of Grosseteste as a commentator? Modern hermeneutics has isolated a small number of preconditions of all interpretative work, and these can usefully be made into headings for a particular examination of the dominant

---

[136] '. . . virtute enim intellectiva, qua potest intelligere omnia perfecte et abundanter, et in omnibus deum et verbum incarnatum, et ordinem ecclesiasticum, et ordinem vitae agendae superat, et praepotens est et dominatur, et quasi potestative tenet omnia post caelestes substantias. Est etiam homo superativus, non solum secundum virtutem intellectualem, qua absque ratiocinatione comprehendit prima principia, vel quaevis alia, sed et superdominatione seu superpotentatu secundum scientiam rationalem ratiocinatione adquisitam . . .' (*Comm. in Hier. Cael.*, ed. McEvoy, pp. 162-3).

[137] See, for instance, *Comm. Myst. Theol.*, ed. Gamba, p. 68; *Comm. in De Div. Nom.*, ch. 2 (MS Florence, *Laur. Plut. XIII dextr. 3*, fol. 35ᵈ), and the important statement edited by Dondaine, "L'Objet et le 'medium' . . .", in *Rech. Théol. anc. méd.* 19 (1952), pp. 124-5.

[138] See the passage in Grosseteste's *De Cessatione Legalium*, beginning, 'Item Augustinus in libro de differentia animae et spiritus . . .', printed by Unger in *Franc. Stud.* 16 (1956), p. 12.

[139] *Comm. in De Div. Nom.*, ed. Ruello, pp. 154-5, 171.

characteristics of Grosseteste's commentaries on the Pseudo-Dionysius.

In the first place, any interpreter must possess what we may call a pre-understanding of what he has to interpret; it is difficult to see how in the lack of this he could manage to crack the basic code of an alien language or thought. The medieval Scholastic, interpreting Dionysius as one Christian another, brought to the understanding of the texts a broad frame of reference centred upon the Bible and its traditional four levels of meaning, an entity universally accepted as the basic expression of Christian identity. This shared universe of meaning was further extended by the ancient principle known as the analogy of faith, a universal presupposition to which Grosseteste gave unusually resolute expression,[140] asserting that the Fathers of both Greek and Latin traditions must be held to have taught the same doctrine, however great the apparent divergences in their expressions of it might occasionally seem to be. Grosseteste, of course, assumed that in the writings of Pseudo-Dionysius he would find no other doctrine on the Unity and Trinity of God, the Incarnation and the Redemption, the sacramental order and the Church, than that taught by Ambrose, Jerome, Augustine, and Gregory. To the wide *Vorverständnis* thus constituted, we may add the large dose of Greek, and especially of Neoplatonic ideas, which the medieval Schoolmen inherited from Augustine in particular, and which rendered the Dionysian corpus something of a complement to what they already accepted concerning such prominent themes as light and illumination, order and participation.

In every work of interpretation the pre-understanding serves only as an indispensable guide to a new understanding which is acquired through the process of assimilating the strange elements in the text, and an interpreter attains success only in the measure that his appreciation of the differences existing between the text and his own starting-point, or pre-understanding, leads him to enlarge and, if need be, to modify the latter. In this respect Grosseteste, like most medieval commentators, met with a degree of success, which in his case can be adequately estimated only if we look beyond the actual

---

[140] See my article, "Robert Grosseteste and the reunion of the Church", in *Coll. Franc.* 45 (1975), pp. 65-6, and the long extract from the *De Cessatione Legalium* on the theme of the *concordia patrum*, edited on pp. 83-4, ibid.

commentaries that he composed. When we compare his writings up until *c*.1240 with those after that date, not a few notable and novel elements in the latter are seen to derive from his reading of Dionysius. No complete assessment of Pseudo-Dionysius's contribution to Grosseteste's thought has been attempted as yet; we cannot embark upon a full-scale enquiry here, but must limit ourselves for the present to recording the main ideas in question.[141]

Grosseteste gave a very large measure of approval to the concept of hierarchy which derived from Pseudo-Dionysius in its application both to the angelic choirs and to the visible Church. He accepted the taxiarchic law of subordination with reference to handing down the illuminations. The principles thus validated enabled him to place the salvation of souls, the end of hierarchy, at the very centre of the Church's mission, to posit a clear distinction between divine task and human arrangement (legal constitution and processes) in Church life, to support the *plenitudo potestatis* of the pope, and to stress the powers of the bishop and the role of lower ministers, while ruling out all exemptions of lower from superior powers. Much of his practical programme of reform could be brought under this idea,[142] and justified by the principle that the ecclesiastical hierarchy must imitate the heavenly regularity and order of the celestial. Within this new vision, the angels became a corporate exemplar of human activity rather than an individual one.

The *Divine Names* also had a strong influence on Grosseteste, and his interpretation of its doctrine remained close to the spirit of Dionysius. While the name 'good' is privileged above all the others, none of the usual divine attributes is held to designate God as he is in himself, but only as he reveals himself in his actions. Human efforts at assigning a rational order to the divine names are governed by the same fundamental limitation.[143] On this subject, Grosseteste's reflections

---

[141] The works of Grosseteste chiefly drawn upon here are: *De Confessione, Praeambulum* (the hierarchy of creation); *Letter no. 127*, to the Dean and Chapter of Lincoln (Church hierarchy); Sermon, *Ex Rerum Initiatarum* (bishop, priest, and deacon); Lyons Dossier (parallels between the celestial and ecclesiastical hierarchies; hierarchy and pastoral duty); *Letter no. 57* (the monastic state); *Letter no. 36* (the Pope and cardinals); Sermon, *Ecclesia Sancta Celebrat* (mystical ascent).

[142] See Powicke, in *Scholar and Bishop*, xix-xxi.

[143] Ruello, *Les 'noms divins' et leurs 'raisons' selon saint Albert le Grand commentateur du 'De Divinis Nominibus'*, ch. 2: "La raison de nom divin selon Robert Grosseteste", p. 169.

in his *Commentary on the Celestial Hierarchy* on symbolic knowledge of the divine and the spiritual, are in profound agreement with Dionysius.[144] The *Mystical Theology* provided Grosseteste with a description of the mystical ascent and union which, despite the incorporation of Augustinian elements, represents an honest and largely successful assimilation of Dionysius's thought.[145]

It might be suggested that some at least of these ideas are in reality no more than novel expressions of doctrines and beliefs that Grosseteste already held on other grounds; the papal *plenitudo potestatis* is a plausible example which might be alleged in this connection. This view would, however, amount to a serious underestimation of the debt owed, not by Grosseteste only, but by Western Scholasticism in general, to the Pseudo-Dionysius; in the case in point, the West had acquired, largely as a result of the eleventh-century reform movement, a network of custom and practice, laws and institutions, for which its traditional theological sources offered few deep foundations. Grosseteste believed, I think, that he had discovered their deepest support in the theology of hierarchy. The specificity of the language he employed in presenting his ideas concerning the Church's structure and mission is sufficient to leave one in no doubt that, if he made such extensive and punctiliously exact use of the language of Dionysius, this was because 'hierarchy' said more to him than 'order' or 'obedience', 'hierarch' more than 'bishop', 'analogy' more than 'status', 'deificatio' more than 'gratia', and so on. His understanding had been so expanded through commerce with Pseudo-Dionysius that it had to bring with it into Latin the very words that enshrined it, and it would have felt itself betrayed by a simple transcription into the traditional Western terms. Grosseteste's fidelity to Dionysius's language is evidence of the real degree of reciprocity that established itself between his presuppositions and the 'strange' text he interpreted. He was in fact prepared to allow the understanding he brought to the work of interpretation to be affected by the encounter, and was ready to question the text in a way that related it to

---

[144] I argue this when discussing Grosseteste's theory of knowledge; see Part IV, ch. 5 (pp. 354–68).

[145] See the introduction to my edition of the conference, *Ecclesia Sancta Celebrat*, in *Rech. Théol. anc. méd.* 47 (1980), pp. 131–87.

his own situation as theologian and churchman. Even where he rejected Dionysius he was forced by a genuine perception of difference into a reappropriation of his own Latin tradition, and a clearer statement of his beliefs was the result. In these ways, the work of interpreting Dionysius opened up for Grosseteste a new depth of self-interpretation.

A good interpreter is a sympathetic listener, capable of calling upon an affinity of interest to aid the work of understanding, and of employing an appropriate form of expression in his interpretation. In this respect medieval commentators generally erred on the side of excess rather than of defect, through their willingness to identify themselves with the quest of their authorities for the truth, and through over-generosity in evaluating their ideas. Grosseteste had no difficulty in arousing himself to an intense interest in the themes of all four Dionysian works, and even in their method of discussion and the style in which they were handled. From this came his inflexible determination to recapture the fullest possible measure of meaning, and his sincere and repeated protestations that the thought of St. Paul's intimate is more profound than is his own capacity to follow it. Grosseteste was far more successful than the common run of medieval commentators in achieving that more indefinable sympathy with the atmosphere of his author which constitutes the finesse of exegesis. For example, he retained the older method of commentating on the entire text in lemmata, rather than adopt the Scholastics' division of the text and pursuit of questions. Again, alerted by a note of Pseudo-Maximus, Grosseteste uses most commonly the words *'intellectus'* and *'angelus'* to designate the celestial beings,[146] in order to conform to scriptural and patristic usage; such expressions as *'intelligentiae'*, *'primae (supremae) substantiae'*, *'substantiae separatae'*, *'motores'*, all of them terms resonating with strong Arabic and Aristotelean overtones, are avoided in his commentaries, though of constant and normal use in his philosophical works. Finally, the only reference to the angels as movers of the spheres, a belief which he certainly held on philosophical grounds, occurs in a discussion which

---

[146] Grosseteste's version of the note reads: 'E. G. [meaning, 'from the Greek']: Intellectus vocant et qui apud gentiles philosophi intellectuales, id est angelicas virtutes. Vocat autem ipsos et scriptura nomine hoc in Ysaia intellectus magnus princeps babylonis, hoc est dyabolus.' (ed. McQuade, p. 6a).

is carefully segregated from exegesis and pursued explicitly for his own satisfaction, and he is careful not to attribute the belief, or its underlying cosmological preoccupation, to Dionysius.

It is only to be expected that such acute sensitivity to language as Grosseteste manifests should be reflected in the interpretation of thought. We find his sharp consciousness of different atmospheres clearly evidenced when the commentary raises a question to which only a microcosmic view could give a satisfactory answer (Why man alone and not the angel is designated 'image of God' in Genesis), but leaves it to the perspicacious reader to solve, due to the writer's awareness that Dionysius never includes microcosmic themes in his work. Again, Dionysius presents himself in the *Divine Names* as the theologian of the revealed names of God, and Grosseteste, despite his tendency elsewhere to distinguish between natural and supernatural orders, adheres closely to this perspective, withholding himself from introducing speculative elements drawn from the metaphysical theology which was rapidly developing in the Schools of his time. His dissent from Dionysius on two points did not lead him to an unfaithful or tendentious interpretation of the latter's thought; as we have seen, he allowed him his different view. It is also to Grosseteste's credit as a commentator that the excurses in which he develops his personal thought are almost always separated from his exposition of the text and are sometimes explicitly announced to the reader. All this must be positively evaluated, while on the negative side the intrusion of some scholastic language and the consequent conflation of Dionysian by (especially) Augustinian categories of thought must be regarded as having impaired his achievement as an interpreter.

Over and beyond that proximity to his author which sympathy with his programme and companionship in his undertaking in some measure assure, an interpreter requires to possess in almost equal degree the ability to achieve a distance sufficient to permit an objective, and when necessary a critical, evaluation of the ideas present in the text. If he is over-anxious to identify himself with his author's meaning, the commentator will inevitably be guilty of the very fault that his commitment is meant to avoid, namely, of reading into the alien, difficult or repugnant aspects of the ideas he encounters the meaning which seems natural to him, but which, however, is 'eisegetic' rather than exegetic, coming as it does from the native cultural space from which the interpreter has not managed to free himself sufficiently. It was here that the main failure of medieval hermeneutical practice undoubtedly lay. The facility with which writers of that age found an undisquieting solution to problematic texts—transmuting gods into God, and Sybils into prophets—testifies to their inability to place their own ideas within parentheses and to exercise a rigorous *epochē* over their own cultural presuppositions. The habit of systematic *Ausschaltung* of their cultural *a priori* was one that they developed but little, because they were never able to attain to the perception of their own culture as a limited whole, comparable to other cultures under the signs of identity and difference. The motive force driving what we may call their assimilative tendencies in interpretation knew only forward and reverse gears; it was incapable of idling and could not be put into neutral.

It must be acknowledged that Grosseteste made a much more successful effort at hermeneutical *kenosis* than did most interpreters of his age. He rejected from the outset the notion that Dionysius's text is easy to understand. For a start, he stressed that it was written in a language whose difference from Latin he almost celebrated, and whose superiority to it he extolled; hence it could not be read as though it were a product of the Western Church. He knew that words are the key to unlock thoughts, and that in consequence all study must begin with the Greek text, which alone can help resolve the ambiguities the reader comes across. Even the syntactical elements of Greek and the literary style of the writer he thought of as pertaining to the meaning that must be recaptured in its very quality of foreignness. At the level of ideas too, Grosseteste made a genuine effort to avoid facility in his interpretation. Now and then he had to rest content with a probable interpretation of the meaning; sometimes he was forced to acknowledge defeat and to allow the reader to choose from among several possibilities the one which might seem best to him. Undeniably, however, his predominant feeling of difference from his authority was of a reverential rather than a cultural or philosophical kind: he thought of Dionysius as a profoundly learned and holy man, and a disciple of the Apostle Paul;

Grosseteste felt himself to be not only inadequate, but unworthy in his presence. For all that, he tended to assume too
much common ground in ideas as between himself and the
great Christian writer; he obviously could not relativize him
to any degree, in the way that a historically-conscious age
would do, regarding him as a Neoplatonic, Christian harmonizer of philosophy and theology. As a commentator, he was
in several respects less observant, or at any rate less willing to
share his impressions with his reader, than either Albert, who
explicitly remarked upon the gulf between hierarchic subordination and mystical individualism in Dionysius, or the mature
Aquinas, who unhesitatingly pronounced him a Platonist.
Medieval writers generally, and Grosseteste in particular, had
little sense of proportion when it came to reading and writing,
so that, to offer only one example, the relatively slight attention paid by Dionysius to Christology and salvation-history,
as compared with the conjugation of his metaphysical triads,
never led them to a radical suspicion of what Dionysius's enterprise was about; none of them bruited the suspicion that he
might be more a Neo-Platonist than a Christian, or that he
was not in fact the disciple of Paul he professed to be.

The technical element required by a work of interpretation,
namely, the bringing to bear of all the available objective means
for the recovery of meaning, was generally neglected in the
Middle Ages, but it is precisely here that Grosseteste's outstanding merit is most manifest. We look upon him today as a
founding father of scientific philology—the Erasmus of the
Middle Ages, as Miss Smalley has called him. Of course, philology is no more a completely autonomous art nor an all-suf-
ficient hermeneutical instrument than is any other discipline,
and to the extent that it creates an ineluctable demand for
continuity with historical information in order to limit sheerly
subjective constructions, Grosseteste's success with it was far
from total and, when compared with the achievements of later
centuries, was clearly lacunary. He did, however, use such
material as was available to help him trace the rootedness of
Dionysius's language in an ancient culture and dialect, adding
whatever he could find (as distinct from invent) to explain
allusions to historical places and facts such as the Acropolis,
the Areopagus and the Attic dialect. If his reference works—
dictionaries, the *Lexicon* of Suda and the scholia—were of

comparatively meagre help, his intention went beyond the limitations imposed upon him by the circumstances of his age, and what he envisaged was more significant than what he achieved—surely a testimony to his intellectual greatness.

A good interpreter of a document or of another mind always achieves more than the sum of the five elements we have enumerated. In the case of the greatest interpreters, charismatic and highly personal flair, the spark kindled from a leaping fire, fans a flame of inner empathy that manages to intuit the very selfhood of the author, drawing out his fundamental intuitions from behind the written word of his text, but construing them on the basis of the latter. Even the very judgement itself as to whether this quality is present in a commentator is probably subjective to a high degree. It must suffice to say that Grosseteste had no more complete success in lifting the veil with which the obscure author of the Pseudo-Dionysian corpus covered himself, than have had scholars ever since. The notorious (and probably deliberate) obscurity in which the author of the Dionysian corpus still remains shrouded has prevented historians from attaining the kind of intimacy that is sometimes possible between commentator and author. Grosseteste, in any case, committed all of himself to the recovery of Dionysius's meaning, the moralist and preacher just as much as the metaphysician of light; and in the firm and vital connection he made between the ideas of hierarchy and pastoral care he may well have come as close to the spirit of Pseudo-Dionysius as any subsequent historian has managed to do.

The last part of our task, before we pass on to the discussion of the actual ideas found in the commentary, is to formulate some basic ground-rules for disengaging the purely personal elements in Grosseteste's commentaries from their contexts.

In the first place, we can safely assume that Grosseteste was in broad agreement with Pseudo-Dionysius: the number of disagreements he registered is small, as we have seen, and he passed on to his readers a body of doctrine which he regarded as basically sound and pious. If it can be assumed with safety that he agreed with everything from which he did not overtly distance himself, however, it is still not the case that all the themes and doctrines of the treatises came under his approval and caught his interest in equal degree. He had not the same enthusiasm for absolutely rigorous subordination as had Dionysius, nor was he convinced that the latter's system was true in all details. Fortunately, valuable external evidence comes to our aid in the shape of numerous passages in other works of Grosseteste composed after *c.* 1240 and showing a Dionysian influence; the ideas borrowed or influences manifested point clearly to those places where the weight of his approval was concentrated. Secondly, in the personal and independent passages of the commentaries themselves, we hear Grosseteste speaking in his own voice, without any diminution of authority. In the remainder of the commentary—by far the greatest part—where he stays closer to the text, we can listen for tones that are more muted, but never the less distinctly his own. When his explicit reservations and his measurably different perspective from that of Dionysius are borne in mind, a familiarity with his work (and with that of Dionysius too, of course) allows access to Grosseteste's own contribution. Awareness of the general characteristics of his work and of the hermeneutical situation of his age is here the first and indispensable guide. Whenever, following its taut straining after the full sense of Greek words, his language flexes itself momentarily and becomes fluent and familiar, the reader is warned of a possible surpassing of Dionysius's meaning on the part of Grosseteste. This almost always turns out to be the case when the language of psychology and of interiority appears: *interior homo, affectus, amor, voluntas.* Where anthropological themes are introduced by Dionysius, extension of them by the commentator can at once be suspected, and *e converso*, where contrasts between angelic and human nature are found in the commentary, it is well to verify what, if any, foundation they have in the text. Again, where Dionysian concepts have a Western cognate form, and especially an Augustinian one, contamination is always a real risk, one that preys upon a host of philosophical and theological ideas such as light and illumination, grace and union, symbolism and exemplarism. Words of the Greek text referring to cognition and the mental faculties are especially prone to comment in terms of a more differentiated consciousness than was that of Dionysius himself. Finally, even the innocuous-seeming, and oftentimes helpful, pedantry by which Grosseteste consistently brings order and logical sequence into Dionysius's thought as he progresses, must always be scrutinized with care; for it represents an intervention of a kind which transformed two-thirds of the *Mystical Theology* into a contemplative ascent through the levels of reality to the chiaroscuro of the *'locutio cum deo'*, thus aereating and lightening the darkness which envelops the summit of Dionysian rapture.

Chapter 3
The Teaching of the Commentator

I THE NATURE OF HIERARCHY

ORIGEN remarked that the Church's belief in angels had not led it to define their nature.[1] One aim of his practice of the Christian pursuit of gnosis was to define the participation of the angels as being that of higher creatures in the Logos, in order to remove from the gnostics one of the main attractions of their world-picture, namely, the emanation of a series of aeons from the first principle. By linking participation to creation, subordinating the angels to the Word, and conjugating the themes of graduated participation and subordination among three levels of angelic being (the six first-born angels, the archangels and the angels), Origen introduced into Alexandrian theology the notion of hierarchy, and centred it upon the communication of the knowledge of God down from the highest angel to the human hierarchy. This idea had no native roots in the Latin West,[2] and had to await the wide diffusion of the works of Pseudo-Dionysius from c.1150 onwards before it could establish itself.[3] The attention paid it by Grosseteste is a gauge of its success, and of its profound attraction for the mind of his age: 'hypothèse totale', as Père Chenu remarked, when drawing the analogy with the modern theory of evolution.

The opening passage of Chapter Three of the *Celestial Hierarchy* sets out a definition of hierarchy in terms of three components: sacred order, knowledge, and operation. Grosseteste regards this as the material definition; he comments upon each of its elements, adding on his own initiative a fourth, love. All four components of hierarchy, Grosseteste explains,

---

[1] *Handbuch der Dogmengeschichte*, Bd. 2, Fasz. 2b: 'Die Engel' (Tavard), p. 36.
[2] See art. 'Anges', in *DTC* I, 1, p. 1210, for St. Gregory's very limited, practice-bound account of the nine choirs, probably derived from Pseudo-Dionysius.
[3] Chenu, 'L'Entrée de Denys', in *La Théologie au XIIe siècle*, p. 130.

are governed by the idea of the sacred or the holy. He advances a family of interrelated notions which serve to define the holy: precious, honourable, unblemished, wonder-arousing, mighty, pure, chaste. He is well aware that the holy lies at the very centre of the Greek concept of hierarchy which he is attempting to explicate. He locates the central strand of the notion of the holy or sacred in the reference which sacred actions or things bear to God as the Creator, consummator, and perfector of all things; thus he underlines the transcendental reference of the idea by relating it to the creation and to the return of all things to their last end. All ancient and medieval ideas of order, both natural and social, make at least implicit reference to the sacred, and Grosseteste, who englobes the Augustinian definition of order within the more complex and specific Dionysian notion of hierarchy, makes the sacral element in ordering quite explicit when he points out that 'the order (*dispositio*) of homogeneous and heterogeneous elements, in which each receives its due place', is not simply an empirical pattern, but is the result of the divine constitutive action: God arranged (*disposuit*) all things in number, weight, and measure, and the concept of due place therefore refers back to the justice which informs the divine wisdom.[4] Order such as is conferred by God upon the Church and its ministers, to organize those entrusted to them, 'giving each his due', has a 'protestative' meaning; the actual sacredness of order consists in its being directed to God and leading to him as end.

The second note of hierarchy is knowledge (*scientia*), that is, the understanding of truth performed under the sign of the sacred. The knowledge of the divine mysteries has, like hierarchy itself, a transcendental character, since it is reflected in the knowledge of all things when seen as traces of God's nature, plan, and activity. *Scientia* thus defined is really synonymous with St. Augustine's use of the word *sapientia*, for it makes reference to the fullness of the latter in the contemplative vision of heaven, and it implies the unity of contemplation and action: the *ordo rerum*, when brought under the form of the sacred, becomes the *ordo vitae agendae*, and for each rational nature this is further specified by the *ordo* within which it is

---

[4] On the very different meanings and associations of '*hierarchia*' and '*ordo*', which the Scholastics endeavoured in a measure to fuse, see Chenu's interesting exploration, ibid., pp. 129-30 n.

placed as a member of the Church, of earth or of heaven.[5] The guide to the sacred quality of life and the orders (social and cosmic) subtending that is the illumination of God found in the Scriptures, so that Grosseteste can describe these, to which he refers in his very literal translation as *eloquia*, as the very substance of hierarchy.[6] The third component of hierarchy is sacred activity, directed to God as its end, or directing other things to him. Grosseteste passes immediately to the fourth element (presumably because it contains the key to the third), that is, love, ordered or hierarchized according to its objects. All creatures are to be loved in proportion to their place in the order of being; God is to be loved as the end of all. Though all four components of hierarchy belong intimately together, so that love, for example, is ordered according to the objective value of its objects, is informed by knowledge, and is productive of activity, Grosseteste finds that a certain priority must be granted to *amor* because it comes as much from the formal as from the material definition of hierarchy, the form and purpose of which is the assimilation of the creature to God: its divinization, as far as that is possible for a given nature.[7] The notion of deiformity, the word used in this context to define the goal of hierarchy, has a threefold sense. It can mean God himself as *per se* beauty, comeliness, or form; or his operation of these as effects in all creatures; or the goal to which rational creatures strive in hierarchy, and

[5] 'Altera pars est scientia, comprehensio videlicet veritatis, quae tunc sacra est cum in rebus vere comprehensis, sicut in creatoris vestigiis et speculis, ipsum trinum et unum et invisibilia divinitatis ad fruendum speculatur, aut ipsum in se omnino puro intelligentiae aspectu contemplatur, aut verbum incarnatum, per quod solum est reductio et revocatio ad patrem, in rebus vere intellectis aut in se ipso comprehendit, ut per ipsum ad patrem accedat; aut cum in rerum ordine aut in seipso apprehendit ordinem ecclesiasticum seu militantis seu triumphantis ecclesiae, ut in suo stet firmiter ordine, non per pusillanimitatem se deiciens, nec per elationem se supra se efferens; aut cum in rerum proprietatibus aut in seipso ordinem vitae agendae conspicit, ut secundum ordinem et honeste vitam agat, ut sic ad dei contemplationem pure accessum habeat, aut cum deum pure contemplans, in eternis rationibus omnia videt.' (*Comm. in Hier. Cael*, ed. McQuade, pp. 114-15).

[6] 'Adhuc, hierarchia ecclesiastica secundum beatum Dionysium est omnium sacrorum ecclesiasticorum ratio, negotium et adornatio; cuius substantia est a Deo tradita eloquia, cuius finis est salus animarum, ad Deum videlicet ut est possibile assimilatio et unio . . .', *De Iure Eiusque Apparatibus*, § 7 (part VI of the Lyons Dossier), ed. Gieben, *Coll. Franc.* 41 (1971), p. 382; cf *Eccl. Hier.* I, 3-4.

[7] '. . . forma quae perficit ierarchiam, scilicet assimilatio ad deiforme prout possibile, quae forma est reductio seu sursum ductio ad dei imitationem, imitantem eum proportionaliter ad inditas ipsi divinitus illuminationes.' (*Comm. in Hier. Cael.*, ed. McQuade, p. 115).

the degree to which they attain that goal; in other words, their measure of spiritual renewal and likeness, in love, knowledge, and operation, to God himself. Deiformity means sharing in the thearchic light, and therefore in beauty, simplicity, and goodness. This participation is changeable in the first creation of every rational creature, but is confirmed in the faithful angels, and made indestructible.[8] Of course, all creaturely likeness to God is purely imitative and never approaches equality, being governed by the finite capacity or analogy of the creature, and by the infinite gulf between thearchy and its creatures. The aim of hierarchy, in summary, is the union of creatures with God, so far as is possible.

From the definition of hierarchy thus formulated in terms of its attributes and *differentia*, Grosseteste proceeds to deduce the fundamental laws presiding over hierarchical operations—a very Aristotelean procedure. Assimilation to God and union with him remake rational creatures in his image. It follows that, being likenesses of the Good, they will imitate his expansive outpouring. As creatures, on the other hand, they have no influence to pass on to others, save what they themselves have received from above. Thus we have a first approximation to the law of teletarchy: each creature will hand on to those beneath it a share in the light it has received from above, to recall those lower beings to the origin of light and to raise them up towards it. A full statement of this law follows swiftly: the teletarchic law is, that whatever is done hierarchically and in salvific mystery ($\tau\epsilon\lambda\epsilon\tau\acute{\eta}$) is done in order to recall a lower creature to God by the mediation of a higher agent, who neither usurps powers which his place does not accord him, nor neglects to do what is required of him, nor hands on what is above the other's capacity to receive, nor refuses what befits its receptivity. Grosseteste employs examples to draw attention to the universality of the field of application of this law, which governs both hierarchies (celestial and ecclesiastical) and both dispensations (the Old and the New).

---

[8] 'Est autem haec formatio in creatura rationali ab initio suae conditionis transmutabilis, sed in angelis qui in veritate steterunt, lucifero et suis complicibus in veritate non stantibus, facta est per confirmationem intransmutabilis. In sanctis autem hominibus dum in hac carne degunt ideo dici potest intransmutabilis seu intransalienabilis, quia nulla vis potest eam auferre aut a possidente eam alienare, licet ipse possidens eam sponte possit eam abicere, quae post hominum deo adhaerentium confirmationem omnino erit intransmutabilis.' (ibid., p. 119).

Having inserted a long morality concerning the indissocia-
bility of the love of God and of the neighbour, Grosseteste
sets about exploring the ultimate ground of hierarchy. God is
its efficient, formal, and final cause, since all that is in it,
constitutively, generatively, and teleologically, comes from
him, who is head and leader of all ordering.[9] More profoundly
still, however, hierarchy, which derives from the order of grace
and of *bene esse*, as a pure gift, is based upon the innermost
divine nature, for if hierarchy as universally defined embraces
both the human and the angelic orders, it refers beyond these
orders which it englobes to the ordering principle itself, creative wisdom, which disposes all things in harmony (Ecclus.
16:27), in accordance with the divine ideas.[10] There exist
therefore three hierarchies, the trinitarian, the angelic, and
the human; but the term can be applied univocally only to
the two created ones. The angels contemplate the hierarchic,
eternal laws in the divine mind by which the created hierarchies
are constituted and their dynamism is regulated, and according to which God himself will govern the first hierarchy, then
the first will govern the second, the second the third, and
finally the human hierarchy will be governed by the hierarchy
of angels immediately above it.[11] Is it possible to discover of
what these laws themselves are the expression, and thus attain
to the ultimate understanding of the structure of the created
world? Grosseteste thinks he can distinguish the ultimate
reason for the gradation and order in creation; it resides in

[9] 'Ex hoc autem quod ipsa [sc. *supereminentia*] est per se teletarchia, sequitur
quod ipsa etiam sit omnis ierarchiae causa, tam efficiens et formalis quam finalis.
Ab ipsa enim habet esse, in ipsa manet, pulcrificatur, et formatur, ad ipsam et in
ipsam apprehendendam et amplexandam et non deserendam, semper inhiat, ut in
suum proprium et summum et perfectivum bonum . . .' (ibid., p. 147). 'Sic omne
purum et illuminatum et perfectum necesse est a primo et per se puro et illuminato
et perfecto, et non aliunde, puritatem, illuminationem et perfectionem suscipere.'
(p. 145).

[10] 'Licet enim tres dicantur esse ierarchiae, divina videlicet, quae est sapientia
attingens a fine usque ad finem fortiter et disponens omnia suaviter, et angelica,
et tertia quae in hominibus est, cum nihil positive dictum possit de creatore et
creatura univoce dici, manifestum quod ierarchia univoce dicta non potest universalius dici quam de angelica et humana.' (ibid., p. 137).

[11] 'Eisdem namque legibus eternis in mente divina eternaliter scriptis, et ibidem
lectis et intellectis, gubernatur et dirigitur, et ad superprincipale principium et
finem reducitur prima ierarchia a deo, et secunda ierarchia a prima, et tertia a
secunda, et humana ierarchia a tertia, moderatis tamen eisdem legibus secundum
gubernatorum et reductorum differentes possibilitates.' (*Comm. in Hier, Cael.*, ed.
McEvoy, p. 8).

what, following Pseudo-Dionysius, he terms the law of taxiarchy: God brings about an order that is always good order,
and that consists in a beauty that imitates his own nature—
'*omnis pulchritudo in ordine consistit*'.[12] All hierarchical relations among superiors, equals and inferiors, exhibit agreement and analogy, from which harmony results; and harmony
and concord are no less than synonyms of the divine nature
as revealed to us.[13] God, who is described as *supersubstantialis
harmonia*, wished all things, but especially the deification of
his rational creatures, to be done in the highest harmony, in
accordance with his own nature. Not only the division of the
three celestial hierarchies into first, middle, and last, but the
distinction of first, middle, and last orders within each hierarchy, and the similar gradation of powers within each individual spiritual being (making each a microcosm of the whole
arrangement of 'trinal triplicities'[14]) can, Grosseteste thinks,
be deduced from the same rule and principle, '*ad plenitudinem
harmonicae concordiae*'.[15] In common with most of the thirteenth-century Scholastics, Grosseteste extended from the celestial hierarchy to the entirety of creation the principle that the
divine beauty could be expressed only in a harmonious diversity of products, and with it the idea of order, reinforced by
overtones of hierarchy. The immense diversity and the differentiated self-communication of creative beauty make up, he
remarks, a universe of more perfect beauty and utility than a
single uniform product could achieve; and with multitude and
variety included as essential attributes of order, the subordination of lower to higher participations follows necessarily.[16]

[12] God is pronounced to be 'finis et consummatio omnis bonae ornationis,
consistentis in pulchritudine deiformitatis . . .' (ibid. p. 6), '. . . quia omnis pulchritudo in ordine consistit, et ideo ab effectu dicitur bene ornata, licet ipsa in se sit
per se bona ornatio, fit etiam quaelibet harum reductionum in harmonia, hoc est
convenientia, et concordia divina, et analogia, hoc est comproportionalitate ex
qua resultat harmonia.' (p. 7).

[13] 'Sic itaque omnia fiunt analogice et armonice in illis caelestibus ierarchiza-
tionibus, quia deus, qui est omnis harmonia et plenissima sui ad se concordia,
omnem harmoniam causans, principians et terminans, volens omnia secundum
singulorum possibilitatem in harmonicis concordiis sibi assimilari . . .' (ibid. p. 9).

[14] 'Vel potest intelligi haec triplicitas etiam trina in unoquoque intellectu ad
similitudinem trinae triplicitatis in toto angelico consortio.' (ibid., p. 15).

[15] 'Et non solum ierarchiarum quamlibet tripharie divisit, sed et quemlibet
ordinem cuiusque ierarchiae discrevit eterna harmonia, similiter tripharia divisione
ad plenitudinem harmonicae concordiae.' (ibid., p. 12).

[16] 'Sicut enim sunt essentiae diversae simpliciores et compositiores eius simplicitati propinquiores et ab ea elongatiores, sic sunt ipsius secundum maius et minus

Although Grosseteste made it plain that he considered hierarchy to belong to the order of grace, it would be wrong to assume that he would have placed the angelic substances on a single level of participation in the order of natural gifts, or would have envisaged hierarchization to have begun only after the confirmation of the good angels.[17] The metaphysical grounding which he gave to the ideas of creation and order would imply that the hierarchy established and perfected by the order of grace is itself founded upon a gradation that already held good among the angels, from the very first moment of their creation.

The noteworthy patristic tendency to reserve unqualified spirituality of being to God alone became evident very early in the theological tradition, with Origen, who attributed a body of heavenly or starry ether to the angels.[18] With some exceptions, such as St. Irenaeus, the patristic tradition followed Origen. Augustine himself was embarrassed by the question, but after some hesitation he inclined to the opinion that the angels possess a spiritual body similar to that in which men are to be resurrected, one of air and fire.

Although Pseudo-Dionysius was not the first to propagate it, the belief that the angels are pure spirits nevertheless commands the whole of his work, and was to exert through it a notable influence on later theology.[19] Perhaps it was under his influence that St. Gregory defended the spirituality of the angels, a belief which did not establish itself with firmness in the West until the high Scholastic period. Even the Fourth Lateran Council did not intend to settle the matter, and the clarity of its relevant statement is in fact more apparent than real.[20]

Grosseteste appears to have defended the immateriality of

communicabiles, et est in hac rerum summa varietate et summae pulchritudinis varia communicatione universitatis rerum pulchritudo perfectior et utilitas maior . . .' (*Comm. in Hier. Cael.*, ed. McQuade, p. 169). See the close parallel in the prologue to his *De Confessione*: 'Multa quidem et varia fecit, quia non erat unum ex universis quod capere posset totum suae pulchritudinis.' (ed. Wenzel, *Franc. Stud.* 30 (1970), p. 240). Note the quotation of a similar idea from Hugh of St. Victor's *Expositio in Hierarchiam Caelestem*, made likewise in this prologue.

[17] As St. Bonaventure apparently did; see Bougerol, 'S. Bonaventure et la hiérarchie dionysienne', *Arch. Hist. Doctr. Litt. M.A.*, 36 (1969), p. 148.
[18] Tavard, op. cit., p. 33.
[19] Roques, op. cit., p. 154 n.
[20] Tavard, op. cit., pp. 64–5.

the angels even before the influence of Pseudo-Dionysius on his thought became marked, and he apparently even believed that it could be established by reason.[21] The work of commenting on the *Celestial Hierarchy* presented him with the perfect opportunity to reflect on the consequences of a view to which he was already committed. His many partial discussions of the referential value of symbols and attributes to God and the angels crystallized in his commentary on Chapter Two into a little treatise on the divine and angelic names. In this he attempted to set forth the fundamental principle of the logic of attribution of names to spiritual realities. He enunciated it in a double form, to cover positive and negative attribution. The principle is formulated as follows. The negation of all things is true of God; the negation of all material things holds true of the angels. All affirmative attributes literally applied are false of God, while the class of those that affirm some material property cannot be applied literally to the angels.[22] From this principle there follows his consistent effort, renewed throughout the commentary, and especially evident in its fifteenth chapter, to remove all residue of material understanding stemming from the use of earthly symbols. He sees very clearly the danger that symbolism derived from the more immaterial cosmic realities—light and the heavens— may present a greater occasion for the popular, anthropomorphic imagination to set to work than does the symbolism derived from natural life; for while the ordinary believer can readily appreciate that the angels are not formed of the earthly matter that is the source of 'unlike symbols', he may be less able to purify the class of 'like symbols' from all material associations.

Grosseteste as commentator both finds and makes opportunities to explore the concept of pure spirit, even though (and this is significant) his explorations are always pursued by comparing and contrasting the pure spirit with the incarnate,

[21] *De Finitate Motus et Temporis*, ed. Dales, *Traditio* 19 (1963), p. 264.

[22] '... negationes quidem etiam proprie dictae verae sunt, ut praedictum est, de deo quidem negationes omnium, de angelis vero negationes omnium corporalium et materialium, affirmationes autem proprie dictae inconvenientes seu inconcordes occultationi seu absconsioni arcanorum de deo et angelis, quae archana ut supra tactum est abscondenda sunt ab immundis, quod fieri non potest per affirmationes proprie dictas, sed symbolice et ekfantorice sumptas; dictae enim proprie de deo falsae sunt omnes, de angelis vero falsae omnes quidquam materiale praedicantes.' (*Comm. in Hier. Cael.*, ed. McQuade, p. 64).

human spirit. Man and the angel are both rational creatures made in God's image, and our experience of spiritual action is the best basis we are given on which to construct a discourse on the spiritual. Affirmative statements attributing to the angels characteristics which we share must all be qualified, before they hold true of the pure spirit. For instance, the angels are endowed, so Pseudo-Dionysius tells us, with a *logos* superior to ours.[23] Now '*logos*' can be used in three ways, for it can mean the interior reason of which we are conscious, or the exterior, audible word, or the combination of these two, made as the reason produces a verbal manifestation of its interior meaning for the purposes of communication. Now since the latter two meanings of the word evidently cannot be applied to the angels, our own internal experience of meaning remains the only valid clue in the inquiry into the nature of purely spiritual life. Only when an angel has assumed a body for a special ministry can the two operations involving material expression be attributed to it. The angel is to be described as a *logos* or reason of a purely intellectual kind. Its use of reason transcends ours, in that it does not presuppose a corporeal instrument. Positively expressed, the superiority of the angel lies in the mutual compenetration of angelic beings through unclouded mutual love and intellectual knowledge, a form of being and action which materiality excludes. Grosseteste has none of the reservations commonly experienced by the following generation of Scholastics concerning this simple and beautiful idea.[24] The angels, he maintains, manifest their acts of will and understanding to each other '*limpidissime*' and directly, without that need for exteriorization which only matter posits. The presence of one angelic mind to another is quite direct and unmediated; it does not know the distance that separates knower and known, where knowledge has to

[23] 'Logon vero secundum quod significat absolute rationem quae tantum intellectualis est et corporeo non utitur instrumento, non privat ab eis hoc nomen alogia, sed significat supereminentiam rationis ipsarum ad nostram rationem, quae corporeo utitur instrumento et per sermonem exteriorem suos conceptus depromit audienti. Est autem haec supereminentia, quod amore mutuo et aspectu intellectivo sese mutuo penetrantes absque aliquo corporali seu exteriori adminiculo, suas voluntates et comprehensiones adinvicem limpidissime et certissime declarant.' (ibid., pp. 91–2).

[24] Such reservations can emerge on the occasion of the question as to whether one angel knows another *per essentiam suam* or *per speciem seu similitudinem*; see St. Thomas, *Summa Theologiae* Ia, q. 56, a. 2, ad 3.

be mediated through the senses. Grosseteste's theory of the supereminence of angelic knowledge is, of course, built upon the Augustinian belief that the direct apprehension of the eternal reasons affords the angels a clear and certain knowledge of all created being.

Over against Grosseteste's repeated assurances that the angels are immaterial in both nature and activity, there stands a single remark which seems to contradict the general principle. In his commentary on Chapter Four he makes the brief, but at first sight disconcerting, remark that, although a distinction is often made between material and spiritual substances, still, by comparison with the absolute simplicity of God, all creatures are material, since they are formed by his gift of being.[25] The reader is immediately tempted to posit a connection between this remark and the theory of universal hylomorphism, which Avicebrol's influence did so much to spread among the thinkers of the thirteenth century. Nowhere else in his writings does Grosseteste evidence any sympathy for this idea, so far as I can tell. A simpler interpretation of this isolated statement emerges, however, when its likely sources are checked. I think it most probable that Grosseteste quarried the idea in Damascene's treatise on the angels, where it is expressed in almost identical terms. The angel is an intellectual and spiritual substance; immaterial and unembodied compared to us, it is—like all creatures—gross and material by comparison with the incomparable Godhead, for he alone is essentially spiritual.[26] This idea goes back to Origen and to the Apologists, and it finds expression even in writers like St. Gregory, who, as we have noted earlier, defended the immateriality of the angels.[27] Clearly, for many of the Fathers, and of the Scholastics too, the concepts of matter and body were not co-extensive, and they could adopt the idea in question without denying the unembodied nature of the angels. However, its repetition by Grosseteste, though perfectly innocuous, and testifying to a

---

[25] 'Etsi enim quandoque fiat distinctio inter substantias materiales et immateriales, comparatione tamen ad simplicissimam divinitatem omnes sunt materiales, eius formarum largitione et prout eis est possibile communicatione formatae.' (*Comm. in Hier. Cael.*, ed. McQuade, pp. 168-9).

[26] Damascene, *De Fide Orthodoxa*, ch. 17, ed. Buytaert, p. 69: 'Incorporeus et immaterialis dicitur [i.e. the angel] quantum ad nos, sed comparatus ad Deum, corporeus et materialis invenitur.' This text was a classic; see St. Thomas, *Summa Theologiae* Ia, q. 50, a. 1, obj. 1.

[27] St. Gregory, *Moralia in Job*, II, 3.

respect typical on his part for the patristic tradition he revered and imitated, does indicate his share in the general intellectual embarrassment as to how to express the participation of a purely spiritual nature in the divine being. Even in the case of pure and simple spirits, some kind of composition must be posited which separates them as finite beings from the infinite Spirit, and so finally removes the danger of pantheism. The difficulty lay in developing concepts adequate to this task. Grosseteste did not succeed in finding a solution. Aquinas's theory of the participation of each finite act of *esse* in *ipsum esse subsistens*, and the distinction of *esse* and *essentia* in all finite things which it entailed, was the first which overcame the difficulty satisfactorily.

The spirituality of the angels implies their simplicity of nature. Like his authority, Grosseteste never tires of reminding his reader that simple, spiritual being is only very inadequately signified by the multiple and material nature of the symbolic knowledge to which mankind is limited by its present epistemic condition.[28] The intelligible beings are simple or incomposite in themselves, but this does not mean that they all participate equally in the communication of being, for the degree of participation of each being in unity is defined by its own *analogia*. In a graduated hierarchy, those members are more simple in being who are closer to the One, and those relatively more composite who are further removed from that primal simplicity.[29] The simple nature of the angels gives rise to a single act, *intelligentia*. This is not to be thought of as effecting a coldly intellectual form of contemplation, but rather as being the faculty of union with the divine nature, so that we have to think of something that is in itself simple *unibilitas* as having two inseparable aspects, namely, knowledge and love. The equivalent to this total capacity for responding

---

[28] 'Dicit autem hic compositiones communiter formas et figuras materiales et corporeas, quae omnes sunt aliqua participantes compositione, significantes immaterialia et incorporalia intelligibilia . . . in designationem intelligibilium simplicium in seipsis, et ideo existentium nobis adhuc infirmis ignota et incontemplabilia . . .' (*Comm. in Hier. Cael.*, ed. McQuade, p. 40).

[29] '*Est enim hoc proprium causae omnium, . . . vocare entia ad sui ipsius communicationem*, non tamen ut omnia entia ipso equaliter communicent, sed sic, ut *unicuique entium definitur* seu determinatur *a propria analogia*, hoc est proportionalitate suae susceptibilitatis ipsius. Sicut enim sunt essentiae diversae simpliciores et compositiores eius simplicitati propinquiores et ab ea elongatiores, sic sunt ipsius secundum maius et minus communicabiles . . .' (ibid., p. 169).

to God is in our composite human nature only one power among others, although it is the highest. Pure intellects are not hindered by passion, which is a darkness set over against the light and energy of reason in man, and their knowledge requires none of the strategies by which mankind increases its stock of ideas (experience, reason, and analytic science). There is no trace in the pure spirit of the plurality of intellectual powers by which men grasp the first principles of theoretical and practical knowledge, know spiritual being, and reason from known to unknown. The mode of existence enjoyed by the angels is best thought of by us as the complete coincidence of the two forms or activities, life and intellect. In all that they are, they are living, and that life is none other than a purely intellectual life, consisting in the loving contemplation of God and of all things in him.[30] Now life is inseparable from growth; though we cannot represent their intellective process under the form of time (which presides over our reasoning from premiss to conclusion), we can think of it as a reciprocal motion of *aspectus* and *affectus*, that is to say, of the willed urge to conformity with God sustaining and being renewed by contemplation, in an unceasing circulation that is nothing other than their capacity to receive the influence that comes from above.[31] Doubtless it is the angels' capacity for ever-increasing assimilation to the divine mystery which justifies our attributing duration to their lives. Being bodiless, they do not live in cosmic time, the bearer of the seeds of decay; being finite, they cannot enjoy the total simultaneity of eternity. Eternal life is the limit and ambit of their form of life, circumscribing from beyond what can best be called their sempiternity.[32]

[30] 'Ipsae namque totum quod sunt, vita sunt, nec alia vita quam intellectualis vita qua deo fruuntur, ipsum et in eternis rationibus in ipso omnia contemplantes, et toto amore ipsum amplexantes, et singula in ipso amoris mensura et modo quibus convenit et decet; et econverso, in omnibus creatis ab ipso ipsum speculantes et in amorem ipsius assurgentes, et quae ad nutum voluntatis ipsius agunt non corporeis seu materialibus instrumentis agunt, sed intellectualis virtutis imperio, nec agentes, avertuntur ab ipsius fruitione, nec fruentes, impediuntur ab actione. Talis itaque est caelestium spirituum vita intellectualis.' (ibid., p. 182).

[31] 'Sunt itaque caelestes spiritus naturaliter ex libero arbitrio conantes in conformitatem ad dei imitativum, et ex conatu aspicientes et ex aspectu iterum conantes, nec est cessatio huius circulationis; et est haec circulatio aptitudo ad suscipiendum divinas influentias gratuitorum donorum, quibus susceptis couniuntur deo, et in fortitudine amoris iam formati ex donis susceptis ad datorem inflexibiliter extenduntur, et extenti semper suscipiunt, et secundum suscepta ordinati totam ut dictum est vitam habent et ducunt intellectualem.' (ibid., pp. 182-3).

[32] 'Ipsam enim durationem sempiternam substantiarum caelestium divina

The hierarchical activity of the intelligences, conceived of as a participation in the divine life, is characterized by the three moments of purification, illumination, and perfection, whose omnipresence in the text of Pseudo-Dionysius and in the commentary renders their summary exposition extremely difficult. The metaphysical background presupposed by what is in effect a theology of revelation and grace is, of course, the metaphysics of light. God is the very substance and essence of spiritual light, and no light of the spirit can shine save by participation in the uncreated light. Light streaming from a source tends to form a hierarchy of diminishing power, and the created lights in their varying degrees of participation imitate the nature of the source of light, each shining or reflecting upon the next the light which it has itself received from above. The guiding principle of the whole conception is that of unity. There is a single infinite fountain of light, whose unity remains perfect even when its ray is participated in by many. The generation and extension of light downwards from this spiritual sun leaves it in itself unchanged, undivided, and still transcendent to the hierarchy formed by its effusion.[33] Although totally present to all it shines upon, it loses nothing of its interiority in its outgoing. Since all light comes from one single source, downwards from the Father of Lights, the work of the assimilation and transmission of light within the grades of the hierarchy, down even to the lowest, is as much the activity of the *lux suprema* as of the lower beings themselves.

The effect of the reception of light is described by Grosseteste the commentator in a global way, before being analysed into its triple hierarchizing function. The intention with which light is granted is, that its recipients may be raised and led back to the fontal origin of all light or being.[34] This 'raising'

sapientia comprehendit et ambit, ac per hoc incomprehensibili modo terminat.' (*Comm. in Hier, Cael.*, ed. McEvoy, p. 121).

[33] 'Sicut autem haec prima ierarchia de qua praetactum est reducitur ad deum ab ipsa deitate per immediatam luminis dationem ab ipsa, sic proportionaliter secunda ierarchia ab hac prima, et tertia a secunda, et humana ierarchia a tertia angelica hierarchia gubernatione reducitur ad superprincipale principium et terminationem, ut finem et consummationem omnis bonae ornationis consistentis in pulchritudine deiformitatis; et est huiusmodi reductio posteriorum ierarchiarum a prioribus facta secundum eandem legem observatam in omnibus hiis reductionibus, legem dico taxiarchiae, hoc est dei, secundum quod ipse est principium taxeoos, id est ordinis seu ordinationis . . .' (ibid., pp. 6-7).

[34] 'Finis autem largitionis luminum a patre luminum est, ut ipsa reducant

(*reductio*) has two effects. It polarizes in an upward direction the finite spirit, which is, as it were, stretched out horizontally and emitting its energy outward and downward. The unifying power of light thus concentrates the spiritual light on which it shines, rescues it from forces of contraction and dissipation, and produces growth and expansion within it. The effect of this recall from activities directed to a multiplicity of objects is the unification and simplification of spiritual energy, or love, in imitation of the first light.[35]

The hierarchical activity of the reception and handing on of light has as its aim the assimilation of the rational creature to God, or deification: what Grosseteste calls becoming '*luciformis a luce prima*'. The first moment of the process corresponds to the creature's need for purification in view of the reception of light, for light and shadow are opposites, and light can be neither received nor reflected by a distorted or impure mirror.[36] In hierarchic or thearchic activity, however, all things are done according to the state of the receiver; the only absolute constant is the generosity of the giver. Purification in the human and the angelic hierarchies differs somewhat, for in the former it is a temporal process of catharsis, the removal of impurities that are already there, whereas the good angels are purified by preservation from sin, which they never actually contracted.[37] Purification strips the intelligence of all that makes it unlike the divine simplicity, that is to say,

recipientia prout possibile est ad ipsum fontanum lumen quod ipse est.' (*Comm. in Hier. Cael.*, ed. McQuade, p. 4).

[35] 'Sumus autem nos primo per varios variarum rerum amores distracti, divisi, et multiplicati, et ipsa distractione et multiplicatione non expansi nec maiorati, sed contracti et corrugati et minorati. Sed luminis apparitio quaecumque praedicto modo in nos proveniens velut unifica virtus, ordine converso quo in nos venit et retendendo nos secum ad suam originem unde venit, ab huiusmodi contractione et corrugatione et minoratione nos expandit, dilatat et maiorat, et avertens nos a multitudine prius amatorum, convertit nos ad patris congregantis et colligentis nos a divisione et multitudine unitatem et deificam simplicitatem.' (ibid., p. 5).

[36] 'Necessaria est autem primo rationalis creaturae purgatio. Impura enim non suscipit illuminationem, et tenebrosa non potest habere operis perfectionem quia qui ambulat in tenebris, nescit quo vadit, et qui ambulat per viam tenebrosam nescit ubi corruat.' (ibid., p. 140).

[37] 'Necesse est igitur, si impura sit, ab impuritate mundari, quemadmodum est in hominibus. Si vero impuritatem ab initio non contraxerit, hoc est ei purgari, ne incidat in impuritatem conservari. Sic purgati sunt ab initio sancti angeli, non ab impuritate contracta mundati, nullam enim umquam contraxerunt, sed a lapsu in impuritatem conservati.' (ibid., p. 141). This doctrine is stated repeatedly: pp. 154, 288, 341, 315, 350.

of all mixture contracted by disordered love; since love always tends both to unite the lover with the object of his love and to make him resemble it, inordinate love introduces multiplicity into the spirit, whose gold then becomes alloyed through the admixture of baser metals. Pollution, disharmony, and confusion of selfhood are the result. In the human hierarchy, catharsis must bear upon all our loves, sensible and intelligible; in the angelic, of course, it will work on intellectual love alone.

Illumination is more noble than purification, which it presupposes. Grosseteste, like Pseudo-Dionysius, regards the latter as being the negative side of the former. It is the specific role of illumination to transmit the knowledge of God and of divine things. As such it is a communication of God's self-knowledge, conferred in the degree appropriate to the receiver. In the case of the angel, it is undeclining contemplation (*theōria*) of God, without either shadow of darkness or distortion of vision, whereas in man it englobes both active and speculative powers, and all knowledge of art and science directed to their best end.[38] Whether it is conferred directly or indirectly, it is a participation in the beatitude of God himself. It is a light which, as regards *homo viator*, touches and transforms, not the intellect alone, but the entire *mens*, in both thought and action, and in *aspectus* and *affectus* alike.

Perfection is the summit of the process begun by purification and extended through illumination. It is a state of union with God in which the illuminations are present, not as a multiplicity of knowing acts, but as a unified field, a lived unity centred upon God. It consists in the exercise of the best virtue, the speculative or contemplative, whose expression in the perfect life is happiness, and whose object is, of course, the highest being, the divine unity and Trinity.[39] The perfect are transposed from the region of partial knowledge and hesitant love, to be confirmed in constancy of knowledge and action.[40]

[38] 'Illuminari autem est per virtutes et scientias, vel simpliciter per virtutes, quia sub nomine virtutis generaliter dictae ad activas et speculativas comprehenditur omnis cognitio artis et scientiae ad suum finem optimum directa.' (ibid., p. 141).

[39] 'Felicitas autem est operatio secundum virtutem vel secundum optimam virtutum in vita perfecta. Felicitas vero perfecta est operatio speculativa seu contemplativa optimorum, hoc est unitatis et trinitatis ipsius deitatis.' (ibid.).

[40] 'Oportet namque ad hoc ut sint perfecti, primo ab imperfectione transponi et educi, et sic participes fieri sacrorum conspectorum per scientiam a lumine divino susceptam ad perfectionem promotivam.' (ibid., p. 153).

These three moments must be grasped in their unity, as phases of a single process. Thearchic science, the knowledge of God communicated in degrees to rational creatures, drives out all ignorance from the intelligence and all weakness from the will. Beyond this preservation or liberation from darkness, illumination positively conceived is the manifestation of the divine nature and of all other things as known in it. As this knowing becomes habitual and is confirmed, it approaches perfection.

When spoken in their fullest sense, purification, illumination, and perfection apply to God alone; it is only by participation in him that rational creatures enjoy them. God is pure, because his being is a completely simple harmony. He is eternal light in whom there is no darkness. He is perfection in his eternal act of self-contemplation. The purifying, illuminating and perfecting of all creatures belongs to him, who is all these things *per se et primo*, the unique source of being and light. He is therefore the *teletarcha*, the first cause of hierarchy and hierarchical operations. Each order of the angelic hierarchy participates in his eternal act by being made the illuminator of the order beneath it. Of the first hierarchy of angels alone is God the immediate teletarch. Though the three operations can be attributed one to each of the three orders within every hierarchy, each order accomplishes all three functions, though in different degrees according to its place within the hierarchy. The law that governs this divinely-ordained cooperation of created teletarchs is that they act as creatures, receiving the light according to the fullness of their being, assimilating it in sinless, rational freedom, and transmitting it in its undiminished fullness to the grade beneath them.[41] The weakening of light that accompanies its descent is the inevitable result, not of any ungenerosity in those handing it on, but only of diminished receptivity at each lower grade. The Dionysian principle of analogy is an expression of the thearchic laws, which rule the system and ensure the extension of the benefits of the divine goodness and beauty. Thus does God provide for all things in his love;[42] and love, Grosseteste

[41] '. . . sicut speculum lumen quod suscipit in oculum inspicientis ipsum transfundit, nil per hoc de suo lumine sibi imminuens, totum tamen lumen suum alii tradens, licet forte in suscipiente sit minus quam in tradente propter suscipientis minorem aptitudinem.' (ibid., p. 155).

[42] 'Lex igitur teletarchica et thearchica est, ut quidquid ierarchice in teletis

concludes, is something which of and by itself is a source of joy and delight.[43]

II *QUAESTIO*: ARE THE ANGELS ALL OF THE SAME SPECIES?

The diverse grades of activity, on the one hand, and the different distribution of the functions assigned by divine providence to the angelic choirs, on the other, raise the question whether the angels are of distinct species rather than all belonging to the same one. The question arises very naturally from the text of the *Celestial Hierarchy,* Chapter Five. Pseudo-Dionysius laid down that all nine orders can be denominated from the last, the angels, since the superior grades of a hierarchy of participation share in all the properties of the lower ones, while the reverse is not the case. He hesitated none the less before another type of consideration, which, working in the reverse direction, suggested that all titles, even those of the higher orders, therefore, are common to all hierarchies and orders, extending even to the lowest, but that the properties they designate are shared in unequally. In effect, the question thus posed became the introduction to the whole section of the *Celestial Hierarchy* following it, because it marked the transition from the definition of hierarchy in general to the description of the orders and the exegesis of the meaning of their biblical names.

The commentator felt the need to satisfy his own curiosity immediately, perhaps because the question as to whether the angels are of one species or many had assumed a place of some importance in the growing systematic treatises on theology ever since the appearance of Master Hugh's *Sentences.* Grosseteste did something that was very rare, if not unique, in his

agitur fine revocationis et reductionis in deum agatur, nec usurpet agens quod supra se est, nec omittat quod sibi conveniens est, nec tradat alii quod est supra suam receptibilitatem, nec deneget ei quod suae congruit susceptibilitati.' (ibid., pp. 126-7).

[43] 'Caliditas enim est naturaliter inclinativa subiecti in dilatationem et diffusionem sui in omnem partem. Gaudium autem est animi diffusio. Amor autem est res quae per se et sola est gaudiosa et delectabilis. Est igitur amor res per se dilatationis et diffusionis participativa. Quapropter et inclinatio ad actum amoris est inclinatio ad actum dilatationis et diffusionis. Unde convenienter haec inclinatio nomine calidi est expressa.' (ibid., pp. 280-1). Compare the praise of love in the *De Mandatis*: 'Hic amor [viz. ordered love] est res suavissima, quia nulla nisi per amorem sunt suavia, et sine illo omnia sunt amara.' (MS *Laud. misc.* 524, fol. 110[vb]).

Dionysian commentaries, when with the words '*Dubitabit itaque hic aliquis*', he raised a question on the text. I have found only a half-dozen questions in his later theological works, for he did not adopt the generalized use of the *quaestio* favoured by the majority of his contemporaries. The structure of this discussion proves to be quite loose; he left himself the characteristic freedom to follow the intrinsic pattern of the theme without being straitened by a rigid form of dialectic.

Are the names of all nine orders held in common? The question is immediately reformulated, in what amounts to a transposition of level: are the nine orders really nine different species? Now, if there are nine *species specialissimae*, the commentator points out, it will follow that no two orders can share a name which signifies an ultimate specific difference, or a property convertible with a single species. It could be argued, he admits, that if the concept of hierarchy is to be saved, then the angels must be of different species. Uniformity is the law of all species, and it must apply with particular force to the angels, since there can be no differentiation among them from the side of individual merit. If there is only one angelic species, then all of its members must in the nature of things be equally close to God, so that if one receives divine light without any mediation the same must hold true for all. Now Grosseteste is fully committed to the concept of hierarchy, for he regards it as reposing on divine revelation itself.[44] It seems at first sight as though he must posit a multiplicity of species among the angels, and in fact he notes that the philosophers have given reasons for thinking that the angels have such proper and specifically different acts as would assign them to different species, as movers, namely, of the heavenly spheres. Such acts would be incommunicable beyond the particular species.

A further possibility with which Grosseteste finds he must reckon is that the specific differences of the celestial substances are not ascertainable by us, as the scriptural names may well refer, not to their differences, but to the properties they all have in common, however unequally they share in them. He reminds us that Pseudo-Dionysius himself at one point

---

[44] 'Si itaque unius sint speciei specialissimae, ac per hoc unius naturae, videtur quod omnes dicta ratione sint deo eque propinquae et ab ipso suscipientes et ad ipsum reductae immediate, quod non solum est contra hunc auctorem, sed et contra alios theologos, et contra ipsa eloquia.' (*Comm. in Hier. Cael.*, ed. McQuade, p. 237).

introduced the hypothesis that the names signify properties held in common, but subject to gradation in the order of intensity. Thus the seraphim, for instance, are named from their burning love; but it goes without saying that love is a universal activity of spirit as such. In this case we would be dealing with two distinct classes of predicate, the one derived from the specific differences of the orders (if these can be ascertained at all, that is to say), and the other from the common predicates found in the theological tradition.

Up to this point in the discussion, now momentarily blocked to progress, Grosseteste has allowed the assumption that hierarchical order implies the existence of many angelic species; at this point he turns his dialectic upon the assumption itself. Despite the counter-indications thus far met, could it not be maintained, he asks, that all the celestial substances belong to a single *species specialissima*? After all, a common definition can be arrived at, based solely upon the idea of spirit or intellectual nature, and the essential properties implied in it. Angels are unembodied intellects; they are capable of distinguishing between true and false, good and evil, and are ready to cleave to the good and avoid the evil, since they have clear knowledge of the consequences of both; they are possessed of judgement and endowed with the power of command that executes what right judgement enjoins; they are more powerful than bodily things, and hence able to produce effects that lie beyond any power of nature; as intellectual natures they are equipped to communicate with their fellows; and their understanding embraces all intelligible objects. Now if all the angelic intellects can understand all intelligible objects, then no distinction can be made among the angels as intellects, which is essentially what they are. Considered as minds, they are all of the same species.

This turn of discussion represents in fact something of a *tour de force*, since Grosseteste has managed to deduce the properties of the nine orders from the concept of spirit. As we shall see, he makes full use of the deduction in his solution to the question.

Grosseteste turns at this point to the philosophically-based argument for the division of angels into species. The argument runs that, since each planetary sphere has a motion that is visibly its own and unique, it must follow that the movers differ from one another in species, in a way that parallels the differences among the spheres they move. Grosseteste is not impressed by this line of reasoning. Only if the angels were conjoined movers of the spheres, regulating them in virtue of a quasi-personal union with them (as the soul's intentions govern the body by virtue of the union of both), could we argue back from the specific differences of the spheres to parallel differences among their moving powers. In fact, however, the angels are not conjoined movers but *substantiae abstractae*, quite unembodied substances. Hence they do not require specifically different moving-powers, even though each sphere is a species to itself.

The philosophical difficulty thus disposed of, Grosseteste returns to the refinement and defence of the position which, it is now clear, he favours. The common belonging of all the angels to a single species does not, he claims, render all differentiation among them unthinkable. Distinctions could prevail among them on the basis, not of their cosmic operations, but of different vocations, reposing therefore on the divine will. As Jacob was preferred to Esau, so also the higher orders might receive light more immediately from God by being placed in greater proximity to him, but the lower only through the intermediary of their superiors, due to their greater removal. If this is granted, none of the arguments yet encountered is sufficient to prove that the differentiation effected by unequal degrees of participation in being amounts to a difference of species. In fact, the only specific difference falling within the genus 'rational creatures' would appear to be that between angelic and human spirit; and if the soul and the angel differ by a single specific difference, namely, the aptitude and desire of the former for union with a body, then it is hard to see how the angels themselves could belong to several different species.

The defence of specific differences among the angels could only be conducted, concludes Grosseteste, on the basis of an elaborate hypothesis associating the three possible indices of differentiation. This strategy would link, in the case of the seraphim, their natural capacity for the immediate reception of light, their concomitant power of love, and their moving of the sphere of the fixed stars; and so on for the other eight orders, where in each case their specific place in the hierarchy, the concomitant biblical attribute, and the cosmic function assigned to them, would together constitute a specific difference. Grosseteste does not offer a criticism of this hypothesis, though it plainly does not have his sympathy, and his exploration of it is purely dialectical. In a rather characteristic aside to his reader, he confides that he finds the whole problem difficult of resolution. He knows that better minds before him have been baffled by it; he mentions Damascene and recalls the embarrassed agnosticism of his conclusion, that the God who made them is alone in knowing what the *species* and *terminus* of the angels may be, whether they are equal in essentials or different from each other.[45]

Having said this, however, Grosseteste immediately re-embarks upon the question. Resuming the deduction of the angelic attributes from the notion of pure intellect, he now shows to his own satisfaction that the nine attributes exhibit an internal continuity, and indeed flow from each other. Love, the highest reality, receptive, responsive, and unitive, is the condition of knowing; from the union of both derives just judgement; this issues in action through the virtue of control, which in turn requires resolution, and strength to resist all hindrances. Finally, upon these six virtues there follow three that are concerned with providence and instruction. These nine virtues, he suggests to his reader, are so interlocked that each is derivable from and mediated by the preceding one, and the whole system is therefore the product of orderly mediations. Furthermore, they fall into three classes of three members each. The first class englobes spiritual acts of a contemplative nature, while the three last are directed to action; the middle class contains a mixture of action and contemplation. The coherence of the deduction enables Grosseteste to uphold the community of angelic properties by showing reason why they are possessed by all nine choirs, the different measures of whose participation in the One are sufficient to account for the gradation or hierarchy that obtains among them.

Grosseteste is now in a position to give solidly-grounded rules for the attribution of names to the orders. And first, the names of the lower orders can be attributed to the higher

---

[45] Both quotations come from ch. 17 of the *De Fide Orthodoxa* (ed. Buytaert, pp. 69, 72). The first changes a word of Burgundio's version, something typical of Grosseteste's *correcta translatio*.

without any evacuation of their meaning, which is reinforced but not altered; however, if the names of the higher were to be applied to their inferiors, a weakening of signification would progressively evacuate their sense. It follows that the name assigned to a particular order is given, not from the power in which that grade excels all others (for the cherubim have less knowledge than the seraphim, though their name signifies 'knowledge'), but from the power and act which are pre-eminent in the beings of that order, in a sense that is triply specified: with regard to a special office; in virtue of its relative degree of proximity to the One, its place, therefore, within the universal order; and in virtue of a capacity which is subserved by the other properties in its possession, or upon which these are focused.

The two theses affirmed at the close of Grosseteste's discussion are that the angels belong to a single species, and that the nine choirs form a true hierarchy. The latter saves the data of Scripture and authority to Grosseteste's satisfaction.

On reflection, the influences which underlie Grosseteste's approach to this question, and indeed to angelology as such, are seen to be very diverse. In the first place, the notion of hierarchy, which he upheld out of honest loyalty to what he regarded as the firm evidence of Scripture and the teaching of respected Church Fathers, lies as an *a priori* form under the entire discussion. The second influence came from Aristotle's philosophy and was twofold: on the metaphysical side, it urged assigning the separate substances to different species on the basis of their functions as movers; while in logic, it provided the important, clearly-defined concepts of genus, species, and *differentia*, as set out in the *Analytics* of Aristotle. Needless to say, these two systems of ideas are non-coaxial, and the alignment of their parameters proved difficult. The Dionysian conception had its roots in a Neoplatonic notion of participation and a philosophy of spirit, whereas the Aristotelean was grounded in observation and natural classification. Grosseteste was sensitive to their divergences, showed himself conscious of the specificity of their conceptual products, and proved unwilling to make facile equations between the idea of species on the one hand, and, on the other, those of *ordo*, *adornatio*, and *hierarchia*. He evidenced no enthusiasm for the artificial alignment of theological and cosmological ideas.

The third world of ideas which, though itself largely invisible, yet underlies the entire discussion and gives it a measure of cohesion, is the Augustinian *civitas spirituum*, the city of created spirits. Augustine gave it a democratic constitution to bind together its inhabitants. The only difference he admitted was the temporary distinction between the angel and the human soul: the latter is, for a time, of lower status than the former. The basic constitution, however, admits of no ultimate or radical difference among the citizens, and the future state of the society of rational spirits will be completely classless, for the destiny of all finite spirits, the unmediated vision of the face of God, is identical. For Grosseteste and all his Latin contemporaries, whose theological culture and religious sensibility were saturated by this outlook, the unity in species of the angels could have been nothing less than a matter of the commonest evidence. Indeed, the dismay which Aquinas occasioned to his colleagues at Paris, when he argued that each angel is a species apart from the others, can only be appreciated when the novelty of this assertion is measured against the venerability of the tradition.

Of the three influences of which I have spoken, the last it was that marked Grosseteste's native bias. Its gravitational attraction was sufficiently strong to pull the first system, that of Pseudo-Dionysius, into orbit around it; but the second, the Aristotelean, remained largely outside its field of influence.

PART THREE

THE LIGHT OF NATURE

Chapter 1
Creation and the Cosmology of Aristotle

'You cannot really read, or admire or be in a room,
unless natural light is there. We are actually born out
of light, you might say. I believe light is the maker of
all material. Material is spent light.'

(Louis Kahn, interviewed for *Time Magazine*,
15 January 1973.)

IT was Clemens Baeumker's studies on Witelo and on medieval
Platonism, published in the early part of this century, that
established and documented the view which sees the metaphysics of light as being of exclusively Neoplatonic origin and
character, even in its application to nature.[1] This thesis won
broad acceptance among students of Grosseteste's cosmological ideas. Ludwig Baur initiated the study of the sources of
Grosseteste's light-speculation and found that Saints Augustine
and Basil were the chief inspirers of his characteristic theoretical positions; he borrowed also from a number of Arabic and
Jewish Neoplatonic thinkers, and Aristotle supplied a logical
framework, the doctrines of causality, of potency and act, and
a useful terminology, all of which Grosseteste employed to
underpin his essentially Augustinian or patristic Neo-Platonism.
However, the two streams of Aristoteleanism and Augustinianism flowed through his thought without any profound inter-
mingling.[2] Pierre Duhem expressed himself in agreement with
these findings when, with customary energy, he described

[1] Baeumker, *Witelo, Ein Philosoph und Naturforscher des XIII. Jahrhunderts*
(Münster, 1908); *Der Platonismus im Mittelalter*, BGPM, Bd. XXV, H. 1-2: *Studien
und Charakterisken zur Geschichte der Philosophie insbesondere des Mittelalters*
(Münster, 1927), pp. 139-79; reprinted in Beierwaltes (ed.), *Platonismus in der
Philosophie des Mittelalters* (Darmstadt, 1969), pp. 1-55.

[2] Baur, 'Das philosophische Lebenswerk des Robert Grosseteste, Bischofs von
Lincoln (m. 1235)', in *Görres-Gesellschaft*, 3. (Cologne, 1910), p. 82: 'Man kann
im allgemeinen sagen: in metaphysischen und allgemein philosophisch-spekulativen
Fragen ist er Augustinianer und nimmt aus des Aristoteles Philosophie nichts
herüber ausser der Syllogistik, die er auf den Unterbau der augustinischen Erkennt-
nislehre aufsezt, die Kausalitätslehre, Potenz und Akt und einzelne terminologische
Ausdrücke', cf. Baur's *Die Philosophie . . .*, pp. 170-95 for a summary of the chief
Aristotelean influences on Grosseteste's ideas.

Grosseteste as 'a mind in rebellion against the sway of Aristoteleanism'.[3] Alexandre Koyré held Grosseteste responsible for
the attitude of neutrality, or even hostility, to the Aristotelean
philosophy of nature, which prevailed among the mathematical
scientists of Oxford; in Grosseteste's most outstanding successors, Peckham and Bradwardine, mathematics formed a way
of thinking that was essentially anti-Aristotelean, in contra-
distinction to the strict peripateticism of Paris, which Koyré
thought was maintained up until the time of Nicole d'Oresme.[4]
Indeed, the view which regards Grosseteste's mathematical
cosmogony as at best unrelated to Aristotelean physics may
be said to have become general, to the point where even those
scholars who acknowledge a strong influence of Aristotle on
other aspects of Grosseteste's philosophy—his methodology,
for example—tend to assume that the sources of his metaphysics of light are wholly accounted for by Neoplatonic
writers.[5]

It will be the aim of this chapter to fill what I regard as an
important gap in studies of Grosseteste's cosmological views
by exploring the relationship between his world-system and
that of Aristotle. The interest of such an inquiry is manifold.
In the first place, it is of importance to know how high
Aristotle's authority ranked with the founder of the Oxford
tradition, and to reach a correct assessment of the degree of
welcome or reserve which he brought to his reading of the
Philosopher. In the second place, we may hope to discover
Grosseteste's basic aim as a physicist and metaphysician, the
intention that commanded his philosophical options in these
areas of speculation. Finally, Aristotle's was the most complete
and authoritative physical system of the ancient and medieval
worlds, and that being so the historian is justified in measuring a medieval thinker's originality in terms of his personal

---

[3] Duhem, *Le Système du monde*, vol. 3, p. 278: 'Le mépris de R. Grosseteste
à l'égard des ouvrages d'Aristote ne fut peut-être pas aussi complet que Bacon le
prétend . . . Il n'en est pas moins vrai que Grosse-Teste nous apparaît comme un
esprit rebelle à l'emprise péripatéticienne, et singulièrement original lorsqu'il
expose ses idées sur les diverses questions de la *Métaphysique* ou de la *Physique*'.
cf. vol. 5, pp. 341 ff: 'Le Platonisme de R. Grosse-Teste'.

[4] Koyré, 'Le vide et l'espace infini au XIVe siècle', in *Arch. Hist. Doctr. Litt.
M.A.*, 17 (1949), p. 52.

[5] Crombie, *Robert Grosseteste and the Origins of Experimental Science, 1100–
1700*, Introduction, pp. 4 ff., 13, 104–6, on the importance of the Augustinian-
Platonic tradition in the science of the Middle Ages.

criticism of it; one thinks of Middleton, Buridan, Oresme, and Cusa, among many others, as thinkers of later medieval times who breached the Aristotelean system at one point or another and helped prepare the later collapse of the whole.[6] This yardstick of originality, namely, the deliberate deviation or *Distanzierung* of a thinker from important aspects of the dominant world-view, may shed new light on the much-disputed question of where precisely Grosseteste's original contribution to scientific thinking is actually to be found.[7]

I wish to set forth an outline, the briefest compatible with intelligibility, of the *De Luce*, one of the few scientific cosmologies, and perhaps the only scientific cosmogony, written between the *Timaeus* and early modern times.[8] After that I shall offer some comments on its importance and attempt to situate its author within the history of scientific ideas.

I *DE LUCE*, GENESIS AND ARISTOTLE

In the first of the four parts into which the *opusculum* falls, Grosseteste attempts to demonstrate the thesis—original with him—that the first form of corporeity is to be identified with light.[9] The three-dimensional extension of matter follows necessarily upon first form or corporeity, where both matter and corporeity are in themselves simple substances possessed of no dimensions. Since matter and form are inseparable, form could introduce dimensions into simple matter only by its self-generation and instantaneous self-diffusion drawing matter with it. But the only thing that behaves in this way is light; therefore corporeity either must itself be light, or must effect its self-diffusion by participating in light. But corporeity is the

---

[6] For a discussion and justification of this viewpoint see G. Beaujouan, 'Medieval Science in the Christian West', p. 496, in *Ancient and Medieval Science*, ed. R. Taton (London, 1963).

[7] In this chapter I take for granted the chronology of Gorsseteste's scientific works proposed in Appendix B.

[8] Many authors have offered summaries of the *De Luce* (Baur, Duhem, Sharp, McKeon, etc.), and at least three English translations have been made (Riedl, Wallis, Terrell). The nature of the present chapter makes a summary indispensable, and in any case most of the above-named précis are inaccurate in important points; this applies particularly to Baur (*Die Philosophie . . .*, pp. 85–93) and Riedl (the introduction to her translation). Baur's text is uncritical; I offer some corrections in footnotes to this chapter. All references are given to Baur's text, by page and line.

[9] The four parts are: light as first corporeal form (pp. 51.10–52.16); cosmogony (pp. 52.17–57.8); cosmic motion (pp. 57.9–58.7); the perfection of the universe (pp. 58.8–59.2).

first form, and so cannot be the principle of three-dimensional extension in virtue of any subsequent form; wherefore light is not a form supervening upon corporeity, but is corporeity itself.

This conclusion is verified by the dignity and nobility of light, excellences which place it beyond all corporeal and material things and liken its nature to that of the separate forms or intelligences, in the view of all *sapientes*.

The entire world-machine was created in the beginning from first form and first matter. Light multiplied itself from a single point infinitely and equally on all sides to form a sphere, and extended matter into the dimensions of the actual universe. Two questions arise: why was an infinite multiplication of light required? And if infinite, why did it produce a finite universe? To the first: a simple thing multiplied a finite number of times cannot generate quantity, by the authority of Aristotle in *De Caelo et Mundo*.[10] To the second: a simple thing multiplied by infinity can produce only a finite quantity.[11] One simple thing cannot exceed another infinitely, but a *finite* thing exceeds a *simple* thing infinitely, whereas an infinite quantity would exceed a simple thing by infinite times infinity. Light, therefore, a simple thing infinitely multiplied, must extend matter (similarly a simple thing) into *finite* dimensions.

Light extends matter 'gradually and continuously'. Grosseteste prepares this conclusion by enunciating a series of what appear to be original mathematical propositions on relative infinities.[12]

(i)   The proportion between one infinite aggregate and

---

[10] *De Caelo* I, 5,271b,15-25, seems to be the passage Grosseteste has in mind, but the axiom in question here does not appear to be explicitly stated by Aristotle.

[11] Contrast Sharp, *Franciscan Philosophy at Oxford*, p. 20 n.: 'It never occurred to Grosseteste to ask why the diffusion of light in the first instant stops at a particular circumference and there forms the firmament.'

[12] Baur's exposition of the mathematical section is confused and at times misleading. Alexander Birkenmajer, in his article 'Robert Grosseteste and Richard Fournival', *Med. Human.* 5 (1948), p. 39, n. 15, gives a symbolic formulation of the mathematical propositions, which, as he points out, 'are not to be judged from the standpoint of Algebra nor from that of Cantor's Theory of Sets'. Σ signifies the summing up of n = 1 to n = ∞.

(i) The proportion $\Sigma a : \Sigma b$ can be any rational or irrational number (*De Luce*, p. 52.32-4).

(ii) $\Sigma n > \Sigma 2n$ is equivalent to $\Sigma n - \Sigma 2n = \Sigma(2n-1)$ (pp. 52.35-53.3).

(iii) $\Sigma 2n = 2\Sigma n, \Sigma 3n = 3\Sigma n$, etc. (p. 53.4-12).

(iv) The proportion $\Sigma 2n : \Sigma(n + a)$ is an irrational number (p. 53.13-26).

another can be any rational or irrational number.

It follows that a relation of greater and less can obtain between infinities.

(ii) The sum of all numbers is infinite, and double the sum of all even numbers.

(iii) The sum of all doubled numbers from one to infinity is twice the sum of the halves corresponding to those doubles; the sum of all trebled numbers from one to infinity is three times the sum of the corresponding thirds, and so on.

The second and third propositions refer to relations between two infinite sets expressible in terms of a rational number.[13] The fourth proposition instances a relation of two infinite numbers according to an irrational number.

(iv) The infinite sum of all doubled numbers will no longer be related to the infinite sum of the corresponding halves minus a finite number (for example, 1) in the proportion 2:1, but by an irrational number (no finite number being an aliquot part of an infinite sum, its subtraction from an infinite sum cannot leave the latter's proportion to another infinite a rational one).

The conclusion to be drawn is, that the infinite self-generation of light extends matter into finite dimensions of greater and lesser extent related to each other according to any (*sic*) rational or irrational numbers.[14] For example, if the infinite multiplication of light extends matter to a size of two cubits, twice that infinite generation will yield four cubits, and one half of it, one cubit, and so on, according to other rational and irrational proportions. The extension of matter is evenly-graduated and continuous through the whole space of the universe; there is no void.[15]

[13] As Grosseteste says, 'It is clear that there can be a relation of one infinite class to another according to any rational number': I read *infinitum ad infinitum* in p. 53.12, with MSS *R,D,P,V,* as against MSS *M,a,* Baur, and Riedl.

[14] *De Luce*, p. 53.27-30: 'His ergo ita se habentibus, manifestum est quod lux multiplicatione sua infinita extendit materiam in dimensiones finitas minores et dimensiones finitas maiores secundum quaslibet proportiones se habentes ad invicem, numerales scilicet et non numerales.'

[15] There follows in the text a brief digression. Grosseteste claims that his theory of infinite expansion recovers the insight of those philosophers who asserted that all things are composed of indivisible *atomi*, bodies from surfaces, surfaces from lines and lines from points. Aristotle attributes this view to Plato's *Timaeus*, presumably thinking of 54d (*De Caelo* III, 1, 299a, 2-300a, 19), and introduces the Pythagoreans into the subsequent discussion. Taking Aristotle's criticism of Plato to heart, Grosseteste proceeds to show how his own theory does not contravene

The first cosmogonic principle was the even extension of matter into a finite, three-dimensional sphere by the infinite self-generative energy of light; the second yields the mechanics of the unequal distribution of matter in terms of rarefaction and condensation (*De Luce*, p. 54. 11 ff.). Though the propagation of light and the consequent expansion of matter, beginning from the primordial point, take place equally in every direction, of necessity the outermost reaches of extended matter are more sparse and rarefied than are the inner, which remain capable of further rarefaction. The farthest limit of extension is reached when no further rarefaction of matter is possible;[16] the ultimate capacity of matter being realized, the area immediately bounded by the outer spherical surface is incapable of further physical change. A perfect body has come into being, the firmament, having in its composition only first matter and form. The most simple body in essence, it is the greatest in quantity and the container of all subsequent bodies.

The first sphere is supremely active by its form, and its matter is so distended and subtle that it is more akin to spirit than to corporeality, in which grossness and density brake and diffuse the action of form. The light of the first body necessarily diffuses itself in straight lines, from each point on its inner circumference to the centre of the universe.[17] *Lux* as form is, of course, inseparable from matter; what is diffused by the first body is *lumen*, a spiritual body (or bodily spirit) so subtle that it does not divide the body it passes through, and its passage is instantaneous. Like *lux*, it multiplies itself infinitely, and therefore its 'motion' to the centre is not the local motion of one single entity from place to place, but the

the Stagirite's dictum, 'magnitudo solum ex magnitudine componitur'. He gives geometrical examples of rational and irrational proportions (a line and its half, a diameter and a non-aliquot chord, the infinitesimal angle of contingence occurring between tangent and curve, compared to a right angle), which merely illustrate his thesis on rational and irrational proportions and enlarge the notion of 'part' and 'whole' and their mathematical relationship.

[16] That is, presumably, when any additional rarefaction would produce the beginnings of a void. One is entitled to suppose that Grosseteste is thinking of matter as composed of contiguous, indivisible atoms.

[17] 'De necessitate' (p. 54.34), because, of course, light propagates itself in straight lines, and therefore each point on the inner surface of the sphere will generate light at right-angles to the tangent touching the surface at that point; also, a premiss which does not require airing, light propagates itself from the sphere's outermost limit *inwards* towards the centre, because outwards is nothing.

propagation of a self-generating thing. The result of its passage is the collection of corporeal mass towards the centre, a focusing action which condenses and concentrates matter progressively as the lines of force converge upon the geometric centre of the universe. Now matter cannot be drawn from the perfectly rarefied first sphere; it is the region immediately below this that is rarefied by the action of light, and when the limit of possible rarefaction is reached, a second sphere is formed; nature's abhorrence of a vacuum does the rest, by determining that its outer surface be contiguous with the inner surface of the first sphere. Like the first, this is incapable of further impression; but the second is derivative from the *lumen* of the first, and *lux*, in the first sphere simple, is in the second doubled.[18]

Just as the *lumen* generated by the first body completes the second sphere, leaving a denser mass beneath,[19] so in turn the *lumen* of this sphere perfects the third, leaving the underlying mass still more condensed. The same process of condensation continues uninterrupted until the nine celestial spheres are completed and below[20] the ninth and lowest one lies a thick mass, which is to form the matter of the four elements.[21] The action of the lunar sphere conforms to that of the higher, but the gradual diminution of power in the descending hierarchy leaves it too weak to rarefy the elemental matter of the region immediately beneath it, whence the denser mass of this area remains imperfect and susceptible of further condensation and rarefaction. The matter of this region is rarefied sufficiently to become fire, but remains elemental matter.[22] The light of fire in turn produces a sphere of air, fairly rarefied and light, below it,[23] and this differentiates water and earth by condensation and rarefaction. So little of the latter process is possible at this stage that the sphere of water remains with the ponderous earth.

Grosseteste now draws some general qualitative conclusions

[18] Reading 'Et sic est completum secundae sphaerae lumen, quod gignitur ex prima sphaera; et lux . . .' (etc.), with two of Baur's MSS, instead of the garbled text he prints (p. 55.19–21).

[19] Reading *infra* (with 3 MSS), instead of Baur's choice, *intra* (p. 55.24).

[20] Reading *infra* (with 6 MSS), instead of Baur's choice, *inter* (p. 55.29).

[21] The verbal moods ('donec complerentur . . . et congregarentur . . . quae esset . . .') seem intended to convey the planned necessity of the process.

[22] Reading 'materia elementaris' (with some MS-support) for 'materia elementorum' (p. 56.3).

[23] Omitting (with Ms *V*) 'et sic produxit . . . ipsius disgregatione' (p. 56.5–8).

from his description of the cosmogonic process. Of the thirteen spheres forming the sensible universe, nine are heavenly, removed from generation, corruption, and alteration, being perfect; the lower four are their opposite, being subject to generation and corruption, growth and alteration. Every higher body in this universe is the form (*species*) and perfection of that succeeding it and generated from its *lumen*, and the first body contains all potentially, as unity does every number. The focusing of the light of all other spheres makes the earth the representative body of the universe: Pan—all things[24] —or Cybele, fertilized by light to become the mother of all the gods.[25] The middle spheres, he concludes, also represent the extremes; each taken in respect to the lower ones is like the first to all, and compared to those preceding it, is like the earth to the heavens.

Continuity becomes a universal law of physical being. Light, the form and perfection of all bodies, is more simple and spiritual in the upper spheres, more multiple and corporeal in the lower, and the bodies are no more of identical species than are numbers, though they proceed by multiplication from unity. Light is the principle as much of unity and perfection as of differentiation and multiplicity.

It is likewise from the participation of lower in higher (and ultimately in the supreme) that Grosseteste deduces the principal motions of the universe.[26] Sharing in the form of the higher bodies, each sphere receives motion from the same incorporeal moving-power as they, namely, the intelligence or soul which imparts the diurnal motion to the first sphere. If they have all of them a single mover,[27] they have not for all that a single rate of motion, for in each lower sphere the first bodily light is weaker and less pure than in its superior, and

[24] 'Propterea ipsa est quae a poetis Pan dicitur, id est totum.' (p. 56.24-5). Grosseteste's source is presumably the *Etymologies* of St. Isidore (VIII, xi. 81): '. . . unde et Pan dictum est, id est omne.' The resort to such etymological and linguistic lore was, of course, widespread; for a parallel, see William of Conches, *Glossae in Timeum*, ed. Parent, pp. 149-50.

[25] Grosseteste gives an etymology of 'Cybele': 'quasi cubile, a cube, id est soliditate, nominatur' (p. 56.25). This derivation is not found in Isidore, but the latter defines a cube as 'figura propria solida', with the three dimensions. Grosseteste plays on the idea of 'solid' and transposes it into the physical density of the element earth, 'quia ipsa est omnium corporum maxime compressa'.

[26] In the third part of the *De Luce*.

[27] The spheres have in addition their proper motion and movers; cf. his *De Motu Supercaelestium*.

hence both less receptive of the diurnal rotation and slower in following it.

The elements participate in that first light which is the first heaven's form, but at such a distance that they do not share the diurnal motion of the celestial spheres; their form is weak, and their matter—the principle of resistance and disobedience —too dense and heavy. This is an opinion not universally admitted, for some allow the heavenly circular rotation to extend to fire and even water, causing the motion of comets and tides respectively. However, all real philosophers know that the earth at least is exempt from rotation.

A second heavenly motion is caused by the eighth sphere (the second in order of descent), the lower ones obeying it because they participate in its form.

To sum up on motion: the perfection of the heavenly spheres precludes their undergoing further condensation and rarefaction (i.e. rectilinear motion of substantial light to or from the centre), but leaves them capable of circular motion, the bodily expression of the moving intelligence's thought.[28] The natural motion of the elements, on the other hand, is up and down, rectilinear therefore, as the light within them seeks to condense matter by drawing it to the earth's centre, or to rarefy it by moving it away.

That this universe is a complete and harmonious thing is easily proved. In the most simple body there are four things to be found: form, matter, composition, and the composite. Form is totally simple and corresponds to mathematical unity. Matter is the dyad, due to its binary qualities of receptivity and divisibility.[29] Composition corresponds to the number three, for in it are informed matter, immattered form, and the property itself of composition. 'Four' comprehends whatever the composite is beyond these three. The aggregate of these numbers is ten, contained in the quaternity of the first body (which virtually contains all the others), and mirrored in the number of bodies in the world—for the four elements form together a single terrestrial body. Manifestly, ten is the perfect number of the universe and is possessed by every whole and

---

[28] Something of this kind is presumably meant by the following: '. . . sed soiummodo motus circularis a virtute motiva intellectiva, quae in sese aspectum corporaliter reverberans ipsas sphaeras corporali circulat revolutione'. The MSS share our confusion at this point.

[29] Reading 'divisibilitas' (with 3 MSS), instead of Baur's 'densitas' (p. 58.13).

perfect thing. Clearly, too, only the five proportions found in the first four numbers are adequate for the composition and harmony that sustain every composite being; they are the foundation of harmony in musical sound, gesture, and rhythm.

a) The Sources of the De Luce

In many other less important and less original works than his *De Luce*, Grosseteste gives rather copious references to the authorities on his book-shelves; in his little masterpiece, on the other hand, a single reference to 'Aristotle, *De Caelo et Mundo*' is allowed to suffice. The *De Luce* is a work which can be read and admired for its intrinsic brilliance, unity, and clarity; beyond that, however, the identification of its sources serves as an indispensable clue to the intention of its author.

First among the several inspirers of the work we may place the biblical account of the creation of the world. Already in the opening phrase of the cosmogonical description the reference made to 'light, which is the first form in the first matter created', seems to pass through a metaphysical sense of 'first' ('foundational') to a temporally first product of the primordial act of creation which posed the absolute beginnings of the *mundi machina*, '*in principio temporis*' (p. 52.17-21). When the first sphere is formed Grosseteste identifies it as the firmament; but his gloss is unnecessary, for we are all in the secret from the very beginning: the heavens and the earth of which the *De Luce* describes the formation are the work of the first three biblical days; created *ex nihilo*, they were differentiated out of formless matter by the divine command, '*fiat lux*'.

Grosseteste reads Genesis through the tradition of hexaemeral commentaries, and we can easily trace the influences of Augustine and Basil, in particular, on his thought. Augustine had concluded that the light which God created is present in the work of all six days, hence the repetition of '*dies*', the time of light, in the biblical account.[30] The light first created can be interpreted as a material light, and this may have differentiated day and night (even before sun and moon were made) by rotation or by *contractio* and *emissio*.[31] Physical

---

[30] St. Augustine, *De Gen. ad Litt.*, IV, 21, 38: '. . . restat ut praesentia lucis illius, quem diem Deus fecit, per omnia opera eius repetita sit, quoties dies nominatus est . . .'. For a summary of Augustine's light-theory, see Baeumker, *Witelo*, pp. 372-7.

[31] Augustine, *De Gen. ad Litt.*, ibid., '. . . quodam modo lux illa quae dies

light, or ether, and air, the upper elements, are the most subtle and rarefied and, therefore, the most active and dynamic bodies in the universe: '*Praecellentia sunt corpora*', Augustine proclaims, and adds that they are more akin to spirit than to gross corpulence, and are the fitting instruments of the soul in administering the body.[32] It can be amply shown from others of his works that Grosseteste assimilated these ideas on physical light from the great Latin Doctor.

The influence of St. Basil's *Homilies on the Hexaemeron* on Grosseteste's cosmogony is real, but should not be exaggerated.[33] In Basil's view, the first word of God produced light, which shone, however, only to illuminate the heavens already formed; it was not actually the first thing made.[34] The ether and the heavens are made of a subtle and transparent stuff not identified as light, and *lux* filled them by instantaneous self-diffusion.[35] The primordial light of the first three days divided day from night, and for Basil, as for Augustine (who may be following him in this), the periods in question are a rhythm of effusion and contraction. There is no need to see here more than a verbal analogy with the process of condensation and rarefaction described in the *De Luce*; a vague conception of emission and withdrawal of light in the heavens is all that can reasonably be found in these Fathers, as it were an alternation of dawn and dusk without a sun. Even when St. Basil declares light to be a simple and homogeneous body (*P.G.* 29, 47), it is evident that he is thinking still of physical light, not of dynamic energy. And finally, to complete the distinction of the author of the *De Luce* from his source, Basil's lovely remarks on the aesthetic aspect of light are relayed more by the *Hexaemeron* of Grosseteste (which quotes expressly from them) than by his *De Luce*, whose aesthetic is more cerebral and mathematical than sensuous.

If the combined influence of Augustine and Basil by no

appellata est, vel circuito suo, vel contractione et emissione, si corporalis est, vices diurnas nocturnasque peregerit . . .'. Augustine's probable source for this notion is the passage of St. Basil referred to in n. 34 below.

[32] Ibid., VII, 15, 21; 16, 22; 19, 25 (*P.L.* 34, 365); cf. XII, 16, 32 (*P.L.* 34, 466).

[33] As I feel it has been by Baur; see *Die Philosophie* . . ., p. 85.

[34] St. Basil, *Hom. II in Hexaemeron*, 7 (*P.G.* 29, 45-6). Grosseteste quotes the passage in question in his *Hexaemeron*.

[35] Ibid.: 'Talis est enim ipsius natura, tenuis scilicet et pellucida, adeo ut lux per ipsum transiens nulla temporis mora indigeat'.

means determined the *De Luce* even in its general lines of thought, it was never the less of germinal importance. It encouraged Grosseteste to apply the command *'fiat lux'* to a pure physical light preceding the creation of the luminaries, and suggested a composite picture of a substance embracing heavens and earth in its action, a substance simple and homogeneous, diffusing itself instantaneously through space, uniquely active, transcending the other elements in the direction of spiritual being, and conferring beauty upon the world-machine. All this, however, falls far short of the *prima forma* and its mathematical behaviour, as Grosseteste was to conceive of them.

The first corporeal form and matter of *De Luce* have immediate roots in Arabic and Jewish speculation. Avicenna, Algazel, and Averroës developed a theory of common corporeity, or that in virtue of which every material thing is extended, according to which first matter became the possibility of extension and the form of corporeity the active generator of three-dimensionality, i.e. the first and common form of all bodies, to which *forma specialis* remains still to be added.[36] The influence of Avicebrol on Grosseteste is probably even stronger than that of Avicenna. In his voluminous treatise, known to the Latins under the title of *Fons Vitae*, universal matter and form are represented as two substances differing in essence but united by the will of God.[37] Matter is closest to the One and receives the influence of its light or virtue. Universal form is diffused like light over matter (p. 313); it is thought of as something more spiritual, which becomes corporeal in matter as what Avicebrol calls *'prima forma substantialis'*.[38] Like Avicenna, Avicebrol too considered that the first corporeal form extended matter into three dimensions. This view, adopted by Grosseteste, gave the possibility of

---

[36] Avicenna, *Sufficientia* I, 2 (ed. Venice 1508, fol. 14). cf. *Metaphysica* V, 3 (fol. 88).

[37] See ed. Baeumker, *BGPM* Bd. 1, H. 2-4; H. 4, pp. 318-9.

[38] Ibid., H. 3, p. 161. Grosseteste restricted the application of Avicebrol's universal hylomorphism to the material world, and refused to apply it to spiritual being. If Avicebrol really is the source of *De Luce* p. 52.10-16 (concerning the assimilation of *prima forma corporalis* to the Intelligences), then it is possible that Grosseteste thought of him as a Christian author, a common enough assumption in the early thirteenth century. Grosseteste is not given to applying the term *'sapientes'* to pagan authorities without some sort of qualification, such as *'huius mundi.'*

conceiving the most universal aspect of material beings—pure extension—in geometric terms, before other substantial and accidental forms (even that of determinate quantity) are added. Avicebrol regarded the simplicity and nobility of the first form as a near approximation to the spiritual world of intelligences ('*lux spiritualis*'), an idea that could be brought into connection with the Augustinian description of light as a quasi-spiritual substance. No doubt, too, the Neoplatonic theme of emanation, conceived of as a cascade of light descending from the One, which is paralleled in the cosmogony of *De Luce*, reached Grosseteste through Avicebrol and the *Liber de Causis*.[39] Analogies aside, however, it would appear that Grosseteste's identification of light with *prima corporeitas* was original to him.

The Arabic astronomers contributed to the theory of motion adopted in the *De Luce*. The reform of the Aristotelean system proposed by Alpetragius fixed the number of the spheres at nine and attributed the diurnal motion of all the spheres to the action of the first. The spheres become increasingly less simple in proportion to their distance from the first, so that the apparent movement of the planets against the heavens can be accounted for by their relative tardiness in obeying the highest motion. Grosseteste's respect for Alpetragius is manifested in many of his works, right up to the *Hexaemeron*,[40] but he does not accept the whole system of the *De Motibus*; in the *De Luce* he (somewhat hesitantly perhaps) rejects its suggestion that the elemental spheres share in the diurnal motion, and he consequently treats Alpetragius's attempted explanation of the motion of comets and tides as purely hypothetical. Following Thābīt ibn Qurra al-Harrānī, Grosseteste accepts that the motion of accession and recession attributed to the eighth sphere is transmitted to the seven lower ones.[41]

While Grosseteste's argument concerning relative infinities seems to have been his own invention, the derivation of the contrasting arithmetical symbolism of the last section of *De*

---

[39] Bardenhewer, *Die pseudo-aristotelische Schrift über das reine Gute* (Freiburg-im-Breisgau, 1882), pp. 184–5.

[40] See my remarks on the *Computus Correctorius, De Fluxu*, and *De Motu Supercaelestium* in Appendix B.

[41] *De Luce*, p. 57.30–33. Thābīt was one of Grosseteste's principal authorities in astronomy; see my remark on the sources of the *De Sphaera* in Appendix B, p. 506.

Luce present no special difficulty. No doubt it had many sources, for the perfection of the number ten was as much a commonplace in the Middle Ages as it had been in ancient times, ever since Pythagoras. Unity as containing all numbers in germ had been a received idea in Chartres, as had been the identification of matter with the Platonic dyad,[42] or *alteritas* (Grosseteste's *'divisibilitas'*).

The claim that the density of matter is the cause of resistance and disobedience[43] is probably influenced by the *Timaeus*. In Plato's cosmogony the intractable, shifting, chaotic, pre-existing material resists the ordering action of the World-Soul. Plato's application of mathematics to cosmogonical speculation probably justifies our regarding the *Timaeus* as having influenced Grosseteste's light-theory, although in a general way only.

I have reserved to the last an influence which is more pervasive than any yet mentioned, that of Aristotle.

II GROSSETESTE'S SYNTHESIS OF GENESIS AND THE PHYSICAL SYSTEM OF ARISTOTLE

As I have already remarked, Aristotle's *De Caelo et Mundo* is the only book referred to by Grosseteste in his *De Luce*. The Philosopher's authority is invoked to lay the first basis for the mathematical discussion of infinity, namely, the principle that the multiplication of a simple thing cannot generate a quantity.[44] Grosseteste then asserts on his own initiative that an infinite multiplication of a simple thing will produce a finite quantity. The high authority of the Philosopher is again in evidence when Grosseteste defends the compatibility of his theory of infinite aggregates with Aristotle's principle to the effect that a magnitude can be produced only from another magnitude (*De Luce*, p. 54.2).

Among the analogies that could be drawn between the *De Caelo* and the *De Luce* one is particularly striking, namely, the doctrine that earth must congregate in a sphere around the centre of the universe. It forms part of Aristotle's proof of the sphericity of the earth: 'We must imagine the earth as in

---

[42] Parent, *La Doctrine de la création dans l'école de Chartres*, pp. 78–9.

[43] *De Luce*, p. 57.24.

[44] A thesis which recurs in the *De Caelo*, III, 1, 299a, 25–30. Grosseteste asserts that the combination of weightless things could have a quantity of weight, which seems odd to say the least.

process of generation. . . . If all particles move to one point the resulting mass must be similar on all sides and therefore a sphere.'[45] Reading this, Grosseteste may have noted that the same spherical form would result as much from a motion of particles *from* as *towards* the centre.

But it is unnecessary to appeal to analogies of this rather speculative kind in order to establish the dependence of the *De Luce* on Aristotle's physical system and particularly on the *De Caelo*, for the world produced by the expansion and unfolding of light-energy is in its essential result the world of Aristotle's physical system.

In the first place, it is a finite cosmos, unlike that of the ancient Atomists. Grosseteste attempts to prove mathematically that it must be so, but his proof succeeds in recovering only half of his assumption: granted that an infinite multiplication of a simple thing can produce only a finite quantity, Grosseteste still has not given an *a priori* reason why light could not be multiplied *infinities infinite*.[46] It is plausible to suggest that Aristotle's objections to the possibility of an actual infinite form a hidden premiss to the whole argument.

Secondly, in this light-filled world there is no vacuum. Only once is Aristotle's principle concerning nature's abhorrence of a vacuum explicitly invoked, but the principle presides over the entire unfolding of the universe. Light is generated evenly from the centre, so that no void arises; a sphere is formed at that precise point where further rarefaction of matter would produce the beginnings of a vacuum. The positioning of the spheres is determined quantitatively by the attainment of maximal rarefaction, but the flawless contiguity of the inner surface of the containing with the outer surface of the smaller sphere could be assured only by a metaphysical *horror vacui*.

In the paucity of generative causes invoked by the light-cosmogony it may not be fanciful to detect an echo of that efficiently parsimonious nature of the peripatetic school, whose economy in the employment of causes Grosseteste frequently extols in his later scientific works, with explicit appeal to Aristotle's authority.

[45] *De Caelo* II, 14, 297a, 13-26.
[46] In *De Luce*, p. 52.28-9 Grosseteste states that 'quantum infinitum infinities infinite excedit simplex', but treats this only as a part of his proof of the finiteness of the world.

Rarefaction and condensation are twin concepts of which Grosseteste would have learned in Aristotle's account of the Presocratic cosmogonies. Of course, his extension of their efficaciousness to the formation of the heavens is original with him; Aristotle confines that mechanism to the elemental world. Significantly, it is above all in the qualitative and not in the mathematical properties of the universe that Grosseteste is in overwhelming agreement with Aristotle. Somewhat unexpectedly, in view of the Neoplatonic kind of continuity that links all thirteen spheres of the universe, he manages to recover the full Aristotelean contrast of supralunary and sublunary worlds. The qualities of the nine celestial spheres—inalterability, insusceptibility to growth, generation, and corruption —are identically the properties of the peripatetic quintessence, and the attributes of matter within the four elemental spheres are their opposite in every respect: *existentes modo contrario* (*De Luce*, p. 56.17). The elements have the Aristotelean qualities of light and heavy, and their tendency to a natural place is determined by their relative density. The motion towards that place is rectilinear, as opposed to the perfect, circular motion of the heavens.

We are fortunate in being in a position to trace something of the progress of the quintessence in Grosseteste's writings before the *De Luce*, for the development that becomes apparent is instructive concerning the atmosphere of thought from which this work derives.

In his *De Sphaera* (to be dated probably to the period just before 1220), Grosseteste is more interested in the geometric properties of the quintessence or ether, which, forming the entire body of the heavens, is one, endowed of necessity with circular motion around a fixed axis, and spherical; otherwise its rotation would cause a vacuum.[47] In the *De Generatione Stellarum* we learn more of its physical properties. It is noble and unchangeable,[48] a simple body (p. 33.8), more transparent to light than the higher elements (p. 34.32), and considered a 'new matter' by comparison with theirs.[49] Thus far with Aristotle; but one cannot suppress astonishment upon

[47] *De Sphaera*, ed. Baur, pp. 11.21, 12.31, 13.25. Grosseteste's reference to 'Philosophus' on p. 12 is probably aimed at the *De Caelo*.

[48] *De Gen. Stell.*, p. 36.2.

[49] Ibid., p. 35.31: 'Sed caelum est factum ex tota vel nova materia, ut dicit Aristoteles, primo de Caelo et mundo.' (*De Caelo* I, 2, 269b, 13 ff).

learning that the heavenly bodies, far from being quintessential, must be composed of the elements. Arguments are taken from astronomy (the colour of the stars, the density of the moon), and 'experience' (actually, from Albumazar) to show that the luminaries differ specifically from each other. Now we can scarcely assume that Grosseteste propounded this doctrine in blissful ignorance of Aristotle's teaching; if he knew as much as he appears to have done about the quintessence, he must have known that for Aristotle it is the substance as much of the stars as of the spheres. Why then did he reject this teaching of the Philosopher?

The answer to this intriguing question can be found only, I believe, in the loyalty Grosseteste still felt (*c.*1220) to the whole syndrome of astrological and alchemical beliefs which were scarcely to be distinguished from astronomy—the earliest object of his scientific passion, so far as can be ascertained. In the early *De Artibus Liberalibus* the three interests are inseparable. Grosseteste describes how the qualities of the planets combine to produce metals. When these qualities are listed, they prove to be as elemental as the effects they induce: Saturn is earthy, cold; the moon is cold, the sun hot, Jove humid, Mars dry, etc. In *De Impressionibus Aeris* (written between 1215 and 1220, probably), a more refined list of qualities is offered, and these are certainly thought of as inhering in the nature of the planets.[50] The possession of such qualities is incompatible with the attributes of the quintessence.

The *De Cometis* (1222-24) marks an advance in the direction of a purer Aristoteleanism. 'Since there is no change but of *situs* in the supralunary region', Grosseteste's argument begins, 'it follows that the comet cannot be a new star or new supralunary phenomenon'; it can only be a new occurrence '*sub globo lunari*'.[51] Grosseteste says nothing more on the nature of the heavenly bodies, but the strictness with which the Aristotelean principle is enunciated should mean that it applies both to the spheres and to the bodies they bear. Grosseteste, however, still wishes to retain astrology, even within a world shaped by fundamentally Aristotelean parameters: the comet is fire sublimated by the virtue of a star or planet and assimilated to its proper nature (ibid. p. 39.6); its appearance can be used to predict certain future events (p. 40.6 ff.).

[50] *De Impress. Aeris*, ed. Baur, p. 42.   [51] *De Cometis*, ed. Baur, p. 36.17.

The *De Luce* stands, then, at the end of an evolution in Grosseteste's ideas. His early interest in astronomy, astrology and alchemy produced several astronomical works in the period *c*.1215-20. It gave rise to an interest in the physics of the heavens and, of course, of their effects upon the earth, which explains why we have found the *De Caelo et Mundo* cited more than any of Aristotle's other physical works in Grosseteste's philosophical writings up to and including the *De Luce*; and even the latter work itself gives no evidence of any extensive study of the *Physics* or the *Metaphysics*. While the *De Luce* does not discuss the *luminaria*, its conclusions concerning the substance of the heavens are fully Aristotelean, and Grosseteste seems at pains to underline this fact.

The doctrine of circular and rectilinear motion is likewise peripatetic, at least in its effect. The heavens are incapable of further condensation and rarefaction and therefore of any motion to or from the centre of the world; circular motion is their only possibility. The elements have not reached the limits of the possibility of their matter, and so motion in lines up and down from the centre is natural to them.

The doctrine of celestial motion contained in the *De Luce* came from Alpetragius, as I have already stated. Interestingly, a passage in the *Computus Correctorius* of Grosseteste (written before 1230), in opposing Ptolemy's mathematical account of motion to Aristotle's physical one, attributes to the latter the same teaching on motion that is upheld in the *De Luce*, where it is laid out in a more detailed fashion.[52] Towards the end of the passage, Alpetragius is introduced and his effort to save the appearances '*per modum Aristotelis*' mentioned. The only conclusion possible is that for some period of time Grosseteste was not able to distinguish what was proper to each in the physics and astronomy of Aristotle and Alpetragius; for certain purposes at least they formed a single doctrine which in his own mind he opposed globally to that of Ptolemy.[53] The *De Luce* falls with every likelihood into this period, just before

---

[52] *Computus Correctorius*, ed. Steele, *Opera . . . Rogeri Baconi*, fasc. 6 (Oxford, 1926): '*Secundum ipsum*', that is, according to Aristotle, there are nine concentric spheres, each moving on its own axis and poles, but all sharing the diurnal motion caused by the first sphere, only more slowly.

[53] The *De Motu Supercelestium* marks the end of this period, for there Grosseteste could comment on Aristotle without interference from Alpetragius's reform; he mentions Alpetragius only at the close. See Baur's edition, p. 100.33.

Grosseteste applied himself seriously to the study of Aristotle's *Physics* (and, to a much lesser extent, of the *Metaphysics*), between *c*.1228 and 1232. It is difficult to resist the conclusion that the theory of celestial motion which it advocates was adopted by Grosseteste in fidelity to Aristotle, and that he considered Alpetragius a true disciple of the ancient physicist.[54]

I conclude in the light of this exposition that the author of the *De Luce* aimed consciously at producing a synthesis of the cosmogony of Genesis and the cosmology of the *De Caelo*. Taking the light produced by the first word spoken by Yahweh and the amorphous matter, the state of *tōhûwābōhu*, or formless void (Gen. 1:2), with which it is creatively joined, Grosseteste attempted, in Kepler's celebrated phrase, to 'think God's thoughts after him', and to imagine how the omnipotent mathematician planned the process which gave as its result a world-machine Aristotelean in its essential features. Grosseteste set himself to do for Aristotle's physics what the masters of Chartres of a century before him had tried with Plato's *Timaeus*; one wonders whether he was conscious of following in their footsteps. Perhaps we can see a parallel between his attempted deduction of the Aristotelean world from the higher truth of revelation and the next major undertaking to which he turned his attention, for in his *Commentarius in Libros Analyticorum Posteriorum* he attempted to circumscribe the Aristotelean scientific understanding within the entirety of revealed truth, where it has its due place under *sapientia*, as *intellectus* has in the soul under *intelligentia*. Both projects are open compliments to the Philosopher, yet reminders of the partial nature of his synthesis and exaltations of the greatness of God's gift of revelation.

III NON-ARISTOTELEAN ELEMENTS IN GROSSETESTE'S COSMOLOGY

The world which Grosseteste deduced from a handful of cosmogonical premises was Aristotelean in its fundamental

---

[54] Just as there is evidence that he looked upon Avicenna as a representative of Aristotelean psychology. Our suggestion above raises the possibility that other Neoplatonic doctrines of the *De Luce*—first matter and form, notably—may also have been thought of by Grosseteste as genuinely Aristotelean. In general, his whole generation lacked a differentiated grasp of the conflicts between purely Aristotelean doctrine and Neoplatonic infiltrations into it and interpretations of it.

contours, but the process of deduction itself and the principles inspiring its conception transposed and sublimated the elements of the resultant cosmos. It could not have been otherwise. When Grosseteste composed the *De Luce* his acquaintance with the Aristotelean physical corpus was not extensive, amounting in all to little more than the *De Generatione et Corruptione, De Caelo et Mundo*, and *Meteorologica*. In an era when the assimilation of these works was effected through a screen of patristic and Neoplatonic ideas, together with Arabic reforms of the initial system, the net result could not be pure Aristoteleanism. (The same holds true of all his contemporaries, and indeed in some degree of all medieval authors.) Some of Grosseteste's deviations from Aristotle's system of the universe were conscious ones, as, most obviously, when he rejected the assumption that the world had no beginning in time; others, such as his valuational heliocentrism, may have been quite unconscious; while as regards the idea of space and the aesthetic approach to nature, it is almost impossible to say how he saw them in relation to the Aristotelean corpus. I propose to examine a series of positions in which one can discern a clear difference between Grosseteste's ideas on the physical universe and those of Aristotle. I hope thereby to determine the factors which gave him distance from the imposing world-system of his greatest ancient authority, and to assess the possible importance of his positions for the history of cosmological ideas.

a) The mathematical God and the geometric world

The evangel of mathematical truth receives its only properly kerygmatic proclamation in Grosseteste's *De Lineis, Angulis et Figuris*:

There is an immence usefulness in the consideration of lines, angles and figures, because without them natural philosophy cannot be understood. They are applicable in the universe as a whole and in its parts, without restriction, and their validity extends to related properties, such as circular and rectilinear motion, nor does it stop at action and passion, whether as applied to matter or sense . . . For all causes of natural effects can be discovered by lines, angles and figures, and in no other way can the reason for their action[55] possibly be known.

This is the first statement of a claim which was to prove

---

[55] That is, the reason *propter quid* as distinct from *quia*. This passage is found in Baur's edition, pp. 59–60.

fundamental to the metaphysics of early modern science. The unusual degree of self-consciousness with which it is made suggests that Grosseteste was not unaware of his originality. It amounts to a demand that the underlying mathematical structure of the physical world be laid bare, where by 'mathematical' is meant Euclidean geometry to the practically complete exclusion of arithmetic.[56] Compared with the unqualified claim he made, of course, the actual success which met his efforts at the mathematical explanation of nature was exiguous, and his original contribution to scientific knowledge was meagre enough: it reduces itself on critical examination to the theory that the rainbow is caused by refraction, and not by any degree of complexity in the reflection, of light.

Grosseteste's faith was greater than his works, but if we regard the latter more leniently and benevolently than the chronicler of the successful advance of scientific ideas strictly speaking can perhaps afford to do, we can discover in them an originality and an importance which in the long run did have a bearing on science, in so far as science came into being in dependence upon certain metaphysical beliefs.

From this perspective, the most attractive aspect of Grosseteste's attempted scientific explanations of natural phenomena is the way in which they manage to simplify physical realities (such as the climates, heat, and the rainbow) and to confer upon their analysis something of the abstract intelligibility of geometry. In the *De Natura Locorum*, as well as elsewhere in his scientific writings, Grosseteste acknowledges that natural philosophy must account for the variations of nature in the different parts of the world (ed. Baur, p. 66). The first rule that can be employed concerning the effects of lines of force emanating from a natural body is that a short straight line is the most effective agent. It is easily proved.[57] The briefer line is more fit to carry the virtue of the agent than the longer, and the straight line coming immediately from the surface of the agent to the patient, being the shortest possible, is a stronger transmitter of power than the oblique or reflected one. Nature is sparing and finds simpler means better. As for figures, the

[56] Grosseteste probably knew little more of arithmetic than was contained in Jordanus's *Algorism*, of which he is known to have had a copy made, perhaps *c.*1215 (see Thomson, *The Writings*, p. 30). The only practical scientific use he envisaged for arithmetic was in making astronomical calculations.

[57] *De lineis*, p. 60.30.

pyramid represents the strongest and best action there is (p. 64-13). Compared with individual, straight lines of force radiating from every point on the surface of the agent to the nearest point on the patient, the pyramid, whose base is the entire surface of the agent,[58] has greater effect, as it focuses its whole *virtus* into a cone which finds its point on the patient's surface, and consequently acts on it with greater force. Moreover, since it is in reality always surfaces rather than points that are involved, each point on the patient will be the cone of a pyramid based upon the entire surface of the agent. Now shorter pyramids are capable of transmitting greater power than are longer ones, the distance from the source to the patient being smaller; and the angle formed by their rays at the cone, being more acute than in the case of longer pyramids, will produce more united and intense action.

Grosseteste can now apply this highly intelligible and extremely simple *a priori* mathematical scheme to the complex problem of temperature variations, to prove that mountain heights essentially receive more heat from the sun's rays than do valleys; *accidentaliter*, he admits, they are mostly colder because other causes are in fact operative as well, and only when the complexity of the latter is taken into account can the variations experienced be accounted for. Examples of such factors are the blowing of the winds, the extent to which the mountain reaches into the upper, colder interstice of the air, and the fact that the sun's rays remain longer on peaks. The solution to the problem is reached by abstracting from a multiplicity of possible causes—winds, atmosphere, motion—and searching out a single aspect, namely, the force of the sun's rays, which is capable of geometrical treatment and allows a perfectly intelligible, because simple, account of the cause of differential temperature. The other relevant physical causes then receive attention as components of the total phenomenon.

This same ability of Grosseteste to simplify the physical phenomenon has been remarked on by Eastwood in his study of Grosseteste's theory of the rainbow.[59] Grosseteste, Eastwood points out, searches for elements of an explanation that

---

[58] Evidently Grosseteste is really thinking of a cone.

[59] Eastwood, 'Robert Grosseteste's theory of the Rainbow. A chapter in the history of non-experimental science', in *Arch. Intern. Hist. Sciences* 19 (1966), 313-32.

can be assimilated to the simple optical phenomena with which his authorities have familiarized him. The essence of his solution is the idea that sunlight passes through a cloud and is refracted to another cloud, that, namely, in which the rainbow appears to an observer. To render the phenomenon fully permeable to the laws of geometrical optics, Grosseteste describes the cloud as spherical; he has already learned how light-beams act in passing through a water-filled glass sphere.[60] Reaching towards another geometrically-simple shape, Grosseteste extends the concavity of the cloud to the earth beneath to form a pyramidal cone, which is precisely the shape taken by the layers of rare and dense (lower) moisture between the cloud and the earth. He now has four refracting media (atmosphere, cloud, rare moisture, dense moisture—a plane common surface is supposed between the latter two), through which the light must pass consecutively, and from the multiple refractions that ensue he can extract an explanation of the bow and its colours. As Eastwood says:

> The use of such unsubstantiated notions as a spherical cloud, a pyramid of moisture, and macroscopic refraction is for Grosseteste [not purely *ad hoc* but] quite proper. His idea of scientific explanation is founded on application of basic mechanisms to more complex phenomena. The mechanisms are taken from the available literature. The relationship between the basic mechanism and the complex phenomenon is considered to be simple and direct. He sees the rainbow, not as a complex phenomenon, but as a composite phenomenon (Eastwood, art. cit., p. 323).

Eastwood relates Grosseteste's procedure convincingly to the respect in which he held the principle of parsimony. This respect gave rise to a conviction in him that reality should reflect the simplest complete explanation.

Something additional is at work, though, for simplicity and simplification are not identical, and Grosseteste undoubtedly tried to search out among the phenomena to be explained one or more basic formalizable patterns, being guided in the search by his geometer's eye and by his faith that mathematical explanations attain to the true reality of natural phenomena. It is in this respect that Grosseteste's scientific use of mathematical explanation is to be referred back to his cosmogony, where the same fertile geometric imagination is at work. The behaviour of visible light is nothing more than a reflection at

[60] *De Nat. Loc.*, ed. Baur, p. 71.

the sensible level of metaphysical *lux* or corporeity, and it yields a privileged glimpse of the inner workings of a mathematically-functioning reality. Three foundational principles of *De Luce* can be singled out as lying at the heart of the matter.

The first foundational principle is the coincidence of the physical with the geometric centre of the universe. It was doubtless the belief that reality is mathematical which prompted Grosseteste to prefer Aristotle's and Alpetragius's system of homocentric spheres to the non-concentric system of Ptolemy, with its postulated distance between the physical centre of the world, namely, the earth, and the several geometric orbit-centres which circle around the stationary earth. Grosseteste yielded to this preference, however, only in his physical works, for he continued to employ Ptolemy's system for astronomical purposes, and never managed to arrive at a final decision as to the comparative merits of the two systems.

Now it may seem somewhat paradoxical that the physical system should have prevailed over the mathematical one for reasons that have to do with mathematics, but the paradox disappears with the realization that Grosseteste's option was meant to prevent all possibility of a breach between mathematics and reality. A breach of just this kind had been introduced through the positivism of the sophisticated Hellenistic astronomers, who had succeeded in saving the appearances of the heavens only at the cost of a double renunciation: of the effort at physical understanding, and of faith in mathematics as the key to nature's real functioning. Grosseteste rejected this product of resignation, and with it the implication that mathematics is a technique of purely notational value for a description of the physical world that is imposed on it from without. To the pure astronomer, his choice would have amounted to an unscientific rejection of the evidence favouring the Ptolemaic system (and its reformulations); to the historian of science, on the contrary, it is of profoundly positive significance, for early modern science found one of its foundational options in the rejection of the same defeatism and the affirmation of an unrestricted metaphysical faith in mathematical realism, albeit a realism in some respects specific to that age.[61]

[61] See further the important article of Koyré, 'Die Ursprünge der modernen Wissenschaft. Ein neuer Deutungsversuch', in *Diogenes* 4 (1957), pp. 442-3.

The second notable presence of the mathematical imagination in the *De Luce* resides in the identification of space with extension posited in the very first moment of the universe's existence. The pains which Grosseteste does not spare himself to prove mathematically that light expands matter from a primordial point evenly and continuously (because according to any relative infinities), are justified by the requirement that three-dimensional space be simply the product of extended matter. In this first moment, space is relatively homogeneous, co-extensive with physical reality, and one; it is the indissociable partner of a material energy programmed to perform according to the laws of geometry. Despite its kinship with geometrical occurrences, however, this space of extension of the universe in three dimensions falls short of the Euclidean in two important respects: it is finite and bounded, and it is not neutral or isotropic, for it has two 'natural' places, centre and circumference.

In the second moment of the cosmogony, light initiates from the circumference of the world-body that series of developments in which the one becomes many and the differentiation between simple and more complex, the distinguishing out of highest and best from derivative and lowly, is effected. On the completion of this process, the *Stufenkosmos* of Aristotle is present, with its order of rank among bodies determined by proximity to or distance from the first and noblest. In this world, whose graduated variety is emphasized in *De Luce* by the Neoplatonic emanation of many from one, space can be little more than the aggregate of the places differentiated according to their dynamic function: to each place a specific kind of body is allotted by nature, and a characteristic 'pull' is granted to attract that body to its natural site. In the heavens, the 'place' of each sphere relative to the first mover determines its responsiveness to universal motion, its speed of rotation, therefore, and degree of regularity, in the case of planets.[62] Now the contrast between the 'space' of these first and second moments of Grosseteste's cosmogony is real and evident, and is a measure of the distance that divides his ideas from orthodox Aristoteleanism. The final product is physically

---

[62] This holds true, not only for Alpetragius, but for Aristotle also, who explains it teleologically (*De Caelo* II, 12, 291b, 24-292a, 9), as Grosseteste was to become aware; see *De Motu Supercelestium*, ed. Baur, p. 100.21.

indistinguishable from the place-continuum of Aristotle, which ensures that the latter's theories of natural and violent motion will be upheld, and no novelty introduced into dynamics. But what a difference there is at the metaphysical level, where for Grosseteste a unified space gives rise to the multiplicity of places, and physical space presupposes a more neutral geometric space within which it was originally formed and shaped, just as each physical body is pure extension in three dimensions before it receives a specific form, mineral or organic. The difference is indiscernible for the practising physicist, for unification does not imply uniformization; but the imaginative superimposition of a real and unifying space upon the layered world, the reduction of multiplicity to unity, is of importance for the metaphysician, and not without interest for the historian of science.[63]

The third and last important witness to the mathematical foundations of reality is the attempt made in the *De Luce* to derive the qualitative properties of the Aristotelean world-system from concepts of a more quantitative nature. Firstly, the principle that no void can form in the universe received an attempted partial mathematical justification in the description of the mathematically-programmed extension of matter by light's infinite multiplication, according to any infinite proportions.[64] In the second place, the mechanism of rarefaction and condensation which determines so much of the world-formation is arguably more quantitative than qualitative, being based upon the distribution of matter according to degrees of density; the variations involved are functions of the distance of extended body from its generative source. The qualitative differences between heavens and earth, as also among the elements of the latter, heavy and light, are consequences of relative rarity and density; and circular and rectilinear motion, as well as the differential rates of the former, are deduced from the varying intensity of light in the bodies of spheres and elements, where intensity again is the result of differences in degree of condensation and rarefaction.[65]

[63] See Koyré, *From the Closed World to the Infinite Universe*, ch. 2, pp. 40 ff., for the development by Bruno of the concept of an infinite Euclidean space.

[64] The finiteness of the process depends implicitly upon the impossibility of a void, as I have remarked earlier in this chapter.

[65] I do not wish to maintain that all the elements in the deduction of motion are quantitative, nor that the quantitative ones could be assigned arithmetical

'Geometric' might seem too narrow a qualification for the powerful mathematical imagination which produces this cosmogony, if we considered that the intricate laws of the even expansion of matter from a point to a sphere were arithmetical ones, unrelated to 'lines, angles, and figures'. On examination, however, the arithmetic of *De Luce* reduces itself to a geometrical basis. Grosseteste's illustrations of the production of proportionate finite qualities by infinite multiplications of a simple, are formulated in terms of lines of two, four, and one cubit. In the postulates of geometry, the infinite multiplication of a dimensionless point yields a line, and the infinite multiplication of a line, a surface. Grosseteste's mathematical deduction of spatial dimensions from a dimensionless simple is rooted in geometry, to which he gave 'real' or physical value together with the possibility of arithmetical formulation, by treating the Euclidean point as a simple but real unit; that is to say, dimensionless, yet capable of yielding dimensions by infinite replication. This led him to sin against the Aristotelean principle concerning the impossibility of an actual infinite. More faithful to Greek geometry, Aristotle would only allow that a continuum is potentially divisible to infinity, that is to say, that division could never end by reaching an infinitely small but indivisible unit. For Grosseteste, on the contrary, every line and every surface is an actually-existing infinite, and the divine mind knows the size of its atomic unit.[66]

We are thus led back from the mathematical cosmogony, itself already the ground of the phenomena of visible light, to the first light and absolute ground of being. The cosmogony of light takes its character from the idea of the God whose word planned and executed it, and in a highly original and significant passage occurring in his *Notes on the Physics*, Grosseteste describes God as a mathematician who established the basic indivisible units of space and time from which the whole

values. Perfection, impurity, and the refractoriness of matter are qualitative concepts to which Grosseteste accorded great importance.

[66] The originality of Grosseteste's treatment of relative infinities and the extent of his influence on later Scholastics have been well brought out by Analiese Maier in various of her works: *Metaphysische Hintergründe der spätscholastischen Naturphilosophie* (Rome, 1955), pp. 399-400; *Ausgehendes Mittelalter. Gesammelte Aufsätze . . .* (Rome, 1964), pp. 82, 84, 295 ff., 311-314; *Zwischen Philosophie und Mechanik* (Rome, 1958), p. 24.

extension and unfolding of the material world is effected.[67] If we follow Grosseteste's exposition, we have the privilege of assisting at the birth of a new idea, one which inflected traditional conceptions of creation and God in a manner that was to prove decisive for the metaphysics of modern science.

The question at issue is the definition of time: is Aristotle's conclusion, to the effect that time is *numerus motus secundum prius et posterius*, an adequate one? Having examined it, Grosseteste concludes that the idea of motion falls obliquely under it, and that Aristotle's conclusion is less a demonstration of the reality of time than the outcome of a superficial discussion (*'levis confabulatio'*) arriving only at the meaning of the word (pp. 88–89). He begins on his own account an examination of the nature of measuring as such, one of the earliest in the history of Scholasticism. Numerical measurement of time or space is done by the addition of pure unity. But whence come the units by which a line, for example, is measured?[68] If there were but one line in existence, which we will suppose abstracted from matter, we could never pronounce it long or short or of one cubit, because, being the only longitudinal dimension in existence, it could not measure itself, nor could its aliquot parts of unknown size yield a measurement of the whole. The parallel with time is exact: if there were but a single motion in existence, let us say the diurnal motion of the first heaven, how could it be pronounced quick or slow, or the time of a single rotation be measured? 'Here', concludes Grosseteste, 'we are in an area of such difficulty that it lies almost beyond the scope of the human mind.'[69]

Credo tamen . . . Grosseteste professes his conviction that there is an ultimate foundation of extension in space and time, a final unit and measure which determines their nature. The line mentioned above can be measured and counted, if it consists of indivisible points whose sum can be reckoned and makes an extension. Only, since their number is infinite, the reckoning cannot be done by a finite mind, but only by a mind to whom even infinite numbers are finite, and related

---

[67] See ed. Dales, pp. 86 ff.

[68] From this point in the text (i.e. Dales's edition, p. 90 onwards) I follow Annaliese Maier's transcription of the Venice *Bibl. Marciana* MS lat. VI, 222, printed in her book, *Zwischen Philosophie und Mechanik*, pp. 24 n.–25 n. Dales's text presents many difficulties at this point.

[69] 'Haec difficilia sunt valde et ad quae vix pervenit humanus intellectus'.

to each other as greater and less;[70] for one infinite number can be related to another in every proportion, both rational and irrational: as double or treble, or as the diameter to a chord.[71] Now God's wisdom is 'without number' (*sine numero*), and infinite numbers are for him no more limitless than is 'two' for you or me. The infinite sum of all equal numbers, of all odd numbers, and of all infinite numbers infinitely divisible, or any other complication you care to invent, is for him a finite number. Just as things in themselves finite are infinite as far as we are concerned, so what is infinite in itself is finite for God. If he 'created all things in number, weight, and measure', he is the *Mensurator primus et certissimus*, who measured the lines he created by infinite numbers, which are, however, finite, so far as his mind is concerned.[72]

That Grosseteste, throughout his discussion, is keeping in mind the actual process of creation as he had reconstructed it in the *De Luce*, is beyond all possible doubt. Besides the evidence of his reference to the mathematical argument of that work, there is at this point an example taken almost verbally from it: it was from a determinate, infinite number of points that God measured and counted a line of one cubit, from double that infinite number a line of two cubits, and from half that number a line of one-half cubit. What is meant by a determinate number is made clear: it is the number of actual points in a given line, and it is such, that it is not found in any line of a different length. There is then an absolute, indivisible point which is the unit of all extension in the universe, and we risk nothing whatsoever in identifying that point with the simple point of light from which the cosmogonic process began, in *De Luce*.

[70] 'Puto quod numero infinito punctorum illius lineae, finito tamen mensuranti, qui numerus punctorum non est in aliqua alia linea, sed in maiori est maior numerus infinitus punctorum, et in minori minor. Per hunc modum mensurandi non potest mensurare nisi cui numeri infiniti finiti sunt, et cui unus numerus infinitus est magnus et alter parvus. Unde si nullo creato est infinitum finitum, nullum creatum sic mensurat.' (ibid.).

[71] 'Et hoc alibi probatum est', states Grosseteste, referring his reader implicitly to his own *De Luce*.

[72] 'Sicut enim que vere in se finita sunt, nobis sunt infinita, sic que vere in se sunt infinita, illi sunt finita. Iste autem omnia creavit in numero, pondere et mensura, et iste est mensurator primus et certissimus. Iste numeris infinitis, sibi finitis, mensuravit lineas quas creavit. Numero aliquo infinito sibi certo et finito mensuravit et numeravit lineam cubilem, et numero infinito duplo lineam bicubilem et numero infinito subduplo lineam semicubilem.' (ibid.).

From this metaphysical basis Grosseteste returns to human measurement. Knowledge of the infinite number of points by which the *Mensurator* measured a line *'simpliciter, verissime et certe'*, is not possible to a finite mind. A line already thus measured we take over as the established standard of measure, without being able to measure it; but it will always be a real line in some determinate nature (a finger, hand, span, forearm, and so on). The same applies to time: the Creator measured all time in eternity before time began, producing all temporal differences of longer and shorter by the infinite multiplication of a determinate and indivisible instant, according to the proportion of one infinite number to another. All we can do, on the other hand, is to settle on one time-span, a determinate motion of the heavens, as the unit for measuring things in time.

The fundamental importance of this passage for the history of ideas has not fully been adverted to. What fourteenth-century Scholastics such as Alnwick, Burley, and Gregory of Rimini, and indeed one or two modern historians, found striking in it, is its teaching on relative infinities and on the reality of an actual infinite. It is clear, however, that Grosseteste would simply not have arrived at his original positions without his belief in the infinite mind. Those Scholastics who later pointed out that these infinite numbers which measure the universe are irrelevant and unhelpful to physics were blind to the sense of Grosseteste's undertaking, for no consideration of any possible practical value of his ideas so much as entered his head. Only because he believed in an all-knowing God did he bother to speculate about infinite numbers, which no human or created mind could ever count, and consequently to acclaim this God as a mathematical designer.

Grosseteste finds it possible and desirable to graft his essentially new and unheralded conception of a mathematical God upon an ancient truth: God 'ordered all things by number, weight and measure' (Wisdom 11:21). This is a favourite text of his; it would be difficult to count the number of times he quotes it in his writings. All the other uses I have observed relate fairly directly to St. Augustine's foundational interpretation, which made number, weight and measure exist from eternity in the Divine mind as *modus, species et ordo*, primary determinants of a creation to which God gives 'limit, form and

order'.[73] The triad refers to transcendental qualities of creation,
its order and gradation (*ordo, gradata distinctio*). In the *Notes
on the Physics*, Grosseteste effects an evident and admirable
glissement of meaning, shifting the emphasis from the qualitative harmony and measure of the hierarchic creation universally, to the numerical value and operation from which the
material universe has received its origin. The *Mensurator
primus* determined upon a definite unit as the basis of all extension and, as *Numerator*,[74] operated on it by multiplication
to produce an infinite series of infinite numbers governing
finite material extension.

Grosseteste has no hesitation in founding his novel conception upon the biblical view of the Creator, whose natural attribute is, after all, the thinking of infinite thoughts. That we
are here very far removed from Aristotle is too obvious to
merit remark. That Grosseteste's conception was never shared
by more than a handful of later Scholastics is something of a
surprise, for it is surely more in keeping with the Christian
idea of God to attribute to him certain knowledge of the infinitely small atomic unit of the world than to deny him the
power to measure the only potentially-infinite constituents
of a continuum, on the grounds that 'there is no actually indivisible thing' (*ens actu indivisibile non datur*), since a creature
infinite in act is unthinkable. Grosseteste achieved a double
breakthrough, first in positing an actual infinite in creation,
secondly in honouring God with the title of mathematician.
The first thesis was heresy to Aristoteleans, the second would
have been meaningless; but both ideas were destined for a long
and fertile career.

This is not the first occasion on which we have traced a
non- or even anti-Aristotelean notion back to its ground in the
conception of an infinite, creative God. We shall meet the same
phenomenon again, and more than once, in Grosseteste's later
writings.

[73] Augustine's basic exposition of the text is found in *De Gen. ad Litt.*, IV, 3,
7-5, 12, *CSEL* 28 (3, 2), pp. 98-101 (*P.L.* 34, 299-301), but see also *De Civ. Dei*,
XI, 15; *De Musica*, VI, 17, 56-7; *De Vera Religione*, VII, 13; *De Trinitate*, VI,
10, 12.
[74] The strictly mathematical meaning of the words is reinforced by the use of
'*mensurare*' and '*numerare*' to describe God's action, whereas '*pondus*' is simply
passed over and remains an undeveloped concept by comparison with the other
two.

It has been our contention that Grosseteste was equipped with a true geometer's eye that allowed him to introduce simple formal structures into the explanation of natural phenomena; with faith that the working and inner nature of material reality are mathematical; with a powerful geometric imagination, fertile enough to support, with the first interesting cosmogony to arise since the times of Plato, his intuition that all is formed from energetic light; and with a belief that God must be a mathematician. It is chiefly in this nexus of ideas, imaginings, and intuitions that his importance for the development of science resides, that is to say, his importance as a rich begetter of the metaphysical principles that bore science into the light.[75]

b) The unity of matter

'The heavens were made of a completely new matter, according to the first book of Aristotle's *De Caelo et Mundo'*, noted Grosseteste, shortly after 1220.[76] It was a doctrine with which he was never satisfied, and his attempts to overcome it led him into the second of what we have termed his major deviations from the Aristotelean physical system.

The Aristotelean doctrine of matter and motion represents in a sophisticated form the universal, primitive perception that heavens and earth differ profoundly from one another. The cyclic and timeless harmony of the heavens, manifested by the regular motion of sun, moon, and stars, received a philosophical transcription in Aristotle's concept of quintessential matter, a simple substance which, having no contrary quality, cannot decay and requires no periodic rejuvenation. A consequence of the strict opposition between the different and opposed matters of heaven and earth was drawn: there is no exchange of matter between supralunary and sublunary worlds, no quintessence on earth and no elemental matter in the heavens. The inner circumference of the lunar sphere marks a strict dichotomy between two different natures, sets of operations, and physical laws.

Several influences rendered Grosseteste either blind to or

---

[75] On the importance of geometry for the development of early modern science see Butterfield, *The Origins of Modern Science 1300-1800* (London, 1956), pp. 13-14, 73-4, 142; Koyré, art. cit., pp. 442 ff.; and *From the Closed World...*, pp. 100-04, 114; Kuhn, *The Copernican Revolution*, (Cambridge, Mass., 1957), pp. 212 ff.

[76] *De Gen. Stell.*, ed. Baur, p. 35.31-3.

dissatisfied with the full consequences of this doctrine and stimulated him to produce a more flexible alternative.

First among them was his early attachment to astrology, unselfconsciously and unambiguously proclaimed in a series of his works from *De Artibus Liberalibus* up to the *De Cometis*. He accepted unreservedly the belief that the stars (including the planets)[77] act upon the world in virtue of qualities—hot and cold, dry and wet—which they share with the elements,[78] and he drew the only possible conclusion, namely, that even though the spheres may be quintessential, never the less 'in the making of the stars there is no making of new matter', *ergo* they are made from the matter of the elements, and proceed from a different act of creation than do their spheres, not being of uniform matter with these.[79]

One might think it possible that the quintessence of the spheres at least is the pure Aristotelean substance, and its mention an index of a growing fidelity to Aristotle on Grosseteste's part; but on further investigation a second factor, namely alchemy, announces itself. The simplicity and immutability of the quintessence, Grosseteste maintains, do not prevent its entering into a mixture with the elements. When it descends into the lower world to function as the container of the elements or the harmonizer of their complexion in a mixed body, it simply changes its nature and becomes mutable. Now if humiliation of the quintessence is possible, Grosseteste sees no reason why the reverse process, namely, the sublimation of essentially mutable, elemental matter to a heavenly state of incorruptibility, should not be equally so (ibid., p. 36.3-11). Sublimation forms the foundation of his explanation of the shooting-star.[80] The *De Cometis*, however, marks a degree of rapprochement with Aristotle: no new phenomenon occurs

[77] '*Stella*', without qualification, is commonly enough used to refer to both stars and planets.

[78] This is a belief he probably never surrendered: the *Hexaemeron* contains (*MS Bodl. lat. th. c. 17*, fol. 204b) a long disquisition on the 'frigiditas stelle saturni, que stella . . . deberet esse ceteris planetis ferventior, nisi aquis illis superioribus temperatus esset eiusdem stelle fervor'. The Arabic influence here is clear: Grosseteste cites Albumazar as showing 'from experience' that the planets differ in specific nature from each other.

[79] *De Gen. Stell.*, pp. 35.26-36.2. It is in this context, curiously, that he appeals to Aristotle's authority for support, quoting from the *De Caelo et Mundo* the idea expressed above.

[80] 'Palam est ergo quod trica est ignis sublimatus separatus a natura terrestri et assimilatus naturae caelesti.' (*De Cometis*, ed. Baur, p. 38.1.)

in the supralunary region, wherefore the comet must be an occurrence *sub globo lunari*, in the region of elemental fire (pp. 36.17, 37.2-5).

From the point of view of strict Aristoteleanism, Grosseteste's attempt in these earlier works to obscure and partially to obliterate the sharp differences between heavenly and earthly matter and processes could only be seen as a senseless deviation, a fragmentary and eclectic assimilation of the Philosopher's true position. Regarded, however, in the light of its sequel, his attachment to the two pseudo-sciences must be given its place as a germinative influence upon his *De Luce* and his metaphysics of light generally. In the rational and scientific cosmology of *De Luce* two basic ideas are enthroned, namely, the continuity of nature and action throughout the material world, and the ultimate unity of matter. Both of these bear some resonance of the half-magical world of astrology and alchemy. In Grosseteste's fully mature thought, the earth's place at the geometrical centre of the universe gives a solid, scientific reason why it is most receptive of the light of all the heavens.[81] Light is the unique connection or influence between heavens and earth and the explanation of all earthly phenomena is found in its presence and action, from the flux of tides[82] to the presence of vegetation on earth.[83] Every natural body on earth has something of celestial light and luminous fire in it;[84] light incorporated in the elements is the explanation of sound and colour,[85] of the generation of heat through the repercussion of matter's particles,[86] of the animal spirits, and of all five senses and their *spiritus*.[87]

Now the analogies that undoubtedly exist between Grosseteste's early and his mature thinking in the present connection, though they seem not to have received the adequate attention of historians, are striking and important. At an earlier stage

[81] See, e.g., *De Luce*, p. 56.23; *Hexaemeron, MS cit.*, fol. 210ᶜ.

[82] *De Fluxu*, ed. Dales, p. 464.150 ff.: 'Item si corpus celeste non agit in inferiora nisi secundum suos radios luminares, et ex hiis radiis luminaribus cum incorporentur aliquo modo cum elementis . . .'.

[83] *Hexaemeron, MS cit.*, fols. 212ᵈ-213ᵇ; see Part IV, ch. 4, i.

[84] *Comm. in Anal. Post.*, Venice 1514, fol. 37ᶜ: 'Omne namque corpus naturale habet in se naturam celestem luminosam et igneum luminosum, et eius prima incorporatio est in aere subtilissimo.' This forms the premiss of the explanation of sound-generation.

[85] 'Color est lux incorporata perspicuo.' *De Colore*, ed. Baur, p. 78.4.

[86] *De Fluxu*, loc. cit.; *De Calore*, p. 81.20.

[87] See Part IV, ch. 4 (pp. 296-7).

Grosseteste was prepared to believe that parts of the quintessence can become humiliated and that these govern and preserve the harmony of elemental complexions, due no doubt to their higher power and their lack of an opposite. Later on he kept returning to the idea that light (an accident of the heavenly bodies, which he considered should not be thought of as opposed to or separate from the pure substance begetting it)[88] becomes incorporated into the elements, where it is the source, apparently, of all causal action. Lying behind both of these notions is a challenge to the Aristotelean dualism of heavenly and earthly matter. It becomes, then, a matter of great interest to inquire how that dualism fared in the *De Luce*.

I remarked when summarizing the *De Luce* that the heavens which result from the cosmogonic action of light are essentially Aristotelean: their only composition is of matter and form, and their matter is not subject to further impressions, wherefore they are incorruptible and immutable, and hence the qualitative opposite of the sublunary world (*De Luce*, p. 56.14 ff.).

This description no doubt coincides very largely with that of the quintessence, but it is suffused with an entirely different feeling for the nature of matter, as the remark immediately following it makes clear: just as unity contains the whole sequence of numbers in potency, so the first sphere is every later one by virtue of the multiplication of its light that produces them. The significance of this Neoplatonic idea is only fully grasped when it is realized that all bodies or spheres of heavens and earth are included in the power of the first. Admittedly, Grosseteste is at pains, a few lines later on, to emphasize that physical light allows for differentiation and gradation: bodies are not all of the same species, even though all proceed from simple or multiplied light, any more than all numbers are of the same species (pp. 56.36-57.4). Light, the *species et perfectio* of all bodies, accounts for both the unity and the diversity found at the level of material being. It is, however,

---

[88] In his *Hexaemeron* (MS cit., fol. 203c) Grosseteste attempts to reconcile the doctrines of Augustine and Damascene concerning the nature of light, by maintaining that light is both a *corpus* or substance and a *qualitas* or accident: 'Cum igitur horum auctorum utrasque sententias credamus esse veras, et sibi invicem non contrarias, dicimus quod necesse est lucem dupliciter dici; significat enim substantiam corpoream subtilissimam et incorporalitati proximam, naturaliter sibi ipsius generativam, et significat accidentalem qualitatem de lucis substantie naturali generativa accione procedentem. Ipsa enim generative accionis indeficiens motio qualitas est substantie indeficienter sese generantis'.

the continuity of the productive process that leaves a lasting impression, and that makes the difference of material heavens and earth no longer the absolute principle it had been for Aristotle, but a derivative and secondary effect which can be reduced to a higher principle of intelligibility, one whose application does not extend only as far as the cosmic watershed of the lunar sphere, but encompasses the entire system: the principle, namely, that each lower sphere participates in the form of the higher in a weakened and more derivative manner, and consequently acts with less energy in the production of its inferior. This law applies to the process of condensation and rarefaction by which the first sphere of pure and simple light produces the second, of 'redoubled' and more multiple light, and applies in exactly the same way to the lunar sphere as the begetter of the sphere of fire; only, in the latter case, the degree of cumulative weakening is such that further rarefaction and condensation are possible, so that rectilinear motion comes to replace circular—only gradually, however; the transition is not abrupt, as it must be in pure peripateticism.[89]

The unity of material energy within Grosseteste's universe is, if anything, even more pronounced than that of the 'common corporeity' and of the *materia et forma universalis* of his Arabic and Jewish predecessors.[90] It goes back to a primordial act of creation in which the matter and form of the universe were not merely generically but also numerically one, proceeding directly from the hand of the Creator, and not derived through the mediation of intelligences. Its direct source is the biblical account of creation; and fundamentally Grosseteste's view of material being is unitary because it is creationist. The idea of unity to which his faith committed him became a first category in his natural philosophy in a way that it never had been for Aristotelean science, and it commanded his rejection of the unqualified duality of matter so fundamental to Aristotle's cosmology. The atmosphere of thought in which the germinal idea flourished was Neoplatonic, and we could bring the whole material cosmos under the rule of a principle applied to natural heat in the *De Operationibus Solis*: every

---

[89] In the *De Luce* Grosseteste dallies with Alpetragius's suggestion that the upper elementary spheres of fire and air have a weak circular motion, and in the *Hexaemeron* he accepts it. This adds to the continuity between the two regions.

[90] Crombie notes the importance of these notions; see *Augustine to Galileo*, vol. 1, p. 88.

plurality that is unified into one collectivity must be traced back to a single root-principle.[91] That radical principle, the *lux* or first corporeal form,[92] made essentially one physical system out of what for Aristotle had been two separate ones, for it abolished the difference in principle whereby the higher universe was thought to be of an essentially different stuff from the earth. In replacing a difference in kind by a distinction in degree of density, perfection, and beauty, the light-metaphysics achieved a crucial turning-point in the history of the correction of Aristoteleanism, and if a long *attente* was still to ensue before Galileo's use of the telescope destroyed the duality of heavens and earth on the level of observation and science, the metaphysical belief in the radical unity of origin of material being was already assured a place in the history of ideas by Grosseteste.[93] The more concrete, one might almost say scientific, gains of the new system are already apparent in his own philosophy of nature, where a series of principles concerning the nature and the mathematical behaviour of light are deduced, principles which, applying equally to earth and heavens, presage the emergence of a unified physical theory in later centuries.

The idea that there exists a continuity between light and the elements is treated as a matter of course in Grosseteste's later writings. In his remarks concerning the relationship of the microcosm to the macrocosm in the *De Cessatione Legalium*, he presents the unity of the human body as being representative of a unified cosmos, in which the heavenly bodies share the nature of light with the contiguous element of fire, as fire and air share the quality of heat, air and water that of humidity, and water and earth, frigidity.[94] Grosseteste draws the conclusion that 'In this way all things are linked together in the most orderly way by natural connections.'[95]

[91] *De Operationibus Solis* § 15, ed. McEvoy, *Rech. Théol anc. méd.* 41 (1974), p. 76.

[92] Ibid., § 2, pp. 63–4. See further Grosseteste's reflections on unity in the extract from *De Cessatione Legalium*, edited by Unger, *Franc. Stud.* 16 (1956), pp. 13 ff.

[93] It is true, of course, that the use of the telescope destroyed the myth of the nobility and perfection of the heavens, beliefs which were just as firmly tenets of the metaphysics of light as they were of the physics of Aristotle.

[94] *De Cessatione Legalium*, ed. Unger, p. 14; see the parallels in the sermon, *Exiit Edictum*, ibid., p. 22.

[95] 'Sic omnia naturalibus nexibus ordinatissime sunt adinvicem complexa.'

What is left of the quintessence? Grosseteste makes no use of the term in the *De Luce*, and he employs it only sparingly thereafter, preferring the biblical word for the heavens, 'firmament'. This, it is true, incorporates some important attributes of the quintessence. Together with the heavenly bodies it contains, the firmament enjoys perpetuity of being, in the sense that, having no opposite quality, its matter lacks a principle of corruption.[96] In discussing these matters, the later Grosseteste prefers to invoke patristic authority in his support: Augustine, he once remarks, derives *'firmamentum'* from the *firmitas inalterabilis essentiae* which the heavens enjoy until all things will be made new.[97] Only once in the later works does he invoke the quintessence by name, in the context of a scientific argument, when he proves that the sun does not generate heat by contact as does a warm body, because the quintessence cannot undergo and could not transmit qualitative change;[98] but this argument too is based upon the incorruptible nature of the heavenly matter and does not indicate any revision of doctrine on its author's part.

Was Grosseteste conscious that his ideas on heavenly matter, and therefore on matter simply speaking, diverged from those of Aristotle? This kind of query is often impossible to settle where writers of the early thirteenth century are concerned, for oftentimes one suspects that even rather fundamental deviations from Greek authorities were unintentional, or were even intended, paradoxically, as faithful representations of authority. Fortunately, a brief passage in the *Hexaemeron* settles the question in the present instance.[99] Grosseteste points out first that the Bible does not say from what matter the luminaries were made. There are three theories: Aristotle's, that they were formed from the body of the firmament, the fifth essence separate from the four elements; that of Plato, approved by St. Augustine and St. John Damascene, that they

From a second parallel passage found in the *Hexaemeron, MS cit.*, fol. 223ᵃ; cf. fol. 233ᶜ. Grosseteste also draws the conclusion that the five senses parallel the five elements of the world, since they are formed by the incorporation of light into the lower elements; see Part IV, ch. 6, v.

[96] 'Substantia corporis celi habet esse permanens, et substantia corporis solis et huiusmodi . . .' (*De Cess. Legal., MS cit.*, fol. 167ᵇ). Cf. *De Operationibus Solis*, § 30, ed. cit., p. 88; and *Hexaemeron*, fol. 199ᶜ.

[97] *Hexaemeron, MS cit.*, fol. 204ᶜ.

[98] *De Calore Solis*, ed. Baur, p. 82.6.

[99] *Hexaemeron, MS cit.*, fol. 214ᵃ.

were made of the elements and ether, but so harmoniously compounded that they are neither heavy nor light and have no contending qualities;[100] or, finally, that they were made from that primordial light which, *secundum quosdam* (for Grosseteste prefers to remain anonymous), materialized to fill the universe during the first three days of creation. Neither here nor elsewhere in the work does Grosseteste undertake to determine the question, which he finds to be of little relevance for the interpretation of the Bible. Of sole interest to us is his manifest consciousness of the philosophical originality of *De Luce* and of its clear divergence from Aristotle's explanation.[101]

As he often does, Grosseteste reserves the best wine to the last: perhaps, he remarks innocuously, none of the three views is anywhere near the truth, and the command issued by the Creator was of an other order altogether!

The influence of astrology and alchemy made it natural for Grosseteste in the earlier stages of his philosophical itinerary to look for continuity of nature and action between the heavens and the earth. At a later stage, his belief in creation and his attraction to a Neoplatonic idea of continuity reinforced his dissatisfaction with the irreducible duality of celestial and elemental matter that lay at the very basis of Aristotle's system of nature. In Grosseteste's metaphysics, material being is one by origin (it is even numerically one) and by nature. Differentiated by a single physical process of production, it remains essentially one in heavens and on earth, the real differences of which are to that extent gradualized and merged into one another.

In asserting the unity of matter throughout the universe, blurring the difference between heaven and earth, and adumbrating the advent of a unified physical system, Grosseteste opened a radical breach in the theory of Aristotle. Later medieval critics, as is well known, were not slow to follow him through. William of Ockham seems to have been the first

[100] The identification is made in fol. 204^c, where the doctrine of the *Timaeus* is alleged, Augustine is said to be on the whole in agreement with it, and Damascene's vehement opposition to the quintessence is recalled. Grosseteste passes over his own original contribution in silence, and dismisses the whole question as being of little relevance for the interpretation of the Bible.

[101] It is likely that Grosseteste regarded his theory as being closer to the patristic view, according to which fire and ether were held to be continuous and light was considered the first body in nature (a doctrine which Grosseteste attributes to Augustine in *Hexaemeron*, fol. 213^b), than to the Aristotelean idea.

thinker of the fourteenth century to deny that the heavens and the sublunary world differ fundamentally in their matter and in their laws, and Jean Buridan enlarged the scope of this challenge by extending the concept of *impetus* from terrestrial into celestial mechanics.[102] Oresme[103] and Nicholas of Cusa further encouraged the homogenization of the cosmos: for the latter, sun and earth have *quasi* the same matter and perfection.[104] If the scientific revolution of later times was much more than a simple extension of such speculations—for it had to overturn the whole Aristotelean world at once—still the subjection of the entire universe to uniform physical laws, as with Galileo,[105] or the indefinitely-extended universe of Descartes, unified by the omnipresence of one and the same matter,[106] do in retrospect owe something to those medieval Schoolmen whose critical distance from the dominant Aristotelean system undermined its authority and encouraged later generations to reflect that there could be other possible, if in some respects less empirical, ways of looking at the world.

c) Divine omnipotence and the limits of human knowledge

In the *Hexaemeron* Grosseteste mentioned three suggestions about the matter of the heavens but concluded his discussion of the question with a warning to the reader that all three may be wide of the mark; God's disposition for the generation of the bodies of the heavens may in fact have been quite other than the hypotheses men have yet constructed. On closer examination, the same work yields a number of such oppositions between divine omnipotence and human speculation. These are no doubt intended in the first place to glorify God's unlimited power and in the second to safeguard the authority of Scripture. Regarded from another point of view, however, their cumulative effect is to highlight the impossibility of a purely deductive theory of the world reposing on principles internal to the divine nature, such as one meets with in emanationist systems.

---

[102] Beaujouan, 'Medieval Science in the Christian West', p. 509; cf. Crombie, *Augustine to Galileo*, vol. II, pp. 82, 85.
[103] F. Fellmann, *Scholastik und kosmologische Reform*, BGPM, N. F. 6, ch. 2, especially p. 36.
[104] Koyré, *From the Closed World to the Infinite Universe*, pp. 19-24.
[105] Butterfield, *The Origins of Modern Science*, pp. 15-16, 30, 60-1.
[106] Koyré, op. cit., p. 105.

As it happens, it is in every case the nature or some attribute of the heavens that forms the context of these remarks of Grosseteste. At the end of his discourse on the heavens in the *Hexaemeron* he returns with full sympathy to a word of Aristotle: the study of the heavens is toilsome and involved; its yield in terms of knowledge of the celestial substances is small, because of their remote nobility.[107] There is a touching sincerity about this admission. Grosseteste has busied himself for more than twenty years with the leading authorities on problems of celestial motion, yet he admits candidly at the end that he has reached no certain conclusions.

This is not the place to tell the full story of the hesitations, new positions, and reconsiderations which mark his advance from the theory of the World-Soul to the apparently certain doctrine of the movers set out in his *Commentary on the Celestial Hierarchy*.[108] I propose simply to examine the implications of the distance which he takes in his hexaemeral commentary from the physical and astronomical theories current in his time, for I believe that his comments are of profound significance in themselves. I shall simply mention here one important moment of his itinerary which is relevant to the general consideration of the relations between Aristotelean science and Christian faith, namely, the conscious and radical effort he made in the *De Motu Supercaelestium* to remove from the heavens any suspicion of divinity, and to secure their finitude as the material creatures of an omnipotent and eternal Creator. Grosseteste expresses the fear that within the premisses of Aristotle's argument for the necessity of a first mover (in both *Physics* VIII and *Metaphysics* XII) there lie the seeds of pantheism. While not directly accusing Aristotle of maintaining pantheism, Grosseteste keeps the accusation hovering around his consideration of the necessarily eternal nature of the world and of motion.[109] He notes that a shift of meaning from the intrinsic inalterability of the heavens—which he, like Aristotle, defends—to their substantial perpetuity, in a positive

---

[107] *Hexaemeron*, fol. 207ᵇ: 'Ut autem dicit Aristoteles in libro de animalibus, sermo de celo est cum labore et difficultate, nec comprehenditur nisi parva scientia substantiarum celestium propter magnitudinem nobilitatis earum.'

[108] The heavens are inanimate and are moved by angels separate from them in nature.

[109] *De Motu Supercaelestium*, ed. Baur, p. 96.19: 'Et difficultas non minima est in tot sermonibus quibus usus est Aristoteles in ostensione ultimae positionis', i.e. the existence of an unmoved mover.

sense indistinguishable from eternal being, would render the heavens infinite in moving-power and act, and would dispense with the need for any more noble principle such as an immaterial first mover: 'a great error, and one for which it is necessary that we devise a refutation'.[110] The refutation which he sets out proceeds in two stages. The substance of the heavens, he admits, is perpetual, having no contrary nor intrinsic possibility of corruption, but their motion cannot be permanent of itself, since it has an opposite, namely, rest (p. 98.8 ff.). Secondly, it is shown that movers who move by will and desire are of finite power: if they effect an unceasing motion, this cannot be of themselves but must derive from a necessary being, one that is *'semper in actu'*, and therefore immaterial and infinite (pp. 97.26-98.7; 98.19-33).

The first reference to celestial motion occurring in the *Hexaemeron* extends the correction of Aristoteleanism by biblical revelation into the area of final causality (fol. 199[d]). To what he calls 'the worthless argument of the philosophers', which explains the circular motion of the heavens in terms of the assimilation of the moving intelligence to the divinity, Grosseteste objects that it is based upon an unworthy conception of God and of his creative will. His refutation is not without a dry, salty humour, as when he remarks that, in moving its body to take up successively every position possible to it,[111] the intelligence has no better prospect of divinization than a man racing over land and sea in futile imitation of the divine ubiquity. He accuses the Philosopher of having a false idea of God, whose true nature consists in perfection and rest; of motion, which is the activity of creatures aiming at self-perfection; and of God's intention in creating. The simple truth is, he states, that the motion of the heavens was willed by God to carry round the luminaries which effect the generation of the human body and supply man's requirements on earth,

[110] 'Aut si hoc corpus est in sua substantia perpetuum, sicut ostensum est in primo de caelo et mundo, tunc non opus est ponere principium eo nobilius ... Et si aliquid est aeternum in sua substantia, videtur quod necesse est potentiam eius moventem esse aeternae motionis, et ita potentiae infinitae ... Si igitur caelum non alteratur in sua substantia, neque accidit ei fessitudo, non est impossibile, ut agat actionem infinitam a potentia existente in ipso, et tunc non erit necessarium ponere motorem separatum ab ipso. Et in hoc sermone non est error parvus, et ideo necesse est ut speculemur in eius destructionem.' (ibid., p. 97.7 ff.).

[111] In imitation, as Grosseteste says, of God, 'qui simul habet omnia quae habere potest'.

'*cum omnia propter hominem sint*': all creatures were made for man. Beyond that function, celestial motion has no significance, and that is why the first heaven, the *caelum aplanon*, lacking bodies, does not revolve, but only imparts motion to the lower spheres. If it is kept perpetually at rest (by itself, or through the agency of an angel) and only moves the lower spheres by the power of its light, in this state it far more resembles eternity in its energy-transferring (*actuosa*) state of rest than it ever could by motion. In any case, the hypothesis that heavenly motion proceeds from an intelligence is not part of our Christian tradition, Grosseteste warns his reader in conclusion.[112] This is presumably a large part of the explanation of his reticence concerning the angel's role in cosmic motion; it is in fact asserted only once, in his late *Commentary on the Celestial Hierarchy*, where it is carefully segregated from the discussion of the Christian author's views and not intruded into the religious themes developed in the rest of the book.

The 'waters above the firmament' were from patristic times onwards a difficult *locus* for commentators on Genesis whose world-view incorporated much that was derived from the Greeks. It is with this classic difficulty concerning the harmonization of two different conceptions of the physical system that Grosseteste begins his major discussion of the nature of the heavens in his hexaemeral commentary. He works through the difficult cosmological question towards the affirmation of a principle whose force is felt in all that follows: the principle that the Creator's omnipotence far transcends the limits of the human mind (fols. 204^a-205^b).

Against those who set themselves to prove the impossibility of there being waters above the starry heaven, since as a weighty element water has a natural place deputed to it, Grosseteste advances a number of explanations, some of them as old as Augustine's time, some inspired by more recent learning, but none of which is of direct interest to the present discussion. Ending his review, Grosseteste quotes from St. Ambrose the judgement that since the word of God is the

---

[112] 'Hanc rationem motus celi multis modis pro nichilo reputamus, quia nec hoc conceditur a plerisque auctoribus nostris, quod intelligencia moveat celum.' (*Hexaemeron*, fol. 199^d).

origin of nature, God's title to legislate for creation is absolute.[113] It should arouse no surprise that God's power and might should suspend the waters above the firmament of heaven, and the creature has no more right to ask why he should have done it in this way rather than that, than it has to inquire why he divided the waters of the Dead Sea to allow the Israelites escape, when he could have delivered them in other wise. Nothing lies outside the power of him who gives power to whom he pleases.[114] Grosseteste maintains that the recitation of the various opinions has been worthwhile. Even though they partly exclude each other and none rallies full consent, they represent several ways in which the author of the Bible can be defended against attack; and the Bible, when reporting the free actions of God both in nature and in history, is of greater authority than the capabilities of the human mind, as Augustine reminds us.

Grosseteste now begins his investigation of the heavens' nature with the significant declaration that though so many thinkers have studied '*scrupulosissime*' their nature and number, he doubts whether any has come upon the truth; if anyone has, he has not, apparently, been able to offer convincing reasons in its support.[115] He finds a good example of the mutual contradictions of the philosophers in the question concerning the substance of the heavens (ethereal fire, according to Plato, the quintessence according to Aristotle), at whose treatment we have already looked. Grosseteste wishes to avoid prolixity and tediousness by sparing his reader all the controversies and arguments involved, and he passes on to consider whether the heavens are animate or inanimate and, if the former, whether they are endowed with a single World-Soul, as Plato maintained, or with many, as did Aristotle (as Grosseteste

---

[113] 'Sed quid in hiis querimus naturam, cum dicat Ambrosius, cum sermo Dei ortus naturae sit, iure usurpat legem dare nature, qui originem dedit.' (fol. 204^b).

[114] Grosseteste's conclusion is not very convincing, even when taken on its own terms, for the idea of intelligible natural law had gained a strong hold on his mind, forcing him to regard the waters above the heavens as a kind of permanent miracle, 'Contra et supra solitum nature cursum', as he says. This type of solution could not stand for long against the Scholastics' demand for rationality; cf. St. Thomas Aquinas, *Summa Theol.*, I, q. 68, a. 3.

[115] 'Multi scrupulosissime investigaverunt, sed nescio an aliqui veritatem [fol. 204^d] invenerunt, aut si forte invenerint, nescio an eorum aliqui se invenisse veritatem veraci et certa ratione deprehenderint. Scribunt enim super hiis philosophi sibi invicem contraria.'

understood him, that is to say). Having examined the opinions, he draws attention to their variety and uncertainty, and excuses himself from offering a solution to a problem for which neither philosopher nor Church Father could bring forward a definitive answer.[116]

The third topic concerns the number and motions of the heavens, and this one interests him more, although he confesses from the start that he has no answer to it; he could cite multiple opinions both of astronomers and of natural philosophers, but there is not one of them that he can conclusively verify or refute, for *'non nisi ambiguitatem nobis relinquunt'*, he complains. For who knows whether the fixed stars and planets move in paths of cycles and epicycles under a motionless firmament, as Ptolemy maintains, when, after all, his entire mathematical account of celestial movements remains an intelligible description even if the spheres do not move?[117] Grosseteste announces casually that it would not be difficult to suppose that a conscious power, or even a bodily agent, moves the stars along the circuits determined by Ptolemy or by his predecessors and successors on the basis of 'observation with instruments': St. Augustine was well aware that the stars could make their circuits even if the firmament stands still.[118]

Uncertainty extends further still. The widespread view fixing the number of the spheres at no more than nine (one for each planet, then the stellar sphere and the anastral sphere), which was used in a fragile attempt to support the traditional number of the angelic choirs,[119] inspires no confidence. If indeed the stars do not have a motion of their own but are borne

---

[116] A form of what we may term literary humility is at work here, for within a year or two Grosseteste felt himself in a position to rule out animation-theories and pronounce the heavens to be purely material bodies, moved by angels (1239–40, in his commentary on the *Celestial Hierarchy*).

[117] 'Quis enim scit an non moto firmamento . . . moveantur stelle fixe et planete secundum vias circulorum et epiciclorum, quemadmodum videtur sentire Tholomeus de motibus eorum? Totum enim quod ipse dixit et demonstrare se credidit de siderum motibus imaginari potest, absque motibus celorum, licet ipse ponat firmamentum eniti contra stellas, et stellas contra firmamentum.' (fol. 205ª).

[118] 'Non enim dificile est ponera aliquam virtutem intellectivam, vel virtutem etiam corporalem, que stellas moveat per illos circuitus, quos experimentis et instrumentis adinvenit Tholemeus, sive astronomi qui ipsum precesserunt, sive qui ipsum subsecuti sunt. Unde Augustinus ait, si autem stat firmamentum, nihil impedit moveri et circuire sidera.' (ibid.).

[119] '. . . ex quorum numero quidam credunt se probare numerum ordinum anglicorum' (*sic*: the Bodleian MS is of English provenance!).

by the heavens, it has a certain validity. On the other hand, the fixed stars have a motion of accession and recession that is not accounted for by the simple diurnal rotation imparted by the first moving heaven, so that the hypothesis of an anastral sphere has had to be invoked: can we be certain, then, that the compound motion of each planet does not require one or more heavens beyond that which actually bears it, in order to explain its variations? The consequence would be a multiplication of the number of heavens admitted by science. In any case, comments Grosseteste in another of his throwaway asides, no planetary orbit has so far been observed throughout its whole course. The implication of this remark is, clearly, that the empirical basis required for such speculations has not been adequately laid.[120]

But more is yet to come. No one can give a reason, remarks Grosseteste, why there cannot be numerous planets which we cannot see, but whose activity is in fact necessary to sustain the balance of nature, that is to say of growth, in the lower world. When one considers that the stars which constitute a galaxy are optically indistinguishable, yet are all required for the effects they produce on earth (otherwise why would God have created them?), it is manifest that only by a divine revelation could we learn whether or not there are numerous planets which the human eye cannot discern, each adding its sphere to the number of heavens which the unaided reason postulates.[121]

There can be but one conclusion: no one can profess to have the truth about the nature, number, motion or motors

---

[120] 'Praeterea, sicut secundum ipsos celum applanon et anastron . . . necessarium est ut moveat inferiores celos et stellas eorum motu diurno . . . sic forte cum quilibet planeta plurimos habeat motus, et forte plures quam adhuc sint deprehensi, quia nullius eorum motus adhuc plene deprehensi sunt, eget quilibet planeta preter proprium celum in quo est [fol. 205^b] alio celo, vel aliis celis, qui suis motibus eius varios motus efficiant.' (I have interchanged the order of this argument and the following one for purposes of clear exposition).

[121] 'Sed unde scietur quod non sint plures stelle erratice nobis invisibiles, generacioni tamen in inferiori mundo necessarie et utiles? Dicunt enim philosophi galaxiam esse ex stellis minutis fixis nobis invisibilibus. Unde igitur sciri posset, nisi divina revelacione, an non sint plurime *huiusmodi* stelle invisibiles nobis, quorum quelibet suum habeat celum movens ipsam ad profectum generacionis in mundo inferiori? Stelle enim que galaxiam constituunt licet indistinguibiles sint secundum visum, non carent effectu generacionis et profectus in mundo inferiori.' (fol. 205^a). I interpret 'huiusmodi' as referring back to the erratic stars, or planets; it can scarcely be thought to apply to the fixed stars.

of the heavens. The philosophers of this world are foolish boasters if they think they possess the truth, for their fine-spun reasonings are more fragile than spiders' webs.[122] It is a pity we do not know these things, but we can at least be grateful for the divine voice which gives us the final cause of the firmament, namely, to divide the waters above from those below. And besides, there are things which nobody doubts, such as that the firmament with its stars affords a great help to generation and growth; for of course the light of the heavenly bodies favours vital heat. In the following folio too, and without qualifying his scepticism with regard to much of the current science of the heavens, Grosseteste feels able to describe certain properties of the heavens which he considers certainly to belong to them, and others which, though less certain, are supposed true by the commentators on Genesis and made the basis of spiritual interpretations. He expresses the hope that his *'lector parvulus'* will find plenty of material of this type both here and in other chapters (on the earth, sun, moon, and stars).

Taken all together, these pronouncements of Grosseteste on the heavens, despite, or even mostly because of their largely negative or sceptical character, merit a place of some importance in the pages of thirteenth-century philosophy. That it is in his biblical commentary rather than in some purely scientific work that they are voiced might suggest that they are not to be treated with full seriousness, being perhaps the effort of a believer to divert his readers from too much *curiositas* about nature and to recall them to the purity of revealed truth; but the objection thus formulated has its own presuppositions, for its implication that hostility or indifference determine the believer's attitude to science is largely anachronistic. Grosseteste's utterances are sincere, as I think their coherence with his properly scientific writings proves. For more than thirty years he had read all the books he could find on astronomy and natural questions about the heavens; he had studied, compared, revised, and adapted their contents, but at the end of it all he remained unsatisfied. The contradictions existing

---

[122] 'Cum igitur haec ita se habent, nullus potest de numero celorum, aut eorum motibus aut motoribus aut ipsorum naturis aliquid certum profiteri, licet de talium scientia inaniter se iactent mundani philosophi; ratiocinationes enim quas de hiis contexerunt aranearum telis fragiliores existunt.'

between the chief rival conceptions, those of Ptolemy and Aristotle, were too profound to be levelled, yet neither taken separately satisfied strictly scientific criteria of truth. What comes to expression in this passage of the *Hexaemeron* bears little or no relation to latter-day Christian fundamentalism, with its unwillingness to budge beyond the letter of the biblical text. It is the frustration of a well-informed and highly-intelligent critic of scientific hypotheses, who is criticizing from a position within the scientific movement of his own times, and allying his methodological criticisms with the demand of reason itself for coherence among the various truths which it accepts from different sources. Grosseteste, it may be said, longed for the discovery of a system of the world-machine which would satisfy all the criteria of truth. He admitted that the Bible has given only precious hints concerning the nature of the heavens, and he considered that science must consequently be constructed by what he thought of as a very fallible human *ingenium* on the evidence of the senses, with their inherent and, as it must have seemed to him, insuperable limitations. We did, it is true, detect a form of literary humility —the property of a genre—at work in the earlier part of the chapter; but the contrast which its expression makes with the reasoned criticism which follows it is too evident to allow us to take it at face value.

One re-reads the propositions in which Grosseteste's treatment issues with a sense of astonishment that increases on the way. His rejection of lingering traces of divinity in the Greek view of the heavens is expected, but is not on that account of negligible value. It forms part of a long campaign that began early in Christian intellectual history, as the beliefs in divine transcendence and immanence, and in a temporal creation, went to war against Greek polytheism and pantheism, and effected the desacralization of the heavens, by ridding them of the pantheon of Greek philosophical belief. Grosseteste's criticism of the Aristotelean deduction of circular heavenly motion goes in this direction; it is a substitution, not simply of one teleology for another, but of a Christian theology for a pagan one. At the heart of the conception according to which the material cosmic processes serve the divinization of a spirit, there lies the notion of a necessity which constrains any possible divine freedom: the whole interlocking system of matter and divinized intelligences simply must be so and could not be otherwise. Grosseteste's faith makes it natural for him to ridicule the pagan idea and to begin reflection, not with matter, but with the unconstrained freedom of the Creator. God freely decided to create spirit and matter, the latter to serve the former. The modalities of that service possess, on this view, a merely instrumental importance, being interesting as objects of speculation, but not central. God can freely choose the means to turn the stars; we can be sure only that, whatever arrangement he has devised and implemented, it is one which adequately secures the conditions required for the life of embodied spirits on the earth's surface. Grosseteste's biblical teleology commits him to no single hypothesis of celestial mechanics, indeed it ensures his distance from all those known to him from scientific authorities. He himself draws the consequences of divine transcendence: the intelligences may or may not be one or many; the stars need not necessarily be physically anchored in the bodies of the spheres, they could be moved in free orbit by an angel, or just as easily by a *virtus corporalis*, a material power, should God so will. In the event, Grosseteste was to opt several years later for what seemed to him to be on balance the simplest solution (that angels move inanimate material spheres carrying the stars); but with what reservations about its truth we can well guess on the basis of his remarks in the *Hexaemeron*. Certainly he never risked inflating the speculation with a religious value which it simply did not possess. His reminder that the hypothesis is to be distinguished from religious belief found echoes till the end of the Middle Ages and beyond, though more slumbrous spirits than his made a virtue out of oft-repeated habit and seemed to regard the angel's cosmic functions as a religious truth.

In the course of his discussion in the *Hexaemeron* we have seen Grosseteste advert to the possibility that the motion of the heavens may be due to a bodily or physical cause. It might prove of interest to trace the fate of this idea in the thirteenth-century schools, for we find St. Bonaventure taking it very seriously indeed. In inquiring whether the motion of the heaven derives from its own form or from an intelligence, the Franciscan Doctor pronounces the first to be '*modus satis catholicus*', that is, acceptable to the theologian. The connection of the idea with the tradition of light-speculation becomes clear from the nature of one of his arguments, to the effect that the power which belongs essentially to light as a self-propagating agency could be postulated to be the moving-power of the heavens, a hypothesis which would render the usual appeal to spiritual movers redundant.[123] The defenders of the theory, he admits in his summing-up of the question, can answer the physical objections to it easily enough, although he is inclined to doubt whether they can account for the eschatological station of the heaven without making it into a form of violent motion, since they invoke a special *influxus divinae potentiae* to bring the heavenly revolutions to rest. Bonaventure himself thinks it simpler to assign the governance of celestial motion to the angels. The terms in which he speaks imply, however, that the physical hypothesis has already won support and been defended. My impression is that some author or authors had developed the idea in the intervening period of twenty or so years between Grosseteste's *Hexaemeron* and Bonaventure's *Sentences*.

However that may be, one finds the elements of the physical hypothesis present in successive stages of development in Grosseteste's writings. In the *De Luce* already (*c*.1226), though the first sphere admittedly receives motion from an incorporeal intelligence or soul, it is by participation in the form of the first that the lower spheres are moved in the diurnal rotation. In a passage occurring in the *Hexaemeron* (and already adverted to), there is envisaged the full possibility that heavenly motion could be caused by the light of the first sphere even without any angelic mediation. Grosseteste wishes to show that rest is a better imitation of divinity than is motion, and he illustrates his thesis with the following consideration. If the first heaven is kept always in a state of rest, '*sive ab angelo, sive a natura propria*', and transmits something of its light-power to act on the lower spheres, its active state of rest is more like eternity, the fullness of divine being, than any form of motion, even circular, can be.[124] This is sufficient to confirm

---

[123] *II Sent.*, d. 14, p. 1, a. 3, q. 2, opp. 3. (*Opera Omnia* vol. II, 347). It is in fact incorrectly termed an objection, for Bonaventure admits both hypotheses, although he prefers the second.

[124] 'Quapropter si celum istud primum perpetuam habeat quietem, sive ab angelo sive a natura propria, et quietum manens aliquid de virtute luminis sui transfundat in inferiora et agat in illa, multo verius assimilabitur statui eternitatis per huiusmodi actuosam quietem, quam per aliquem modum motionis' (*Hexae-*

that the later side-reference to a possible material cause of celestial motion is not a thoughtless fantasy but a well-considered possibility unfolded from his central cosmological positions.

The idea thus tentatively launched found, perhaps, few supporters during the thirteenth century, but its sophisticated restatement by Jean Buridan has been recognised by contemporary historians of scholastic mechanics as an important attack on Aristotle's presuppositions.[125] Buridan was doubtlessly original in several respects in applying a theory of impetus to celestial motion, but two of his basic starting-points had already been coupled by Grosseteste a century previously: angel-movers are no part of religious belief; and none of the traditional philosophical hypotheses has been able to establish itself with conviction, wherefore the mind is free to explore and speculate in this domain.[126]

The scepticism which emerges in the purely scientific aspects of Grosseteste's criticism had, of course, scientific rather than theological grounds. It was born of his conflicting loyalties to two mutually incompatible accounts of celestial motion, the physical, and the geometric or astronomical. His critical remarks, however, went behind both accounts to reveal the limits of what passed for knowledge in the scientific circles of his time. He pointed out that the calculation of the number of the spheres represented a series of speculative guesses, none of which stood up to criticism. He insisted that theories depended upon observations, and that the latter were only as good as the eyes that recorded them; moreover, there was no principle that obliged God to make all features of his creation evident to the senses, and there might, therefore, be heavens which no one would ever see.

Grosseteste's second major criticism was of a less *a priori* character. He remarked that the observations that had been made were lacunary in every case. Since, in the history of astronomy, not a single planet had been observed throughout

meron, fol. 199ᵈ), since God, as Grosseteste adds, 'Stabilis manens dat cuncta moveri' (Boethius, *De Consolatione Philosophiae* III, 9).

[125] Maier, *Zwei Grundprobleme der scholastischen Naturphilosophie*, pp. 211 ff; Crombie, *From Augustine to Galileo*, vol. II, pp. 80–86; Beaujouan, 'Medieval Science in the Christian West', p. 509.
[126] See the texts published by Maier, op. cit. pp. 211–2, 213–14, 223, and translated by Crombie, op. cit. pp. 82–3, 85.

its orbit, whole segments of the motion of the planets might, for all anyone knew, have gone unremarked. This complaint was to be repeated two centuries later by Tycho Brahe, who himself took the first steps to rectify the situation.[127]

Now while it is the case that Grosseteste's criticisms reveal his grasp of scientific and philosophical principles, there is no denying that they also presuppose a world-view whose distance from the Greeks is the whole breadth of revealed religion. The adage *'omnia propter hominem'* implies a functional view of the heavens' turning: the needs of humanity as envisaged by the Creator determine the number of heavenly bodies made. Looked on from this religious perspective, the limitations of astronomical speculation are evident, and Grosseteste is capable of formulating those limitations in terms which the scientist would have to acknowledge and accept, once stated. The point is, that from within other possible world-views they might not have been stated at all. This is a consideration of some importance and one to which I shall return in the following chapter (section ii).

d) Valuational heliocentrism

T. S. Kuhn numbers among the non-astronomical factors giving rise to the Copernican revolution the need for astronomically-based methods of navigation which the voyages of discovery created,[128] the demand for a reform of the calendar,[129] and the widespread belief that the sun is the source of all vital principles and forces in the universe.[130] However correct he may be in tracing this last factor to Neoplatonic and not to Aristotelean currents of thought, it is undeniable that its rise

[127] Kuhn (*The Copernican Revolution*, p. 200), describes Brahe's innovation in the following words: 'Most important of all, he began the practice of making regular observations of planets as they moved through the heavens rather than observing them only when in some particularly favourable configuration'; and p. 201, 'His observations provided a new statement of the problem of the planets, and that new statement was a prerequisite of the problem's solution. No planetary theory could have reconciled the data employed by Copernicus.'

[128] See, however, the wise cautions of Beaujouan on this subject, 'Medieval Science in the Christian West', p. 525.

[129] Copernicus's complaint, that the manifold uncertainties surrounding the science of planetary motion rendered impossible an accurate estimate of the constant length of the seasonal year (Preface to *De Revolutionibus*, printed in Kuhn, p. 137), echoed that of medieval reformers in a line whose origin is traceable back to Grosseteste; see *Computus Correctorius*, ed. Steele, pp. 215-18, where the various methods of computation are outlined.

[130] *The Copernican Revolution*, pp. 124-32.

must be placed somewhat further back than the Renaissance humanism in which Kuhn finds it expressed. The theme of the sun's cosmic centrality finds, in fact, its natural place in our present study of the non-Aristotelean elements in Grosseteste's cosmological thought. It is a natural and evident consequence of his metaphysical views concerning light, one to which later humanists contributed nothing essentially new, and which Grosseteste himself regarded as a pure extension of patristic thought, the thought of Ambrose, Augustine, Basil, and Chrysostom (i.e., Pseudo-Chrysostom, Scottus Eriugena, author of the sermon, *Vox Aquilae*).

Among the reasons given by Kepler for preferring the heliocentric hypothesis to the traditional systems of the heavens, is the following:

Of all the bodies in the universe the most excellent is the sun, whose whole essence is nothing else than the purest light, than which there is no greater star; which singly and alone is the producer, conserver, and warmer of all things; it is a fountain of light, rich in fruitful heat, most fair, limpid and pure to the sight, the source of vision, portrayer of all colours, though himself empty of colour, called king of the planets for his motion, heart of the world for his power, its eye for his beauty, and which alone we should judge worthy of the Most High God, should he be pleased with a material domicile and choose a place in which to dwell with the blessed angels ... For if the Germans elect him as Caesar who has most power in the whole empire, who would hesitate to confer the votes of the celestial motions on him who already has been administering all other movements and changes by the benefit of the light which is entirely his possession? . . . Since, therefore, it does not befit the first mover to be diffused throughout an orbit, but rather to proceed from one certain principle, and as it were, point, no part of the world, and no star, accounts itself worthy of such a great honour; hence by the highest right we return to the sun, who alone appears, by virtue of his dignity and power, suited for this motive duty and worthy to become the home of God himself, not to say the first mover.[131]

With the exception of physical centrality, all the items in this list of the sun's attributes are to be found in some form or fashion in the *De Operationibus Solis* of Grosseteste, in which Neoplatonic axioms of unity are made to yield the sun's central causal action in the production of worldly phenomena.[132]

---

[131] This passage, taken from a fragment of an early disputation on the motion of the earth, is translated by Burtt (*The Metaphysical Foundations of Modern Physical Science*, 2nd. ed., London, 1932, p. 48) from Kepler's *Opera Omnia*, ed. Frisch, (Frankfurt and Erlangen, 1858 ff.), vol. VIII, p. 688. See Kuhn, op. cit., p. 130.

[132] The same ideas are present in part in Grosseteste's *De Calore Solis* and in

Fountain of all light, the sun is generous in distributing light to the moon and the stars (§ 29); it is the principle and heart of all heat (§ 15), the purest and loveliest of things and *'specialiter opus excelsi'* (§ 12), the cause of all colour, the source of visibility and of the visible spirits in the eye (§ § 6, 7, 11); it has the simplest motion of any planet (§ 32), is the chief begetter of corporeal forms, which are impressed in its annual cycle, and it controls growth (§ § 9, 12). The classic metaphors for the sun (heart and eye of the world), traceable to ancient roots and present throughout the patristic tradition, occur in both authors, Grosseteste and Kepler. Interestingly too, the idea that the sun is the begetter of all 'motions and changes' retains in Kepler's text the broad Aristotelean sense of *motus* found in Grosseteste,[133] while its context looks forward to Kepler's later, more specific view that the sun is the *anima motrix* whose forceful light-rays must account for the velocities of all planets in their own orbits.[134]

There can be no doubt that it was the continuous tradition of light-symbolism and light-metaphysics that can be traced in Christian history through patristic and medieval times to the Renaissance, which allowed Copernicus himself to present his novel system of the heavens as being in one respect acceptably conservative, that is to say, as a physical theory fully in harmony with the central place occupied by the sun in traditional thought. In the High Middle Ages it was Grosseteste who took the initiative and stimulated what we may call a valuational heliocentrism, founded not merely upon symbolism, but upon the whole nature of reality as he portrayed it in metaphysical terms. The scholastic metaphysics of light, of which he was the greatest and best-known exponent, was popularized by Pico della Mirandola, Marsilio Ficino, Pietro Pomponazzi, and

---

the passage on the goodness of the sun in the *Hexaemeron*, fol. 217[c–d]. The references which follow are given to my edition, published in *Rech. Théol. anc. méd.* 41 (1974), 38-91.

[133] 'Emissio namque radiorum solis plurimorum motuum causa esse videtur', *De Operationibus Solis* (§ 25). See also Grosseteste's brief treatise, *De Motu Corporali et Luce* (ed. Baur, p. 92), which ends thus: 'Et in hoc patet, quod motio corporalis est vis multiplicativa lucis.'

[134] Kuhn, op. cit., p. 214. It should not be forgotten that for William Gilbert also the sun was the 'chief inciter of action in nature' and the strongest magnet in the gravitational system of the universe; see Butterfield, *The Origins of Modern Science*, p. 128.

Francesco Patrizzi, to find its way into the circles of thought out of which modern astronomy came forth.[135]

e) Science and aesthetics

Grosseteste's metaphysical aesthetics have been well studied[136] and the generality of his ideas on beauty is of no direct relevance here.[137] There is, however, a connection between his aesthetics and his philosophy of nature which is worthy of mention, because it is something of a novelty and constitutes another aspect of his distinctive contribution to the origins of science.

If the ideas on aesthetics which underlie the *De Luce* are of a predominantly musical character, based on a Boethian and Augustinian arithmetical notion of proportion, there are already in that work intimations of another point of departure for the consideration of beauty in itself. The *species et perfectio* of bodies universally is said to be *lux*, and the essence of light is simplicity: arithmetical unity represents the unity of first corporeal form in itself, and is a transcription of metaphysical identity or simplicity. The metaphysics of *De Luce* evolved under the aegis of a notion of perfection, not of beauty; but perfection itself is associated indistinguishably with simplicity and purity of nature in a way that strongly suggests an influence of St. Basil's *Hexaemeron* and that looks forward to Grosseteste's later and more self-conscious effort

[135] For the history of light-speculation during the Renaissance period see Goldhammer, 'Lichtsymbolik in philosophischer Weltanschauung, Mystik und Theosophie vom 15. bis zum 17. Jahrhundert', in *Studia Generalia* 13 (1960) 11, pp. 671-5; compare with Koch, 'Über die Lichtsymbolik im Bereich der Philossophie und der Mystik des Mittelalters', 653-70 in the same number; Mazzeo, 'Light metaphysics, Dante's *Convivio* and the letter to Can Grande Della Scala', in *Traditio* 14 (1958), 191-229 (prints texts of Ficino); Kristeller, *Eight Philosophers of the Italian Renaissance* (California, 1964), ch. 7: 'Patrizzi', pp. 110-26; Maechling, 'The Metaphysics of Light in the Natural Philosophy of Francesco Patrizzi da Cherso (1529-1597)' (unpublished dissertation for M.Ph., University of London, Warburg Institute, 1977); *Le Soleil à la Renaissance*, Colloque international (Université Libre de Bruxelles, 1965), 584 pp.

[136] De Bruyne, 'Grosseteste et l'esthétique mathématique', pp. 72-120 in *Études d'esthétique médiévale*, t. III: *Le XIIIe s.* (Bruges, 1946); *L'esthétique de la lumière*, ibid. pp. 3-29; *L'Esthétique du moyen-âge* (Louvain 1947), pp. 70-9, 80-2; Pouillon, 'La Beauté, propriété transcendentale, chez les scholastiques (1220-1270)', *Arch. Hist. Doctr. Litt. M.A.* (1946), pp. 285-91, and texts on pp. 320-23.

[137] For a very brief summary of Grosseteste's ideas on aesthetics, see McEvoy, *Rech. Théol. anc. méd.* 41 (1974), pp. 53-4.

at harmonizing Greek and Latin aesthetic concepts within the metaphysics of light.

When considering the nature of the firmament, Grosseteste makes his metaphysics issue for the first time in an aesthetic of a geometrical character.[138] Light is *per se* beautiful, rendering beauty a transcendental presence in the material world. Its physical perfection is found in the perfect beauty of the sphere of the first heaven, where the simplicity and unity of the firmament is the highest that can be found in bodily nature, since the composition is of first matter and first form alone. The identity of proportion which establishes harmony between its matter and form manifests itself in the equality of every part with the whole and in the uniformity of its motion.

Light, the form and beauty of heaven, generates or propagates itself (beauty and self-diffusive goodness being only formally distinct, whether in the first creative light or in his creation), and from every point directs radial lines of light-energy to the earth at its centre. The concurrence of these lines from all sides towards the centre of the perfect geometric figure 'composes', says Grosseteste, 'a texture of indescribable beauty' that is the most revealing symbol of divine glory in the material world, for in it the invisible reveals itself to our understanding.

With this passage we associate the references to the '*mira-bilia*' and the '*pulcherrima*', Grosseteste's descriptions of the geometric operations of light-phenomena, such as the rainbow, whose wonderful aspect consists in the effects of refracted rays.[139]

Grosseteste's assertions concerning the loveliness of the *textura* of light-rays acting in the world as a whole are the expressions of an aesthetic attitude to nature which goes through and beyond the purely artistic, sensuous delight in colour and splendour—a commonplace in the Middle Ages— to the unqualified beauty of the abstract formal and geometrical simplicity of things, seen when their essence has been grasped in its unity by the *oculus interior*. It is of a piece with his metaphysics of beauty, which finds in the aesthetic object unity and simplicity, identity of part and whole, or, in the graduated outflow of beings, the Beautiful and Good gathering

---

[138] Ibid., pp. 62-5, where the Latin text is printed.
[139] *De Natura Locorum*, ed. Baur., pp. 69.33, 71.6.

all things into unity in response to the mutual complementarity of their appetite and function: *'alioquin non essent unius universi partes'*.[140] The geometrical aesthetics of nature seems to have been Grosseteste's own discovery (it is almost certain that he had not read Alhazen's *Optics*, its main source in the generation after him), and it cannot be claimed that in its pure form it gained many adherents. For the stricter kind of Aristotelean it could have little meaning, and in any case its roots in the light-metaphysics would have raised the suspicions of all who regarded that current as merely a form of philosophical poetry which debased the currency of language and inflated metaphors into metaphysics. The seventeenth century, with its quasi-mystical reverence for the marvellous, simple and abstract geometry of things, would have its own impatient, visionary answer to Aristotelean prose.

[140] Pouillon, art. cit., p. 321 (the text in question comes from Grosseteste's *Commentary on the Divine Names*).

Chapter 2
Grosseteste's Place in the History of Science

I EXPERIMENTER, MATHEMATICIAN, OR METAPHYSICIAN?

SOME British historians of the later nineteenth century thought to have discovered in the figure of Roger Bacon—a prophetic, heroic, and persecuted individual as they depicted him—an outstanding harbinger of modern empirical science; they displayed in some regards a *naïveté* quite in proportion to Bacon's own credulity on many points. The gradual realization of Bacon's manifold dependence on the ideas of Grosseteste, for which the main credit was due to Baur,[1] has tended to displace the interest of twentieth-century historians from the *Zauberlehrling* back to his reputed master.

The arguments advanced by scholars for regarding Grosseteste as a founder of modern science have differed. Baur himself saw the essence of his scientific greatness in his attempt to ground the philosophy of nature upon mathematics and experiment and hence to introduce an empirical interest into medieval Scholasticism.[2] The chief and best-known defence of Grosseteste's claim to scientific greatness, however, has been set out by Dr Alastair Crombie in his scholarly work, *Robert Grosseteste and the Origins of Experimental Science 1100–1700*. To his central thesis, which is that the experimental method devised and accepted within Grosseteste's school at Oxford marks the beginning of the modern tradition of experimental science,[3] Crombie added a rider, to the effect that the set of methodological procedures so evolved was applied with a degree of success to the problems of optics.[4]

Crombie's valuable and interesting work is yet another striking proof that there can be no history-writing without presuppositions. In the history of science, above all, beyond

[1] Baur, 'Der Einfluss des Robert Grosseteste auf die wissenschaftliche Richtung des Roger Bacon', in *Roger Bacon, Essays* . . . ed. Little (Oxford, 1914), pp. 33–54.
[2] Baur, *Die Philosophie* . . ., pp. VII, 92–3.
[3] Preface, p. vii; p. 1.
[4] Ibid., p. 14, and ch. V, 'Mathematical Physics' (pp. 91–127).

properly historical presuppositions are philosophical ones too, concerning the very nature of scientific inquiry. On the one hand, Crombie holds unswervingly to Duhem's view that medieval and modern science represent an unbroken continuity, and on the other he advances the proposition that logical and methodological revolutions must be held to be the real initiators and the decisive factors in scientific progress. Now the view for which Duhem is famous, and to the demonstration of which he devoted his vast historical erudition, is open to the objection that it rests upon a conventionalist philosophy of science, i.e., it represents the conviction that science cannot and should not hope to be anything more than an approximate ordering of phenomena in function of non-hypothetical principles, whose justification lies precisely in the complex process of their historical maturation; and Duhem failed to rally the majority of philosophers of science to his support. Equally, Crombie's principle favouring methodological innovations is open to the theoretical objection that in scientific progress it is advances in actual methods of inquiry rather than in abstract methodology that are of determinant value, and to the historical objection, that if a broadly adequate methodology be regarded as a necessary and quasi-sufficient basis for scientific advance, then it becomes at once essential and at the same time embarrassingly difficult to account for the relative scientific sterility of the late medieval and early Renaissance period, if one grants—and this is the force of Crombie's main thesis—that the requisite methodology was already available from around 1240 onwards.[5]

These lines of criticism lead us to consider first the validity of Crombie's subsidiary thesis, namely, that Grosseteste made a direct scientific contribution of some value when he applied his theoretical principles of methodology to geometrical optics, using experiment and observation to verify or falsify his hypotheses.[6] Now it is true that Crombie himself is acutely aware of the limitations of medieval natural philosophy in the domain of observation and experiment;[7] yet he maintains

---

[5] Koyré makes a number of pertinent criticisms of Crombie's work in an article already referred to, in *Diogenes* 4 (1957), pp. 421-48.

[6] Dales agrees ('Robert Grosseteste's scientific works', p. 382): 'I have found the role of experiment in Grosseteste's scientific works to be somewhat more extensive than was described by Dr Crombie.'

[7] *Augustine to Galileo*, vol. 2, pp. 126-9 contains excellent remarks *à propos*.

that experimentalism in something of a modern sense was introduced and practised by Grosseteste as a central element in his science of nature. However, when the wider context (linguistic, literary, and historical) of the *experimenta* to which Grosseteste appeals in his works is examined in more detail, the impression of modernity evaporates without trace. Eastwood has no difficulty in showing that in Grosseteste's optical treatises 'experiment' means 'experience', in a sense at once perfectly Aristotelean and unashamedly commonsense, but bearing no relationship to the controlled experiment which was devised in later times, and which included as its essential component the careful, methodical segregation from eclectic and un-selfcritical experience which is implied in its very notion.[8] Furthermore, many of what Grosseteste refers to as '*experimenta*' are drawn directly from his Greek and Arabic authorities, with the assumption that human experience is for all practical purposes a constant.[9] The totally undifferentiated state of the notions involved is well illustrated by Grosseteste's repeated remark in his *Commentary on the Celestial Hierarchy*, that the good angels, unlike humanity, have no 'experimental science' of sin, but do possess '*scientia experimentalis*' of the divine essence, which we in the state of pilgrimage lack.

Naturally enough, the failure of an impressive attempt to convince other scholars that Grosseteste occupied a central place in the emergence of experimental science has led to a major reassessment of the nature of Grosseteste's contribution to the natural philosophy and science of his own time, and even, most recently, to the expression of a distinct scepticism as to whether he made any scientific contribution at all.[10]

[8] Eastwood, 'Medieval Empiricism. The Case of Robert Grosseteste's Optics', in *Speculum* 43 (1968), 306–21, p. 321; 'Robert Grosseteste's Theory of the Rainbow. A Chapter in the History of Non-experimental Science', in *Arch. Intern. Hist. Sciences* 19 (1966), 313–32, p. 331.

[9] A good example of the coincidence of meaning of '*experientia*' and '*experimentum*' with authority occurs in Grosseteste's *Notes on the Physics* (p. 42.23), where Grosseteste glosses Aristotle's conclusion, 'Casualia esse quae nec semper nec frequenter uno modo accidunt', with the remark, 'Quod autem talia sint quae nec semper, nec frequenter uno modo accidunt, docet sensus et experientia, et est propositio et principium experimentale sensibile.'

[10] Eastwood allows that Grosseteste's theory of the rainbow shows a modest advance beyond the sources available to him. Beaujouan states that 'though the actual discoveries of Grosseteste were therefore negligible, we must remember that without him there might never have been an Oxford School'. ('Medieval Science in the Christian West', p. 491).

While admirers of Grosseteste may feel that such negative estimates as these are too soberly demythologizing to represent the full truth of the matter, they in turn have to struggle with the workings of an inferiority complex that affects all historians of medieval science *vis-à-vis* their counterparts in the ancient classical or early modern periods. A healthy self-criticism on the side of the interested parties is called for, if the repeated alternation of the inflation and devaluation of the medieval scientific tradition is to be avoided in the future.

As it happens, the criticisms made of Crombie's thesis, far from putting an end to the discussion, have in some respects established it on a broader base. This is doubtless a tribute to the stimulus which Crombie's work gave to Grosseteste-studies.

When Fr Alessio's main assertion has been separated from the tissue of observations, analogies, and parallels that cloak it from view, it reads somewhat as follows: the Creator envisaged by the author of the *De Luce* was a geometer who constructed his universe by simple and economical mathematical means that dispensed him from further intervention in the working of the world-machine.[11] Whether this perfectly acceptable contention is compatible in strict logic with Alessio's other belief, that in Grosseteste's eyes the mathematical idea of the world expressed in his masterpiece was meant to be nothing more than a non-realistic, symbolic or conceptual fiction, is a matter of some doubt. I at least can find little worth in the second idea.[12]

[11] 'Storia e teoria nel pensiero scientifico di Roberto Grossatesta', in *Riv. Crit. Stor. Fil.* 12 (1957), 251-92; p. 273: 'In effetti il Dio creatore del *De Luce* è un grande geometra euclideo che si limita a porre un solo assioma all'atto della creazione.' The substance of Alessio's contention is to be found also in the author's *Mito e scienza in Ruggero Bacone* (Milan, 1957), pp. 115-74. See the review of this in *BF* XI, n. 1506. For his criticisms of Crombie, see 'Studi e ricerche su Roberto di Lincoln', in *Riv. Crit. Stor. Fil.* 12 (1957), 231-7.

[12] 'Storia e teoria', pp. 285-6. Alessio seems not to offer evidence from Grosseteste's texts in support. Indeed, this criticism could be extended to much of his theorization. There is no evidence that the mature Grosseteste aimed his researches at the practical exploitation of nature through science (pp. 262-3); or that he shared that other Baconian myth, the recapturing of a lost Golden Age of technical science and power (pp. 260-3); or that the fundamental inspiration of his scientific work was Franciscan (pp. 268-9); or that the influence of the *Didascalion* and Grosseteste's 'religious fundamentalism' combined to inflect his science in a 'purely anti-speculative direction'. And surely one can agree that the total conditions of the England of that time combined to produce Oxford and its masters, without yet regarding, say, the *De Luce* as a sort of codicil to the *Magna*

Crombie's tendency was to find both the origins of the scientific revolution and the originality of Robert Grosseteste in a combination of empiricism and logic of procedure. Alexandre Koyré on the other hand would award Grosseteste the palm for making light-metaphysics the foundation of geometrical optics and thereby taking the first step towards the establishment of a mathematical science of nature.[13] Grosseteste's effort to reduce physics to optics may have been premature, his convictions may have won negligible following in the Middle Ages, and optics may have remained marginal to the scientific revolution of the seventeenth century; but for all that it was the Platonic love of mathematics that was to inspire the mathematical science of nature and to turn minds away from unprofitable Aristotelean empiricism.

It will be recognized that this view is in total agreement with the conclusions that have emerged from our analysis of Grosseteste's texts in the previous chapter. It seems to me in no way paradoxical that, although Grosseteste's logical theory was fully Aristotelean and he was neither an experimenter nor yet an observer of distinction (not even in astronomy),[14] and although he was not a great mathematician, he yet took a decisive step in the direction of mathematical science and that he may, indeed, be regarded as in some respects its progenitor. His intuition led him to the conviction that mathematics, far from being an abstraction from aspects of the physically real, is the very internal texture of the natural world, presiding over its coming to be and controlling its functioning; that, in the words of Kepler, '*Ubi materia, ibi geometria*'. Of course, this faith was metaphysical; but then so too was much of the high-level inspiration of scientists in the seventeenth century. It

Carta (p. 262). Alessio's transcription of the matter of *De Luce* is without value (p. 283), and the various parallels he draws between the *De Luce* and Newton's *sensorium*, ether, and the theory of evolution, dubious, to say the least of it.

[13] Koyré, 'Die Ursprünge der modernen Wissenschaft', p. 440: 'Crombie ist durchaus im Recht, wenn er sagt, dass die Lichtmetaphysik Grossetestes, die dieser überdies zur Grundlage der Optik machte, der erste Schritt zur Konstituierung einer mathematischen Naturwissenschaft war. Auch diese Auffassung teile ich durchaus. Denn ich glaube, dass gerade hier die eigentliche Originalität Grossetestes zu suchen ist.'

[14] His calendar is computed from observations based on the Paris meridian, both in his *Kalendarium* and in his *Computus Correctorius*. Duhem supposed that the latter work must have been written while Grosseteste was a student at Paris. A simpler and likelier explanation is that he used tables that were to hand, rather than constructing his own.

was abstract, because the mathematical structure of reality is not given to the senses, but intuited or believed in by the mind. What it afforded was not so much scientific results as delight in the pure understanding of the essence of things, and, what Grosseteste valued most of all, a glimpse beyond the beauty of the harmonious *textura* of things to the mind of the *primus numerator*, the *lux prima et inaccessibilis*.

II GROSSETESTE'S CHRISTIAN FAITH AND HIS CRITICISM OF SCIENTIFIC AUTHORITY

The question as to whether Christian belief has been a favourable, a neutral, or a pernicious factor in the rise of the natural sciences is not one which may be blundered into, nor is it one that can be completely avoided by the historian. Part of its attraction—I abstract from the evident seductive power which it held and holds for controversialists—may perhaps lie in the vertiginous thrill the mind feels in asking it at all: the precipice, the declivities before which the inquirer stands offer a perilous sense of giddy freedom; the wide horizons appearing from the lofty crag give a momentary illusion that the vantage-point offers an unsurpassable view of the contours and folds of the historical landscape. It is only in the safe descent to lower levels that the sense of buoyancy is left behind, the vista is replaced with an ever-changing series of perspectives, the eyes attain a sharper delineation of individual features, and the choice of the advantageous path becomes pressing.

When two such very general phenomena as religious faith and science themselves undergo the processes of time and unfolding, each successive generation can only hope to establish some comparison between relatively impure essences of each of them, and must rest content with an estimate of their relations which will be honest, but in the nature of things not final. The historian in particular will attempt to distinguish such 'essences' in the configurations they assume at different periods (and periodization itself is not absolute but represents a projection of the mind, a likely story, an hypothesis), and will hope to arrive at limited observations of each great form, and of their interweaving by the discourse of time. A serious effort at estimating the relationship between science and faith in the High and Late Middle Ages (where do the latter end and the Renaissance begin—in art, music, science, philosophy, religion, politics?) would have to determine what, not Christianity in general, but medieval Christianity was like, in life and reflection, and what science and its applications represented for that same age. In asking the question more narrowly still: what was the relationship between his religious belief and the specific characteristics of Grosseteste's natural philosophy, it would be similarly inappropriate to begin from an *a priori* idea of the essence of Christianity and superimpose that upon his beliefs, as though there could have been for him some sort of absolute segregation of philosophical and cultural elements on the one hand (like exemplarism or essentialism, with Platonic roots, or arithmology, with Pythagorean origins, or the spirit of other-worldliness, or all the other threads that were woven into the hexaemeral tradition to make up its texture), from pure Christian faith on the other. The historian, with his distance and perspective, may be conscious of the heterogeneous derivation of ideas and inspirations which, for even the cultured believer of a given epoch, all seem essentially bound together.

I shall simplify the question, then, in the hope of reducing it to answerable proportions: did Grosseteste's belief in an omnipotent and good Creator influence those original aspects of his metaphysics of nature which, as has been argued, had a bearing on the emergence of science? It is now up to anyone who pleases to object that the question thus framed is of a leading—or misleading—nature, or to argue that this or that given aspect of Grosseteste's idea of the Creator or creation was not really Christian at all, but something quite different, Greek, or Platonic, or Stoic.

Grosseteste inherited the belief of ages, that the Creator constructed the universe with intelligence and purpose and that the world-machine therefore exhibits number, weight, and measure. That this belief does not produce scientific knowledge and the attitudes that germinate it, is clear; but that other images of the divinity could, on the other hand, prevent the emergence of scientific curiosity is, for me, equally clear. If there were many gods locked in feud over the provinces of the world; if there were but a finite deity unable to bend things to his conception; if the divine omnipotence were conceived of in a way which pulverized the continuity of time into atomic instants, each unrelated to the others, and no prediction were possible from one state to the next . . . and so on.

The conviction that the universe had an absolute beginning with time opened up to Grosseteste a certain imaginative space that an Aristotle or an Averroës did not have. He felt his imagination freed for the attempt to reconstruct the cosmic generation and so come to guess at the shape of the divine command that initiated the whole process. In this he had the very literal help of divine authority in Genesis, which told him what God did first, what he did next, and what last. Now that is a different kind of order—even granted the common medieval belief in simultaneous creation—from the Greek ordering of things, which thought of them as having always been in their natural places. It is an order of timed, tensed events moving in a linear sequence, an order of emergence, in contrast to an order established but ungenerated; it is like the difference between a master-builder, to whom the architect's sketch recalls the building of a house, and the occupant of the third generation who experiences the house as space and material long since differentiated by family tradition for living in, never having been put in the way of thinking that the whole is the result of an assembly of the parts sketched in exploded form in the master-plan. Since the plan Grosseteste possessed (the Genesis narrative and its patristic exposition) was strong on outline sequence, but vague in detail and light on colour, there remained, within fidelity to its contours, a possibility, for the innovating imagination, of replaying the narrative in slow motion and imagining its articulating-points, filling in the gaps and perceiving new relations between the days and nights of the first week. The process really took place; if one pondered the plan with curious love, who knows if one might not just possibly see how it may actually have happened? The medieval imagination was everywhere lively, realistic, and detailed.

An infinite mind was at work in the production of the world; Grosseteste was, as far as I can discover, the first figure in the Judaeo-Christian tradition to find a real corresponding infinity in the world itself. Here we have no need to ask whether or not the influence of his Christian faith was real and direct, for Grosseteste himself tells us that the mathematics of infinite aggregates which programme the expansion of the world could have no meaning for any finite mind and could only have been employed by a mind to whom infinite numbers are as though finite. Without his belief in God's infinity Grosseteste would not have played with the conception of numbers that for a human mind are theoretical and unthinkable, and which he would have looked upon as useless in fact. Granted an infinite mind, however, an element of actual infinity in the creation becomes a thinkable possibility, and something which Aristotle would have rejected on axiomatic grounds becomes for the first time plausible, even in a certain way congruent, as a fuller expression of unbounded creative wisdom and power. Of course, between the bare idea of infinity and the unboundedness, the *indefinitas* of the universe of later thinkers such as Nicholas of Cusa, Giordano Bruno, and René Descartes, there is a world of difference. For Grosseteste, quite certainly, no less than for St. Augustine, order and harmony implied boundedness in the universe; he would have agreed wholeheartedly with the pronouncement Kepler was to make: 'It is not good for the wanderer to stray in that infinity'. The fact is that the fourteenth century recognized the radical nature of Grosseteste's innovation but found no conclusive refutation of it. The axiomatic character of Aristotle's abhorrence of an *ens infinitum in actu* was eroded by Grosseteste's speculation, in a way that even he himself did not grasp, and the repugnance before the infinite and the indefinite was removed by his innovation from the domain of logic to that of psychology.

The novel aspect of Grosseteste's world-system goes back entirely to his conception of God as the great calculator. For the first time, it would appear, in the history of Christian belief, God is addressed as a mathematician whose ideas for creation are mathematical operations realizable in matter and form. On the other hand, this new divine attribute is only an extension of two traditional ones: infinite power and wisdom, the latter referring to the order and intelligibility of his creative action which founds *numerus, pondus et mensura* in the world. A traditional, and indeed a scriptural, idea is then filled with a novel content: novel, because '*numerus*' and '*mensura*' had a more determinate content in the culture of the early thirteenth century than they had had in earlier Christian ages, struggling to master the abacus rules of thumb and reserving their admiration for a symbolic or mythical numerology which offered itself as a relief for the mind (and perhaps for the thumb), a satisfaction gained without the toil of operations, an invitation to bound associations canonized by the reading of the breviary and the ecclesiastical authors of antiquity. If the patristic numerology generated any mental expectation about the nature of divine action, then it might have been that it would reveal a coherence under the regulative concepts of congruence and harmony. The novelty of Grosseteste's idea of God as *Numerator* becomes clear by contrast: this image of the deity generates expectations of an altogether more mathematical, both arithmetical and geometrical, type. This is the kind of expectation that fulfils itself in the world of science, when an inner need for meaning and form sharpens the eye and encourages it to read upon the screen of ideal reality the after-image of the programme that determined the innermost structure of things. In scientific inquiry, evidence turns up to answer inner needs of the questioning mind, if they are insistent enough and sufficiently clear and coherent; for in this respect nature is not parsimonious or ungenerous: she is ample enough to suit different tastes.[15]

It appears to have become something of a dogma among contemporary historians of science that the origins of modern science are Platonic rather than Aristotelean, since, it is widely held, the association of mathematics with the philosophy of nature is a Platonic—or Neoplatonic, or even Neopythagorean —heritage, but not an Aristotelean one. The truth of this assertion, it seems to me, lies in the antithesis it posits between Platonism and Aristoteleanism, and cannot be retained by the thesis alone (i.e., by the assertion, 'Platonism associated mathematics with the philosophy of nature'); the historical reality is more complex than that. It may be true that all Platonists recognized in mathematics the key to natural reality, but their belief would never of itself have advanced natural science in the modern sense by a single hair, had not both terms (mathematics and reality) undergone extensive and independent modification. Two examples of what I mean may be offered. In the *De Musica* of St. Augustine, numbers of a certain kind (Augustine himself refers to them as *numeri progressores, occursores, judiciales*) are represented as holding the secret of

---

[15] See the fascinating (and for the present writer, at least, convincing) reflections on the sources of change in world-pictures and the growth of knowledge contained in Lewis's *The Discarded Image* (Epilogue).

a whole branch of knowledge, namely, how the senses make contact with their objects, and how the soul's judgement makes sense of the flux of sensation. This is a piece of true Platonism as I understand the term, for mathematics is regarded by Augustine as the key to knowledge and thinking, to the inner world and the permanence of the ideas. The effect of this belief upon attitudes to nature and science is, of course, nil. My second example refers to Grosseteste's *De Luce*, where the author demonstrates the perfection of creation by associating natural numbers with aspects of natural reality (matter, form, composition, and the composite) and, by adding together the numbers assigned to each of these, arrives at the number of the world, ten, the perfect number. In the same work he rests the construction of the universe on the mathematics which the infinite mind posited in creating extended material substance. The difference between the two kinds of number-game is salient *and was so for Grosseteste himself*. The second kind responds to a rational and scientific search for the principles of natural extension and action, the first to an aesthetic of arithmological symbolism which is purely Neoplatonic, an inheritance from Boethius, Augustine, and other Fathers of the Church. This one is not of the slightest imaginable relevance to Grosseteste's scientific insight, but is rather something like a canticle sung in an ancient and revered language after the creation's completion, a transcription into conventional notation of the harmonious compact of the cosmos, or an illuminated decoration of the *liber creaturae* which leaves the text unglossed but indulges the scholarly reader's eye. The second consideration on the other hand is sober stuff and supposes criteria, not of symbolic congruence, but of coherence and truth invoked. It has a clear source in the Christian idea of divine infinity, and I think it has no affinity with Neoplatonic mathematical utterances.

The feeling for the unity of the world and for its action as a unified whole was inspired in Grosseteste by his biblical faith that all material reality has a single central function: the service of man. However, its relevance for the history of science is only very indirect: it removed Grosseteste from assumptions that pervaded Greek astronomy and natural philosophy and allowed him a standpoint from which these lost their claim to absolute validity and were relativized. This belief did not stimulate him to produce a rival system of the heavens; it did, on the other hand, encourage him to put his scientific imagination to work on a theory of how vegetable life is formed by the action of the light of the heavenly bodies.

One aspect of Grosseteste's belief in the unity of the universe is of particular importance for the history of science, namely, the conclusion that it is not two irreducibly different kinds of material being that are given in the universe, but one, created from the insertion of a single point of energy into a matter numerically one in the first instant of time. Again it is a Christian belief which Grosseteste is here transcribing into other terms and placing at the foundation of a metaphysics of unity.[16] Here, however, there is a deal of Neo-Platonism at work as well, filtered through the pastristic tradition. Grosseteste was aware that his original philosophical realization of the Genesis narrative ran in this respect against Aristotle, but was in sympathy with the Fathers, and even with Plato.

Grosseteste's religious belief in an anthropocentric, divinely-made plan for nature and history led to a certain secularization or materialization of the Greek heavens; that is to say, it led him to contribute his part to a continuing critique by patristic and medieval theologians of the divinity of the heavens. Reverence for God's jealousy of idols urged him to deny to the spheres and their movers all latent aspirations to divinity of the Greek kind. The negative side of the Christian concept of divine infinity is at work here. The rotation of the spheres he considered to be a pure function of the divine arrangement for the balance of nature; how it is effected was a matter of relative indifference to him, since God is omnipotent and free, and may employ whatever instrumental means he chooses, whether angels or material energy. There was room, he considered, for any number of speculations as to the mode of operation, number and so forth of the heavens, since the actual form which God freely chose to realize in them has not been revealed to us. Observations were so lacunary, he felt, that little or nothing could be said with confidence concerning celestial motion until reliable and more complete data were available. Was he conscious that his Christian scepticism

[16] I recall the wise conclusion of Baeumker: 'Aber bedeutungsvoll bleiben doch zwei Grundmotive [i.e., of the metaphysics of light] : der ästhetische Sinn für die Natur und die Vorstellung von einem die ganze Natur durchziehenden Zusammenhang.' (*Platonismus im Mittelalter*, BGPM 25, H. 1-2, p. 165).

vis-à-vis Greek theories left the way open to progess in astronomy?

If it is demonstrable that in all these ways Christian beliefs lay at the basis of most of Grosseteste's convictions about the world, and that they *de facto* stimulated him to revise whole aspects of the inherited world-picture in a way that was to prove healthy for natural philosophy, and even anticipated some attitudes and frameworks of early modern scientific thinking, might it not none the less be felt that his literal belief in aspects of a rather primitive Semitic world-view (the waters above the heavens, for instance) was just as harmful for his own scientific or rational investigation of the world as it proved itself to be in later centuries? On the whole the reverse is, I consider, the case. The Genesis account of the creation of the world, as expanded by the whole panoply of the hexaemeral tradition, and in general by the Christian view of God, man, and the world, formed a firmly-grounded world-picture whose detailed cosmological tenets were few, and which could accept much supplementation from pagan science. The very tenacity with which the letter of the Bible was held decreed that there would be unavoidable clashes of authorities between certain of its elements and Greco-Arabic philosophy. So long as only one world-picture (the biblical-patristic one) was available to them, medieval scholars not surprisingly showed no initiative in matters of cosmological thought. As it became evident to them, however, that the Bible and Aristotle said different and sometimes contradictory things, then initiatives were called for which could result in something as lovely as the *De Luce*, in which a roughly Greek and Aristotelean world was circumscribed within principles of the higher level of generality implied by a creation-belief. It is difficult to see how otherwise original thought could have begun, if not as a result of clashes between authorities.

The foregoing examination of the role of Christian belief in the genesis of Grosseteste's thought has been too microscopic to encourage historico-philosophical extrapolation to the general question of faith and science. However, I do wish to risk a cautious step beyond it.

It seems to me that no deduction of scientific method is possible whose sole premiss is the divine nature, the Creator. The objections to such procedure on orthodox theological grounds alone would be insuperable, and the only attempt yet made (so far as I know) at offering a deduction of scientific method from the historical Christian belief in a Creator must, I think, be pronounced an interesting failure.[17] If, however, there is not, as has been seriously claimed, 'a relation of strict reciprocal implication between theology, philosophy of nature and scientific method',[18] neither is there, historically speaking at least, a total disjunction between any pair of the triad. A concept of God's nature, of the world's action and principles, and of methods of exploration, all three co-existing in time and place and in a continuous cultural matrix—which is to say, in the same human minds—can achieve a degree of mutual interaction and interpenetration. It is doubtful whether empirical-mathematical science could have developed in the way it did apart from the Judaeo-Christian faith in the Creator, to which its protagonists almost universally gave their assent; historically, it is certain that it did flourish in Western Christian Europe as nowhere else. The belief in a Creator could never of itself be induced to yield scientific principles, but it could and did allow scientists a freedom that other more jealous divinities might have denied them. Besides, and more importantly, the image of God which they carried in their minds gave rise naturally to convictions of a metaphysical character which are by no means the natural, spontaneous, and universal products of the human spirit, but are bound to a particular culture. Without belief in the goodness and the reality of matter, the intelligibility of things, regularity and a degree of uniformity of natural action, contingency under law, beauty, simplicity, and, no doubt, many other such ideas, it is difficult to see how science could have come about. All these principles could easily be brought under the attributes of the Christian God (or Yahweh, or Allah), whereas those of them that existed in the world of classical Greece could not be given the same degree of grounding in metaphysics and religious belief.

If I do not say that such principles could have been deduced from the notion of a Creator-God, that is because the God-concept is itself to some extent part of the cultural process,

[17] Foster, 'The Christian doctrine of the creation and the rise of modern natural science', in *Mind* 43 (1934), 446 ff.; 'Christian theology and modern science of nature', in *Mind* 44 (1935), 439 ff.; 45 (1936), 1 ff.

[18] Foster formulates in these terms 'the principle upon which I depend throughout' [his proof] : *Mind* 45 (1936), p. 5.

and it changes and develops too; we have seen one concrete example of such development in Grosseteste's idea of God as *Numerator*. Giordano Bruno constructed an ontological type of argument from the possibility to the actuality of infinite space, the possibility being established by the idea of divine infinity;[19] while Kepler, his contemporary, argued that God could not but create (if create he did) a mathematically-ordered, harmonious—and therefore finite—world (pp. 58-9).

Descartes could argue that God created the world by a pure act of will, and that it is preposterous for us to scrutinize his aims; whereas teleological consideration of nature in the Middle Ages felt itself supported and encouraged by the belief that if God willed ends they must, often at least, be important enough for us to explore them and evident enough to be uncovered. In the Leibniz-Clarke correspondence, finally, important differences between the physical systems went back to different conceptions of the Creator (pp. 235-272).

In the cases mentioned, the disagreement of the attributes derived from the idea of a Creator are presumably due to the fact that the content of the concept 'Creator-God' is not necessarily indistinguishably the same even for men of the same generation, but is liable to contamination by other cultural factors present in a given age or mind, such as basic metaphysical principles like plenitude and sufficient reason, the ideas of perfection and harmony, the belief in mathematical realism, the desire to construct a theodicy, or to obstruct the secularization of knowledge, etc. (the possibilities are numerous). The relationship, therefore, between the two ideas 'God' and 'world' is a complex one; neither is a concept capable of direct definition and abstract comparison with the other, nor is either an empty concept, to which the other may at will be commanded to lend content. Ultimately our concept of God—which is never more than an approximation to his inaccessible Light—is filled with ideas drawn from our experience of the world and the other person, and our representation of his action is likely to reflect concepts of order or beauty formed by reflection on the world—not only, nor even principally, by individuals, but by all the cultural forms bearing individual life: language, work, association, and communication.

[19] Koyré, *From the Closed World . . .*, p. 40.

Yet behind the differences in the Christian concept of a Creator-God as between one generation and another is a common element, residing in the attributes of infinite power, wisdom, goodness, and freedom; and historically, these attributes acted above, or even through, the configurations into which they were drawn by philosophers, and, in many cases, left their mark on the metaphysical ideas furthering emergent science. Divine infinity suggested that the Greek understanding of the finitude of the world and its contents is not axiomatically true, and that mathematical infinity, in multiplication of atoms, or extension of space, could *a priori* have a certain reality for the divine mind, even if not for the human. The freedom of the divine will demanded that no hypothesis concerning the nature of reality be accorded an absolute grounding in 'the' divine nature and be thus canonized by an umbilical attachment to the sacred.

It has frequently been claimed in the past that the Middle Ages were dominated by the authority of the ancient world in political and intellectual life. Few historians would be prepared to deny that that authority counted for much more than it was to do in later times, yet fewer still would today endorse the older claim in its absolute form. If the scientific authority of the ancients was never accorded absolute value by medieval thinkers, or permitted to determine the whole world-picture, this was due to three main factors: those authorities disagreed among themselves, and from their friction the spark of enquiry was generated; authorities were sometimes found to disagree with facts of experience; and authority was clearly seen to be in frequent contradiction with Christian beliefs and traditions. We have instanced these factors in following Grosseteste's ideas in this chapter; we have seen him on occasion outrightly deny some presuppositions of Aristotle, reduce accepted theories to probabilities or possibilities, or outflank and disarm some of his authorities by a movement of thought directed by his Christian outlook. This critical assimilation of ancient ideas preceded him by at least a century and continued long after his death. When the Bishop of Paris in 1277 condemned certain attempts to place limits on the infinite power of God (such as the theses that God could not create an infinite world or a plurality of worlds, or a vacuum within the world, nor cause the world to move linearly), he quite unintentionally fostered the critical spirit among theologians, for it was the concept of an infinite Creator free from all natural necessity and constraint upon his will that kept open whole areas of the late medieval *Weltanschauung*, areas which Greek and even Arabic science had historically tended to seal.[20] The same idea lent positive encouragement to some fourteenth-century Scholastics in their assault upon the finitude of the Greek cosmos and upon its congenital attachment to the divine, and prompted them to evolve elements of a world-view more congruent with their creationist metaphysic. The idea of a contingent creation, freely affirmed by a sovereign will out of a multitude of possibilities, does not imply a scientific method of explanation, still less does it admit of the deduction of actual laws of functioning; but it did incline the mind awakened to wonder to seek the limits of truth in commerce with the world itself, a world from which the gods had flown and their idols been chased.[21]

[20] For a more recent, and on the whole sympathetic, evaluation of Duhem's classic thesis to this effect, see Beaujouan, 'Medieval Science in the Christian West', pp. 500–01. The articles condemned by Étienne Tempier have been studied individually and placed in their historical context by Hissette, *Enquête sur les 219 articles condamnés à Paris le 7 mars 1277* (Louvain-Paris, 1977), 337 pp.

[21] Beaujouan, p. 502: 'Empiricism and metaphysics, and even a very definite kind of metaphysics, the creationist, are closely linked together. What other means, indeed, but observation and experience can we possibly use for the study of a world freely created by an infinite God?' A similar argument in favour of the Newtonian method in science was used by Ralphson, the first theologian of the Newtonian system.

PART FOUR

THE LIGHT OF INTELLIGENCE

Chapter 1
Grosseteste's Psychology in its Historical Setting

I OLD THINGS AND NEW

ROBERT Grosseteste devoted himself to the study of philosophy and theology over a period of some sixty years. Although it is probable that almost all his extant work dates from after 1215, it cannot be forgotten that his uniquely long career spans the turn of the thirteenth century, and covers therefore a period of intense interest in the intellectual life of the Middle Ages. It can be assumed that Grosseteste was active from around 1215 as a professional theologian, so that for the ensuing twenty years of his life in the Schools he was committed to lecturing on the Bible as his main task, and to holding disputes on theological questions. That he achieved so much more than his academic duties commanded is a testimony to his wide interest in the newly-discovered Greek and Arabic works which were winning notoriety just at this period. This new scientific literature attracted his attention and drew from his pen a series of commentaries on Aristotle's logic, treatises on motion and other natural problems, and writings on metaphysical topics. By the end of the period in question Grosseteste was as well acquainted as any man of his time with the new natural philosophy, whose bases were the physical works of Aristotle and the astronomical writings of Ptolemy. He had absorbed the chief metaphysical ideas, like matter and form, act and potency, and causality, which derived from Aristotle and were filtered through Greek and Arabic Neo-Platonism. This body of literature was welcomed by him because it filled a gap in the traditional theological world-view within which he had grown up. It had been no part of St. Augustine's concern to develop a systematic philosophical doctrine of astronomy or physics, and the sciences in the Latin tradition, of which his thought had always remained the mainspring, ploughed only a tiny furrow on the edge of theology's broad acres, as when Bede wrote on the calendar for practical ecclesiastical purposes, or treatises *De Naturis Rerum* were deposited by the mainstream of exegesis.[1]

With the arrival of the new physical works, which despite their prohibition at Paris in 1210 began to exert a traceable influence on theology after *c*.1220, there also came a new series of treatises on human nature. In the Western theological tradition there had until then been no place for a philosophy of man rivalling the theological doctrine. Here once again the influence of St. Augustine had proved determinant, and if he had rejoiced in opposing Christian wisdom to pagan speculation, it had been far from his mind to develop side by side a theology based upon the scriptural data and an autonomous philosophy of man. Of course, Augustine's anthropology, though centred upon the biblical view of man, owed much to Neo-Platonism and Stoicism, but the elements borrowed from those philosophies were not allowed to form an autonomous department within theology; they were rather re-thought, and integrated into the relevant biblical ideas to form a single whole. The Middle Ages followed Augustine, in that the only teaching on human nature given in the schools up until the mid-thirteenth century formed part of the course of theological studies. It is consequently no surprise to find that the new sources deriving from Aristotle and his Arabic adherents, and containing teaching on ethics (*Ethica Vetus, Nova*) and psychology (the treatises *De Anima* of Aristotle, Avicenna, and Algazel), found a welcome and an influence in the theological faculties, long before they were made part of the syllabus in arts. At Paris, for example, the years 1220 to 1240 witnessed a considerable interest in philosophical problems on the part of theologians, while as for the faculty of arts, the obscurity in which the period is still shrouded none the less permits us to ascertain that only the logical and ethical works of Aristotle were taught until 1252. The *Metaphysics* and the *Libri Naturales* were not widely read until after that date, and did not become an official part of the syllabus of studies until 1255, when the new constitutions of the faculty imposed the reading of all the main works of Aristotle then recovered.[2]

---

[1] Alexander Nequam used the plan and themes of Ecclesiasticus as the framework of his cosmological work of that title.

[2] Van Steenberghen, *La Philosophie au XIIIe siècle* (Louvain-Paris, 1966) p. 359.

It has not always been realized that the first theologians to welcome the new philosophical literature (the Paris teachers, William of Auvergne, William of Auxerre, Philip the Chancellor, Alexander of Hales, and, at Oxford, Grosseteste) were in effect taking a new step in introducing elements of Aristotelean and Neoplatonic philosophy into their theological syntheses. Not that they managed to achieve at once a perfectly consistent harmonization of traditional theology and philosophical motifs, nor that they produced an overnight change in the spirit in which theological discussion was conducted; on the contrary, their first efforts at assimilation were feeble and eclectic by comparison with the achievements of the following generation, and the spirit animating their efforts continued to be theological and Augustinian, posing no formal distinction between theology and philosophy, and subordinating the latter to the former. The novelty of their project lay in the fact that they formed a theologico-philosophical movement by borrowing from Aristotle and his *sequaces* whatever seemed to promise something of utility for their own theological interests.

The problem of the soul has a special interest, because the challenge involved in the appropriation of psychological ideas by theologians was greater than in the case of physics and astronomy. Theology had inherited from Augustine a conception of the soul which was sacral and religious in character, as had been that of the whole Neoplatonic movement to which he was extensively indebted. The spirituality of the soul and its independence of the material instrument which it rules as mover were the main emphases of Plato's heir, and both beliefs invited immediate baptism into Christian faith, largely, we may suppose, because of their ethical implications. The *De Anima* of Aristotle on the other hand, despite its Platonic elements, represented such a difference in method, doctrine, and mood that those who read it in the early thirteenth century often despaired of understanding it. They had to assimilate at once a new method, a new doctrine, and a strange vocabulary. They were forced to acquire categories of thought which lay at a far remove from their traditional and essentially religious concept of the soul, if they were to absorb Aristotle's ideas, with their strongly biological basis. Starting from a simple division of the soul into rational, irascible and concupiscible parts, the whole inspiration of which was ethical and religious, they had to move to the Aristotelean scheme of vegetable, sensitive, and human souls, a scheme whose study embraced all living things, and set man in a naturalistic context far removed from sacrality.[3] No wonder so many despaired of understanding the Philosopher, and turned for guidance to the easier and more religious work of Avicenna; and it is scarcely more remarkable that, with Avicenna established as the trusted exegete of Aristotle's psychology, no theologian of the first half of the thirteenth century perceived the outright contradiction between the conception of the soul as a spiritual substance using a material instrument, and as the form of an organic body. Not until Averroës had opened their eyes to the Platonism of Avicenna did it occur to the Latin thinkers that in this respect they had been misled in their interpretation of Aristotle.

Scholars who have studied Grosseteste's views on the soul and knowledge have been hampered by a number of difficulties, such as the disputed authenticity of the *De Anima*,[4] confusion as to the sources he employed,[5] and the lack of a reliable chronology of the relevant writings. Yet, surprisingly perhaps, they have reached a considerable measure of agreement concerning the general features of his doctrine, regarding his thought as being overwhelmingly Augustinian in its strict dualism of soul and body,[6] its clear reliance on the illumination of the mind,[7] and its opposition to all the characteristic posi-

---

[3] Of course, Aristotle's theory of soul was not completely lacking in religious content. The *nous* is anything but a merely biological agent, it rather comes 'from outside' and possesses something of the divine. And, on the other hand, the opposition between Aristotle's and Avicenna's psychology should not be exaggerated; the latter's vocabulary remains in large part Aristotelean.

[4] The *De Anima*, regarded by Thomson (*The Writings*, p. 90) as authentic, is now generally rejected; see Appendix A, no. 33, p. 484.

[5] For instance, Sharp (*Franciscan Philosophy at Oxford*, pp. 9–46) held that Avicebrol was a major influence. In fact there is no evidence to suggest that Grosseteste favoured the universal hylomorphism advocated by the Jewish thinker.

[6] Zedler, in particular, exaggerates the element of dualism; see her article in *Proc. Amer. Cath. Phil. Assoc.* 27 (1953) pp. 144–55.

[7] Accepted by all who have written on the subject: Gilson, "Pourquoi S. Thomas a critiqué S. Augustin", in *Arch. Hist. Doctr. Litt. M.A.* 1 (1926), pp. 90–99; Callus, "Introduction of Aristotelean Learning to Oxford", in *Proc. Br. Acad.* 29 (1943), pp. 28–9; Lynch, "The doctrine of divine ideas and illumination in Robert Grosseteste, Bishop of Lincoln", in *Med. Stud.* 3 (1941), pp. 172–3; Miano, "La teoria della conoscenza in Roberto Grossatesta", in *Giorn. Metaf.* 9 (1954), pp. 60–8. The positions adopted by these authors will be discussed below at appropriate places in chs. 2 to 5.

tions of Aristotle (the abstraction of universal ideas from sensation, the distinction of active and passive intellects). It will be the task of the following four chapters to put this consensus of scholarship to the test by studying Grosseteste's writings on the soul and knowledge, putting them into strict chronological sequence, and identifying his sources with accuracy.

We plan to trace Grosseteste's study of the soul through the greater part of his writings, attempting to view the evolution of his thought as a whole. It will be made clear that his development parallels in many ways that of the Paris masters who were his contemporaries. His initial preoccupation with psychological questions is of a theological character, and makes sense only in an intellectual world in which the traditional theology is the sole point of reference, its centre being the doctrines of St. Augustine. As time goes on, however, his analysis of the soul's functions betrays a strong imprint of the newly-available philosophical literature; there are clear signs that he has grappled with the *De Anima* of Aristotle, though like most of his contemporaries he read it through Avicenna's treatise. The evolution we find present in his thought does not mean that by *c*.1240 he had as good as renounced his Augustinian starting-point; rather he had retained all the Neoplatonic elements present in traditional thought, while none the less managing to incorporate whole areas of the new teaching where he found it serviceable. His psychological doctrine does not amount to a synthesis, either in the sense that it resulted in a systematic treatise of psychology, or that the scattered elements of theory can be drawn together to compose a well-defined and consistent philosophical doctrine; but there is real interest, never the less, in seeing both the extent and the limits of the welcome he accorded to the new ideas.

II A CHRONOLOGICAL LISTING OF GROSSETESTE'S WRITINGS ON PSYCHOLOGY

De Artibus Liberalibus In all probability the earliest extant work of Grosseteste, this treatise on the utility of the arts cannot well stem from the years after the great dispersal of 1209, when Grosseteste presumably went to Paris to study theology. Since on the other hand he was *magister in artibus* already by 1189, it may have been composed very early indeed. Its interest for us lies in a passage in which Grosseteste discusses sensation, using ideas that owe everything of importance to the *De Musica* of Augustine, and revealing no traceable influence of the newer psychology.[8] The work thus marks the point of departure for the evolution of Grosseteste's anthropology.

De Veritate and *De Veritate Propositionis*[9]   The *De Veritate* represents an attempt by Grosseteste to define the concept of truth in both its ontological and noetic meanings, and *De Veritate Propositionis* extends the discussion to a particular difficulty, namely, that of future contingents. In style, spirit and content, both treatises are steeped in Augustinianism, Augustine and Anselm being the only real sources. The only Aristotelean note struck is the remark with which the *De Veritate* concludes: the concept of truth (*intentio veritatis*), like that of being, is ambiguous, for on the one hand it is one in all truths, and yet is diversified in each by appropriation (p. 145).

Baur was inclined to consider the treatise a late work, composed *c.*1240 (p. 104*). I would place it a decade or more earlier, since it bears none of the characteristics of Grosseteste's late interests (knowledge of Greek and wider use of Greek sources), and finds its natural affiliations with the group of brief metaphysical and theological treatises which are much more likely to derive from the decade 1220-30 than from any earlier or later period. We know with reasonable certainty that the *Commentary on the Posterior Analytics*, Grosseteste's most extensive work on knowledge, was finished by around 1231, and probably composed for the most part around 1228, though of course it may have incorporated notes and comments made by Grosseteste in the earlier years of the decade. In that commentary he was preoccupied with completing Aristotle's conception of scientific knowledge with ideas deriving from the Christian tradition, and in particular from St. Augustine. It is possible that his *De Veritate* represents a reassertion of the Augustinian theory of truth as against Aristotle's lacunary ideas, much as the *De Finitate* was directly occasioned

[8] See ed. Baur, pp. 3-4.
[9] Ibid., pp. 130-45.

by his grappling with the Aristotelean *Physics*, and was meant as a Christian correction of the pagan conception of the eternity of the world. If this symmetrical relationship between the two Aristotelean commentaries and the two original treatises be admitted, the *De Veritate* could be considered contemporaneous with the *Commentary on the Posterior Analytics*. Against this conjecture stands the fact that the *De Veritate* wastes no time in countering Aristotle, but dedicates itself to a calm and reasoned re-assertion of the doctrine of Augustine and Anselm. On the other hand, it could be argued that, since Grosseteste did not find Aristotle's theory of knowledge shocking (as he did the notion of the world's eternity), but only incomplete, there is no reason why he should not have written a treatise on the Christian concept of truth and its relation to the supreme truth and light of the mind, without referring to Aristotle, who had quite simply nothing to say for or against that conception. What evidence there is suggests a date of composition *c.* 1225–30 as the most likely, and nothing of which I know counts against this.

De Intelligentiis and *De Statu Causarum*   The *De Unica Forma* and *De Intelligentiis* form a single epistolary treatise, written *c.* 1228 and devoted to the relationship of the spiritual to the material world: God (the exemplary form of all things), the angels, and the soul.[10] In the second part of the work the soul's relationship to the body is made a qualifiable model on which the angel's relationship to matter can be imagined, so that there is a deal of psychological doctrine in the piece.

With this work we can associate a passage from the *De Statu Causarum*,[11] where Grosseteste describes the soul as a substantial form and discusses its origin. The passage is brief, but important for the development of his psychological doctrine.

The *De Statu* is almost contemporaneous with the *De Intelligentiis*. Grosseteste evidences considerable knowledge of Aristotle, quoting from the *Posterior Analytics*, the *De Generatione*, and even the *Metaphysics* (twice). He can scarcely have read so widely in Aristotle before the decade 1220 to 1230. Close doctrinal parallels occur between the two works: thus how God can be *forma prima omnium* is acknowledged to be

---

[10] See Part II, ch. 1 (pp. 52–8).
[11] See ed. Baur, pp. 124–5.

a difficult problem (p. 125.25); it is in fact tackled in the *De Unica Forma*, which may, then, be slightly posterior to the *De Statu*. Similarly, the relationship of intelligences to matter is brought into association with soul-body relations in both *opuscula*. The psychological doctrine of both treatises gives no reason to postulate a long period of time intervening between their dates of composition. All things considered, a date between 1225 and 1228, just before the *De Unica Forma* and contemporaneous in a general way with the group of metaphysical and theological treatises, would be very suitable.

Commentarius in Libros Analyticorum Posteriorum Aristotelis[12]   Grosseteste's sole finished commentary on Aristotle is of central importance for his theory of knowledge, particularly in those passages of the work where he departs from the exposition of the text in order to offer what he considers a more complete view of human understanding. Crombie proposed that the work was written *c*.1220.[13] Dales, on the other hand, has made the inherently more likely suggestion that the commentary was composed *c*.1228, and given a number of reasons in support of his view.[14] While no absolute certainty on the matter is at present available, we can safely take it that the commentary is a work of the decade 1220–30, and probably of its later years.

Commentarius in VIII Libros Physicorum Aristotelis   Most of these notes were probably written between 1228 and 1232. The only passage of relevance is one which develops a contrast between intellect and sense, acknowledging the *De Musica* of Boethius as its source.

De Operationibus Solis   Written around 1232, this little work contains important references to the spirits active in sensation, the *species visibilis*, and truth.[15]

---

[12] I have used the Venice edition of 1514 (Nachdruck Frankfurt/Mainz, Minerva, 1966), to which all references are given.

[13] *Grosseteste and the Origins of Experimental Science*, pp. 46–7.

[14] "Robert Grosseteste's scientific works", in *Isis* 52 (1961), pp. 395–6.

[15] I have published the text in *Rech. Théol. anc. méd.* 41 (1974), 38–91.

De Cessatione Legalium This biblical treatise can be assigned to *c*.1235.[16] The section on the Incarnation develops a philosophy of human nature based upon microcosmism, which is of depth and importance. Its systematic nature provides a strong framework into which scattered remarks of a microcosmic nature occurring in other works can to some extent be built. The *De Cessatione* also contains some interesting remarks concerning the union of soul and body.

Hexaemeron It can be said of Grosseteste's finest exegetical work (finished *c*.1237) that it contributes something of interest to all the main anthropological questions of the age, and is the chief source for his views on the vegetative soul, the process of sensation, and, of course, on the soul as the image of God.[17]

The commentaries on Pseudo-Dionysius Composed between 1239 and 1243. I have drawn on all but the *Commentary on the Ecclesiastical Hierarchy*. Grosseteste not infrequently adds comments which reveal a continuing interest in questions of psychology and epistemology. Of particular importance is the passage in the *Commentary on the Celestial Hierarchy* in which he defends the traditional Augustinian position on the beatific vision.[18] Symbolic knowledge is a theme common to all the commentaries. The hierarchical vision underlies Grosseteste's fully mature views on ecclesiastical and social order.

Three late works, two sermons and a treatise on confession, must be considered for our purpose as a group, since they all derive from the period after 1240 and devote considerable attention to the powers of the soul. All three come under the clear influence of Avicenna, Pseudo-Dionysius, and the Pseudo-Augustinian *De Spiritu et Anima*.

a. *Ex Rerum Initiatarum.*[19] The very tangible Dionysian influence on the thought and vocabulary of this clerical conference places it certainly after the year 1240. It includes an important statement on the powers of the soul.

---

[16] See Appendix A, no. 77, pp. 489–90.
[17] The section dealing with the soul as image has been edited by J.T. Muckle, in *Med. Stud.* 6 (1944), pp. 157–74.
[18] See ed. Dondaine, in *Rech. Théol. anc. méd.* 19 (1952), pp. 124–5.
[19] See ed. Gieben, *Coll. Franc.* 37 (1967), 100–41 (see Appendix A under 'Sermons', pp. 498–9).

b. *De Confessione.*[20] The influence of Pseudo-Dionysius dates this treatise to after 1240. Its chief interest for our purpose lies in a passage where Grosseteste, expounding the commandment to love with the whole soul and heart and mind, resorts to Avicenna for an analysis of the soul's powers (pp. 260-1). Also, the prologue contains a passage on man as the image of God, as *minor mundus*, and as lord of creation.

c. *Ecclesia Sancta Celebrat.*[21] The wide scope and the detail which mark the classification of the powers of the soul contained in the opening pages of this sermon make it the most valuable surviving witness to Grosseteste's psychology. The close verbal parallels between its closing section and Grosseteste's *Commentary on the Mystical Theology* suggest a dating around 1243.

III AUTHORITIES ACKNOWLEDGED BY GROSSETESTE

It is helpful for our purposes to list those works of predominantly psychological interest from which Grosseteste quotes, or to which he refers by name, and to add the date of the reference as exactly as possible. Thus we hope to effect a reconstruction of his library of psychological works, and to trace the development of his interests by noting the first occasion on which a particular book is mentioned by him. Clearly, in this undertaking, one cannot argue from silence; but one works on the assumption that a book mentioned or quoted from at least once was consulted, and is thereby alerted to the possibility of doctrines derived from that same source being adopted elsewhere without express acknowledgement.

St. Augustine No author was more read and quoted by Grosseteste than the great Doctor, as a glance at the *Concordance of the Bible and Fathers*, compiled by Grosseteste and Marsh, attests.[22] Works of psychological interest in this extensive list include *De Duabus Animis, Epistolae* (no.187 is quoted in the *De Intelligentiis*), *De Genesi ad Litteram, De Immortalitate Animae* (quoted in the *Hexaemeron* and elsewhere), *De Libero Arbitrio* (often referred to by Grosseteste), *De Musica*

---

[20] See ed. Wenzel, *Franc. Stud.* 28 (1970), 218-93 (see Appendix A, no: 80, pp. 492-3).

[21] See ed. McEvoy, *Rech. Théol. anc. méd.* 47 (1980), 131-87.

[22] See Thomson, "Grosseteste's concordantial signs", *Med. Human.* 9 (1955), pp. 47-8.

(quoted in the *De Intelligentiis*, and a source for the *De Artibus Liberalibus*), *De Trinitate*, etc.

Aristotle The *De Sensu et Sensato* and *De Anima* are both quoted in the early (1220?) *De Generatione Stellarum*, and both by name. The *De Anima* is also named in the *De Motu Supercaelestium* (*c.*1229). Quotations from and references to the *De Anima* are infrequent in Grosseteste's later works. The latest is probably the reference to '*Philosophus*' in the sermon *Ecclesia Sancta*, where *De Anima* is clearly meant.

St. Ambrose Ambrose was not a major source for psychological doctrine in the Middle Ages, but Grosseteste quotes from his *De Dignitate Conditionis Humanae* in the *De Intelligentiis*, when discussing the soul's presence in the body.

Boethius '*In capitulo quinto primae musicae ubi dicit Boethius*', is the only relevant reference I have found. Occurring in the notes to ch. 1 of the *Physics* (before 1232), it is used to develop a contrast between sense and intellect.

Algazel In his unedited *Expositio in Galatas* (1233-35), Grosseteste quotes from an unnamed work of the Arabian philosopher an obscure passage concerning the impression of the soul worked through its *affectus* upon bodies, which impression is called fascination.[23] I have not found any other reference to Algazel in the context of psychological doctrines.

Avicenna We read, once again in the *Expositio in Galatas*, '*Avicenna autem philosophus dicit in libro suo de anima quod multociens anima operatur in corpore [alieno] sicut in proprio, quemadmodum est opus oculi fascinantis et estimacionis operantis . . .*', and the quotation continues for several lines more.[24] Strangely, this casual reference to Avicenna's *De Anima*, made in an excursus in the middle of a scriptural commentary, is the only occasion on which the work which laid the basis of Grosseteste's mature psychology is mentioned by name.

[23] Oxford, *Magdalen College MS 57*, fol. 9ᵃ.
[24] Avicenna, *De Anima* IV, 4 (ed. Van Riet, p. 65).

St. Anselm   Grosseteste's knowledge of Anselm's works was wide, as appears from his *Concordance*, and while they contributed nothing definable to his classification of psychological powers, they did exert an influence on his *De Veritate*, in which he twice cited Anselm's book of the same name, and upon aspects of his doctrine in the *De Libero Arbitrio*, where he made frequent and acknowledged use of the *De Concordantia Praedestinationis Liberi Arbitrii et Gratiae*, and on other works.[25]

Pseudo-Dionysius   I cannot think of any directly psychological doctrine of Grosseteste upon which the Areopagite left his mark, but in the period after 1239 his influence upon Grosseteste's epistemology is real. Dionysius reinforced, without, however, notably clarifying, the doctrine of illumination, and encouraged Grosseteste to describe the angelic hierarchies and the human soul in terms of spiritual light. This influence was, however, more theological than philosophical, for the thearchic ray which hierarchizes the intellectual creature is not the light of natural knowledge, but a salvific light of grace, the self-communication of the divine knowledge, and as such had more direct influence on Grosseteste's ecclesiology than on his philosophy. It is clear, however, that Dionysius represented for Grosseteste the mystical theologian *par excellence*, and he adopted much of his thought in the sermon *Ecclesia Sancta*, to make the crown of psychology and anthropology.

In the second place, the exercise of translating the Dionysian corpus became for Grosseteste the occasion for reflecting on God-talk, and the commentaries on the *Celestial Hierarchy* and the *Divine Names*, especially, contain much that is of value, particularly on symbols and the knowledge they mediate.

Plato   The *Timaeus* is the only Platonic work named by Grosseteste.[26] On one occasion he quotes Plato's definition of the soul as '*substantia seipsam movens*'.[27] Aristotle and Augustine were the main sources for his knowledge of Plato, whose opinions and even errors as known to him Grosseteste

[25] *De Veritate*, ed. Baur, pp. 132, 135; *De Libero Arbitrio*, ibid., pp. 151, 165, 168, 176, 178, 199, 217, 231.

[26] In the notes on the *Physics*, ed. Dales, p. 147; *De Cessatione Legalium*, MS *Bodl. lat. th. c. 17*, fol. 158; *Hexaemeron*, ibid., fol. 196[C].

[27] *De Motu Corporalium*, ed. Baur, p. 91.

is often inclined to treat with indulgence and understanding. It cannot be claimed that Plato had much influence on Grosseteste's psychology, except through the intermediary of St. Augustine. The concept of the soul's *eros* as *'recordatio visae pulchritudinis, ut Plato'*, is noted in the *Commentary on the Celestial Hierarchy*.[28] Grosseteste took this piece of information from the *Lexicon* of Suda.

Pseudo-Augustine In the *De Cessatione Legalium* Grosseteste quotes a long remark from the *De Differentia Animae et Spiritus*, a work which, like most of his contemporaries, he attributed to St. Augustine.[29] The same work (and with the same title)[30] is listed among Augustine's works in the concordance of Grosseteste and Marsh, which probably indicates that it had been in Grosseteste's hands for some time before 1236; and indeed he seems to have used it in the *Commentary on the Posterior Analytics*. The influence of the little treatise on his developed psychological theory was profound, crystallizing as it did in the division of the rational cognitive powers into *ratio, intellectus*, and *intelligentia*. Its authorship is now frequently attributed to Alcher of Clairvaux.

Pseudo-Andronicus Grosseteste quotes Andronicus by name in the *Commentary on the Celestial Hierarchy*, reproducing the definition of *passio animae* and *eros* from the *De Passionibus*, which must therefore be reckoned among the minor sources of his psychology.[31] He thought highly enough of the work to translate it, *c*.1237.

Averroës '*Ita dicit Commentator super secundum de anima'*. This reference is found in the *De Lineis* (1231-35), and is the only one to Averroës's commentary on the *De Anima*. Grosseteste claims his authority for the proposition that every agent multiplies its virtue in a sphere, but he does not quote, and it

[28] See Part II, ch. 2, p. 86.

[29] See ed. Unger, p. 12. The editor has misread the title: *divisione* instead of *differentia*.

[30] Its title varied considerably in the Middle Ages. In Migne's edition (*P.L.* 40, 799-832) it is called *Liber de Spiritu et Anima*.

[31] See n. 28 above. Only the latter borrowing is acknowledged; cf. Franceschini, *Roberto Grossatesta*, p. 115, and Tropia, *Aevum* 26 (1952), 97-112.

is impossible to know which commentary he consulted, and whether he already possessed a full text of it. This must be one of the earliest quotations from Averroës on the *De Anima* in the philosophical literature of the period. It can be said at once that none of the Commentator's characteristic psychological doctrines surfaces in Grosseteste's works, nor yet any protest against them. One would have expected an explosion similar to that in the *Hexaemeron* (against the philosophers who believe in the eternity of the world), had Grosseteste realized the true nature of Averroës's teaching. But for this solitary and insignificant reference one would have assumed he had not read Averroës on the *De Anima*.

Also in the *De Lineis* is found a reference to the '*Commentator super tractatum de sono*', again, so far as I am aware, the only one to this work in Grosseteste's writings.

The *Concordantia*, while it contains references to some pagan works, refers to no ancient or Arabic book of psychology, though Aristotle's *De Animalibus, De Vita et Morte*, and *De Somno* are noted.

Chapter 2
The Theological Definition of Human Nature

It might be considered that a study of Grosseteste's philosophical ideas could afford to dispense with insisting on the theological character and setting of his concept of man. Could not the expositor rest content with a glancing reference to the theological framework of his thought and pass quickly on to the examination of its more strictly philosophical content?

Aside from the general objection to this procedure, arising from the risk of falsifying the total historical perspective and presenting a fragmentary analysis of what was in the eyes of its author a unitary whole, a number of important considerations oblige the expositor of Grosseteste's anthropological thought to be especially attentive to the whole, before analysing the parts. In the first place, it will emerge that his preoccupation with the defence of the vision of God *facie ad faciem*, which in the Latin theological tradition was unanimously considered to be the last end of human existence, governed and triumphed over his consistent effort to absorb the contributions of Greek and Arabic philosophy into the science of man. The Latin Christian viewpoint became in effect the criterion which governed his judgement on pagan philosophy, influencing on the one hand his adoption of some psychological, epistemological, and ethical theories (drawn from Aristotle and Avicenna, in particular), and on the other, his decision variously to criticize, reject, reinterpret or neglect others. Granted this, it would help but little towards the understanding of his intellectual options if we were to confine our attention to his borrowings, and thus fail to bring out the concrete nature of the choices he made.

Two examples of this are particularly pertinent. Grosseteste's later works offer a complete description of the powers of the soul, a description taken in part from Avicenna, and therefore based ultimately upon Aristotelean science. However, his borrowings concentrate almost exclusively on the material functions of the soul, its vegetative and sensitive powers. Of the rich Avicennian psychology of the intellect, the detailed and often edifying analysis of the properly human powers of understanding and contemplation, Grosseteste adopts as good as nothing. Instead, he offers an analysis of the intellectual functions in terms of the Pseudo-Augustinian triad of *ratio, intellectus,* and *intelligentia.* Into this schema he can still manage to fit Aristotelean themes, but the triad itself is not adopted simply in order to facilitate their assimilation, but to express the gulf existing between *scientia,* whose object is the created world, and *sapientia,* which is turned upon God alone, both here and hereafter. The beatific vision responds, he maintains, to the deepest level of desire implanted in human nature; but of this the pagan scientists knew nothing.

The theological end of man is not less present to the mind of Grosseteste the epistemologist. Commenting on Aristotle's sharply anti-Platonic thesis, that no intellectual knowledge can be attained which does not begin from sensation and develop by abstraction ('one less sense, one less science'), Grosseteste seizes upon the occasion to draw attention to the limits of the Aristotelean doctrine. There is in fact, unknown to Aristotle, a kind of intellectual knowledge *sine sensus adminiculo,* to which the human mind, considered in itself and without reference to a particular state of man, is ordained. In the beatific vision, namely (which is clearly what Grosseteste has in mind in the context),[1] the soul's *suprema pars,* intelligence, will be irradiated by the divine light, and will receive knowledge of God and of his creation seen in its eternal causes—a form of knowing which is complete, leaves nothing to be desired, and yet owes nothing to sense. This remark has been misunderstood, because insufficient care has been taken to interpret it in the light of the Augustinian theory of the vision of God, to which it never the less makes an explicit reference of unmistakeable clarity. The suggestion has wrongly been made that it was formulated as a criticism and rejection of Aristotle's theory of knowledge and a defence of the philosophical doctrine of illumination deriving from Augustine; Gilson's conclusion, that Grosseteste was so passionate an Augustinian as to draw no distinction between the illumination of the blessed, of Adam, and of man after the fall, is based partly on this text.

---

[1] Grosseteste refers quite explicitly to the state of the soul after death, and the knowledge it will have 'cum anima erit exuta a corpore'.

I THE VISION OF GOD: LATIN AND GREEK DIFFERENCES

In 1241 the Theological Faculty of Paris University condemned the proposition, *'Divina essentia in se nec ab homine
nec ab angelo videbitur'*.[2] Grosseteste made his main contribution to the theology of the beatific vision (in his *Commentary
on the Celestial Hierarchy*, 1239 to 1240) by attacking at its
chief source the doctrine rejected a short time later at Paris.
The importance of his intervention in this vital question lies
only partly in the fact that it provides the first available evidence of a reaction to the current of thought in question; its
chief value is intrinsic to itself. Grosseteste set himself not
merely to contest and reject what he looked upon as an error
of great theological consequence, but to restate the traditional
belief in terms which gained definition from the progress of
the debate, and to draw from the teaching thus formed its
most important implications for the theory of human nature.

The background to the doctrinal condemnation of this proposition has been brought out of obscurity by the researches
of recent years,[3] and requires only a brief summary here,
sufficient to sketch the context of ideas in which Grosseteste
wrote.

The question whether it is possible for the human soul to
see God is posed to faith by the data of revelation, which provide the double assurance that God is inaccessible light whom
no man sees nor can see (I John 18, I Tim. 6:18), and that we
shall yet see God face to face, as he is (I Cor. 12:12, I John
3:2). Christian theologians found themselves obliged to seek
a balance between the two poles of this mystery, and in their
search were led to invoke a range of noetic conceptions which
their philosophical culture placed at the disposal of their doctrinal reflection. Acting under diverse impulses, the Greek and
Latin traditions evolved with different emphases, whose opposition remained concealed only for as long as their isolation
from each other persisted. The meeting of the two streams in

---

[2] *Chartularium Univ. Par.*, t. 1, n. 128 (Paris, 1889). Nine propositions were
included in the condemnation.

[3] Chenu, "Le dernier avatar de la théologie orientale en Occident", in *Mélanges
Pelzer* (Louvain, 1947), pp. 159 ff; "L'homme, la nature, l'esprit. Un avatar de la
philosophie grecque en Occident au xiiie siècle", in *Arch. Hist. Doctr. Litt. M.A.*
(1969), pp. 123-130; Dondaine, "L'objet et le *'medium'* . . .", in *Rech. Théol. anc.
méd.* 19 (1952), pp. 60-131.

the Latin Middle Ages occasioned a work of assimilation, and very soon of self-defence, on the part of the Western theologians. Robert Grosseteste presents us with the first evidence of a reaction to the encroachment of the Greek influence at Paris.

As was so often the case, it was in the thought of St. Augustine that a formulation of teaching crystallized which became the received doctrine of the medieval schools of theology. That the direct vision of God is the hope of supernatural beatitude inscribed in the depths of the rational nature was a belief inculcated throughout the great Doctor's writings, and not distinguished in certitude from the truth of revelation itself, although, as was oftentimes the case, the doctrinal statement achieved by a writer of seminal but unsystematic genius was clear at its centre, but diffuse and uncertain at its outer reaches. God, the supreme truth and the light of the spiritual universe, has made us for himself, and our destiny is a pilgrimage from partial truths and flickering lights, through personal dialogue with their source, towards full enjoyment of the vision which overtakes faith, and replaces the mirror's image with the direct sight of God's face.[4] Faced with the apparent difficulty of the Old Testament theophanies, where God was revealed, indeed, but only as the invisible one 'whom no man can see and live', Augustine responded that God is indeed invisible to our eyes of flesh, but can allow himself to be seen when and by whom he wills. The *mundicordes* of the beatitude will see him: '*nec in loco videtur, sed mundo corde; nec corporalibus oculis . . .*'

St. Gregory, on the other hand, while he reinforced the Augustinian theology of vision, was already obliged to react against a different tradition, which denied the direct vision of God to the angels and the blessed;[5] perhaps he was alerted to this current by his contacts with Greek theology. His wise conclusion entered the deposit of the later Middle Ages by being incorporated into the Gloss: '*in futuro reperietur omnipotens per speciem, sed non ad perfectum, quia eius essentia a nullo plene videbitur*'.

The formulation of the Greek tradition by the Cappadocians

---

[4] See the texts quoted from Augustine by Dondaine, art. cit., pp. 62-3 nn., and especially *Epist. 147, Ad Paulinum de Videndo Deo, P.L.* 33, 613.

[5] *Moralia in Job* I, 18, c. 54; *CCSL* CXLIIIA, 950-5 (*P.L.* 76, 91-5).

took place under different impulses from those which pre-occupied the Latin Fathers. The reaction against Arianism demanded that the Word Incarnate alone be endowed with a knowledge of the essence of God, of which in principle no mere creature is capable. This doctrine entered Western theology mainly through three writers, whose ideas require a brief summary.

The writings of Pseudo-Dionysius, reinforced by the *Scholia* of Maximus and John of Scythopolis, contained throughout their length an insistence on the absolute unknowability of the divine essence by any created intellect. The knowledge which the angels and saints have of the thearchic substance is communicated under the form of theophanies, i.e., created powers or energies adapted to the finite knower. Through these, the actions of God's providence and the attributes mentioned in Scripture can be known and loved, but not the hidden *ousia* which transcends them. The Neo-Platonist in Dionysius wished to preserve the transcendent nature of the One inviolate and removed from the finite understanding, which of necessity breaks up the undifferentiated unity into a plurality of judgements and attributes.[6]

St. John Chrysostom commented the text, '*Deum nemo vidit unquam*', in his fifteenth homily on the Fourth Gospel, a homily widely diffused in the Latin translation of Burgundio.[7] His respect for the mystery of the divine essence, and his emphatic defence of the exclusive privilege of the Son of God, gave rise to the assertion (which the thirteenth-century theologians tried hard to qualify after the condemnation of 1241), that no one save the Word knows the *substantia* of the Father.

St. John Damascene reiterated the doctrine of Dionysius almost to the letter: the *substantia et natura* of God are totally unknown to creatures.

It was, of course, in John Scottus Eriugena, whose discovery of Greek theology converted him from the Augustinian Neo-Platonism of his native Latin tradition to that of the disciple

---

[6] On the adaptation of the Neoplatonic One made by Pseudo-Dionysius, see above all Koch, "Augustinischer und Dionysischer Neuplatonismus und das Mittelalter", in *Platonismus in der Philosophie des Mittelalters,* ed. Beierwaltes, pp. 329 ff.

[7] Grosseteste contests the teaching of Pseudo-Dionysius alone, without reference to Chrysostom's *Homilies.*

of Proclus, that the influence of Dionysius first became tangible in the West. He was led by his mentors, Pseudo-Dionysius and Maximus, to place the divine unity on a plane beyond the possible attainment of created minds, and to fill the infinite gulf between them with finite mediations emanating from the One who is beyond being. Eriugena, admittedly, employed the traditional language of the vision of God more freely than his Greek models had done, but he inscribed the vision within their theology of mediation. Though he considered it impossible for man or angel to see God as he is in himself, Eriugena held none the less that *'quasdam factas ab eo in nobis theophanias contemplabimur'*.[8] The theophany which bears whatever is visible of the unknowable divine essence is nothing other than the divinization effected by grace, which, by purifying and illuminating the intellect, enables the latter to 'see God' in the glory and beauty it has acquired. Eriugena concluded his treatment with a light-metaphor which was to enjoy considerable popularity among thirteenth-century writers. As the light of the air illumined by the sun appears bright, even though the sun itself is invisible, so the nature that is united to God shows forth his divinity, but without ceasing to be a creature. Eriugena's specification of the mysterious Greek theophany was to be found at the very centre of the condemnation of 1241, though less perhaps in his own formulation of it than in its diffusion by Alan of Lille.[9]

Eriugena's solution did not go unchallenged. Long before the beatific vision had become a matter of controversy, Hugh of St. Victor found in his *Expositio in Hierarchiam Caelestem* an opportunity to denounce Scottus's theory, on the grounds that it imperilled the Augustinian teaching. If all we can ever reach, he protested, is an image of the truth, then the truth itself is not seen; but nothing more nor less than the unclouded vision can satisfy the desire of our hearts.[10]

Hugh's outburst raised no echo in the early decades of the following century.[11] So far are the sources of the period leading up to the condemnation from perceiving any conflict between

---

[8] *Periphyseon,* ed. Sheldon-Williams, vol. 1, p. 50 (*P.L.* 122, 448C).

[9] *Distinctiones Dictionum Theologicarum, P.L.* 210, 791C. Alan elsewhere adopts a less daring approach to the question; see Dondaine, art. cit., p. 67.

[10] *Expositio in Hier. Cael.* I, 11; *P.L.* 175, 955A.

[11] See Dondaine's pages on the efforts at assimilation of the Greek theology, art. cit., pp. 74–88.

the traditions, that they can tranquilly discuss problems at the opposite remove from the Greek challenge. Thus the *Summa Aurea* enquires why it is that we cannot enjoy the vision of God already in this life, since God is present within us by grace.[12] Augustine had so emphasized the continuity between faith and understanding, and also the intimacy of the divine presence within the soul, that he had left his disciples poorly equipped to ground the disparity between faith and the vision.

Dondaine has identified three works in which an attempt is made to assimilate the Greek teaching on the absolute transcendence and unknowability of the divine nature. Alexander of Hales maintains in his *Gloss on the Sentences* that we shall see God's essence, not *per se* but through his glorious presence within us acting as a *species*, and he cites the light-metaphor of Eriugena in illustration.[13] With greater firmness, Hugh of St. Cher, commenting on St. John's Gospel under the influence of Chrysostom, denies that the blessed see God in his *substantia*, allowing them only the vision of his attributes. The anonymous author of q. 9 in the Douai *MS 434* not only limits the beatific vision in imitation of Eriugena, but like the latter postulates the indispensability of a medium between the soul and the divine essence: *virtus veritatis et amoris*, which is described as a *similitudo interior* through which we see the *species* that reveals God.[14]

Other contemporary masters may, for all we know, have taught similar doctrines in a more controversial manner, but it is not clear what exactly it was that precipitated the condemnation. Few details of the *déroulement* of events at Paris are available, but it would seem likely that the condemnation was not preceded by a long period of doctrinal tension such as was to occupy the decade before 1277. The three masters cited, the two identified being among the most respected of the age, evidence no awareness of a tension between Greek and Latin traditions, and could claim the support of authorities who were of irreproachable orthodoxy. We can, however, distinguish four closely interrelated questions which were

---

[12] William of Auxerre answers that the immensity of God is disproportionate to the present state of our intellect, which advances to its perfect state only by degrees.

[13] For the text, see Dondaine, art. cit., p. 79.

[14] The circumstance of the condemnation makes it probable that the question derived from a period before 1241.

emerging in various degrees of clarity during the years immediately preceding 1241. The first and most important problem
to be raised by the assimilation of the Greek teaching was that
of whether the divine essence is the object of the beatific vision
on the part of the angels and the blessed in heaven. Following
closely upon that, and complicating it, are the further questions, of the *medium* posited between the mind and God, and
the nature of the theophanies granted to the prophets and to
St. Paul. Finally, resonating above the more audible issues
there is the fundamental and ever-recurring problem of the
knowledge of God as such, introduced this time by the distinction between the divine essence and attributes. What do the
attributes of God revealed in Scripture tell us of him? Do they
admit us into the mystery of his nature, or are they in the last
analysis but names we are forced to use for the whirlwind?

II GROSSETESTE'S DEFENCE OF THE LATIN TRADITION

The passage which occasioned Grosseteste's defence of the
Augustinian theory of vision occurs in Chapter Four of the
Celestial Hierarchy, where Pseudo-Dionysius illustrates the
mediatorial action of the angels between God and mankind.
The Pseudo-Areopagite is content to give a brief reference to
the biblical data which served as the basis for his metaphysical
superstructure: the law was given through angels (Gal. 3:19),
and angels led the patriarchs to the divine, both by bringing
moral instruction concerning the divine will, and by manifesting visions of supermundane mysteries, sometimes of a prophetic nature. The scholion accompanying the text illustrates
these different interventions by reference to the cases of Joshua
and Gideon (instruction), Cornelius the centurion (conversion),
Daniel, Ezechiel and Isaias (the mysteries of the angelic orders),
St. Paul (vision in the third heaven), and St. John (the apo-
calyptic prophecy). Grosseteste expounds the text without
changing its sense. The actions of the angels are instrumental,
he reminds us, and therefore more truly God's than their own.
He expands on the mysteries revealed, extending them to the
universal ordering of creatures 'in number, weight and measure'
(the orders of heaven, of the Church, of the university of
creatures, and of the exemplar world), and to the mysteries of
the divine nature (the Trinity and Incarnation).

Dionysius hastens to restrict the extent of the revelations given by introducing his concept of theophany. The Scriptures forbid us to believe that theophanies have occurred without mediation, for no man sees or will see the hidden nature of God ('*occultum Dei*', Grosseteste renders it), but only a vision of his likeness (*assimilatio*), 'as it were, by a formation of the things that are formless . . . since divine enlightenment takes place through it to the beholders, and they are taught something of things divine'. Once again the scholion reinforces the sense of the text: all theophany is granted by the mediation of the angels, and occurs, not through God's appearing and showing what he is, for that is impossible, but by the saints' being granted divine enlightenment through sacred visions adapted to their level, and effected by the instrumentality of the angels.

When confronted by an authoritative assertion of a doctrine he found himself unable to accept, the medieval expositor was placed in an embarrassing situation. His sense of responsibility to his reader forbade him to pass on the mistake uncorrected, while on the other hand overt confrontation with an ancient and acknowledged authority was to be avoided, if at all possible. The popularity of the benign interpretation was a natural consequence of severe embarrassment seeking escape from an unwelcome dilemma. It is instructive to observe the solution proposed by the influential Dionysian scholar, Thomas Gallus, Abbot of St. Andrew at Vercelli, to the very difficulty under discussion here. In his *Extractio* from Chapter One of the *Celestial Hierarchy*, composed only a few years after Grosseteste's commentary on the same work, he is twice forced to resort to an anodyne paraphrase of Dionysius's outright denial that God can be seen by his intellectual creature. On the first occasion he modifies his author's contention by inserting the reassuring clause, '*in hac vita*', and he explains the indirect character of our knowledge of God in the present life in terms of its derivation from the sensible world.[15] A comparison between the *Extractio* and Grosseteste's translation of what he considered to be the most offending passage in the Dionysian works makes clear the reason why readers

[15] 'Non est enim possibile aliter nobis *in hac vita* supersplendere divinum principalem radium, nisi circumvelatum variis velaminibus sensibilium formarum.' Both passages of the *Extractio* are quoted by Dondaine, art. cit., p. 69.

of the *Extractio*, such as Richard Rufus, could take no scandal in it, even though they may have been alerted to Dionysius's teaching by certain other passages.

Grosseteste	Gallus
Si autem quis dicat et per se immediate infieri quibusdam sanctorum theophaneias, discat et hoc sapienter ex sacratissimis eloquiis, quod *ipsum quidem quodcumque est dei occultum nullus videt neque videbit*. Theophaneiae autem sanctis factae sunt secundum decentes deo per quasdam utique sacras et videntibus analogas visiones et manifestationes.	Quod si quis putet aliquos sanctorum patrum vidisse deum immediate, attendat quid dicit scriptura: *deum nemo vidit umquam, et non videbit me homo et vivet*. Dei autem apparitiones sanctis manifestatae sunt per quasdam formas visibiles vel intelligibiles, deo quidem utcunque congruentes et videntibus cognoscibiles.

By substituting for Dionysius's explicit words a combination of John 1:18 and Exod. 33:20, the Abbot of Vercelli is enabled to restrict the unqualified assertion that the immediate vision of God is impossible, and to deflect its meaning in the direction of the present life, whose noetic conditions impose limits upon our knowledge of God. The banality of the statement thus modified is an inevitable consequence of its disarmament by the device of benign interpretation.

Grosseteste, on the other hand, decides that the discomfort and risk attendant upon a resort to manoeuvre are too great, and that he is dealing with a position which is in irreducible opposition to the Latin tradition, which he upholds. His acknowledgement of the clear doctrinal difference which is at stake stands in contrast to the alternatives of ignorance and innocent approbation of the Greek thesis, which divided his contemporaries in roughly equal measure, up until the very eve of the condemnation. What was it that awakened his sensitivity to the possibility of a conflict? The question does not admit of a certain answer. Perhaps he had Hugh of St. Victor's warning in mind as he read Dionysius, but if that was the case he made no reference to it. It may be that the answer lies in his knowledge of Greek, which conferred upon his judgement of the text a quality of independence which was lacking in his contemporaries, who were not similarly equipped.

So scrupulously faithful is Grosseteste in his translation and exposition of the text that his reader is given no forewarning of the sharp opposition that is to ensue. The *occultum dei* of Dionysius is given its full, original weight when elaborated in the relevant comment: '*occultum dei, hoc est ipsam divinam essentiam inaccessibilitate occultam et invisibilem*'.[16] Grosseteste does, however, correct Dionysius's scriptural references in favour of greater exactitude; '*nullus videt neque videbit*', is qualified: '*sicut ait beatus Johannes in evangelio, et in canonica, "deum nemo vidit unquam", et Moises in exodo ex persona dei, "non videbit me homo et vivet"*'.[17]

The long, personal passage which follows, and in which Grosseteste develops his own thought in independence of the text occasioning it, is unique in his Dionysian commentaries, amounting as it does to a brief treatise on the immediate vision of God as the last end of human nature. Its systematic character suggests a divison into six articles as follows:

a The supreme power of human nature, *intelligentia*, is limited in its activity by the conditions of our earthly life.

b Human nature has an innate capacity for the immediate vision of the divine essence in itself.

c No exhaustive knowledge of God is possible to a created mind.

d The direct vision of God is not limited to the afterlife, but is a privilege accorded to a few chosen souls already in this one.

e The definition of integral human nature is made by reference to the possibility of the immediate vision.

f On theophany.

Grosseteste's tactics are revealed in the first member of the discussion. If he is to maintain, in accordance with the Latin tradition, that the vision of God is an intrinsic possibility of human nature, then respect for the scriptural evidence, asserting that our actual condition as men precludes the enjoyment of the vision, will necessitate showing the reason why this is so. Grosseteste sees no difficulty in supplying such. Original sin has for its effect the disintegration of the natural hierarchy

---

[16] I give references to the text published by Dondaine, art. cit., pp. 124-5, up to the point in the commentary where this leaves off, and thereafter to McQuade's edition.

[17] Dondaine, art, cit., p. 124. The references are to John 1:18, I John 4:12, and Exod. 33:20.

of human powers, so that what was made to serve and follow tends to rebel and initiate inordinate action: the body rebels against the leadership of the soul, and the affections of the soul itself contend against the highest faculty, that of vision. There can be no question of the vision's becoming effective in such a state, for the human power that is destined to see the divine essence in itself and without an intermediary can do so only if it is unhindered in its natural action.[18]

The reader who bears in mind the nexus of emergent theories of the beatific vision present in Parisian writings during the years before 1241, cannot fail to be struck by the decisive nature of Grosseteste's expressions. The Parisian followers of Dionysius, Scottus, and Chrysostom sail a little flotilla, which fluctuates in a variable rhythm as it pursues its single course: '*non ipsum Deum per semetipsum videbimus*'; '*non quid est, sed solummodo quia est*'; '*a nullo mortuorum videtur in substantia sua*'; '*videbimus Deum, non tamen essentiam*'; *per medium, speciem, virtutem*'. The precision and insistence of Grosseteste's language as he formulates the Western reply— '*virtus humanae mentis nata comprehendere divinam essentiam sine medio sicuti est*'—tempt us to seek the source of his exactitude not merely in immanent terms of his own urge to lucidity, but in the light of theological actuality. It is very likely that his doctrinal formulation must be placed in a relationship of direct opposition to the developments in the Paris Faculty.

The second part of Grosseteste's little treatise seeks to prove that the human mind has of its nature an innate capacity to see the divine essence, as it is and without a medium. His thoughts turn at once to Christology. When the Word assumed flesh, he took upon himself a created nature, to which nothing beyond the fullness of human being was added; and though no doubt the angels ministered to him (Grosseteste has not forgotten the Dionysian context in which his discussion took its origin), it was as a man using a human intelligence that he knew the divine nature in itself. No more is needed in order to show that the fullness of human nature (*integritas naturae*) is possessed of the power of understanding the essence

---

[18] 'Dum enim mole corruptae carnis aut inordinatae affectionis aggravatur suprema mentis humanae intelligentia, non est possibile quod Dei essentiam sicuti est in se videat immediate. Necesse est enim quod virtus humanae mentis nata comprehendere divinam essentiam sine medio sicuti est omnino sit ex omni parte impraepedita et irremissa cum in actum ad quem nata est prorumpet.' (ibid., p. 124).

of God without intermediary, and that it will require, after death, only the divine light in order to be activated.[19] Since the conditions which here and now prohibit its exercise, and which all derive from the fall of mankind, will no longer obtain in that state, what is perfect will have emptied out what is in part, and we shall see no longer in a mirror darkly, but face to face, knowing even as we are known.

We cannot fail to observe, as a mark of the specificity of the two theological traditions, Greek and Latin, that Christological considerations both inhibited the Cappadocians' attributing to creatures a capacity for the immediate vision of God, and proved capable of reinforcing the contrary doctrine in the Latin atmosphere.

If the Greek theology of the vision made inroads into the Latin tradition after *c*.1220, that was only because it asserted and safeguarded a fundamental Christian value, to wit, respect for the transcendence of the infinite God over his creatures. St. Gregory the Great, in whose writings, as we recall, the Greek idea of the limitation of the vision of God first found a warning echo, saved this value by recognizing the limits inherent in the creature's direct vision of the divine essence: '*In futuro reperietur omnipotens per speciem, sed non ad perfectum, quia eius essentia a nullo plene videbitur*'. Grosseteste reveals himself as the faithful disciple of Augustine and Gregory, as well as the admirer of Greek theology, when in the third article of his discussion he stresses the limits which surround the vision. It is clear from what he says that he considers the Latin theology which he upholds to be nothing other than a transcription of the biblical data concerning the life of heaven. To see God 'as he is' and 'face to face', and 'to know him even as we are known'—such expressions neither assert nor imply that the created intelligence is capable of penetrating the entirety of what God is (*'ipsam totalitatem quidditatis divinae essentiae'*), for this does not lie within the powers of anything finite. They mean no more than that the inaccessible light which God

---

[19] 'Quod autem habeat humana mens naturaliter sibi insitam potentiam seu virtutem comprehendendi divinam essentiam sicuti est sine medio, patet ex hoc quod dei verbum non assumpsit nisi naturam humanam, supra naturae humanae integritatem nullo addito, et quod ipse homo humana intelligentia sine omni medio (licet angeli ei ministraverint) intellexit ipsam divinam essentiam sicuti est in se. Habet igitur homo in suae naturae integritate potentiam intellectivam divinae essentiae sicuti est in se sine medio . . .' (ibid., pp. 124–5).

inhabits is visible only by the purified understanding capable of functioning without a phantasm, and then only in such a way as to remain transcendent to the understanding of the intelligence. God's 'dwelling in unapproachable light' implies that the true and direct knowledge of his essence still does not reach through to the totality of what he is, though an exception must, of course, be made for the human intelligence of Christ, which is theandric.[20]

This is the best statement of a doctrine to which Grosseteste remained faithful throughout his Dionysian commentaries, and which, with the minimum necessary qualifications, he adapted to the discussion of the divine names. In this statement, the *'sine medio'* of the first article is expanded slightly to become *'nude sine imaginibus materialibus et immaterialibus'*, which is not without an element of obscurity.[21] Clearly, it presupposes an epistemology in which intellectual knowledge is always accompanied by an image drawn from sense experience; but what can be meant by an 'immaterial image'? Two possibilities suggest themselves.In the Dionysian theology, names are applied to God according to his theophanies in material and spiritual creation, and in the negative moment the whole hierarchy of names is negated in an effort of transcendence, which culminates in the cloud of unknowing. The *'imago immaterialis'*, when used in this context, might mean the image accompanying our thought of spiritual substance; but the whole tendency of Grosseteste's commentaries is to say that the idea of spiritual being is reached only by negating

---

[20] 'Notandum quod videre Deum sicuti est . . . et huiusmodi non asserunt nec insinuant quod intelligentia creata penetret vel penetrare possit ipsam totalitatem quidditatis divinae essentiae, hoc enim est impossibile; sed quod nude et sine imaginibus materialibus et immaterialibus lucis quam inhabitat Deus et quae ipsi est inaccessibilitas, ab omnino puris intelligentiis est perceptibilis et visibilis *eo* quod ipsa vere excedit et exsuperat in infinitum omnis intelligentiae comprehensibilitatem. Et quia ipse Deus est lux inaccessibilis, et lucis inaccessibilitas . . . cum sine imaginibus nude sic videtur, sicuti est videtur, et facie ad faciem videtur, et vere cognoscitur, et eius essentia cognoscitur; non tamen totalitas suae quidditatis penetratur.' (ibid.). Dondaine's edition omits the word *'eo'* here italicized, and his punctuation alters the sense slightly.

[21] A parallel passage in the closely contemporary sermon *Ex Rerum Initiatarum* (ed. Gieben, *Coll. Franc.* 37 (1967), p. 121), also mentions the imageless character of the vision, but without distinguishing between different kinds of image: 'Et in quantum rationalis, habet potestatem naturalem et virtutem contemplandi Deum unum et trinum, veritatem nec fallibilem nec fallacem, iustitiam et bonitatem, sicuti est, absque imaginibus, et in rationibus aeternis et causalibus omnium in ipso, quae sunt sapientia genita, cognoscendi omnia.'

the material characteristics of physical symbols. It is more likely that Grosseteste is here thinking of the powerful revelatory quality of images derived from light, because light in its highest physical manifestation was thought of as approximating to the lowest grades of spirit, and as being itself almost immaterial.

Up until now Grosseteste has spoken as though the beatific vision were confined to the angels, the saints, and Christ. He makes an extension, however, when he turns to the visions of God spoken of in Scripture. God singled out Moses for his faithfulness, and spoke to him 'openly, face to face'; and Paul was rapt into the third heaven. The intimacy of these visions (which Grosseteste extends to the Virgin Mary), distinguishes them from the innumerable other visions recorded in Scripture, and he accounts for them by assimilating them to the beatific vision. It emerges from his remarks that, for the divine essence to be perceived, it is only required that the inglorious conditions of fallen nature should cease to impede the action of the intelligence, or *suprema pars*. It still remains true that 'no man has seen God', because in the moment when such a vision was accorded the subject no longer lived as a man, that is to say, under the weight of sin's impediments, which define the usual meaning of the word 'man'.[22] What is human nature, and how is it to be defined? The problem arises naturally out of the preceding stage of inquiry, and its solution furnishes an excellent summary of the little treatise. '*"Homo" multis modis videtur dici*', but its highest meaning refers to the integral, perfect, and incorrupt human nature, such as it was in its first condition, made in God's image. 'Man' is also used of our condition after the fall, and of mankind as living according to the political and speculative virtues. In its most noble sense the word 'man' refers to the person who lives according to wisdom, the supreme speculative virtue, the knowledge of God. Among such, it applies with fullest force to the one who contemplates God openly, and not darkly through a mirror, for he has attained to the end for which humanity was created.

---

[22] 'In momento autem visionis huiusmodi, qui videt non vivit homo, quia non praegravatus nec praepeditus praedictis molibus et impedimentis ab actu supremae potentiae naturae humanae, quam aggravationem et retardationem importat in se hoc nomen homo communiter dictum, sed magis vivit super hominem.' (Dondaine, art. cit., p. 125).

The contemplative of this kind is said to live superhumanly, because his ecstasy raises him above the community of men practising the political virtues, and really he shares the life and contemplation of the angels.[23]

Grosseteste's conclusions amount to a definition of human nature's fullness in terms of its future destiny, which is to say, in terms of the activity of the *intelligentia*, variously called *suprema pars (facies) mentis*, or *apex mentis*, but always thought of as directed towards wisdom as its object. This highest faculty is present in every state of man, but can be rendered inactive by the disorder of sin. Its purification, however, remains a possibility for man even in this life—under grace—and in the measure of the progress he makes, the highest faculty, that element in man which is most fully human, is freed for the knowledge of God. It is therefore not only in the tiny handful of actual mystics, in the plenary sense of that word, that the intelligence is active, nor after our death merely, but rather in all men who seek the light that alone assures the flowering of their human nature at its best.

It only remains now for Grosseteste to determine the meaning of the difficult term '*theophaneia*', which was making such a theological career at the time, and to give it an acceptable sense within the scheme of ideas already developed, if that proves possible. He succeeds in forging a connection with the previous subject. Save in the case of a tiny group of privileged saints, he claims, human nature in its present condition does not enjoy the direct vision of God, and Dionysius, he tells us, wishes to warn us not to take the theophanies of the Old Testament to be direct, non-symbolic encounters of man with the divine essence. Theophanies are manifestations of things divine, made to the saints through symbolic expressions which fulfil two requirements: they befit God, their giver and revealer, and they are adapted to the level of attainment of the person receiving them. The scriptural use of the term refers, as Dionysius claims, to a vision containing a symbolic likeness

[23] 'Maxime autem et excellentissime dicitur homo, qui vivit secundum supremam virtutum speculativarum, quae est sapientia, id est ipsius divinitatis cognitio; et inter viventes secundum hanc virtutem, ille est summe homo qui palam, non per enigmata vel figuras, sed immediate contemplatur divinitatem; hic namque est in actu et fine propter quem conditus est homo, et tunc unumquodque propriissime et maxime dicitur quod dicitur, cum est in actu completo et fine proprio propter quem conditus est . . .' (ed. McQuade, p. 197).

of the reality, which is formless; and the word itself indicates that such vision leads us to God. Grosseteste cannot resist an etymological gloss, which aims at showing that the 'appearance of God' signifies at once the raising of the soul to contemplation of him as revealed, and the splendour of God entering the illuminated mind. Grosseteste is forced to hesitate between two interpretations of the 'reality which is formless': either the phrase refers to immaterial reality, so that the symbolic element in a theophany would rest upon a *similis similtudo* of the divinity, one, namely, which is derived from incorporeal realities such as reason, intellect, light, or life,[24] and not from grosser material things; the other possibility is, that the 'reality which is formless' is the divine being, the final cause of symbolic discourse, which can be represented, albeit inadquately, by any created form. Grosseteste admits as an afterthought that Dionysius does not restrict himself to the former use, but allows his usage to fluctuate.[25]

Grosseteste attempts to formulate a definition of theophany based on Dionysius's remarks, as a vision or intelligible form impressed on the understanding, manifesting a likeness to God, and being derived from immaterial reality.[26] With this he has achieved something of note, in interpreting 'theophany' as representing a limited and indirect, symbolic vision of God clearly distinguishable in its characteristics from the beatific vision, to which it can no longer threaten any interference, since it is affirmed of a lower level of religious experience. He has thereby managed to disarm Dionysius and even partially to win him over, leaving himself in consequence entirely free to adopt the Neoplatonic theology of mediation, including the role of the angels as secondary causes between God and the human hierarchy, and the theology of the divine names

---

[24] '. . . similis formatio est *ex informibus*, id est incorporeis, ut est ratio, intellectus, lux et vita . . .' (ibid., p. 199).

[25] Grosseteste's *notula* to this effect refers to ch. 7 of the *Celestial Hierarchy*, where the variation in usage occurs. That it was an afterthought is indicated by the uncertainty of the copyists as to where best to place it; the most faithful MS puts it at the end of our present section. See McQuade, p. 200.

[26] 'Est itaque theophaneia visio, id est forma intelligibilis aspectui intellectui impressa et unita, inquantum ipsa manifestat intellectui assimilationem divinam scriptam in ipsa intelligibili forma ut in formatione et figuratione sumpta ex informibus, id est immaterialibus.' (ed. McQuade, pp. 199–200). Grosseteste's free use of Avicennian and Victorine vocabulary may be regarded as a mark of his approval.

and their signification. These acceptable and useful theological elements can no longer confuse nor undermine the Western tradition concerning the direct vision of the divine essence, which is the very kernel of the definition of human nature in both its present and its future state.[27]

[27] It is very unlikely that Grosseteste's *prise de position* concerning this doctrine exerted any influence on the condemnation of 1241 at Paris, for his Dionysian commentaries were slow to enter the Parisian scene, and were not widely circulated there even at a much later period.

Chapter 3

The Soul and its Relation to the Body

I THE EVOLUTION OF GROSSETESTE'S PSYCHOLOGICAL DOCTRINE

NEITHER the *De Artibus Liberalibus* nor the *De Generatione Sonorum* sets out to develop or retail a theory of soul, but, although each has a limited objective, we can garner from passing references some bearings that are indicative of Grosseteste's psychological teaching as a young Master of Arts.[1]

It is their eminently practical aim that makes the seven liberal disciplines to be arts and not sciences, for they exist in order to purge mankind of error. Now the works which lie within our power consist of knowing, willing, moving the body, and the desires effecting movement.[2] That which is known is judged to be fitting or harmful, whereupon the will (*affectus*) tends to embrace the suitable or to retire within itself in order to avoid what harms. Now it is the science of moral philosophy which teaches us what is to be pursued and what avoided, and here the art of rhetoric proves its utility; as a good *ministra* of moral philosophy, it moves the concupiscible power to the pursuit of the fitting, and the irascible to the avoidance of the harmful.

Natural philosophy also gains from the ministry of the arts. Musical modulations, medical authorities believe, can be used for healing, for the reason that illness is cured by ordering and tempering the spirits. Even the deaf can be treated by music, for the following reason: the soul follows the body in the latter's affections, and the body follows the soul's actions. When, therefore, the body is affected by sounding numbers,

---

[1] There is every reason to take both works together, as they are similar in doctrine and spirit. Dales places the *De Gen. Son.* in 1221–2, but gives no argument for such a late, and precise, dating. The only sources quoted are Priscian and the *Etymologies*, both of them part of the stock-in-trade of the Master of Arts, which Grosseteste was until 1209. The work exhibits no knowledge of Greek, and its only other apparent source is the *De Musica* of Boethius.

[2] 'Opera enim nostrae potestatis aut in mentis aspectu, aut in eiusdem affectu, aut in corporum motibus, aut eorumdem motuum affectibus omnia consistunt.' (ed. Baur, p. 1.12 ff.).

the soul draws out of itself numbers which are of the same proportion, and the spirits adjust the proportions of the numbers to agreement. The wise doctor must therefore have a knowledge of the due proportion of the body as impressed on it by the stars, and must be acquainted with the proportions which induce concord among the elements and the humid parts of the principal spirits, and between the soul and the body. When these proportions are expressed in terms of musical sound, upon the numbers' reaching the soul everything in man returns to a proportioned state. The doctor must also have studied the behaviour of the spirits prevailing in different emotional states, such as in joy, when they dilate, and sadness, when they contract; for the states of the soul too can be affected by the knowledgeable employment of musical sound (ed. Baur, pp. 4.35-5.20).

The numbers present in sound and in the soul, through which the due proportion between body and soul is restored, are described as '*progressores et occursores*', and this nomenclature draws our attention to a passage in the same work where the process of perception is outlined (pp. 3.30-4.9). Again the example is derived from musical sound.

When a sounding object sends forth a *numerus in progressione* to the ear, the soul sets to work in the air inside the ear a number which meets that received by the sense. This number then progresses from the external sense to the memory, following which a certain number is adapted within the whole soul, one fitted to the number stored up in the memory. The result is pleasure within the soul if the sound struck was harmonious, and pain if it was dissonant. In all this process reason has had as yet no part to play, but the soul can now apply the numbers of judgement, by which it can distinguish and pronounce upon the rest. Music and harmony are therefore not limited to the study of sounding numbers, but extend to *progressores* and *occursores, recordabiles, sensibiles* and *judiciales (numeri)*.

The teaching of the Master of Arts, while it reveals an interest in medicine and astrology which derived from Arabic science, is in all points of psychological doctrine thoroughly Augustinian. The use of *affectus* and *aspectus mentis* is based on a remark of Augustine.[3] The division of the powers of the

[3] *Soliloquia I*, 6 (*P.L.* 32, 875-6).

soul into *ratio, concupiscibilis,* and *irascibilis* is perfectly traditional, and defended as the doctrine of the Fathers in the
Speculum Speculationum of Nequam: '*Vis rationabilis ut
discernat bonum a malo, vis irascibilis ut detestetur malum,
vis concupiscibilis ut eligat bonum.*'[4] Concerning the Aristotelean tripartite division of the soul, the *De Artibus Liberalibus*
is silent. The soul is the noble and spiritual being of Augustine,
which knows, wills, and moves the body, which cannot itself
be affected by anything ignoble or material, but which provides the active element in the sensation taking place in the
body's organs. Sensation is but the occasion for the soul to
draw from within itself the material of knowledge, the numbers. In all of this, as in his enumeration of the numbers of the
soul, Grosseteste is the faithful disciple of Augustine's *De
Musica.* If we add the spirits which operate as a medium between soul and body, mediating between the numbers of both,
and the proportion between a soul and a body which are conceived of as different realities, we have a doctrine that is fully
Augustinian in every important respect. Of the new psychological theories there is not a trace.

In this respect the *De Generatione Sonorum* represents an
advance on the *De Artibus,* and may with certainty be considered posterior to it. In a *resolutio* aimed at reaching the
causes of sound, Grosseteste divides the *virtus motiva* into
intrinsic and external agencies. Now the only cause of sound
that is intrinsic to the sounding object is soul; inanimate
nature cannot produce sounds as an intrinsic moving agent.
And since the movement which produces sound is not continuously present in that which possesses soul, it cannot derive
from the vegetative soul, but only from the sensitive, when
this is moved by a voluntary motion presupposing imagination
or apprehension of an object. Thus the definition of *vox* can
be formulated: a sound formed by a first motive cause possessed of imagination. We can distinguish further to reach the
articulate human voice (*vox litterata*). What gives a voice its
kind and perfection is the formation of the vocal instruments
and of the moving spirits which affect them; only the perfect
formation of both results in articulation.[5]

---

[4] Callus, 'Introduction of Aristotelean learning . . .', p. 24.
[5] 'Primum autem motivum talis motus non potest intraesse nisi ipsa anima,
quia natura non potest esse principium primum talis motus. Et cum non sit talis

A new triadic division of the soul (vegetable, sensible, and rational) has here replaced the traditional schema, and betrays the influence of Aristotelean philosophy. Typically, Grosseteste's new source is made to take its place side by side with the old; the soul still acts upon the body's organs (which are adapted to its grade of being, sensible or rational) through spirits, whose main characteristic is their mobility (p. 8.30 ff.).

If our chronological argumentation is sound, the next evidence concerning the development of Grosseteste's psychology dates from the period 1225 to 1230, that is, some twenty years after his earliest extant compositions. Though the nature of the soul is approached from two different angles in the *De Statu Causarum* and *De Intelligentiis*, the doctrine of both treatises is similar, and betrays the fact that his reading of the new psychological literature has caused Grosseteste to add an important element to his concept of the human soul.

The *De Statu* is devoted to a discussion of the Aristotelean four causes, but breathes a Neoplatonic atmosphere throughout. Grosseteste is aware that Aristotle's metaphysics is faulty, that, in particular, it has no systematic theory concerning the relations of the first cause to the secondary causes. If there is no first cause, he argues, then there is simply no causality, as each lower causal agent derives its activity as cause from a higher one, not merely that next to it in the hierarchy of causes, but from the one above that again, and ultimately from the first.[6] The real problem concerning causality lies in proving that secondary causes have real, and not simply instrumental, causal efficacy.[7] One suspects that the influence of the *Liber De Causis* is at work here.

It is the section on formal causality which interests us, since Grosseteste is forced to situate the formal causality of

motus continue *in habente* animam, non erit talis motus ab anima vegetativa, sed a sensibili motiva motu voluntario, *quem* necessario praecedit imaginatio vel apprehensio. Ergo sonus formatus a primo motivo, in quo est imaginatio, vox est. Sed cuidam voci dat speciem et perfectionem ipsa figuratio actualis instrumentorum vocalium, et figuratio motus spirituum motivorum instrumentorum vocalium.' (ed. Baur, pp. 7.25-8.6). The first two words italicized are noted by Baur as a variant reading for *habens* in his text, which, however, does not make sense. I have read *quem* for Baur's *quam*, with the support of one MS, because the context demands it.

[6] See ed. Baur, p. 120.7-10.

[7] Grosseteste decides that they do have real causal efficiency, because efficient causes intervening between the first cause and the last effect possess an *intentio propria*, which makes them agents rather than instruments.

the soul, Aristotle's doctrine, therefore, within his review of the causes. He begins his treatment with an Aristotelean division of formal causes into substantial and the nine accidental forms, then of substantial form into absolute and relative (the latter being, e.g., the whiteness of some concrete white thing); thirdly (and less peripatetically), *forma substantialis absolute dicta* receives a triple division into the separate exemplar, the form by which a thing is, and the form which is both exemplar and cause of the object.[8] Real substantial form may further be purely material, impressed upon matter so as to receive a definite place (e.g., the forms of the elements and of mineral bodies, in which each part is of the same species as the whole),[9] or it may be immaterial, i.e., one not localized in the different parts of matter, and therefore a principle of life. All souls are unlocalizable in this sense, but the vegetative and sensitive souls are educed from potency to act by the intervention of celestial virtue and the agency of a form, both of which are in place and matter,[10] whence they are corruptible. These souls are characterized by a higher grade of unity than material forms possess, because their parts are organic, and therefore not of the same kind as the whole, but only of the same operation; by which is meant that every part of the plant has vegetable life, and sensation is present throughout the animal body. The intellective soul, however, which 'comes upon' the vegetative and sensitive, is a form of a different kind. Because it is added directly to an unlocalized form, it cannot have been generated by any material agent, but only by an immaterial form, namely, the first form of all things.[11] Unlike inanimate

[8] 'Forma vero substantialis absolute dicta adhuc dicitur multipliciter. Dicitur enim uno modo exemplar separatum a re et non quo res est, et alio modo quo res est, ita quod sit coniunctum rei et non exemplar; tertio modo dicitur simul exemplar et quo res est.'

[9] Reading *eam* for *autem* at p. 124.30. The *forma impressa* indicates the influence of Avicenna; *De Anima* V. 1, ed. Van Riet, p. 91.

[10] 'Alio modo dicitur forma, quae non est situalis et non recipit participationem secundum situs diversos in partibus materiae, ut est anima, et hoc dupliciter: aut cum non sit situalis, mediante tamen virtute caelesti, quae situalis est, a forma situali educitur de potentia in actum, ut est anima vegetativa et sensitiva.' (ed. Baur, pp. 124.34-125.2).

[11] 'Alio modo est anima intellectiva forma superveniens his, de quibus iam dictum est, quae, quia immediate advenit formae non situali, necesse fuit ut non educeretur in esse *mediante forma situali*, sed immediate a forma non situali, scilicet prima.' (ibid., p. 125.6-10). (The words underlined are given by Baur as a variant reading, but the balanced contrast in the phrase as a whole confirms their genuineness.)

natures, in which the part of a substance is of the same kind as the whole, and unlike vegetative and sensitive things, in which each part performs the operations of the organic whole, the intellectual soul cannot be divided into parts standing in any way for the whole, for no part of a man is the man, and it is not a part of man which thinks, but the whole man.[12] Its perfection, that is to say its unity and simplicity of essence and operation, makes the intellective soul the supreme natural form.[13]

'Substantial form' has two further meanings, for it applies to both the intelligences and God. The 'separate forms' are so called because they are united to the celestial bodies only as movers, and therefore extrinsically; they do not need contact with matter for their perfection (p. 125.16-19). In this respect they stand in contrast to the soul, for though the latter's incorporeal and simple nature has been emphasized, the soul is not merely the mover of the body, but is united to it in such a way that its highest operation, thinking, is impossible without a phantasm, which is the product of the sensible power.[14]

The form mentioned earlier 'which is both exemplar and cause of the thing', is the first form, abstract, simple, and separate.[15]

It has been customary to find in this passage the origin of the theory of plurality of forms, which was to be popular two generations later in the Franciscan school at Oxford.[16]

---

[12] '. . . quia nulla pars hominis est homo, nec aliqua pars hominis intelligit.' (ibid., p. 125.11)—proof that Grosseteste has made some acquaintance with the *De Anima* of Aristotle.

[13] This seems to be the meaning of the rather oblique statement: 'Et propterea, cum haec forma nullam relinquat multitudinem vel diversitatem operationis vel essentiae, in sua perfectione, dicitur quod est ultima formarum naturalium.' (ibid. 125.12-15). Miss Zedler (art. cit., p. 147) arrives at a paraphrase which expresses a sense opposite to ours: 'the diversity of operations and perfections makes (the) intellective soul the highest of natural forms'. Miss Sharp (op. cit., p. 27) had spoken in the same terms. Grammar aside, it is unlikely that the disciple of Augustine would find diversity and not its opposite, simplicity, a virtue in the soul. Simplicity is clearly what Grosseteste intended in building a hierarchy of inanimate things, vegetative and sensible things (possessed of higher, organic unity, with a soul that is active throughout), and intellective soul. It is totally in the sense of the *De Statu* that the higher a being's causal operation, the greater its unity should be.

[14] 'Anima vero rationalis non solum unitur corpori humano sicut motor, sed etiam sicut intelligens, mediante virtute corporea. Intelligit enim non sine phantasmate, quod est actus virtutis sensitivae.' (ibid., p. 125.20-22).

[15] 'Haec est forma prima, quae qualiter sit forma prima, difficile est explanare.' (p. 125.25). Had Grosseteste at this stage already treated the question, he would surely have made reference here to his *De Unica Forma*.

[16] Sharp, op. cit., pp. 28-9; Zedler, art. cit., pp. 147-8.

This interpretation has two disadvantages: it cannot be substantiated on the basis of the text, and it has distracted attention from the very real evolution of thought which is witnessed to here. In language and thought, this passage belongs in its totality to the genre of literature attempting to assimilate the new *scientia de anima*, as the psychology of Aristotle *et sequaces eius* came to be called; and it registers a degree of success. Compressed into a minimum of words is the whole psychophysical hierarchy of forms, beginning in the inanimate world with the forms of the elements and complexioned bodies. That the discussion conducted in terms of *forma situalis—non situalis* presupposes an accurate idea of the categories and the definition of *situs* and *locus*, is confirmed by reference to the parallels in the *De Intelligentiis*.[17] Avicenna's specification of the Aristotelean theory of vital generation is compressed, but is present in all its essentials: the vegetable and animal souls are educed from potency to act by a material form with the cooperation of the celestial virtue, which proportions the elements and brings them to the degree of equality required for the development of an animate soul. The remark to the effect that 'it is not a part of man that thinks' (but the whole man), betrays some acquaintance (though perhaps through a florilegium) with the *De Anima* of Aristotle. The impossibility of the soul's attaining knowledge without reference to the phantasm indicates that Grosseteste has adopted a theory of abstraction, and with it a tendency to regard the soul and the body as somehow united, and not simply juxtaposed. But above all, there is the assertion that the soul is the substantial form, as well as being the mover, of the body. Of course, the simplicity and spirituality devolving upon it as the direct creation of the first form preserve its nobility and distinctness from the purely natural universe, and the soul has lost nothing of its Augustinian grandeur; never the less, Grosseteste sees no obstacle to speaking of it in the language of Aristotle. He was not aware of the irreducible opposition between Plato's[18] and Augustine's

---

[17] See, e.g., ed. Baur, p. 115.13-15.

[18] Grosseteste was well aware that the definition of the soul as self-moving went back to Plato: 'Si loquamur de motu secundum modum Platonis, verum est quod anima est substantia seipsam movens.' (*De Motu Corp.*, ed. Baur, p. 91.12-13). He may have derived his information on this occasion from the *De Anima* of Aristotle, I, 2, 404a, 21; I, 3, 408a, 30; but he had also read the *Timaeus*.

concept of the soul as substance and mover, and the Aristotelean definition of it as a substantial form animating and
perfecting the body.[19] The contradiction remained unperceived
by Grosseteste's whole generation, and was not detected by
anyone before Aquinas.[20] Grosseteste, then, managed (to his
own satisfaction, at least) to wed the traditional theological
conception of the soul's nature and functions to the new
philosophical trend, which he welcomed and flattered in the
most sincere way, by imitation.

As for the question of the plurality of forms, the text does
not allow of any positive pronouncement concerning it. That
the vegetative and sensitive souls are conceived of as substantial forms there can be no doubt, but it is not stated, nor yet
clearly implied, that they remain such after the coming of the
rational soul. Miss Zedler's case for plurality reduces itself to
the affirmation that in the case mentioned, the intellective
soul 'comes upon' (*superveniens, advenit*) the others; but this
expression does not unambiguously imply plurality.[21] It is
better to approach the question of plurality only after a more
detailed discussion of Grosseteste's doctrine on the powers of
the soul.

Are the angels in distinct places, or everywhere at once?[22]
By a deft generalization of the problem referred to him,
Grosseteste manages to produce a response that is valid for the
relations of spiritual to material being as such: of God to the
material world, the angel to the body in which it ministers, or
the region over which it presides, and the soul to the body
united to it. If the theological nature of his inquiry encourages
him to remain within the atmosphere of Augustine's writings,
from which he quotes at length, the psychological doctrine
presented does not differ from that of the *De Statu*, though
it is in some respects more, in others less, explicitly delineated.
The divine ubiquity provides a model of the soul's relationship
to the body: as God is everywhere in the universe, so is the

---

[19] In *De Intell.*, p. 115.33 the soul is described as the *perfectio* of the body,
to which it is united, as well as being its motor.

[20] See Crowley, *Roger Bacon*, p. 191 ff.

[21] Zedler's other argument (p. 148) is based upon the faulty text of Baur,
corrected above by the choice of an alternative reading.

De Intelligentiis, ed. Baur, 112.1–3 (see Part II, ch. 1, pp. 52–8). Unlike many
other thirteenth-century Schoolmen up to and including Aquinas, Grosseteste does
not seem to have used Hugh of St. Victor's discussion of the same question in *De
Sacramentis* I.3, 28 (*P.L.* 176, 224BC).

soul in the body it animates, for the soul is the image of God. With St. Anselm, Grosseteste rejects the popular ancient idea that the soul is located essentially in a single dominant organ, the brain or the heart, from which its power can be exercised in the other parts.[23] Grosseteste's emphatic objection to this theory is clearly motivated by his desire to preserve the essential simplicity of the human soul: its substance is everywhere present where it acts, since it is incorporeal and lacks dimensions (*De Intelligentiis*, pp. 114-15). Not even the light-metaphor, which images the soul as a point of light situated in the heart (or brain) and radiating its *virtus* to the members, has any validity, for, were the soul situated thus its physical distance from any given co-ordinate of place would be susceptible of accurate measurement. As it is, one might as well attempt to measure the distance intervening between the foot and the health, or the complexion.

The soul lacks *situs* and *locus* within the body, as does God in the world. The soul can, however, be said in normal parlance to be *in* the organ in which it initiates the functions under its control, the heart for vegetative and the brain for sentient and motive functions. However, the site here in question is not that of the soul precisely, but of the spring of corporeal motion affected by it.

Two more questions remain to be answered: how do the angels move the bodies they assume, since they are incorporeal? How is it that angels (and souls) can suffer from corporeal punishment? As to the first, the soul is once again the model. Between the extremes in man, namely, the weightier members and the incorporeal soul, there are transmitters of motion, namely, nerves and muscles, which themselves receive movement from the bodily spirits. The spirits are made of the finest material in the body and are close to spiritual nature, being in their essence light. Motion begins to be transmitted when the impulses of the soul (*affectiones*), themselves spiritual, reach the light-energy of the spirits.[24] A continuous process of this kind, beginning in the soul and ending in the members, presupposes a bond between soul and body, whereby a

---

[23] *Proslogion*, c. 13 (*Opera Omnia*, ed. Schmitt, vol. 1, pp. 110-11).

[24] '(Anima) . . . commovet sine medio hoc quod in corporibus magis appropinquat incorporalitati, et hic est spiritus corporeus sive lux, quo medio moto movet consequenter corpora grossiora.' (*De Intelligentiis*, p. 116.18-27). Grosseteste's notion of the spirits is discussed below in section ii (pp. 278-89).

connection is established between the soul's impulse and the body's adapted response. It is therefore difficult to transfer the model to a separated intelligence, related to matter purely as a mover.[25] But in any case it is clear that, though the soul's motion is neither material nor local, its bond with the body enables it to produce a motion that is both, as the hinge allows the door to swing, though it itself remains unmoved (pp. 116.34-117.2).

The second question concerns the bodily suffering of the damned, which Grosseteste tries to understand by considering the relations of soul and body in this life. A single Augustinian principle underlies his entire thinking: incorporeal substances have a higher (*nobilius*) mode of being and activity than material things, which cannot directly act upon them,[26] for the least of spiritual beings is more noble than the best of material things. This principle must somehow be reconciled with the experience the soul has of suffering from extreme sense-stimuli, for Grosseteste is Aristotelean enough to be aware that the soul enjoys moderate sensations, whereas extremes cause it pain.[27] Augustine helps to resolve the difficulty by suggesting that the embodied soul is not directly affected by the body, but yet cannot avoid being affected when the latter receives impressions.[28] The body does not act upon the more noble soul, but it is the necessary occasion of an action within the soul, which, therefore, suffers in relation to matter, without suffering from it.[29] Grosseteste gives an example illustrating

---

[25] 'Sed videtur animam non sic posse movere corpus . . . nisi esset ligata corpori, ut propter nexum colligationis motionem huius sequeretur motio comproportionalis sibi colligati.' (ibid., p. 116.31-4).

[26] '. . . licet substantiae incorporeae possint agere in corpora, utpote nobiliora in minus nobilia, non tamen ut videtur e converso corpora possunt agere in substantias incorporeas, quia ignobilius non potest agere in id quod nobilius est; substantia autem incorporea, etiam informis, quovis formato corpore nobilior est.' (ibid., pp. 118.37-119.4).

[27] 'Sicut anima . . . ad motiones corporis per qualitates sensibiles excellentes patitur et torquetur. . .' (ibid., p. 118.7-9). That the interpretation of the difficult word '*excellentes*' given above is the correct one is confirmed by p. 118.20-24: 'Est enim passio delectabilis, sicut et passio poenalis; in sentiendo enim media delectabiliter patitur anima, sicut poenaliter patitur in sentiendo excellentia extrema.' The classical formulation of this experience is that recorded by St. Bonaventure, *Itinerarium* II, 5 (V. 301a): 'Sensus tristatur in extremis et in mediis delectatur.' The source is Aristotle, *De An.* I, 12, 424a, 18-31, III, 2, 425b, 21 ff., and 426a, 30 ff., III, 4, 429b, 1-5.

[28] *De Musica* VI, 5, n. 9 (*P.L.* 32, 1168).

[29] 'Dixi . . . omnes animas, dum sunt in corpore, ad motiones quorundam

this nice distinction by which the soul's transcendence is saved. The motion of a mirror necessarily occasions a change of direction in the light-rays reflected from it, but it does not for all that account for their movement: they propagate themselves (p. 119.24–26).

As was remarked above, the theological nature of Adam of Oxford's question gave Grosseteste ample scope to expound a psychology derived in essentials from St. Augustine. The soul is conceived of as a spiritual substance which, like its kin, the angel, is nobler than any material thing. As distinct from the separate intelligence, however, it is united to a body as its perfection, and here Grosseteste borrows the language of Avicenna: *colligatio, perfectio*.[30] Both mover and form, the soul is present everywhere in the body without being located in an organ, but it acts directly only through affecting the luminous spirits, the intermediary between spirit and matter favoured by all Neo-Platonists. Grosseteste's remarks on sensation make it clear that his reading of the newer philosophy has not led him to doubt the validity of the sensation-theory based upon numbers, according to which the sense-stimulus constitutes the necessary occasion for the soul to draw out of itself the *numeri iudiciales* corresponding to the number sensed and remembered. We may well wonder how Grosseteste reconciled this Neoplatonic thesis with belief in the necessity of the phantasm and in abstraction, but in any case he found no contradiction between them, any more than in considering the soul as both substance and substantial form. He witnesses to a period of untroubled eclecticism in which the awareness of sharp oppositions was still a thing of the future. There can be no doubt that Avicenna's *De Anima* was partly responsible for blurring the essentially sharp contrasts of language and thought existing between Peripatetic and Neoplatonic psychologies. A scholar of Grosseteste's generation, equipped with an integral theological culture absorbed from Augustine and others of the Fathers, a Neoplatonic guidebook to Aristotle, and the inaccurate translation of the already obscure *De Anima*, could not possibly have achieved the hermeneutical

corporum poenaliter pati, et non dixi eas a corporibus pati . . .' (ibid., p. 118. 34–6, cf. 119.13–24).

[30] See, e.g., *De Anima* I, 5 (ed. Van Riet, p. 100.88); IV, 4 (p. 65.43); V, 7 (p. 160.2). The designation *perfectio* occurs with relative frequency in Avicenna, e.g., I, 1 (pp. 18.10, 16; 19.27; 20.36, 45; 21.4); I, 5 (p. 81.20), etc. etc.

kenosis required in order to arrive at Aristotle's psychology in its purity. Eclecticism and syncretistic accumulation of doctrines were an inevitable first stage of the philosophical renaissance then just beginning.

Grosseteste's unpublished works from the decade 1230 to 1240 attest a degree of development as over against the doctrine already considered, and tend in general to place a new emphasis on the unity of man. It should not be too readily assumed that this evolution was the result of purely philosophical considerations, or of a later, less superficial reading of Aristotle. Its causes were more complex, as I hope to show.

In Grosseteste's time, the remains of the Origenist theory of dual creation had still not been definitively exorcized from what might be called the latent anthropology of Christianity, in which they retained a presence which was, one suspects, more than merely rhetorical. The close kinship between *contemptus mundi* and such negative attitudes to the body is evident in many theological writers and homilists of the period, and formed a pervading atmosphere from which masters in the arts faculties were not exempted. Numerous passages in the early works of Roger Bacon, for example, describe the body as something ignoble, the soul's prison-house, by confinement in which the soul is hindered in its flight, and soiled.[31] We have seen that Grosseteste has been accused of sharing similar attitudes, and in his *Commentary on the Posterior Analytics*, and indeed elsewhere, the rhetoric of dualism appears. In fact, however, his later works leave no doubt that his attitude to the body and to terrestrial realities became increasingly positive. The unpublished works stemming from the decade 1230 to 1240 contain a number of passages in which a new emphasis on the unity of the person is apparent. Admittedly, the impulse towards this new accent came largely from theology, but the contribution which philosophical ideas made to it should not be minimized.

The first full affirmation of the natural unity of body and soul in man that I have found in Grosseteste's writings occurs, significantly, in the lovely microcosmic passage of the *De Cessatione Legalium*. It will be argued more fully in a later section that, although the deposit of microcosmic doctrines bequeathed to Christianity by Greek thinkers, from Heraclitus

[31] Crowley, *Roger Bacon*, pp. 123–4.

to Nemesius, originated in a dualism of matter and spirit (which was at its clearest in Plato), it contained, as it were in a seminal reason, an affirmation of the unity of man which was destined to appear late in time, probably not long before Grosseteste's own age.[32] If we translate Nemesius's conception of man as the unique meeting-point of visible and invisible worlds into another culture, where the imperceptible rhythm of thought and sensibility displaced the accent from the contrast between the worlds of spirit and matter, to the uniqueness of man's privileged being, then the affirmation of the consistency and homogeneity of God's entire handiwork cannot but demand the abandonment of dualistic exaggerations. The tragic accents in which the union of an angelic spirit with an irredeemably sordid carnality had been proclaimed must yield before the more benign and sanguine recognition of a specific role and a unique destiny for man, granted to him in consequence of his composite nature.

It is the combination of anthropological and ontological interest that lends Grosseteste's development of the traditional parallels between the microcosm and the macrocosm its philosophical depth, for it derives as much from his concern to ground the unity of the *universitas creaturarum* as from his desire to celebrate man's central place in creation. The two preoccupations are inseparable. The perfection and beauty of the universe are only as great as the degree of unity its different regions possess, and the hope for the desired unification resides precisely in human nature, in which all the structural elements of the entire macrocosm are to be found. Purely spiritual beings share only being with matter, and are incapable of union with it, being perfect in themselves. In this respect the soul, which in its spiritual functions and destiny is the equal of the angel and shares the same level of being with it, stands in contrast to the pure spirit, for although it is of a different nature from the body, it has a natural capacity to join with it and form a unified personal being.[33] Since the body shares in all the elements of the universe which produces it and whose laws govern all matter, the human person is a microcosm of

[32] See below, ch. 6, iii (pp. 376-401).
[33] 'Anima autem rationalis et angelus communicant in natura rationalitatis et intelligentiae . . .', *De Cess. Leg.*, MS cit., fol. 177ᵈ. Intelligence is the faculty of direct vision of God.

creation, representing each and every kind of thing in the cosmos.

The language of soul-body relations employed in this context deserves to be examined. The rational soul is 'by its nature adapted to become the perfection of an organic body, and to be united to it *in unitatem personae*'.[34] The words used derive their significance from the philosophical revival, and owe most to Avicenna, though they do not seem to be a direct quotation. The term *'perfectio'* had, as we have seen, been part of Grosseteste's psychological vocabulary for the preceding decade, and had been associated with the Aristotelean conception of the soul as substantial form. However, this was the first occasion on which Grosseteste, while respecting the soul's essential superiority to the body, clearly stated that the union of both elements is natural, and is one which does not do violence to the spiritual soul, but which produces an original reality, the person, by comparison with whose unity soul and body taken separately are but parts. The phrase *'in unitatem personae'* is significant in this regard. It derives from Grosseteste's Christological vocabulary, and signifies the product of the hypostatic union in Christ, in whom two different natures are joined without confusion to form a new and perfect unity.[35] Its application to the soul-body problem therefore admirably suits Grosseteste's purpose, for the brief formula of words combines indissolubly a reference both to the distinction of spiritual and material natures, and to their full union in one person.[36]

The natural aptitude of the soul for union with the body is further specified in two texts which follow the *De Cessatione* in close chronological succession. The first of these, found in the *Hexaemeron*, is another good illustration of that mixture of traditional ideas and new expressions which was characteristic of the period. The soul, Grosseteste asserts, was created

---

[34] 'Est tamen anima rationalis apta nata ut sit perfectio corporis organici et uniatur ei in unitatem personae.' (ibid.).

[35] This usage occurs on three occasions in the same folio cited above; cf. also the sermon *Exiit Edictum*, ed. Unger, p. 22.

[36] This is only another example of the way in which Grosseteste's anthropology —and in a measure that of his entire century—was developed in fruitful and continuous circumincession with Christology. The context of his microcosmic discussion is the Incarnation.

by God on the sixth day, not as a seminal reason in potency to perfection, but complete, including a natural desire to rule the body in which it is to work out its eternal destiny.[37]

In the *Commentary on the Celestial Hierarchy* is to be found the latest and most satisfactory definition of the union of soul and body. Grosseteste is inquiring whether the intelligences are one or many in species, and if many, what the principle of difference among them is. Is it perhaps their moving different spheres? No, he replies, for their relationship to the spheres they move is part of their operation, not of their nature. He can reasonably claim that the human soul is specified among spiritual substances by its union with an organic body to form a person, since this union responds to a need of its nature, and is not simply a contingent operation, as is the moving of the heavens for the angels.[38] We could not wish for a clearer statement of the soul's specific and non-angelic nature. Grosseteste's doctrine of the soul's natural desire for the body is consonant with that of Avicenna.[39]

If the desire of the soul for the body is sufficient to constitute it as a species, the problem of the individuation of members of that species still remains. A brief remark, occurring, once more, in the *Hexaemeron*, hints at a solution to this question, which seems nowhere to have been explicitly determined by Grosseteste. Arguing that God's creation of the individual human soul does not imply the creation of something new in a sense conflicting with Gen. 2:2 ('On the seventh day God rested from his work'), Grosseteste distinguishes the creation of new species, something which cannot take place after the end of the six days' work, from that of individuals within species already established. In the latter case, he remarks, the individuals differ only accidentally from each other, and are not a 'new creation' in the fullest sense of the word.[40]

[37] *Hexaemeron*, MS cit., fol. 237c: 'Non igitur sexto die facta est materialiter, sed perfecte, cuius naturalis appetitus ad corpus administ[ra]ndum inclinatur, in quo iuste et[*sic*]inique vivere potest, ut habeat vel praemium de iustitia vel de iniquitate supplicium.'

[38] 'Ad haec, rationalis anima non videtur differre ab aliqua caelesti substantia alia specifica differentia quam potentia et appetitu naturali unitionis cum corpore organico in personalem unitatem.' (ed. McQuade, p. 241). cf. St. Bonaventure's statement to the same effect in *II Sent*. d. 1, p. 2, a. 3, q. 2 and ad. 1. (II, 50).

[39] See the introductory remarks of Prof. Verbeke in *De Anima*, ed. Van Riet, vol. 1, p. 36, and the texts to which he refers.

[40] 'Dicuntur enim opera adinvicem altera opera cum differunt specie et natura,

Grosseteste gives no further amplification of the point, but we risk little enough in supposing that the human soul is individuated by an accident, which is its relation to a determinate body (as Avicenna had professed). It is difficult on the evidence to see what other view he could have taken, since he nowhere gave evidence of any sympathy for Avicebrol's universal hylomorphism, which provided an alternative solution to the problem of individuation, one much appreciated by many of his contemporaries.

The philosophical affirmation of the naturalness of the union of soul and body to form a perfect, composite nature, finds emphatic expression in Grosseteste's theology of the death of Christ and the resurrection of the body. When he writes on the sufferings of Christ in the *De Cessatione Legalium* and the sermons of his late period, he betrays a tender sympathy for the humanity of Christ that is a fine fruit of the new Franciscan spirituality. His meditations contain one invariant theme: Christ delivered himself up voluntarily to death, that is, to the separation of body and soul. No pain is so great as the deprivation of the most profound and most natural need. The soul of a man young and healthy in body, vital heat, and spirits, is of its nature inseparable from its partner and beloved, and resists the violence of separation with all its forces. This was the unnatural and horrible death to which Christ voluntarily submitted himself for our sake.[41]

The same need of the soul for the partnership in which alone the person exists in fullness renders the immortal soul unhappy until the general resurrection of the body restores to it the fullness of human being.[42] Grosseteste thought he was

eiusdem autem speciei et nature individua non differunt natura et specie, sed solo accidente.' (*Hexaem.*, MS cit., fol. 232d).

[41] 'Evidentissime autem, ut mihi videtur, manifestavit se esse deum cum moriebatur cruce, supra omnem namque potentiam creatam est, a corpore et corde humano sano animam humanam dividere, cum anima naturaliter appetat coniungi suo corpori, nihilque tam abhorreat quam a corpore suo per mortem separacionem, unde et ipsa naturaliter inseparabilis est dum in corde nondum defecerit calor vitalis.' (*De Cess. Leg.*, MS cit., fol. 180d). See the sermon, *Tota Pulchra Es*, ed. Gieben, *Coll. Franc.* 28 (1958), p. 226; cf. *Ex Rerum Init.*, *Coll. Franc.* 37 (1967), p. 132, and the popular Anglo-Norman *Chasteau d'Amour*, ed. Murray, p. 121, vss. 1151–67.

[42] 'Item cum completa beatitudo non erit ante receptionem [*conjecture* resurrectionem] corporum generalem, non enim plene beata est anima dum adhuc tenetur desiderio corporis recipiendi . . .' (*Expositio in Gal.*, Oxford MS *Magdalen College* 57, fol. 31b). cf. *De Cess. Leg.*, MS cit., fol. 159a.

being faithful to Augustine in offering an enthusiastic welcome to the teaching of the *De Spiritu et Anima* that the unitary happiness conferred upon man in heaven implies, as an accompaniment to the beatific vision, the direct vision of the glorified Christ.[43]

It is clear from the variety of Grosseteste's approaches to the soul-body problem that several influences were at work on him during the period between 1230 and 1240, all of which, by a happy concurrence, resulted in a positive affirmation of the unity of man, which places his later doctrine out of reach of accusations of exaggerated dualism.[44] Concerning the concerted influence of the new psychological ideas and St. Augustine's theology of the resurrection (for Grosseteste never suspected that the *De Spiritu* was a late compilation), Grosseteste must have felt that they displayed a happy doctrinal parallel, sufficiently well grounded to remove his acceptance of their common indications far from all hesitation and doubt.

At the same time as he drew soul and body into closer intimacy and adopted in consequence the more up-to-date language of their union, Grosseteste continued to regard the soul as a substance in its own right and the mover of the body. The two impulses remain as firmly united in his writings of the years 1230 to 1240 as in those of the previous period. Numerous passages refer to the soul as *substantia . . . incorporeus spiritus*,[45] *substantia incorporea*,[46] or *intelligibilis substantia*.[47] How did he combine these Platonic denominations with the Aristotelean designation of *forma substantialis*, applied to the soul in the *De Statu Causarum*? In this area we are without positive information, for it is a problem he nowhere discussed. It has already been pointed out that he did not grasp the implications of the Aristotelean theory. It is not unlikely that he would have endorsed without reservation Avicenna's reasoning to the effect that the human soul is not related to the body as its form; not being an accidental nor a material form, but a

---

[43] *De Cess. Leg.*, ed. Unger, p. 12; cf. *De Spiritu et Anima*, ch. 9 (*P.L.* 40, 785). Grosseteste regarded this consideration as one of the reasons for the Incarnation.

[44] See Zedler, art. cit., p. 152: 'Soul by its nature has no need of the body', and pp. 148 and 154, where misleading words and phrases like 'juxtaposition', 'accidental unity', 'moral integrity rather than metaphysical unity', are used.

[45] *Hexaem.*, fol. 232d.

[46] *Comm. in De Div. Nom.*, ed. Ruello, p. 139 § 16.

[47] *Comm. in Myst. Theol.*, ed. Gamba, p. 62.15.

form which is not located in a subject, it can only be a substance in its own right.[48] This statement illustrates perfectly the contamination of pure Aristotelean metaphysical concepts by Neo-Platonism, for Aristotle had used the same or similar language to affirm the substantial unity of organic body and form, as the two inseparable components of every living thing, man included; whereas for Avicenna and Grosseteste, 'substantial form' referred in the unique case of the human soul to a substantial and separable entity, superior in being to the body which it vivifies, rules, and moves.

Grosseteste's treatment of sensation in the *Hexaemeron* reveals how far he was from qualifying any of the traditional Neoplatonic beliefs.[49] Stimulation of the senses renders the soul attentive to its action of governing the body, and this attention makes it aware of the stimulus received, which awareness is simply sensation. The result of the process is ambiguous. If the soul perceives in sensation a threat to its control over the body it can struggle against its adversary, the *passio corporis*, and with difficulty bring the material body round to its own path, despite the suffering resulting to the senses. If, on the other hand, the body's perception is in harmony with the soul's task, it can easily be guided towards the soul's purpose, which establishes itself in this case without pain.

This Neoplatonic phenomenology of sensation, with its emphasis on the superiority and responsible freedom of the soul, becomes Grosseteste's defence against the exaggerations of the *mathematici* (astrologers), in an emotional passage of the *Hexaemeron*.[50] These 'professors of vanity', relying upon the medical thesis that the soul is affected by the states of the body, teach that it can be affected indirectly by the influences of the stars reaching it through the body. They are ignorant of its nobility, which does not allow it to suffer direct bodily

[48] 'Anima autem humana non habet se ad corpus ut forma', *De Anima* II, 1 (ed. Van Riet, pp. 113-44); and as a corrective and completion, see I, 3 (p. 60. 59-61): 'Ergo animam esse in corpore non est idem quod accidens in subiecto esse; ergo anima substantia est, quia est forma quae non est in subiecto.'

[49] The untroubled juxtaposition of the two very different definitions of sensation in the relevant passage is noteworthy: 'Est autem sensus exterior vis susceptiva et apprehensiva sensibilium specierum sine materia [cf. Aristotle, Avicenna], sive secundum Augustinum, sensus est passio corporis non latens animam; non enim patitur anima a corpore.' (*Hexaem.*, fol. 223a).

[50] A comment on this passage is to be found in Dales, 'Robert Grosseteste's views on astrology', in *Med. Stud.* 29 (1967), 357-63.

action, but only to respond actively in proportion as the body is affected. Grosseteste expands the example (already used in the *De Intelligentiis*) of the self-propagating ray of light whose path is changed by the motion of a mirror. Astrologers claim that the states and passions of the soul fall within the domain of prediction of their pseudo-science. Can they not be made to understand that the body is subject both to impressions of the stars and to the actions of the soul; that when the soul is obedient to God in its proper activities, these allow it to command the body and to overrule the weaker influence of the heavenly bodies; that to Saturn's or Mars's action on the blood, conditioning the moods of sadness or anger, the well-ordered soul can oppose its resistance and, retaining its tranquillity, dominate the physical disturbance of blood and spirits, which, when it so wishes, it can so much more easily inflame to anger, than can a planet's far-off action? If the mind is in order, neither it nor its passions, nor yet the affections of the body, fall within the domain of astrology.[51]

A proof (offered in the same work) that each human soul is created *ex nihilo* and infused into the body, lists the traditional attributes of the soul, and is a summary of the doctrine concerning its nature. The human soul cannot be made from pre-existent matter, because it is a substance. Its mutability confutes the teaching of pantheism, its incorporeal and spiritual nature that of materialism. Its rational endowments prove that it does not come from nature, and its moral responsibility, that there is not merely one soul shared by all men.[52] The only remaining possibility is, that for each individual born a soul

[51] Since the text is much too long to be quoted in full, I reproduce only the central statement: 'Sed etiam istis dicendum est quod corpus humanum velud duobus subiacet motoribus. Recipit enim multas passiones et impressiones a sideribus, et recipit etiam motus et impressiones ab anime proprie accionibus. Et cum anima secundum vim rationabilem subiecta est deo, potens est secundum eandem vim imperare virtutibus inferioribus, et potencior [est *adds Grosseteste in the margin*] in efficiendo corpus proprium quam sint corpora celestia. Unde quantumcumque moveat saturnus vel mars corpus . . . plus potest ratio bene ordinata in contra operando, ut sit in anima gaudium et mansuetudo . . .' (MS cit., fol. 215c).

[52] That this heresy possesses no degree of actuality in Grosseteste's mind is indicated by the demonstrative structure of the argument, which is designed—like Avicenna's similar one—to rule out all possibilities save one. This is confirmed by a reference in the previous column (232c): some have held 'unam et communem esse animam omnium, quibus respondet Ieronimus dicens . . .'. cf. *De Cess. Leg.* 179b, where Grosseteste lists the three errors of the philosophers, who offended against revealed truth without knowing it: the eternity of the world, the transmigration of souls, and the existence of a single soul for all men.

is created *ex nihilo* and is individuated by its relation to the body.[53] If we add to these attributes the essential features of the soul met with in other texts, we arrive at a very complete definition of it, something like the following: a simple and incorporeal substantial form, created individually by God *ex nihilo* and infused into an organic body adapted to its reception, to which it naturally desires to be united as mover and perfection, from which it acquires its individuation, and with which it forms a new unit, the person. Within this definition the different impulses of new and old in psychological theory are joined in a precarious unity. This, however, does not make Grosseteste an exception in his generation, as has sometimes been claimed, for it was characteristic of the theologians and artists of that time to attempt to combine fidelity to the old with receptivity to the new ideas.

On a few occasions Grosseteste lifts the veil and discloses his thoughts concerning the whole man. These generally brief passages merit a moment's attention, for each is revealing in its own way.

Human nature consists of body and soul, and no more than that. But what is man, not merely in his constitution, but as an individual being? The concept of person remained, in medieval anthropological discussions, almost submerged by the more essentialist one of nature. None the less there is an indication that *'persona'* is beginning to emerge, from being a technical term limited to a strict theological sense, into a more general anthropological light. In his *Expositio in Galatas*, perhaps the earliest of his works to evidence a personal knowledge of Greek (1232 or 1233, probably), Grosseteste devotes a brief scholion to the Greek word *prosopon*. The Greeks use it to mean the external appearance and the person, and the concept it represents is very diffuse, he complains. Its poorest meaning is 'individual', without qualification. It can mean the individual of rational nature,[54] and it sometimes even refers

---

[53] See fol. 232d. Grosseteste's argument, that if there were only one soul it must be at once just and unjust, may be a variant of one of Avicenna's refutations of the same theory: the universal soul would have to be either wise or unwise in all men; *De Anima* V, 3, ed. Van Riet, p. 110.9. An influence of the *De Spiritu et Anima* (*P.L.* 40, 784) can also be suspected; the attributes of the soul listed there are much the same: 'substantia rationalis, intellectualis, a deo facta, spiritualis, non ex dei natura . . . ex nihilo facta, in bonum malumque convertibilis'.

[54] A sign that Grosseteste is no longer referring simply to its Greek usage:

to the joint actions of soul and body as projected outwards, and serving therefore to distinguish one man from another in their differences. It is according to this extrinsic appearance that one man judges of another and finds him acceptable or objectionable—for only God knows the hidden will of man. In this sense it differs little, Grosseteste adds, from the word 'appearance' (*facies*).[55]

However shallow and undeveloped we may consider this definition of the person to be, we can see that the concept has moved out from theology to the edges at least of anthropology, and hence represents the emergent possibility of a unified discourse on human action and being.

If we inquire further what man is, beyond being a unity of body and soul externalizing itself in actions that are identifiably individual, then the questioner must be turned inwards upon himself to ask, 'Who am I?' With this new question, Grosseteste is free to return to the native soil of truth, and, in a climate of Neo-Platonism, to affirm that the true man is the interior man, of whom it was written, 'Let us make man in our image and likeness.'[56] The hidden, invisible man stands in contrast with the evident exterior presented to the world, for what is *in* me *is* me: not the hand which belongs to me and is *mine*, nor the whole body, for it is in my rational faculties that I resemble God, as Augustine says.[57] It is only as the interior man models the externals into conformity with what he truly is, in the dynamic encounter with grace, that the whole man becomes an image of God. However, 'man', unqualified, means the interior man, since self-consciousness and thought

this definition comes, of course, from Boethius, *Liber de Persona et Duabus Naturis*, ch. 4 (*P.L.* 64, 1343C); cf. *Hexaem.*, ed. Muckle, *Med. Stud.* (1944), pp. 159-160.

[55] 'Ubi autem nos habemus "personam", grecus codex habet prosopon, quod significat apud illos tum "faciem" tum "personam", et hoc nomen "persona" multipliciter sumitur. Persona quandoque significat individuum simpliciter, quandoque vero solummodo individuum rationalis naturae, quandoque significat collectionem accionum anime et corporis et [*conject.* ad] extrinsecus advenientium, qua potest alius in alio dinoscere; secundum hanc personam iudicat homo de homine et acceptat homo hominem vel reprobat. Deus secundum hanc personam non iudicat nec acceptat, vel reprobat, sed secundum occultam voluntatem, quam ipse solus novit. Et hec persona, secundum quam homo iudicat et non deus, dicitur quandoque facies.' (*MS. Magdalen College* 57, fol. 5ʳ).

[56] In the *Hexaem.* (ed. Muckle, p. 165) Grosseteste follows St. Basil's exegesis of Gen. 1:26 as found in the *De Structura Hominis* I, 6-7 (*P.G.* 30, 16-17).

[57] *Contra Manichaeos* I, 17, 28 (*P.L.* 34, 186-7).

do not depend upon the body. If the members were cut off I should still remain the same person; without hands and feet and eyes I could still affirm, or at least think (if my tongue were cut out), that I am Robert, that I am myself, and that I have suffered the loss of the members I previously had.[58]

It may be of interest to remark that Grosseteste traced the origin of language to the composite unity of human nature. The intelligences or angels share something of trinitarian circumincession as they dwell in each other in immediacy of mutual knowledge and love, while retaining their distinctness in unity. If our minds were likewise manifest to each other, then the exterior signs or words we use would be superfluous, and would no longer need to be employed in communication. As it is, language derives from our incarnate rationality.[59]

II LIGHT AS MEDIUM BETWEEN SOUL AND BODY

THE distance of nature between body and soul was an axiom among all heirs of the Platonic tradition. The attempt to harness the divine steed to its heavy chariot, without depriving it of its winged nature, formed a large part of the soul-body question in Neo-Platonism, and the efforts at its solution established a problematic which was accepted by whole generations, first of Arabic, then later of Western thinkers. The degree of fidelity with which the individual Latin philosopher or theologian appropriated this problematic was a function of his adherence to Augustine, who was its main carrier. It goes without saying that in circles where the Neoplatonic light-metaphysics represented a particular attraction, the firmness of that adherence was not to be called into question.

The resulting doctrinal complex can be distinguished into three component elements, all of which were to be found at a considerable stage of differentiation already in Augustine

---

[58] 'Homo namque simpliciter est homo interior, unde et exterioris hominis partibus detruncatis, non minus remanet unus et idem homo. Manibus enim et pedibus abscissis, oculisque erutis, adhuc vere dicere, et lingua praescisa vere cogitare possum, quod sum Robertus, et quod sum ego, et quod ille ego sum truncatus, qui prius fui integer.' *Epistola* 2, ed. Luard, p. 19 (*c*.1232). The whole example and its use are reminiscent of Avicenna's 'flying man', and are but a corollary of the substantiality of the soul; cf. *De Anima* I, 1, (ed. Van Riet, pp. 36–37), V, 7 (p. 162).

[59] See my article, 'Language, tongue and thought in the writings of Robert Grosseteste', in *Miscellanea Mediaevalia* 13/2 (Berlin, 1981), pp. 585–92.

himself. In the first place, the soul cannot be directly in contact with the body it administers, but must relate to it through
a *tertium quid*, its medium or vehicle. It follows that the same
intermediary will be called upon to transmit the action of the
soul to the grosser matter of the body, as a moving instrument.
A special case of the soul's activity in the body is sensation,
to account for which in turn a specification of the general theory
is required.

The background of the Augustinian approach is of considerable interest. Plato's conception of the link between soul and
body as being number and proportion (*Timaeus*, 33c–34a) was
turned by Plotinus into the innate love of the Good, which
brings all the diverse elements in the universe and in man into
a unity (Enneads III., v. 6). His disciples, however, gave expression to a widely-acknowledged need for a double systematic
emphasis, on the one hand, upon the metaphysical distance
between the Absolute and matter, and between the soul and
the body; and, on the other, upon the need for hierarchical
mediating-links to gradualize the chasm obtaining between
extremes, and to assure at the same time the continuity demanded by an emanationist system. To this need the ether
(which was conceived of as a spiritual, heavenly matter) seemed
well-adapted to respond; in the Stoics it had already served
as a quasi-immaterial locus of divine action.[60] Proclus represented the pre-existing soul as already wrapped around with a
cover of ether, which accompanied it as a protection into the
material body, and which remained with it to become the
vehicle of its returning flight to immortal life. Vehicle ($\delta\chi\eta\mu\alpha$)
and mediating body, cover, protection ($\pi\epsilon\rho\iota\beta\lambda\eta\mu\alpha$), this celestial substance was the servant of the soul's divine and heavenly
nature.

St. Augustine shared with the Greek Neo-Platonists their
desire to emphasize the distance obtaining between the natures
of soul and body, and likewise their willingness to fill up the
gap with a spiritual type of matter: ether, fire, or light. It is
the soul's natural *praestantia* over against all material natures
(comparable even to God's transcendence of his creatures),
which demands that it should not meet directly with matter,
but relate to it only through light (or light and air), whose

[60] Moraux, art. *'Quinta essentia'*, in Pauly-Wissowa, *Real-Enzyclopädie*, Bd. 24,
col. 1251.

active nature contrasts sufficiently with the passivity of the two lower elements to render them akin to spirit.[61] The at first sight contradictory notion of a 'matter that is akin to spirit', can be clarified only within the light-metaphysics, the Western origins of which lie principally in Augustine. The three basic propositions of this conception follow each other with strict consequence: God is light in a more than metaphorical sense; the essence of light is to be sought in spiritual rather than corporeal being; and in the visible world, light is the first, subtlest, and most active of material things, and hence closest to immaterial nature.

Augustine's medieval disciples showed themselves fully aware of the profound motivation of the medium-hypothesis, which lay in the need to assert and protect the dignity of the soul.[62] We have seen the same preoccupation at work in Grosseteste's *De Intelligentiis* and in others of his works, and the identical awareness was prominent in St. Bonaventure's thought.[63] But Bonaventure brought out an element of actuality in the question by attributing the same hypothesis to Avicenna (ibid., ad. 1), revealing himself to be at the confluence of two distributaries of the original doctrine. Turning to the pages of Avicenna, we find the thesis on light as the vehicle of the soul developed in the context of the *complexatio* of the animal body and the production of the animal spirits.[64] Here the Neoplatonic problematic of the union of soul and body was associated with the very different problems (in origin of course Aristotelian and medical) of the preparation of the *corpus organicum* for the reception of form.

An Aristotelean analogy to the *lux-vehiculum* theory was not far to seek. In one of the few passages of his work which is of direct interest to the history of microcosmism, Aristotle

---

[61] *De Gen. ad Litt.* VII, 19, 25 *CSEL* 28 (3, 2), pp. 215–16 (*P.L.* 34, 364): 'Sicut enim deus omnem creaturam, sic anima omnem corpoream naturam naturae dignitate praecellit. Per lucem tamen et aerem, quae in ipso quoque mundo praecellentia sunt corpora, magisque habent faciendi praestantiam quam patiendi corpulentiam, sicut humor et terra, tamquam per ea quae spiritui similiora sunt, corpus administrat.' cf. ibid. VII, 15,21 (*CSEL*, p. 213) and XII, 16, 32 (*CSEL*, p. 401): '. . . illud quod est subtilissimum in corpore et ob hoc animae vicinius quam cetera, id est lux.'

[62] Augustine seems nowhere to have applied the words *medium* or *vehiculum* to light.

[63] 'Body and soul are far apart, and require a mediator which is bodily yet subtle, and therefore light.' *II Sent.*, d. 17, a. 2, q. 2, a. 6 (II, 421b).

[64] *De Medicinis Cordialibus*, in *De Anima*, ed. Van Riet, Appendix, p. 190.

spoke of πνεῦμα (*spiritus*) as being the active element in the reproductive matter, 'which is not fire nor any such force, but is analogous to the element of the stars'. It was further described as being more divine than the elements (though differing in nobility according to the dignity of the corresponding soul), connate with and yet analogous to the element of the stars, and to be identified with the vital heat.[65] Arabic Neo-Platonism was not slow to syncopate this theme with its Proclean analogates, and to evolve a doctrine of the origin of life, according to which the sphere of the fixed stars educed life in all its forms through the influence of its light.[66] The most influential propagator of this view was Avicenna, from whose teaching the thirteenth century saw few convinced dissenters. According to his view, it is the proportion of the active elements (fire and air) to the passive (water and earth) which produces a balance between activity and mass in all elementated bodies. For the production of the more noble organic bodies (those of animals and men), a harmonious balance must obtain between the alternate dominance of the two pairs of elements. The process of complexation, which takes effect in bodies through the influence of light, has the effect of weakening the contrareity of the elemental qualities, since it itself has the nature of a conciliating force.[67] The closer the form of complexion approximates to a true medium—one of justice, not of numerical equality, recalls St. Bonaventure[68]—the more it is adapted to receive a more perfect form of life. In its best-tempered state, when the contraries are perfectly equilibrated, the body is proportioned to the reception of a rational soul of heavenly nature.

According to Avicenna, this aptitude of the body for the soul subsists in the human spirits. Generated from the elements, they are an image of the celestial bodies, since the whole process of harmonizing that has resulted in their perfection is governed and influenced by the light of the heavens. The spirits are of the substance of light, enjoying harmony

---

[65] *De Generatione Animalium* II, 3, 736b, 29 ff. On this difficult and controverted passage see Wiersma, 'Die Aristotelische Lehre vom *Pneuma*', *Mnemosyne* 3. F., 11 (1943), pp. 102-07, and Solmsen, 'The vital heat, the inborn *Pneuma* and the Ether', in *Journal of Hellenic Studies* 78, 1 (1957), pp. 119-23.

[66] cf. Baeumker, *Witelo*, pp. 454-7.

[67] Avicenna, *De Medicinis Cordialibus*, ed. cit., p. 190.

[68] *II Sent.*, d. 15, a. 1, q. 2 (II, 377a).

without the interference of a contrary quality. They are active in sensation (the *spiritus visibilis*, for example, is *'radius et lux'*), and they form the bearer of the soul in the body.[69] When we come to Grosseteste's texts from a background of these two Neo-Platonisms, Augustinian and Arabic, we find that he has read and absorbed them both. In his discussion of the influence of the celestial bodies and their light on the eduction of vegetable forms, his acceptance of the Arabic complexion-theory is evident.[70] At the same time, he accepts Augustine's medium-theory, refines it in certain respects, and unites it with the medical doctrine of the spirits.

The contexts in which Grosseteste speaks of the luminous spirits are those of motion and sensation.[71] He does not concentrate directly upon the nature of light as vehicle or medium; neither word is employed by him in this sense.[72] Never the less, it is clear that he adopts Augustine's and Avicenna's view of light as a *tertium quid*, whose presence is indicated by its activity and operation. The evidence is unambiguous, and is revealed by a moment's reflection upon the qualities attributed to *lux* or *spiritus*, which he describes as being 'bodily, but the closest (material) approximation to incorporeal nature',[73] or again: 'the most subtle thing of bodily nature, and therefore close to the soul, which is immaterial without qualification.' That is to say, the spirits serve as a principle of continuity between a noble soul and a gross body, preserving the former from direct contact with the more ignoble parts of matter. That Grosseteste's theory is more sharply-defined than that of Augustine is evident from his refusal to extend the activity of light to the lower element of air; Augustine's elemental theory, it is plain, has been developed by dialogue with the Aristotelean physics. The fineness or subtlety which is

[69] 'Ipsa enim lux comparata est eius [i.e., the soul's] vehiculo, et confortatur propter eam vehiculum eius, quod est spiritus, cui tenebra est contraria.' (Avicenna, *De Medicinis Cordialibus*, ed. cit., p. 190. 45).

[70] *Hexaem.* fol. 212^d. As will be seen later, Grosseteste restricted the application of the physical explanation of the origin of life to vegetables. After having admitted in the *De Statu Causarum* that sensible forms are educed by natural causes, he declared in the *Hexaemeron* his belief that they are produced directly by God.

[71] *De Intell.*, p. 116.14-30; *De Oper. Solis*, ed. cit., pp. 69, 84; *Hexaem.*, fols. 203^{b-c}, 218^c.

[72] '*Medium*' (*De Intell.*, p. 116.26), designates light as mediating the soul's motion to the body.

[73] *De Intell.*, p. 116.25: '. . . in corporibus magis appropinquat incorporalitati'.

attributed to light and affiliates it to the spiritual, may likewise be thought of as resulting from the passage of an Augustinian concept through the refinery of Avicenna's complexiontheory, because for Grosseteste light is neither simply the elemental fire of Augustine, nor the quintessence of Aristotle. Unlike the former, it is removed from the conflict of contrareity, yet is not confined to the supralunary world, as is Aristotle's fifth essence, for it is present in all matter as the principle of extension and of energy.

It is important that the difference between the light which is the first form of corporeity and the light which the spirits are in their essence, be kept in view. They represent diffuse and concentrated material energy respectively, for we are dealing with a gradation of intensity at the physical level. The spirits are the very highest product of the material cosmos. They are comparable only to the matter of the empyrean. They are privileged even beyond it, since it is through their possession that the organic body is adapted to the coming of the soul. The fineness of the spirits is the result of the most balanced possible complexion of the elements under the influence of celestial light, and this quality of penetrability, the consequence of their rarefied state, allows the corporeal spirits to permeate the form of the mixture and to neutralize the warring qualities of the elements. By producing stability and balance in the body, these render it apt to receive a spiritual soul.

Before examining the spirits' intermediary role (in moving the body, and in sensation), two remarks concerning the foregoing doctrine are called for. Grosseteste showed himself alive to the microcosmic possibilities of the spirits-doctrine,[74] in this being no doubt stimulated by the Augustinian parallel between the senses, operative through light and spirits, and the elements of the world.[75] The presence of light in the human body, composed of the four lower elements, completed its total representation of the material cosmos. The body was therefore seen as being at once the product and the image of the entire universe, sharing each single nature within the graded spectrum of materiality. Now this idea was one that

[74] In the *De Cessatione Legalium* and *Exiit Edictum*.
[75] *De Gen. ad Litt.*, VII, 15, 21 (*CSEL* ed., p. 213) and III, 4, 6 (*CSEL* ed., p. 67).

had to be rejected by all who assented to the Aristotelean physical scheme, wherein the separation between sublunary and supralunary worlds precluded the quintessence from entering into compounds with the four elements. That a nature which had no contrary could not form part of a chemical mixture, was axiomatically evident. In this respect, the contrast between Grosseteste and St. Bonaventure on the question in hand is instructive.[76] St. Bonaventure was equally concerned that the human body should be 'somehow all things', so that the cosmos might find in man a focus of unity, as well as finality and ultimate glorification. If, however, the body were constituted by the four elements alone, it could represent only the lower part of the world, not the quintessential heavens; yet, since the fifth element was ungenerable and incorruptible, it could not in its own nature form part of a sublunary mixture, but could at most be present in the body only in its power and through a kind of conformity.

In contrast to Bonaventure, Grosseteste was not inhibited by close adherence to the Aristotelean teaching. In the light-metaphysics he produced a cosmogony whose chief inspiration was the biblical-patristic tradition, and in which the fundamental unity of the cosmos was ensured by the omnipresence of light as the first form of matter. The continuity of qualities among the elements of this universe meant that the rarefied and more simple light-substance of the heavens was a specification of matter in general, and not an area exempt from all contact with the lower world. Thus Grosseteste could affirm that the celestial bodies shared with elemental fire in the nature of light, as did fire with air in heat, air with water in humidity, and water with earth in frigidity.[77] The mediating link of shared qualities, which in the strict Aristotelean scheme were never dissociated from the generative play of their chemical interaction throughout the sublunary world, became in the new cosmogony a law that reached from end to end of the cosmos and tolerated no regional exemption. In other words, Grosseteste's cosmological ideas preserved direct contact in this respect with the common fund of Greek and Latin

[76] For a development of this point, see my study on 'Microcosm and macrocosm in the writings of St. Bonaventure', in *S. Bonaventura 1274–1974*, vol. 2, pp. 319–23.

[77] *De Cess. Legal.*, ed. Unger, p. 14.

patristic conceptions of the ether, that heavenly fire which was held to be in continuity with the earthly elements. Grosseteste was not forced to raise the question of how light could be present in a sublunary body (the question over which St. Bonaventure agonized), because he was never fully committed to Aristotle's physical scheme, and felt no tension in ignoring the doctrine of the *De Caelo* when loyalty to a patristic view was at stake. It should, however, be noted that he nowhere referred to the 'spirits or light' in man's body as being a fifth element. The human body, he taught, is composed of the four elements; but he could affirm in the same breath that it *therefore* contains all material natures.[78] That he was aware of his separation from Aristotle in this respect is not to be doubted.[79]

A second conclusion emerges from a comparison between Grosseteste's theory of soul-body relations and that of Philip the Chancellor, his exact contemporary.[80] Philip was like Grosseteste, in that he felt the strength of Augustine's demand for an intermediary between the rational soul and the body. The *multimoda distantia* separating body and soul emerges, he suggested, from the contrast of their qualities: the soul is simple, incorporeal, and incorruptible, whereas the body is corporeal, composite, and corruptible. It follows that the soul can only be united to the body provided that certain mediating dispositions intervene, and the vegetative and sensible souls, Philip thinks, fulfil this role:

The sensitive soul is simple and incorporeal, but is corruptible; it has two properties in common with the rational soul and one in common with the body. The nutritive soul has only one property in common with the rational soul and one in common with the body. There are consequently two media—the nutritive and sensitive souls—and in that order.
'Exiguntur ergo ad coniunctionem animae rationalis cum corpore anima

[78] '. . . corpus autem humanum constat ex quattuor elementis, quapropter communicat in natura cum illis, et per consequens cum celestibus corporibus, cum quibus communicat ignis in natura lucis.' (*De Cess. Legal.*, ed. cit., p. 14; cf. p. 22).

[79] Aside from his own studies of Aristotle's physical works, Avicenna's care in expressing the equation of light and spirits in such a way as to preserve the genuine Aristotelean physical doctrine would have alerted him to what was at stake: 'Spiritus vero humanus est quiddam quod generatur ex elementis, et effigiat vultus ad simulitudinem celestium corporum . . . et ideo spiritus visibilis vocatur radius et lux.' (*De Med. Cord.*, ed. cit., p. 190).

[80] See Crowley, *Roger Bacon*, pp. 143-9.

sensibilis et anima vegetabilis, et iterum spiritus qui est a natura corporis superioris, et calor elementaris.'[81]

In this pluralist system, the intellective soul was not united immediately to the body, but only through several graduated and subordinate forms. The interest of this finding for our topic is that Grosseteste, in his concern to preserve the soul's nobility and transcendence of the body, appealed, not to a theory of plurality of forms, but to the mediating presence of the luminous spirits. Is it not altogether likely that, had he accepted a pluralist theory similar in form to that of Philip, he would have used it in a similar manner, as one obvious way of blending Aristotelean with Augustinian thinking? His silence, his willingness to promote the spirits to the single medium between soul and body, argues that he defended rather the simplicity of the rational soul as the only soul in man and the principle of nutritive and sensitive operations, as well as intellectual ones.[82]

The function of the luminous spirits follows their place in the psychosomatic scheme, and their intrinsic nature. Since they are of the most subtle matter and are close to being of spiritual nature, they constitute the first instrument of the soul in its action on the body, transmitting the motion they first receive to the more weighty, terrestrial members.[83] It is by its rational and voluntary desire, which, like its substance, is incorporeal, that the soul moves the spirits, acting directly upon them and producing in them a motion that is proportionate to its own spiritual affection.[84] It is the faculty of perceiving what is fitting or harmful for man, and the associated reaction of desire or flight, that is therefore the prime efficient cause of the body's motion.[85] The spirits transmit the command to the nerves and muscles, for it is within their

[81] Ibid., p. 146. One could go further and say that there are not merely two intermediaries between soul and body, in Philip's scheme, but a series of three or four, depending as he distinguishes or identifies the spirits and the vital heat.

[82] Positive grounds for this conclusion will be adduced below, ch. 4, iv (pp. 312–19).

[83] *Hexaem.*, fol. 203[b]. cf. Augustine, *De Gen. ad Litt.*, VII, 15, 21 (*CSEL* ed., p. 213) and 19, 25 (*CSEL* ed., pp. 215–16); XII, 16, 32 (34, 466). Grosseteste paraphrases the last-mentioned passage in the sense of both former ones: 'Lux quoque secundum Augustinum est id quod in natura corporea subtilissimum, et ob hoc anime, que simpliciter incorporea est, maxime vicinum; et ideo est ipsi anime in agendo per corpus velut instrumentum primum, per quod instrumentum primo motum movet cetera corpulenciora . . .'

[84] *De Intell.*, p. 116.23.

[85] *De Statu Caus.*, pp. 90.26–91.6.

capacity to move the solid, bodily matter forming the nervous system, and it is through the muscular network that the larger members of the body are set in motion.[86] That the soul can initiate in the spirits a movement proportionate to its desires is due to its union with the body. It remains something of a mystery, Grosseteste admits, how the angel, which lacks full personal union with its instrument, manages to transmit its affections to it.

The two qualities underlying the functions of the spirits are luminosity and fineness. To the extent that a thing shares in the nature of light, which the spirits do pre-eminently among material beings, it is characterized by activity; now this is conceived of as a kind of agility, whereby it is easily set in motion.[87] Fineness or subtlety enables the spirits to penetrate the entire body and be effective throughout it.

Sensation is a specification of the soul's motive action in the body, as Grosseteste makes clear in the *Hexaemeron*,[88] where he passes from the bodily spirit, which is the soul's first instrument in moving the body, to that instrumental light through which it acts in the senses. In his doctrine of sensation Grosseteste is clearly inspired chiefly by St. Augustine, but here also Arabic influences are detectable. He believes that light is the active element in all five external senses, and with this affirmation he places himself consciously within the perspective of Augustine. Vision is the privileged one among the senses, because through the eyes pure, unmixed light is diffused, shining forth in rays to perceive objects. Mixed first with the pure, higher air, it acts in the organ of hearing, then with thicker, misty air in the sense of smell, which requires qualities of heat to produce evaporation, and moisture to prevent the steamy substance dissipating through volatility. When light combines with earthy dampness, the mixture of both

---

[86] *De Intell.*, p. 116.31 ff. In the *De Statu* (p. 90.30) the briefest of references is made to the Aristotelean *calor naturalis* as a moving cause of the body, but no details are given. The medieval physiology of nerves, muscles, and ligaments was derived largely from Avicenna's *De Anatomia Nervorum, De Natura Animalium* XII, 8 (Venice, 1508), fols. 47ᵛ-48ᵛ.

[87] Augustine, *De Gen. ad Litt.*, VII, 15, 21 (*CSEL* ed., p. 213). Also, Avicenna's transition from *spiritus humanus* to *spiritus visibilis* (identified as *radius* and *lux*) implies the same (*De Mot. Cord.*, p. 90.40 ff.).

[88] In *Hexaem.*, fol. 203ᵇ, Grosseteste is content to speak through the words of Augustine, *De Gen. ad Litt.*, XII, 16, 32 (34, 466). cf. *Comm. Myst. Theol.*, ed. Gamba, p. 60.15.

forms taste; when light reaches down through the thicker air right to the grosser dampness, it produces the activity that is smell; when finally it penetrates through to earth, with its passive heaviness, it forms the sense of touch.[89]

The only sense concerning which Grosseteste gives us further information is, of course, sight, and here he says enough to reveal an Arabic accompaniment to his predominantly Augustinian theme.[90] In the *De Operationibus Solis* he affirms that the visible spirit of the eye is of the same nature as the sun's light, because the sun is the unifying root of all light and heat in the universe, wherefore anything in the lower world that shares light and heat is connatural with it and dependent on it, as the ultimate source of all material energy.[91] It is through the spirit that the soul acts in sight, for the spirit emits rays through the eyes.[92] Behind the physiological explanation of sight which Grosseteste adumbrates, though somewhat obscurely and from a limited point of view, one can discern the influence of Avicenna,[93] who taught that the *spiritus visibilis* was a transparent, fine, and fluid substance filling the optic nerves, conveying the species to the brain through the 'common nerve', and conditioning all aspects of visual sensitivity and perception. Grosseteste, no doubt, accepted this account.

---

[89] *Hexaem.*, fol. 203b-c. At once clearer and more personal is *Hexaem.*, fol. 218c, where, instead of 'light', he uses its patristic quasi-synonym, 'fire': 'Is quoque ordo elementorum observatur etiam in ordine exteriorum sensuum, sicut docet Augustinus . . . agit enim anima sentiens in oculis per ignem purum lucidum, represso calore eius usque ad lucem eius puram. In auditu vero usque ad liquidiorem aerem, calore ignis penetrat; in olfactu autem transit aerem purum, et pervenit usque ad humidam exalacionem, unde crassior haec aura subsistit; in gustu autem et hanc transit et pervenit usque ad humorem corpulenciorem, quo eciam penetrato atque transiecto cum ad terrenam gravitatem pervenit tangendi ultimum sensum agit.'

[90] See Avicenna, *De Anima* III, 8 (ed. Van Riet, pp. 269.50, 273.15, 277.92, the *spiritus visibilis*), and V, 8 (176.61); *De Anima*. XII, 2 (Venice ed. 45rb).

[91] *De Oper. Solis*, § 6: 'Oportet enim omnem lucem visibilem in una universitate et in uno mundo ad unam reduci radicem, quae radix una nusquam probabilius invenitur quam in sole.' (ed. in *Rech. Théol. anc. méd.* 41 (1974), p. 70). The thought is characteristically Neoplatonic, and closer in spirit to the *De Causis* than to Grosseteste's other sources.

[92] Ibid., § 7: 'Praeterea, spiritus visibilis oculi corporalis creditur esse de natura luminis solaris, et in actione videndi agit anima per lucem huiusmodi puram radios emittentem per oculum, et haec eadem lux magis magisque incorporata perficit perfectione corporali totos sensus corporales; et propter hoc sol per virtutem luminis est in conspectu oculi videntis et in opere cuiusvis alterius sensus corporalis percipientis.' (ed. cit., p. 70; p. 85).

[93] See Baeumker, *Witelo*, pp. 613, 635. It is not very likely that Grosseteste knew Alhazen's *Perspectiva*.

Despite the lengthy and detailed attack made upon the theory of the emission of rays by Avicenna, Grosseteste held on to the latter view, in fidelity to St. Augustine.[94] Refusing to admit with the Aristotelean scientific tradition that sensation is a passive process, he stressed that every sensation is an act of the soul through its instruments: the light or active element within the sense-faculties multiplies itself into a sphere surrounding the organ.

This theory of sensation becomes just one more way for Grosseteste to affirm the dominance of spirit over matter. Though all natures, even material ones, are active, as the light-metaphysics sees things, none the less that which is closer to spirit cannot be acted upon by a baser element. There is nowhere an indication that Grosseteste's acquaintance with the new psychology of Aristotelean provenance ever led him to place the Augustinian conception in doubt. As we shall see, the motivation for retaining the old-fashioned physiology of the visual ray lay deeper than simple loyalty to Augustine: sight was for both men, as Platonists, the privileged model for intellectual understanding.

---

[94] Avicenna calls the emission theory '*absurditas magna*', but still he criticizes it at enormous length; cf. *De Anima* III, 5 *in toto*.

Chapter 4
The Division of the Powers of the Soul

THE interest of this question for the history of the assimilation of the Greek-Arabic scientific literature is very considerable. Psychology and its associated sciences, biology and botany, had received no place among the traditional Latin *artes*, and the little that was known of them was either confined to theology or marginalized by absorption into the medical tradition. Now while theology traditionally included the vivification of the body among the soul's powers, and developed a phenomenology of the imagination and the sense-powers, these aspects of the soul's functioning never really interested theologians for their own sake, but suffered neglect due to the concern to preserve the spiritual simplicity of the soul, a quality more apparent in the higher faculties of understanding and willing than in the lower, through which the soul risks losing its path to salvation by having too much commerce with the variety and multiplicity of matter. So much is apparent from the traditional classification of powers, which was frequently repeated with little development well beyond the threshold of the thirteenth century:[1] reason is given to us to discriminate between the good and the bad, desire to grasp the good, and vigour (*ira*) to repel the bad. The moral inspiration of the division is apparent, as are its Platonic roots; but while it invited a religious phenomenology of the path of salvation, it made no place for a more detached form of scientific curiosity, concerned to explore the underground of the moral personality.

When the more morally neutral and biologically-grounded Aristotelean concept of soul became available through translations, it was not long in convincing such minds as had been awakened to new and less exclusively sacral interests by the naturalism and rationalism of the twelfth-century schools, that it filled a gap in their tradition, the very existence of which had not been suspected by previous generations. The

[1] By Alexander Nequam, for instance, as we noted earlier (p. 259).

history of the psychological movement in the generations following the translating work of James of Venice (the *De Anima* of Aristotle) and Gerard of Cremona (*De Anima* of Avicenna) is that of a growing awareness of a space or vacuum within the traditional theology, as it was realized that the prevailing concept of the soul was like the completed roof of an unbuilt dwelling: it required foundations and walls to carry it. The new literature, then, at once created the perception of a vacuum and supplied the means of filling it. The question as to whether the ground plan laid out originally by pagan architects was at all suitable for the structure raised above it, was one that did not emerge until the building was almost completed. It was left for Aquinas to perceive the ill-adaptation of the several styles and the heterogeneity of the materials employed in the construction. By good fortune he combined the skills of master-mason and architect, and he was not a craftsman in one medium merely.

Already, as a Master of Arts, Grosseteste adopted the classification of the soul's powers as vegetable, sensitive, and intellectual, deriving from Avicenna. These remained the fundamental categories within which his ideas were structured. Our task is now to trace the development within this division from vagueness to sharpness of focus, and to discover his originality.

I THE VEGETATIVE SOUL

A glancing side-reference in the *De Generatione Sonorum* enables us to conclude that Grosseteste accepted the new triple division of soul at a relatively early stage in his career.[2] The same passage in the *De Statu* which proved so informative concerning the psychosomatic problem, defines the vegetative soul as a substantial form which is not localized within the matter it informs but present throughout it by its operation, which is educed from potency to act by the influence of the heavens, and which is corruptible.[3] It is only in the works dating from his episcopal period, however, that we are able to determine the exact extent of Grosseteste's psychological

---

[2] *De Gen. Son.*, ed. Baur, pp. 7.28-8.3.
[3] *De Statu Caus.*, pp. 124-25.

studies. The *Hexaemeron* contains a long passage on the functions of the vegetative soul, and the *De Confessione* a brief
summary of Avicenna's classification of its powers. His debt
to Avicenna is very great,[4] and can be documented even to
the details of the language he employs, but never the less his
restlessness to relate the doctrines which he adopted to his
own more personal thought results in a degree of originality.
We give first his classification of vegetative functions, before
discussing his attempt to explain the origin of plant-forms.

Vegetable life represents already a complex structure within
which four powers can be distinguished, the first of which is
the capacity to attract the nutriment which the plant requires
in order to grow to the average size of its species, and then to
restore the matter lost in the unceasing flux of life. The expanding and stabilizing chemical qualities of heat and dryness
assure the attractive power of efficacy, while it is the balance
of cold and dry that enables the retentive power to coagulate
and unify the new matter to the vegetable substance. The
digestive power then works by warmth and dampness to heat
and soften the nutritive material; this process of 'cooking'
converts the latter into the nature of the plant, whereupon
it is distributed to the parts in the requisite quantities. The
superfluous and harmful matter must then be evacuated, lest
it corrupt the whole or preclude the entry of fresh nutrition;
this is the function of the expulsive power, with its qualities
of cold and wet.[5] The movement of material in the vegetative
cycle takes place through little hollow tube-like canals such
as one can distinguish in trees split open, and even in cooked
meats.[6]

If diverse combinations of the four physical qualities are at
the basis of these four powers, the latter in turn are at the
service of the three fundamental functions of all living things,
nutrition, growth, and reproduction. Nutrition assures the
survival of the specimen and augmentation its full representa-

---

[4] Avicenna, *De Anima* II, 1 (ed. Van Riet, pp. 102–13): 'De certificando virtutes
quae sunt propriae animae vegetabilis', is his chief source, but he refers at least once
to the *De Plantis* of Pseudo-Aristotle (Nicholas of Damascus), from which he
derived information concerning the description and classification of plants and
trees; see *Hexaem.*, fol. 212^d.

[5] *Hexaem.*, ibid., *De Confessione*, ed. Wenzel, p. 260.

[6] A detail which may represent a personal observation on the part of Grosseteste; it is not found in Avicenna.

tion of the species, while reproduction guarantees the perpetuation of the species. In his description of the powers and functions characteristic of vegetable life, as well as of their hierarchization, Grosseteste gives a summary paraphrase of Avicenna, the economy and completeness of which indicate close familiarity with the text of the *De Anima*, and even instance close verbal parallels to it, but manifest no desire to go beyond it.

Where he was original was not in classifying and observing, but in suggesting a hypothesis to explain the functions of vegetative life in terms of material energy, in other words, in relating life to light, and thereby renewing one of the oldest, most sacred and most mysterious associations in the history of man.

Every vegetable nature possesses a teleological principle which guides its formation and growth and controls the production of its nutritional and generative organs. Now life-functions begin from a central point and extend outwards in all directions the matter which they inform. It is as though something analogous to the heart in animals were at work in the plant, as a first, active instrument of life.[7] Now it is a general characteristic of the activity of energy that it works from a central point outwards by rarefying and extending the surrounding matter, and the energy referred to as 'vegetable life' or 'life-force' can be brought under this universal natural law. Whence is this energy derived? Grosseteste's answer is that it is a product of the universe acting as a whole. The shape of the cosmos results in the light, not of the luminaries only but of the entire heavens, being generated in straight lines from the spheres and meeting in the centre of the system, the world (*mundus*, the area below the sphere of the moon), and with particular force on the surface of the earth. When intensified light is incorporated into matter that is apt to receive vegetable life, the energy thus focused in a point begins to move the matter affected. The motion there initiated, however, is no more simple than is the energy which gives rise to it, because the light emanates not from one heaven only but from all nine, and the light generated by each heavenly

---

[7] On the importance of the heart in animals as the first member generated, the prime begetter of the spirits and the seat of the soul, see Avicenna, *De Anima* V, 8 (ed. cit., p. 176). Grosseteste agreed, with few reservations: *Hexaem.*, fol. 212^b-13^a.

sphere retains the characteristic circular motion and speed of rotation of its source. When the complexity of these movements of light is allowed for, the motion of energy at the heart of plant-life must be regarded as a compound movement representing the quotients of all nine component impulses, each of which has a rate of motion constant with regard to itself and its source, but of a different formula from the others (*Hexaem.*, fol. 213ᵃ). While the aggregate speed of the vital force is constant, its efficacy in extending matter is a factor of two variants, namely, its relative force, and the degree of resistance the matter offers.[8]

We must imagine, further, that in the same matter many such quanta of energy are active. Each possesses its own proper speed of circular motion (in imitation of the source which continues to feed it), and reflects in its gyrations the whole complex and harmonious mechanics of the higher universe, where lower spheres have a motion that is a product of their proper turning and the variations imparted to them by the higher moving spheres. These various quanta of energy (in effect tiny microcosms of the upper cosmos) create a number of tractional effects in the matter which they energize, and the aggregate result of their activity is a figure extended according to a complex formula of all the variants and constants involved. Anyone who takes the trouble to look closely at plant specimens will discover that they have been moulded and formed by a revolving motion.

A boldness typical of Grosseteste's later cosmological ideas is certainly not wanting in this extension of the light-metaphysics to explain the production of vegetable life. I have not found a source for it, nor does Grosseteste indicate an authority; it is not improbable that it is his own conception. The influence of the heavens on the formation of organic life was, of course, a generally-received notion, capable, as we have seen, of claiming foundation in Aristotle himself, and developed further by Arabic speculation. That many variants on the Arabic theme of the heavens' influence were in circulation at Paris is not to be doubted. St. Bonaventure quotes one, not without approval (*'opinio satis probabilis'*), which associated the three outermost

---

[8] Grosseteste is aware of the necessity of introducing variables, which represent the only possible way of accounting for the variety of plant life in a purely physical (as opposed to a chemical) explanation.

spheres with the three types of soul: the sphere of the fixed stars conciliates the matter which is to receive the vegetative soul, while the crystal heaven and the empyrean perform the same preparative function with regard to sensitive and human souls respectively.[9] Grosseteste's account, however, is considerably superior to its more vague and exotic rivals. It is related to central principles of the geometrical physics of light, and therefore to a body of ideas of a strongly systematic character. It takes its origin from observation (however rudimentary), and is related to the very sober Avicennian analysis of the powers of the vegetable soul in two ways: Grosseteste attempts to deduce the four powers from his theory; and his speculative explanation stands to the Avicennian analysis of plant life in much the same relationship as does his light-cosmogony to the physical system of Aristotle, i.e., as a speculative discourse constructed at a more abstract level, from which the accepted analysis of phenomena can be to a great extent deduced. Like all inventors of a good hypothesis, Grosseteste was tempted to display the fruitfulness of his light-metaphysics by extending it to embrace a problem-area lying outside its original field of application. The result is a work of admirable intelligence and scientific imagination, which, ill-content to remain at the level of assimilation of and subservience to authority, felt impelled to take flight and test their proper buoyancy.

II THE POWERS OF THE SENSITIVE SOUL AND THEIR OBJECTS

Like Avicenna, Grosseteste divides the sensitive powers first into powers of knowing and desiring, then into the external senses and the internal.[10] So closely, indeed, does he follow Avicenna in treating of these abilities, that we find it sufficient

---

[9] *In II Sent.*, d. 2, p. 2, a. 1, q. 2 (II, p. 74a-b). See also Aquinas, *Summa Theol.* I, q. 76, a. 7, who dismisses a yet more exotic variant with unconcealed impatience; and see Baeumker, *Witelo*, pp. 454–6.

[10] 'Vires sensibiles sunt duae, motiva et apprehensiva ... virium apprehensivarum quaedam est apprehensiva deintus, quaedam de foris.' (*De Confess.*, p. 260).

to spend less time discussing their classification than in drawing attention to the personal elements in his treatment.[11]

a. The external senses

Sight is the privileged one among the senses, forming practically a case apart from the other four,[12] because both its object and its power are light, the pure element, unmixed with lower stuff. Its object is all coloured things, whether in the heavens or on earth, as illumined by 'fiery light', that is to say by the visual rays' meeting with the colours, which are an effect of incorporated light. Hearing has the element of air as its medium, and perceives objects sounding by the vibration of the fine airy element within them. The objects of the sense of smell are active through the motion of air, but this time the lower air, mixed with dampness to form a mist or vapour. The savours which taste apprehends derive from activity in the element of water, and, lastly, the touch senses objects the composition of which is dominated by earth and activated by the pairs of opposed 'tactile qualities' (hot and cold, dry and wet).[13]

The strong microcosmic relationship of each sense to its object is only fully appreciable when its double nature is apprehended. Firstly, the hierarchy of the objects to which the senses are proportioned is constituted by descent from the most noble bodies of light and fire through the middle elements, to the passive earth; and in each object constituted by an element, the activity which produces colours, sounds, etc., is an exteriorization of that in the body which either is light, or is assimilated to light by its active nature. Secondly, on the side of the sense-organ a hierarchy is formed reaching from sight through the middle senses to touch, and the hierarchy of nobility results in this case also from the progressive downward extension of light towards the thick and more passive elements, which weaken the activity of its rays. Thus Grosse-

---

[11] The sermon, *Ecclesia Sancta Celebrat*, has been chosen as the basis of the following exposition, because of its completeness and personal tone. Reference will frequently be made to the *De Confess., Ex Rerum Init., Comm. in Anal. Post.*, and to the Pseudo-Dionysian commentaries.

[12] *Comm. in Myst. Theol.*, ed. Gamba, p. 60.7.

[13] *Ecclesia Sancta*, ed. McEvoy, in *Rech. Théol. anc. méd.* 47 (1980), § 13, p. 176.

teste produces a synthesis of the hints given by Augustine in the *De Genesi ad Litteram*.[14] The symmetry of the resulting doctrine of sensation recaptures and emphasizes wonderfully the Platonic maxim that 'like is known by like'. A body in which a certain element predominates is sensed by an organ in which that same element embodies and modulates the activity of light, the perceptive spirit. While the doctrine of sense-objects has its source in Aristotle and its development in Avicenna, it is Augustine's emphasis on the active nature of sensation which triumphs. Sensation is conceived of, not as the result of the activity of a stimulus upon a passive and receptive organ, but as an act of the soul whose corporeal instrument, light, radiates in a sphere from the senses, though being so weakened by the terrestrial dominance in touch that it requires the physically-conjoined presence of its object in order to perceive.

b. *The internal senses*

Grosseteste's doctrine of the internal senses follows Avicenna's to the details. The internal senses are required in order to grant a degree of stability to percepts and preserve them at the service of the knowing subject.[15] The *sensus communis* is the power in which the five senses are rooted as in their origin and in a higher light.[16] It co-ordinates the data of the individual senses and combines them into a unity of apperception.[17]

---

[14] 'Sunt autem sensus . . . ad similitudinem elementorum mundi in mundo sensibili, sicut supra secundum Augustinum notavimus.' (*Hexaem.*, fol. 223ᵃ). The paralleling of senses and elements under the influence of Augustine was fairly common; it is described with an economy equal to that of Grosseteste in the *Itinerarium* of St. Bonaventure, II, 3 (*Op. Om.*, V, 300a–b), and in the *Summa de Anima* of Jean de la Rochelle (ibid., 227b, n. 5).

[15] See Avicenna, *De Anima*, IV, 2, ed. cit., p. 27.65, and Verbeke's introduction, p. 47, ibid.

[16] 'Non enim visus se videt aut auditus se audit nec aliquis ceterorum sensuum particularium se sentit. Haec tamen lumina sensuum particularium superiori luci cuidam statim manifestantur. Sentit enim sensus [fol. 52ʳᵃ] quidem communis sensus particulares agentes.' (*Dictum 55*, MS Cambridge, *Gonville and Caius 380*, fol. 51ᵛ–52ʳ, cf. *Ecclesia Sancta*, § 13).

[17] '. . . sensus communis, quo iudicatur album esse dulce, aut rubeum odiferum.' (*De Confess.*, p. 260). Here Grosseteste tacitly corrects the mistake he had made some few years earlier (c.1241) in the *Comm. in De Div. Nom.*, ed Ruello, p. 139, § 17: '. . . sensum communem, quo comprehenditur unitas et numerus in corporibus, et quedam huiusmodi.' Since Avicenna carefully admonished his reader not to confound the functions of the *sensus communis* with the perception of the common

Although the external senses apprehend only singular material objects, they do so in a relatively immaterial way.[18] They depend, however, upon the material presence of their objects and cannot function in the absence of the latter, wherefore the soul requires a power which retains the species of sensibles as they were perceived. This *fantasia* or *imaginatio* can also join and divide the percepts at will.[19] It functions as a mediator between sense and intellectual abstraction.[20]

Life in all its degrees desires what furthers and delights it and rejects and avoids that which does not favour it. Grosseteste brings Avicenna's most original discovery, the *vis aestimativa*, under this generalized and metaphysical definition of life.[21] The estimative is a power of animal and human life which judges invisible qualities or 'intentions', like threat, fear and attraction, in what is perceived. Grosseteste retails the classic example of the lamb's instinctive fear of the wolf and love of its mother.[22]

The *memoria*, the ability to recall the past, is located in the farthest part of the cerebral concavity, behind the *sensus communis* and the estimative.[23] In man this sense-memory has a particular importance, representing as it does the condition of possibility of forming universal concepts based upon past experience.[24]

c. *The appetitive capacity*

We turn now to the group of appetitive-motive powers. While here also the influence of Avicenna is paramount, we must

sensibles (*De Anima* III. 8, ed. Van Riet, pp. 281, 283), it is not unlikely that the *De Confess.* postdates the Dionysian commentaries and presupposes an intervening re reading of the *De Anima*.

[18] 'Est autem sensus exterior vis susceptiva et apprehensiva sensibilium specierum sine materia.' (*Hexaem.*, fol. 223ª; cf. *Comm. in De Div. Nom.*, I, ed. Ruello, 137, § 12).

[19] See Avicenna, *De Anima*, IV, 1, ed. Van Riet, p. 6. Verbally close to the text of Avicenna is *Comm. in De Div. Nom.*, p. 138: the imaginative 'format et figurat et fingit et componit et dividit etiam absentia, qualia fuerunt in sensu praesentia'.

[20] *Comm. in De Div. Nom.*, ed. Ruello, p. 139, § 17.

[21] 'Etiam quia vita omnis est appetitiva sibi convenientium et delectabilium et repulsiva contristabilium, et per consequens omnis potentia sensitiva, ideo habet sensitiva animae humanae potentiam iudicativam sibi convenientium et delectabilium et inconvenientium et contristabilium, quae potentia dicitur estimatio.' (*Ecclesia Sancta*, § 14, ed. cit., p. 176).

[22] *De Confess.*, pp. 260-1; cf. Avicenna, De Anima, IV, 1, ed. Van Riet, pp. 7-8.

[23] *De Confess.*, p. 261, as in Avicenna, IV, 1, ed. cit., pp. 9-11.

[24] *Comm. in Anal. Post.*, fol. 39ᵈ.

note from the start an element of originality in Grosseteste's treatment. The concupiscential and irascible powers, ever-present in Neoplatonic psychologies, are brought under a higher capacity, which includes the senses also, namely, the common faculty of desire. This power is the origin of all desire at the sensitive level, much as the *sensus communis* is the root of all perception.[25] It would seem that Grosseteste meant by this extension to bring to light the affective aspects of the senses' relation to their objects.[26]

The *concupiscentia specialis* and the *irascibilis* are the positive and negative affective reactions to the objects of sense-experience. They are accompanied in their reactions by a discriminatory power capable of distinguishing at the emotional level between the fitting and the harmful.[27] The faculty of desire constitutes the governing force of movement towards and from their objects (*vis imperans motum*), and, just as in Avicenna, their commands are effected by the moving-power of nerves and muscles, which, by contracting and expanding the ligaments, move the members.[28]

All these powers man shares with the animals, but he possesses them in a human way, as we shall see.

III THE POWERS OF THE RATIONAL SOUL

The development of Grosseteste's thought is nowhere more apparent than in the analysis of the powers of the rational soul, for his classification of the specifically human powers ranges from the traditional tripartite *ratio-concupiscibilis-irascibilis*,[29] through the Aristotelean distinction of opinion, science, and understanding, to the final, complex scheme expounded in his sermon, *Ecclesia Sancta Celebrat*. It is an interesting reflection to make in passing, that if we did not possess this sermon we should be forced to regard Grosseteste's ideas concerning

---

[25] 'Et sicut hae quinque potentiae sensitivae radicantur in una potentia ex qua hae ramificantur, quae dicitur sensus communis, sic et eis insitae potentiae appetitivae et motivae radicantur in una communi appetitiva, quae dicitur communiter concupiscibilis.' (*Ecclesia Sancta.*, § 14, ed. cit., pp. 176-7).

[26] The common concupiscible power finds no place in Avicenna, nor in Gundissalinus, but the former recognizes that every *virtus apprehendens* is impelled to its object by a form of love.

[27] *De Confess.*, p. 260.

[28] Ibid., cf. Avicenna, *De Anima* I, 5.

[29] See, e.g., *De Art. Lib.*, p. 4.30.

the rational powers as fragmentary and incomplete sketches, for this is, as it happens, the only text where the reader attains a sense of an integral vision of the cognitive life of man. Balancing this reflection is another: had Grosseteste left us a commentary on the *De Anima* of Aristotle, it is not inconceivable that his analysis of the intellect would have been fuller still.[30]

In view of the complexity of the evidence and the need to balance systematic exposition against attention to Grosseteste's sources and evolution, the following method of treatment has been adopted. The discussion follows the systematic scheme of the rational powers adopted in the sermon, *Ecclesia Sancta*, perhaps the latest and certainly the most self-conscious treatment accorded by Grosseteste to the problem-area. The evidence we have been able to assemble concerning the evolution and derivation of the ideas will be built into our remarks on each of the powers discussed. First must come a note concerning his principle of specification of the powers of the soul.

Grosseteste saw the entire range of the capacities of human life as constituting a marvellous unity in which a single dynamic principle is at work, producing an openness to reality, a movement towards being and away from self-destruction and isolation. The natural habitat of this multiple wholeness that is man's life cannot be defined in terms of a restricted environment, as it can in the case of plants and animals; it is Being itself, since all that is can be apprehended by some level of human life, and since the soul in the fullest meaning of the word, the life-principle, 'is in some way all things',—a clear reference to the celebrated dictum of Aristotle found in the *De Anima* (III, 8, 431b, 20).

Grosseteste will now work out the implications of this basic understanding of man by showing how each capacity of the soul apprehends a correlative level of reality.

If human life integrally conceived can only be characterized in terms of its capacity for knowing the totality of being, Grosseteste thinks it only natural and reasonable to postulate that the different regions of reality specify as objects of knowledge a variety of knowing-powers within the unity of the soul.

---

[30] See the reservation expressed at the conclusion of the *Comm. in Anal. Post.*, I (ed. Venice, 1514, fol. 25d): 'De aliis vero viribus anime que non ordinate sunt ad opus demonstrationis non est huius loci pertractare; sed quasdam de aliis pertractat physica in libro de anima, quasdam vero ethica.'

The principle of division of these powers which he adduces, when it is confronted with the results its application gives rise to, is an outstandingly good example of the conscious effort at harmonizing the philosophy of Aristotle with the ideas of St. Augustine which is so prominent a feature of Grosseteste's thought. The principle is firmly Aristotelean, but it is pressed into service as a theoretical basis grounding a classification of intellectual powers which Grosseteste attributed to Augustine; the fact that, like many of his contemporaries, he was mistaken in believing the compilation *De Spiritu et Anima* to be a genuine work of Augustine, however significant it may be for the modern historian, leaves Grosseteste's aim in this respect quite unaffected. Just as Neo-Platonists down the Greek, Arabic, and Latin centuries had dreamed of reconciling Plato and Aristotle, Grosseteste aspired in more than one respect to harmonize the Lycaeum and the Christian Plato. He has no difficulty in laying claim to the authority of the Philosopher for the proposition that powers are specified by acts and acts by objects, so that the natural *differentiae* of the various kinds of being can be used to specify the human powers of understanding constituting in their sum the total intellectual life of man.[31]

A glance at the resulting classification is sufficient to reveal its fundamentally non-Aristotelean character, for Grosseteste reaches a scheme whose essentials are a three-fold division of the speculative virtues into *intelligentia*, the faculty of wisdom and the vision of God, *virtutes intellectiva et intellectualis*, which apprehend respectively the first principles of science and the separate intelligible substances, and the reason, which grasps the objects of the natural sciences and the arts.[32]

Intelligentia, intellectus, and *ratio* is the simplified form of the classification, and its source is to be found in the *De Spiritu et Anima* of Alcher of Clairvaux(?), upon which Grosseteste drew with increasing regularity from *c.*1230 onwards. Alcher

---

[31] 'Et secundum differentias naturales entium apprehensibilium, ut testatur philosophus, differunt naturaliter potentiae partiales naturales apprehensivae vitam hominis integram constituentes.' (*Ecclesia Sancta*, § 3). The reference is to *De Anima* II, 1, 415a, 18 ff. Grosseteste elsewhere extends the same principle of specification to the affective side of human nature: *Comm. in De Div. Nom.*, ed. Ruello, p. 142, § 42.

[32] Besides *Ecclesia Sancta*, §§ 5-8, see *Ex Rerum Init.*, *Coll. Franc.* (1967), pp. 121, 126.

twice catalogues the soul's powers as *sensus, imaginatio, ratio, intellectus,* and *intelligentia,* describes them as five stages on the way to wisdom, and likens their coupling with the four affections of the soul to the nine heavenly choirs.[33] Within this scheme, *intelligentia* is described as the faculty of illuminated wisdom, *intellectus* is the power which knows the immaterial beings (created spirits universally), and *ratio* in a limited sense knows the forms of material things, while in a broader meaning it englobes the totality of the intellectual powers of man. Taken altogether, this analysis of the soul is meant to establish that '*anima omnium est capax*', that the soul is '*similitudo omnium*', a microcosm of reality which '*omnia apprehendit*', is '*capax universitatis*' (i.e., God, angels, and matter).

If Grosseteste's dependence on Alcher's work is manifest, the latter's debt to his correspondent, Isaac de Stella, is complete right down to the very details of allegorization.[34] In turn, Isaac's source for his list of faculties is no longer a mystery after the fine detective-work done by von Ivánka, who has shown that it was derived through an as yet unidentified distributary from the *De Fato* of Proclus.[35] There is no evidence that Grosseteste knew Proclus's work at first hand, nor that he had read Isaac de Stella's epistle, which would surely have raised suspicions in his mind concerning the authenticity of the *De Spiritu*. It is therefore all the more remarkable that his use of the triad *ratio-intellectus-intelligentia* strikingly resembles that of Proclus at two points, neither of which Alcher explicitly represents. In the Latin version of Moerbeke, Proclus distinguishes between on the one hand knowledge reached by the abstractive work of reason (*scientia*) and the knowledge of first principles (*intellectus*), both of which, he remarks, are handled by Aristotle, and on the other the *intelligentia* to which the aforementioned powers are subordinated, and which

---

[33] *De Spiritu et Anima,* among Augustine's works, *P.L.* 40, 781, 782, 784, 786, 787, etc.

[34] Isaac de Stella, *Epistola ad quemdam familiarem suum de anima, P.L.* 194, 1875-90. For the full schema see 1880 and 1884-85; also *Sermo* 4, ibid., 1701-02. On the *intelligentia,* see 1880-81; the *intellectus,* 1880; and *ratio,* 1885.

[35] von Ivánka, 'Zur Überwindung des neuplatonischen Intellektualismus in der Deutung der Mystik. *Intelligentia* oder *Principalis Affectio*', in *Platonismus in der Philosophie des Mittelalters,* pp. 153-7. The only known medieval translation of the *De Fato* is Moerbeke's, and its value is increased by the disappearence of the Greek MS-tradition.

attains to that divine and superintellectual knowing of which Aristotle was ignorant and Plato the great master.[36] Now Grosseteste, as we shall see, likewise builds a strongly Aristotelean content into the lower levels of *scientia* and *intellectus*, while reserving to the intelligence wisdom, or the knowledge of God, and the blessed union of the vision face to face.[37]

The second resemblance is equally striking, namely, the agreement of Grosseteste with the *De Fato* that like is known by like, Plato's principle of the specification of powers we may term it.[38] For Grosseteste, God is known by the intelligence filled with grace, the angelic intellects by *intellectus*, and sensible things by *ratio* abstracting universal forms from the data of sense-knowledge. Proclus does not, however, attempt to infer the scheme of powers from the Aristotelean principle of specification.[39]

Grosseteste, as we have indicated, did not content himself with reproducing a borrowed scheme; he developed it to reflect his own wide reading and informed convictions. We turn now to the examination of the members of his division.

God is the supreme light and the supremely intelligible being, wherefore the natural capacity of the mind to know him, to desire his direct vision, and to enjoy it after death, is the noblest capacity of human nature.[40] The degree of development which *intelligentia* or the capacity for wisdom receives in the sermon *Ecclesia Sancta* emerges best when the history of its use by Grosseteste over the preceding period of twenty or so years is studied. Already in the *Commentary on the Posterior Analytics*, the *intelligentia* was placed above *intellectus* and *scientia* as the supreme faculty of the soul (fol. 17a). In itself immaterial, its natural and destined way of knowing would be through direct irradiation from the supreme light, and not by abstraction from the data of sense; it would know all things in the creative light in the same way as does the angel.

---

[36] See the text quoted from the *De Fato* by von Ivánka, ibid., p. 155.

[37] Proclus made *intelligentia* the faculty of union with the One, see ibid.

[38] See Proclus, ibid.: 'Omnia simili cognoscuntur, sensibile sensu, scibile scientia, intelligibile intellectu, unum uniali,' i.e., the One by the faculty of mystical union.

[39] The evidence quoted by von Ivánka (ibid., p. 157) to prove that Gallus had direct access to the text of Proclus in an earlier translation does not seem conclusive, and is probably weakened by the fact that Grosseteste, working independently, reached conclusions similar to those of Thomas. Two men of the same generation, working with the same problematic and similar sources, often do.

[40] *Ecclesia Sancta*, § 5, p. 171.

This is the kind of knowledge it will have '*cum anima erit exuta a corpore*', and which some privileged souls enjoy already in unclouded moments of this life. Since in the normal state of man this eye of the soul cannot function in its purity, the powers subordinated to it are as it were lulled to sleep, and need sense-experience for their awakening. In this, his first extant treatment of the beatific vision, Grosseteste is anxious to show how completely the Christian anthropology transcends Aristotelean psychology, and he appeals quite naturally to the language of Alcher to make his point.

The distinction of *intelligentia* and *intellectus* reappears briefly in the *Hexaemeron* (*c.*1237), where *intelligentia* is represented as a faculty of contemplation whose object, the Trinity, is grasped in its transcendence, in an immaterial way ('without phantasms'). The *intellectus* is this time the Alcherian capacity rather than the Aristotelean *nous*, for its object is spiritual creatures.[41] To this text may be assimilated some remarks occurring in the *Ex Rerum Initiatarum* (after 1240), where *sapientia* is defined as the natural power of contemplating the triune God face to face and of knowing all other things in him.[42]

An important expansion is given to the notion of *intelligentia* later in the *Hexaemeron* when it is diversified into the *memoria, intelligentia* and *amor* of the *De Trinitate*, and identified with the *suprema facies animae* by which the Trinity is contemplated without the interference of the material phantasm or corporeal instruments.[43] This supreme power of the soul is one and three, and therefore the locus of the image of God in man, and of the restoring work of grace in the soul. It in turn orders the lower faculties of the soul by drawing them into its likeness and imitation, and its influence reaches even the body, so that it stamps the whole person with the seal of divine life.[44] However one wishes to name it, whether *ratio, mens,* or *intelligentia*, it is in this highest power that the image

---

[41] *Hexaem.*, fol. 213ᵃ (Grosseteste unfolds a spiritual sense of the light created on the first day): 'Spiritualiter autem tam in ecclesia quam in qualibet anima facta fit [*sic*] lux cum ipsa rationalis cognicio assurgit in contemplationem trinitatis per intelligentiam a phantasmatibus denudatam, vel in speculatione intellectualium creaturarum et incorporearum per intellectum . . .'

[42] *Coll. Franc.* (1967), pp. 121, 126.

[43] See ed. Muckle, *Med. Stud.* 6 (1944), pp. 164–5.

[44] Grosseteste illustrates the influence of the *suprema pars* on the lower powers by a characteristically lovely light-metaphor. The ether receives the light of the

of the Trinity is to be found. Grosseteste, of course, saw no difficulty in equating 'intelligence' with the Augustinian triad of *memoria, intelligentia,* and *amor,* since he thought both descriptions emanated from the same source. Here already one can distinguish the beginning of a trend which was to achieve its perfect expression only in the *Ecclesia Sancta,* namely, to stress the strongly affective character of *intelligentia,* and to identify its activity with the summit of unitive love.

The doctrine of intelligence acquires this last dimension in the course of the Dionysian commentaries. Suffice it to mention two main emphases from the teaching of Grosseteste's *Commentary on the Celestial Hierarchy* concerning the beatific vision and its earthly anticipations in mysticism, as defining the deepest capacity of human nature: the *intelligentia* can only function at its purest when the soul is released from the body, and the wisdom which *intelligentia* receives admits of varying degrees in this life, ranging through the political and speculative virtues to mystical union.[45] Another important and personal passage from the same work describes wisdom as the highest conceivable participation by a creature in the providence of God, even more, in the very life of God himself.[46] The share in the knowledge of God himself is the specific difference of man and angel over against other creatures, for the natural operation of the angel, as of man, is intelligence.[47]

The most puzzling passage concerning *intelligentia* in the entire corpus of Grosseteste's writings is found in his *Commentary on the Mystical Theology* of Pseudo-Dionysius.[48] I quote

sun immediately and passes it on to each of the lower elements in turn until it reaches the crass earth, when the whole universe is illumined by a single splendour (ibid., p. 164).

[45] *Comm. in Hier. Cael.,* ed. McQuade, p. 194 ff.; cf. Part IV, ch. 2, ii (pp. 245-56).

[46] *Comm. in Hier. Cael.,* ed. McQuade, pp. 173-6. The passage is too long to quote representatively, but the following is its kernel: '. . . rationalia et intellectualia participant eiusdem deitatis sapientia que est ipsa deitas . . .'; cf. p. 197: '. . . sapientia, id est ipsius divinitatis cognitio . . .'

[47] 'Angeli vero, qui est intellectus, operatio naturalis est intelligentia.' (*Comm. in De Myst. Theol.,* ed. Gamba, p. 63.24). In an earlier period Grosseteste used *intelligentiae* as a synonym for the angels—an Arabic influence. In the Dionysian commentaries, however, he consistently used *intellectus,* as more closely representing the usage of the LXX and of Pseudo-Dionysius. *Intelligentia* is, therefore, no longer a general designation of the spiritual nature of the angels (*substantiae separatae*), but explicitly denotes their activity of direct vision. It is not opposed to other faculties in them; their lives are free from the complexity of human nature.

[48] Ibid., p. 63.15-25.

it in full. Dionysius denies that the human language of the faculties of knowledge applies in any way to God:

Rursus autem ascendentes dicamus quod neque anima est, ipsa videlicet omnium causa, *neque intellectus,* id est neque angelus . . . *neque habet fantasiam,* id est imaginationem, *vel opinionem vel rationem* agentes videlicet ut per has sit anima naturaliter agens, *vel intelligentiam,* que est actio intellectus ut per hanc sit intellectus agens, *neque est ratio, neque intelligentia, neque dicitur* verbo videlicet rationis, *neque intelligitur* actione intellectus agentis.

This passage disproves Gilson's claim that Grosseteste never refers to the *intellectus agens,* but its brevity and obscurity enable us to say little concerning the content with which he invests the Aristotelean term. We offer the following tentative interpretation to take account of the evidence.

It is a characteristic trait of Grosseteste as translator and commentator on Dionysius that he attempts to systematize the oftentimes fluid thought of his obscure author, and the strictness he brings to the text frequently tends at the same time to bring its meaning closer to the Augustinian tradition. In the whole chapter of negative theology from which the passage comes, Dionysius offers an ascent which is essentially one of love, rising *not* through the levels of knowledge and being to their source, but in independence of all that is conceptualizable; the jumble of concepts and realities which are denied of God have no coherent order nor meaning for thought.[49] Grosseteste sets out never the less to find one, and succeeds largely, but only at the cost of superimposing a systematic perspective which is not that of his author: he turns a purely affective ascent into an itinerary of the mind through the grades of being.

In the present passage he finds the denial that God is of like nature to soul or angel, followed by a list of the soul's powers, none of which can attain God, nor even serve as an adequate symbol of his hiddenness. Dionysius mentions *fantasia, opinio, ratio,* and *intelligentia;* Grosseteste fills in the gap by reflex: *ratio-intellectus-intelligentia,* the Pseudo-Augustinian scheme of the rational powers. Now in this scheme—of which Dionysius is needless to say completely innocent, despite its origin in Proclus—*intelligentia* is the highest and therefore the most active power, in whose action understanding becomes *intellectus agens.*

[49] See von Ivánka, art. cit., p. 156.

We must assume that Grosseteste, who had read the *De Anima* of Aristotle, did not use the term *intellectus agens* without being aware of its provenance. Speaking, therefore, of the *intelligentia* (which, of course, he knew was not to be found in the Philosopher), of the highest faculty of the soul, its immaterial and immortal part and its most active element, he chose to refer to it by using a term denoting the highest part of the Aristotelean soul, the part, namely, which possesses the same attributes of nobility, separability and activity as the Christian *intelligentia*, though not of course the same function. The conflation thus effected can be regarded as part of a pattern. In order to describe the faculty of *sapientia*, Grosseteste had adopted all the terms he could find in his library for the highest part of the soul: from Augustine, *intelligentia*, *mens*, the triad *memoria-intelligentia-amor*; from Augustine and Avicenna, *facies* or *suprema facies animae*; and others—*acies* (*Comm. Post. Anal.*, fol. 20ᶜ), *apex mentis*, *visus mentis supremus*, *mens casta et sancta*.[50] The only prominent terms he does not seem to have employed are *'affectio principalis'* and *'synderesis scintilla'*, both so beloved of Thomas Gallus and (under the influence of his *Extractio*), St. Bonaventure.[51]

We do not wish to imply that Grosseteste attributed the functions of the Aristotelean *intellectus agens*, with the name, to the *intelligentia*; but it should be noted that, as it happens, the two do have one function in common, aside from their several shared attributes: both stand to the lower power(s) of the soul as active agents of illumination.

The sermon *Ecclesia Sancta* is Grosseteste's last and finest portrait of the *intelligentia*. In correcting the predominantly intellectual character of the intelligence as found in his other works, it testifies to the influence of the *Mystical Theology* on his thought. The affective note attributed to the *intelligentia* appears already in its definition as at once the faculty of searching for and seeing God, and the yearning which lasts as long as his vision remains only the distant object of hope. The desire for the fullness of *sapientia* guides teleologically the entire ascent of mind and love through the material and spiritual world (§ 35). At the height of the ascent it is left

---

[50] *Comm. in De Div. Nom.*, ed. Ruello, p. 141, § 21; p. 143, § 26.

[51] See von Ivánka, pp. 149–50. Further reading in the *inedita* could, of course, reverse this finding.

only for the *intelligentia*, identified with the *amor* which has urged the soul upwards, to still the lower powers and stand beyond all things and outside itself in the darkness of the cloud of unknowing, until the beloved manifests himself face to face (§ 36). In the manifestation, the lower powers receive of the fullness of life from the intelligence, much as was adumbrated in the *Hexaemeron*.

An important conclusion can be drawn from this. It has been shown by von Ivánka that Thomas Gallus, in his *Extractio* from the *Mystical Theology*, breaks with the Victorine tradition to side with Dionysius in the description of the mystical union. The unbroken transition from the dawn light to the noonday brightness of the manifestation (Richard of St. Victor) is replaced by a total rupture between the preparatory activity of reason and intellect and the union of the *pura intelligentia* with the unknown God, which union is beyond thought and language. The decisive influence is, of course, Dionysius, but the Latin expression of the new development sought out the language of Alcher and Isaac de Stella, making the intelligence a *principalis affectio* and emphasizing its supraintellectual character.[52] Now there is no difference that I can see between Gallus and Grosseteste on any of these points, rather an identity of thought. Though Grosseteste does not refer to intelligence as *affectio*, his identification of it with *amor* produces the same result. For the immense influence which this development had upon the history of Western mysticism till the close of the Middle Ages, and even beyond that, Gallus cannot be given the sole credit.

The second capacity of the human mind is *intellectus*, and Grosseteste attributes it a double object, which, according to his principle, specifies two levels of apprehension within it. On the one hand the *virtus intellectiva* apprehends the highest form of abstract intelligibility, the principles of knowledge which are self-evident and immediately known to all men as soon as they begin to think.[53] The *virtus intellectualis* or *supersubstantialis*, on the other hand, the second capacity, has as its object the separate, intelligible substances (§ 6). Let us see how this distinction within the understanding developed, before it crystallized in the very last works of Grosseteste.

---

[52] Ibid., pp. 150-2.
[53] *Ecclesia Sancta*, § 5; *Ex Rerum Init.*, p. 121.

The *Commentary on the Posterior Analytics* provides his most extended commentary on Aristotle's conception of the intellect, which is distinguished carefully from *intelligentia*, as we saw, and also from the lower powers of *scientia* and *solertia*. It is assigned two functions, the first of which is of direct interest here: it grasps the indemonstrable principles of reason and scientific knowledge, and it seizes the universal idea.

As the universal is more worthy than sense-knowledge, so the *cognitio intellectiva quae est dignitatum* transcends even the universal in value (fol. 23c). This is the same *intellectiva* which Grosseteste was to place second only to intelligence in his final scheme of the powers of the soul—surely a remarkable testimony to the consistency of his thought. Now the knowledge of first principles is of a higher order than science, whose principles are not simply first, but depend upon the *dignitates*, than which nothing is more known (fol. 3c-d). A self-evident proposition is one whose necessity and evidence lies in itself and not in another truth, one which must therefore be acknowledged by any mind (fol. 10c), such as that two things equal to a third are equal to each other (fol. 8b). Though the intellect grasps them without reasoning, by simple inspection, just as sight does its objects, knowledge of them is not innate (fol. 10c).

In a somewhat wider sense, intellect is the power of the mind to perceive the first indemonstrable premisses of science, from which scientific knowledge strictly speaking can be deduced (fols. 40b, 25b), for demonstrative science cannot prove its premisses, rather it must accept them as truly first, underived, prior to and more known than the conclusions, to which they stand as cause (fols. 3b-4a).

In a wider sense still, *intellectus* is the power of the mind to abstract universal ideas from the particulars experienced by sense.[54] The universal concept can be arrived at only by induction from the senses (fol. 17a), for universals are not innate (fol. 39d). The intellect is the locus of universals.[55]

[54] 'Sic existente sensu, qui est debilissima virtutum apprehensivarum, apprehendens res singulares corruptibiles, stat imaginativa et memoria, et tandem intellectus, qui est nobilissima virtutum apprehensivarum, apprehensiva universalium primorum incorruptibilium.' (*Comm. in Anal. Post.*, fol. 40a; cf. *De Confess.*, p. 261).

[55] Ibid., fol. 23c. Of course, for Grosseteste their prime locus is the divine mind.

Understanding and demonstrative science apprehend pure essences, abstracted from the confusion of material accidents.[56] From all this it is apparent that the *intellectus* of Grosseteste is essentially the Aristotelean *nous*, rather well recaptured by a careful exegesis of the *Analytics*.

In the *Hexaemeron*, however, another *intellectus* appears, and this turns out to be the Pseudo-Augustinian one: the faculty of cognizing the incorporeal beings who are nearest to God.[57] This usage corresponds exactly to that of Alcher and Isaac, and receives besides a Dionysian intonation when called *'superscientialis'*.[58] This cognitional faculty is on a lower level than that which grasps the principles, for its object is not self-evident like the dignities, but is known *'per media notiora et nota'*.[59] The certainty of such knowledge is based upon the intelligibility of its objects, for the separate substances are the highest created light and changeless in their nature; true knowledge demands a perfectly stable object. The intellect does not have to abstract their species, for they are already free from matter and therefore able to become the act of the intellect without passing through a phantasm.[60]

Below the *virtutes intellectiva et intellectualis* comprising the understanding there stands scientific reason (*virtus scitiva*), with the demonstrative knowledge it constructs concerning the permanent aspects of the material world.[61] Grosseteste places at this level the sciences of nature worked out according to the Aristotelean methodology and ideal, and also mathematics.[62] Science is knowledge of causes demonstrated by deduction from the principles intuited by the intellect (fols. 3^b, 25^c). It is sharply distinguished from opinion because, like intellect, it grasps the essences of its objects, not merely their accidents (fols. 25^{b-c}, 32^c).

---

[56] Ibid., fol. 40^b; cf. 25^b, and *De Confess.*, p. 261: 'Intellectus est acceptatio alicuius absque materia et materialibus dispositionibus,' as distinct from *opinio*, which is 'acceptio alicuius in appendiciis materialibus absque materia'. Once more, Grosseteste is repaying his debt to Avicenna.

[57] *Hexaem.*, fol. 203^a, quoted already in n. 41.

[58] In *Ecclesia Sancta*, § 6, pp. 171-2.

[59] Ibid., cf. *Comm. in Anal. Post.*, fol. 40^b.

[60] 'Similiter subfiguramus eam [i.e., the thearchy, God] per omnia intelligibilia et species ex hiis natas, que coniuncte intellectui faciunt eum actu intelligentem...' (*Comm. in De Div. Nom.*, ed. Ruello, p. 150.45; an Avicennian note).

[61] *Ecclesia Sancta*, § 7, p. 173; cf. *Ex Rerum Init.*, p. 121.

[62] *Comm. in Anal. Post.*, fol. 2^d.

The scientific reason aims at universal and necessary truths and presupposes an object that is immutable. If the highest degree of stability is lacking to the object science is impossible, and the less exacting dialectical method is left to attain probable truth, such as human art must be content with.[63] Art is the fifth power of the human mind or reason.

When reason is turned no longer upon objects independent of the human will but upon voluntary actions, its practical exercise is termed natural prudence, the power to discriminate good from bad. It is not a speculative virtue like the others described, but tends to pass over into voluntary action, seeking the good and avoiding its opposite (*Ecclesia Sancta*, § 9).

Three final powers of the rational soul complete the list. The powers of reasoning, of reasonable assent to fact, and of free choice are enumerated separately because of their circumincessory nature—they are diffused through all the others and cannot easily be assigned a precise object (§ § 11-12). The art of logic is like the life's blood or the nervous system of the whole body of science. There is too a power of the reason which may be called faith, and whose object is singular propositions which are not manifest, nor demonstrable, nor deducible from scientific knowledge, but are held with a certainty deriving from the tradition which transmits them: that this man is my father; that Rome or Jerusalem exist, even though I have seen neither; and the meaning of foreign words (§ 11). Grosseteste's attention had been drawn to this category of proposition by Avicenna, whose example of Mecca he freely Westernized.[64] Finally, human freedom is found in all the powers of the soul, because it was given to man to be mistress of all his capacities and servant to none, so that they may observe due order in their proper acts and cessation of activity (§ 15).[65]

These, then, are the powers whose totality comprises human life at the natural level. The definition of soul which emerges from them from a cognitive point of view can only be constructed in terms of its capacity and desire to know all things,

---

[63] *Ecclesia Sancta*, § 8, p. 173; cf. *Ex Rerum Init.*, p. 121. On dialectic as a rational instrument attaining probability, cf. *Comm. in Anal. Post.*, fols. 11[d], 3[b], 25[c].

[64] Avicenna, *De Anima* V, 3, ed. Van Riet, p. 104.

[65] On the damage done to freedom by original sin, cf. *Ex Rerum Init.*, pp. 124 ff.

and this is a thought to which Grosseteste frequently returns.[66] For someone living within the medieval world view it was a fairly natural one, for the universe mirrored in the Scriptures consisted of God, the angels, the material world and its contents, and man—all of which were held to be in some degree intelligible in principle and also known in fact. By the enumeration of these different areas of being, accompanied by an affirmation of the principle that like is known by like, one could find in the soul a power of knowing each region of reality, and the sum of such capacities, the soul or mind, has then the sum of reality for its object.[67] One could call this idea 'totalization through enumeration of kinds'. To that could be added the certainty that the vision of God would be the full realization of the desire to know all things, for in the divine mind are the exemplary ideas of all that is made.[68] The strength of the body of teachings thus constructed could not fail to attract into its orbit the Aristotelean dictum to the effect that 'the soul is somehow all things'. It must have seemed a matter for wonder to Grosseteste that, once again, his two main authorities, Augustine in theology and Aristotle in philosophy, were in agreement with each other, and he exempted the conflated idea from Avicenna's strictures upon its absurdity.

IV THE UNITY OF THE HUMAN SOUL

The debate on the powers of the soul was intimately connected in the thirteenth century with the question of the plurality of forms, which was bound to be raised once the philosophical division of the soul's powers into vegetative, sensitive and rational had established itself. A third question which asserted its presence at this point was the origin of the soul, in view of the need to define the spirituality of at least the rational part of man. Since the stance adopted with regard to this last question was often determinative of the answer given to the plurality-question, it will be as well to begin our inquiry by examining Grosseteste's remarks on the origin of the soul.

---

[66] See *De Libero Arbitrio*, ed. Baur, p. 238; *Ecclesia Sancta*, § 3, pp. 170-1; *Ex Rerum Init.*, p. 121; *Comm. in Hier. Cael.*, ed. McEvoy, p. 162; *Dictum 55*, MS Cambridge, *Gonville and Caius 380*, fol. 52r.

[67] This method, as we have seen, was adopted by Isaac and Alcher, and followed by Grosseteste.

[68] *De Spiritu et Anima, P.L. 40*, 781, 784.

This appears to be one of those cases where Grosseteste changed his mind, although the evidence for this claim is admittedly not altogether clear. Writing to William de Raleigh in 1236, in the hope of converting him to the reform of the common law of bastardy,[69] Grosseteste argued that if the offspring is legitimated by the marriage of the parents during the earliest stages of conception, when the foetus has not yet received the rational soul which makes it human, it is *a fortiori* legitimated by the marriage of the parents at any subsequent time, when the rational soul is present.[70] Now while it might be possible to claim that Grosseteste was employing this argument as a weapon to bludgeon an opponent whom as it happened he did not very much like, it is more probable that, like Roland of Cremona and others of his contemporaries, he regarded the foetus in the early stages as '*quoddam membrum matris*'.[71] Some years later, however, when discussing the Incarnation, he postulated the assumption of a human nature by the Word from the very first moment at which a human body began to form in the womb of the Virgin, on the grounds that the theological consequences of the opposite view would be unthinkable, namely, the assumption of a fully-formed person.[72] Now what was theologically unacceptable was a relationship of priority and posteriority between the infusion of the soul and the assumption of human nature; Grosseteste's teaching, however, did more than simply outrule adoptionism, it excluded any lapse of time between the formation of the body and the infusion of the soul, in the case of Jesus Christ. It may well be that Grosseteste came round to the view that the rational soul is infused at the moment of conception and not at any later stage.

An examination of the question concerning the origin of the soul in the first half of the thirteenth century suggests a growing cleavage between the theologians and the newly-

---

[69] *Epistle 23*, ed. Luard, *Epistolae*, pp. 86-7. See the useful remark of Stevenson, *Robert Grosseteste*, pp. 169-70.

[70] The *De Spiritu et Anima* may have exerted some influence upon the thesis that the rational soul is infused late into the body (*P.L.* 40, 785).

[71] Quoted by Crowley, *Roger Bacon*, p. 128.

[72] *Ex Rerum Init.*, ed. Gieben, *Coll. Franc.* (1967), p. 130: 'Necesse est . . . [ut] non praecedat decisio materiae conceptibilis eius formationem et figurationem, aut harum altera aut utraque animae infusionem, aut aliqua harum trium assumptionem a divina persona in personalem unitatem; quia, si esset in his praecessio et posterioratio, non esset naturae, sed quod esse non potest, personae assumptio.'

emerging group of philosophy teachers, the former holding in general for the direct creation by God of the human soul with its three powers, the latter asserting that only the intellective soul is directly created, the nutritive and sensible developing *per viam naturae*.[73] Theologians who admitted the late infusion of the human soul (and these were in a majority) had to allow that the embryo possesses vegetative and sensitive principles before the coming of the rational soul, and then maintain either that these preparatory forms disappear with the coming of the human soul (containing all lower powers),[74] or that they remain as *duplex vegetativa et sensitiva*, paralleling the lower powers contained in the rational soul.[75] Those who took this second option could not do other than advocate the plurality of forms.[76]

It would seem clear that Grosseteste must be placed with the theologians, despite Roger Bacon's claim that all English theologians, all philosophers, and indeed the universality of thinkers of *c*.1247, taught the direct creation of the intellective soul only. Grosseteste would have found his closest allies in John of la Rochelle, Roland of Cremona, Alexander of Hales, and Albert, all of whom taught that the soul is entirely created with vegetative, sensitive and rational powers. Indeed, there is firm evidence that he moved from the philosophical viewpoint towards the theological one during the interval between the composition of the *De Statu Causarum* and the *Hexaemeron*. In the former work he adopted the view of Aristotle and Avicenna, that both vegetable and animal souls are products of natural causes; whereas by *c*.1237 he was not prepared any more to admit that the sensitive soul even of animals is the result of unaided cosmic processes. Though their substance is not incorporeal and spiritual like that of the human soul, their dignity is such that they must be created by God's direct intervention.[77] It is a fair assumption that if the souls of animals are directly created, the vegetable and

---

[73] Crowley, op. cit., pp. 124–36.

[74] John of la Rochelle, and later St. Thomas: ibid., pp. 126–7.

[75] An attempt by Richard of Cornwall to reconcile the philosophers and the theologians.

[76] The rejection or non-adoption of this thesis did not, however, preclude an author's defending plurality in another sense; as Crowley shows, Bacon is a good example of someone who did just that.

[77] *Hexaem.*, fol. 221[b]. The statement is unambiguous, and is, so far as I know, unique in the literature of the period.

sensitive powers of the human soul are likewise. We are entitled to speculate that Grosseteste wished to remove the question of the origin of the soul from the danger of naturalism,
and that he extended the protective ditch surrounding the
sacred stronghold to include the outer bulwark, the animal
soul. Few would have had any sympathy with this extreme
position, and we are left to imagine the incredulity and scorn
with which Bacon would have received it.

It is evident that the theory of the total creation of the
human soul together with all its potencies would at the very
least discourage the adoption of any of the several theories of
the plurality of forms in man which were current options in
the early thirteenth-century schools. It is all the more remarkable, therefore, that both authors who have discussed Grosseteste's psychological doctrine should have concluded that he
can only have been a pluralist.[78] A partial cause of this mis-
understanding is, no doubt, the failure to realize that a
thirteenth-century author could defend the plurality of forms
as a general metaphysical theory, without allowing its application to the human soul.[79] Fortunately, there are several passages which leave no doubt concerning Grosseteste's teaching
on the matter.

In the *De Confessione* Grosseteste attempts to give a psychological grounding to the Christian commandment to love 'with
the whole soul, the whole heart, the whole mind'.[80] He appro-
priates 'soul' to vegetable life, 'heart' to animal, and 'mind' to
rational creatures. If, he remarks, vegetative, sensitive and
rational principles act as the substantial forms of different
things, they are termed soul; whereas if they coincide in one
being, they are referred to as powers. He draws attention explicitly to the philosophical usage whereby all three forms are
called 'souls', and are qualified further by their relative level
of perfection as nutritive and so on. In the case of man, however, all three are powers of the human soul, each standing

[78] Sharp, *Franciscan Philosophy* . . ., pp. 28-9. Miss Sharp is also wrong in
finding the basis of plurality in Grosseteste's idea of the nobility of the soul; her
suggestion (p. 28) that it cannot assume vegetative and sensitive functions (which
must in consequence be separate forms) because it is spiritual, is untrue, and her
assumption that *lux* acts as a medium between the soul and the vegetative-sensi-
tive functions is unacceptable, as is her identification of the rational soul with
intelligentia.

[79] Crowley, op. cit., p. 136.

[80] See ed. Wenzel, p. 260.

to its *vires* as whole to part, just as the soul stands to all three powers as a whole to its parts.

It is beyond dispute that Grosseteste here affirms the unity of the human soul as a single substance and a single form, including within itself three powers, each of which in turn has many *vires* or functions.[81] Two other works illustrate the firmness with which he held this doctrine, from at least *c*.1236 onwards.[82] In the *De Cessatione Legalium*, the human soul is said to comprise as a power all that the animal soul possesses as a substantial form, and in addition whatever is found at the level of vegetative life.[83] The sermon *Ecclesia Sancta* regards the soul as a principle of life which animates the body, and it numbers a *pars sensitiva* among its powers.[84]

Concerning the value which Grosseteste perceived as underlying his view of the absolute unity of the human soul, we are not left entirely in the dark. It is plausible at least to suggest that he felt assured of having found the best way to preserve the traditional Christian idea of the soul, while managing also to integrate into it the contribution of the new philosophy. Augustine and Ambrose had emphasized the simplicity of the soul, while still attributing to it a plurality of functions, one of which was to be the life-principle of the body. Damascene, whose influence on Grosseteste's later thought must never be underrated, likewise made the incorporeal soul the giver of life, growth, generation, and perception, once again a plurality of functions deriving from a simple being.[85] Probably also an influence of the *De Spiritu et Anima* must be reckoned with: the unity and simplicity of the soul, in spite of the diversity

---

[81] This view is directly parallel to that of Avicenna, whose language of *potentiae* and *vires* Grosseteste borrows. The similarity of Grosseteste's theory of soul to that later adopted by Aquinas should not pass unremarked because of the differences. For both, the human soul is a simple form totally created by God, and containing three powers or faculties within its unity.

[82] That his position even earlier (*c*.1228) was not radically different can be surmised from a passage in the *De Intelligentiis* (ed. Baur, p. 115), where the soul, regarded throughout this work as a simple substance, is said to exercise the functions of vivification, motion, and perception. The language of *potentiae animae* later adopted can be regarded as an explicit formulation of this conception.

[83] 'Communicat anima rationalis cum anima sensibili brutorum in potentia sensitiva, et cum anima vegetabili plantarum in potentia vegetativa.' (ed. Unger, p. 14).

[84] *Ecclesia Sancta*, § 2, § 14, p. 170 and p. 176.

[85] 'Anima igitur est substantia vivens, simplex et incorporea . . . organico utens corpore, et huic vitae et augmentationis et sensus et generationis tributiva.' (*De Fide Orth.*, ed. Buytaert, ch. 6, p. 112).

of its offices, was one of its main themes,[86] and a genuinely Augustinian one at that.[87]

That Grosseteste taught the simplicity of the soul in the traditional Augustinian sense there can be no doubt,[88] and the reasons why he did so are not far to seek. In the first place, the possession of these attributes was taken as a sign of its divine origin,[89] and as making the soul an image of the unity of God's action in the world.[90] Pseudo-Augustine had expressly repudiated the theory that there are two souls in man, an animal and a human one.[91] Secondly, Grosseteste maintained resolutely on the basis of human experience that everything that is found in man, even powers he shares formally with the animals, is possessed by the person in a human way; in this conviction he was perhaps influenced by the very sound psychology of Avicenna. Thus, he maintained that the senses and the desires can be directed by human reason and governed by prudence, whereas in animals they are tied to immediate objects.[92] Grosseteste managed in this way to re-affirm the possibility of reason's dominating the entirety of human nature, in a context no longer purely traditional but renewed by the more differentiated philosophical account of the sense and vegetable powers. Moreover, he thought out the task of the human soul in the context of redemptive grace, which gradually restores to the highest human power that kingly role which it played in the original institution of human nature. He presented a view of man according to which the love of God could enter as motive and desire into the most quotidian acts of eating and waking.[93] Important for our

---

[86] Crowley, op. cit., p. 135. Bacon regarded its influence as bedevilling and its pseudonymous character as patent. Cf. *De Spiritu et Anima, P.L.* 40, 788: 'Essentia una, simplex et individua . . . simplex substantia est anima', and 794, 803, 807: 'in essentia simplex, in officiis multiplex'. Cf. Isaac de Stella, *Epistola de Anima, P.L.* 194, 1876-7, *et passim*.

[87] Augustine, *De Quantitate Animae*, 33, 70-6; 36, 80-1.

[88] *Comm. in Phys.*, ed. Dales, p. 15: 'Est una simplex et indivisa, tota in qualibet parte corporis.'

[89] *De Spiritu et Anima*, ch. 36, *P.L.* 40, 807: the soul is derived from God, 'simplex a simplici'.

[90] St. Ambrose, *De Dignitate Conditionis Humanae* I (*P.L.* 17, 1135A); quoted by Grosseteste, *De Intelligentiis*, p. 14.

[91] *De Spiritu et Anima*, ch. 48, *P.L.* 40, 814.

[92] *Ecclesia Sancta*, § 15, ed. cit., p. 177.

[93] 'Igitur propter deum sumenda sunt nutrimenta et secundum deum evacuanda superflua; et ad honorem eius membris singulis movendum et apprehensivis

consideration here is the fact that the unity of human nature is made by Grosseteste to depend upon the unity of the soul with its powers. No doubt, the same phenomenology and the same theological view could be recognized and accepted by a proponent of the plurality of ordered forms within a soul only substantially one; but Grosseteste preferred to think that the conception of the soul as a simple form possessed of a diversity of operations was the preferable theoretical basis, both for the description of experience and for the theology of grace.

An area of psychology where Grosseteste may well have made an original contribution is the discussion and definition of life, which appears at its best in the sermon *Ecclesia Sancta*. It will be mentioned here only to the extent that it is relevant to the theme of the soul's unity. Writing in the context of the theology of spiritual life or spiritual resurrection, he attempts to define the natural life of man and to break down human nature into its components. His definition is full of interest:

Vita itaque naturalis humana . . . est illa quae vivificat humanum corpus. Haec inquam vita est potentia apprehensiva, appetitiva et motiva ad habendum in actu apprehendendi sibi naturaliter delectabile, et ad cavendum sibi naturaliter contristabile.[94]

The elements composing the definition are largely Aristotelean ones, but Grosseteste seems to have thought them out and formulated them on his own account. The beauty of the entire first part of the sermon consists in the logical way in which this single definition is made to emerge from the entire range of human powers. From the nutritive powers (which, like higher life-forms, exhibit powers of attraction and repulsion), through the external senses (with their inner demand for their object, their love of light and fear of darkness), the inner senses, to the soul itself, with its capacity for knowledge and love of all things, and its recoil from ignorance and darkness, it is a single dynamic principle of attraction and repulsion which animates all human action, and with which the knowledge and love of God are fused in such a way as to obtain access to every crevice of man's natural being. Is it plausible

utendum, et cunctis affectibus in ipsum aspirandum. Et hoc est Deum diligere ex omnibus viribus.' (*De Confessione*, p. 261).

[94] *Ecclesia Sancta*, § 3; see the introduction to the edition in *Rech. Théol. anc. méd.* 47 (1980), especially pp. 142–4: 'Life'.

to see in Grosseteste the founder of what may be styled the 'metaphysics of life', which is found in representatives of the middle Franciscan School, such as Peckham and Olivi? The question might repay an inquiry which cannot be undertaken here.

Chapter 5
Intellectual Knowledge

WHILE the modern preoccupation with epistemological problems should not be read back into the Schoolmen, there is a deal of interest to be found in inquiring how a theologian of the thirteenth century, writing out of a background of Augustinian ideas, reacted to the theory of knowledge created by Aristotle in his struggle to free himself from Platonism. True, the inquiry is not guaranteed an easy path; it faces among others the difficulty of understanding the theory of knowledge associated with the term 'illumination', which contains elements of obscurity which, it could be argued, were never clarified by its supporters. The modern writer finds it difficult to recover the problematic of a cognitional theory which was still at a relatively undifferentiated level; the distinctions introduced by later discussion have placed him outside the integral perspective from within which illumination-theory could be found not only acceptable but self-evident. The interest of Grosseteste's thought is, that as the first medieval Latin commentator on the *Posterior Analytics* he was forced to confront two different traditions of thought, the Platonic-Augustinian, within which he had been educated, and the Aristotelean. We do not expect a writer of the 1220s to produce a perfectly-defined synthesis of the problem areas, but we look forward to following the honest effort made by a very intelligent mind to interpret, criticize, and understand. That is indeed what meets us in Grosseteste's endeavour, and in much higher degree than has been recognized.

I THE *DE VERITATE*

We begin with a discussion of the doctrine of truth set forth by Grosseteste in his treatise *De Veritate*.[1] Here the disciple of Augustine and Anselm makes no reference to Aristotle's theory of knowledge or any of its attendant problems, but handles the question of the relationship between particular

---

[1] See ed. Baur, pp. 130-43.

truths and the supreme truth in the language and spirit of the two authorities, from whom the arguments for and against are drawn.[2] Of these we retain one which proves to be of particular significance (no. 5 in favour): if only the pure of heart shall see the supreme truth, yet some truth and certainty are available to minds not as yet purified, there must be truth other than the highest truth (p. 131.3).

Grosseteste affirms, with his authorities, that both subject and object require to be illuminated by the supreme light if true knowledge is to come about. The foundation upon which this conclusion rests is exemplarism, a favourite doctrine of his, and one to which he returns at every opportunity and in all his works. Each created reality has a truth which consists in its conformity to its eternal exemplar in the divine mind or creative word of God (p. 135). It follows logically that the mind can only claim possession of complete truth concerning any individual thing when it can see, not merely the object itself, but the exemplar which is its standard (*rectitudo*, *regula*). In other words, created truth can be seen only in the supreme truth, the light of whose eternal reason must be present to the beholder and to the objects ('as Augustine says': p. 137).

Let us take the illumination of the object first. No object, Grosseteste declares, can be seen as true in its created truth alone, any more than a body can be seen to be coloured in the absence of an extrinsic source of light.[3] A created truth, he goes on, manifests what it is, but cannot do so by its own light alone. Just as it is only in the light of the sun that physical bodies become visible, so the light of supreme truth is required to reveal the being of objects to the knower.

Since Grosseteste offers no further analysis of objective illumination it is important that the example from physical sight, repeated in the *De Veritate* and elsewhere, should be properly understood. We find in the *Hexaemeron* a brief statement of the relationship of colour to light.[4] Colour is in fact

[2] The question as posed at the beginning reads: 'An sit aliqua alia veritas, an nulla sit alia ab ipsa summa veritate?' (p. 130).

[3] 'Nec potest aliqua res in sua tantum creata veritate conspici vera, sicut corpus non potest conspici coloratum in suo colore tantum, nisi superfuso extrinseco lumine.' (p. 137.19-22).

[4] *Hexaem.*, MS cit., fol. 213ª: 'Est quoque lux, ut dicit Augustinus, colorum regina, utpote eorundem per incorporacionem effectiva, et per superfusionem motiva. Lux namque incorporata in perspicuo humido color est, qui color sui speciem in aere propter incorporacionis sue retardacionem per se generare non

light incorporated into a medium and, like all light, it seeks to multiply its form into a sphere. Its incorporation prevents its connatural action of self-generation taking place, so that it requires further light to be shone upon it from outside in order to be made *actu visibilis*, capable of affecting the eye. The reader who is inclined to interpret the doctrine of objective illumination as nothing more than an emphatic affirmation of the creature's total ontological dependence on the Creator could find support in Grosseteste himself, as when he affirms that each of the transcendental properties of being, therefore not only truth but goodness and beauty also, leads the mind to the creative source of all things, in whom alone these attributes are found at their fullest intensity. In the *De Veritate* itself Grosseteste asks whether, in view of St. Augustine's identification of truth with being or what is, being can be known only in the supreme being and truth only in the fountain of light.[5] He cites an example: water is of itself fluid and takes its shape from its container, wherefore it would make no sense to try to describe, for instance, water in a cubic shape, without adverting to the cubic form of its container. Now, all being, excepting the Creator alone, is likewise labile, and save for his support would of itself return to the non-being from which it came; therefore, to define the being of a finite thing can only be in the last analysis to demonstrate its dependence on the power of the Word. In other words, the first being is known in each instance of being, just as the first light is seen in the finite truth of things made.[6] Similarly, each thing according to its participation in being and goodness[7] calls out that it is made, and invites the mind to raise itself to its creative source.

potest, sed lux colori superfusa movet eum in generacionis sue speciei actum, sine luce itaque omnia corporea occulta sunt et ignota'. Cf. *De Oper. Sol.*, § 6, ed. McEvoy, *Rech. Théol. anc. méd.* 41 (1974), pp. 69–70: '. . . color connativus est luci visibili. Si igitur lucis visibilis oculis nostris radix est in sole, omnis color habet in sui substantia de luce solari, cui lux superfusa se unit ut faciat colorem actu visibilem.' See also *De Colore*, ed. Baur, p. 78.

[5] *De Ver.*, p. 141.13 ff. The reference is to Augustine's *Soliloquia* II, 5, n. 8 (*P.L.* 32, 889).

[6] Ibid., pp. 141.18–142.5. The example is a favourite; see an identical use of it in *Comm. in Hier. Cael.*, ed. McQuade, p. 146. It is closely related to Grosseteste's fondness for the doctrine of God as *forma omnium*, a very strong expression of the exemplarist teaching, found in Augustine himself and in Scottus Eriugena.

[7] *Comm. in Hier. Cael.*, ed. McQuade, pp. 61–2, 72.

It is not possible, however, to interpret Grosseteste's remarks on created and uncreated truth solely in the light of the other transcendental properties, without doing violence to his own clear intention.[8] In the passage of the *De Veritate* under discussion, Grosseteste has no intention of denying that truth is a transcendental property of finite being (*creata veritas*).[9] He affirms unambiguously that a further illumination is required for the truth of a being to be grasped, and the light-example leaves no doubt as to his meaning.

If the intrinsic intelligibility of finite beings is not sufficient to reveal them to the mind, just as the light incorporated into matter is not sufficiently active to reproduce itself without an extrinsic activator (sunlight), are we to conclude that each human act of knowledge demands a special divine intervention in an occasionalistic manner, in order to supply the lack within the object? Or does the sun of the intelligible world never cease to shine? The question evidently does not occur to Grosseteste, something which indicates that it may spring from a more differentiated way of thinking.

Grosseteste draws an important conclusion from the doctrine of objective illumination: no human being is deprived of the light of the supreme truth (p. 138). Even the *non mundi*, those who are not pure of heart, can know only in the presence of God's light; some of them recognize this presence for what it is (is he thinking of Plato?), while others do not, for an eye as yet too weak to gaze upon the sun itself cannot see the source of all light and is unconscious of its action, believing that the coloured world is itself light enough. With this conclusion Grosseteste satisfies one of his aims, namely, to show how truth, though one and having a single source, can somehow be available even to those not morally qualified to see the Truth itself. Such people are not fully deprived of its vision, but the pure of heart and the *perfecte purgati*, on the other hand, see the light of Truth in itself.[10] Before we explore the

---

[8] If one pursues the form of reduction mentioned, the *creata veritas* being identified with an aspect of God's concurrence with his creature, one reduces objective illumination to subjective.

[9] Indeed, he goes to some lengths to affirm the reality of the finite truth. The light of the supreme truth does not stand to lower truths as does the sun to the other luminaries (occluding them by its brilliance), but as the sun to the colours it activates (allowing them to reveal what they are in themselves).

[10] 'Mundicordes vero et perfecte purgati ipsam lucem veritatis in se conspiciunt,

seeming ontologism of this claim, a word must be said concerning the illumination of the knowing subject.

If the truth of any creature consists in its conformity to its exemplar cause in the divine mind, it follows that the knowing subject must know both the object and the exemplar in order to attain to truth, which can only imply that the supreme truth illumines the finite mind (p. 137.1-19). How can it be known that the object is as it ought to be, asks Grosseteste rhetorically, unless the measure of its being is itself visible to the mind? Moreover, the wholesome mind, *oculus mentis sanus*, seeing the first and supreme light in itself, would see all other things also, more clearly in their exemplars than in their creaturely reality (p. 142.5 ff.). This conclusion follows from first principles: the exemplar must give a more noble, clear, and unveiled knowledge of the object, because the divine essence, *lux lucidissima*, is more intelligible than that of any creature. Grosseteste is ready with an ingenious and singularly lovely light-example to illustrate his point (p. 142.20-30).

Whether, therefore, we consider the light which is cast upon the object or that which illumines the subject in knowing, we arrive at the same conclusion: the immediate source of both is the essence of light, and that *lux lucidissima* can be seen as it is in itself by a created mind in a certain state of adaptation or preparation. No wonder, then, that Grosseteste has been accused of defending the extreme form of illumination-theory that was later to be made the object of criticism by St. Bonaventure, and was not paralleled in full degree by any other author of his time.[11] A theory of cognition which holds that true knowledge of objects is to be had only by direct intuition of the divine mind can only be ontologistic (if it affirms that such knowledge is *de facto* given), or sceptical (if it denies

---

quod immundi facere nequeunt . . . soli mundicordes summam vident veritatem . . . nec etiam immundi penitus eius visione frustrantur.' (*De Ver.*, p. 138.18 ff.).

[11] Gilson, 'Pourquoi S. Thomas a critiqué S. Augustin', pp. 97-8: 'D'abord il apparaît que nous venons de rencontrer enfin la doctrine de l'illumination que saint Bonaventure lui-même considérera comme excessive et que nous n'avions pas encore réussi à déterminer.' The whole passage from which this extract is taken should be studied. Gilson's influence on Lynch, 'The doctrine of divine ideas . . .', *Med. Stud.* 3 (1941), 163-73, is clear. Bérubé and Gieben have traced the influence of Grosseteste's *De Veritate* on the Franciscan School at Paris, through the plagiarism of Guibert de Tournai OFM, the close associate of St. Bonaventure; see the study, 'Guibert de Tournai et Robert Grosseteste, sources inconnues de la doctrine de l'illumination . . .', in *S. Bonaventura 1274-1974*, vol. II, pp. 627-54.

that such is the case); and Grosseteste is demonstrably not of the latter opinion.

Before the final judgement is passed, however, the qualification that is always present in the text must be given its full weight. It is not any mind whatsoever that sees the supreme light which illuminates both the object and the subject, but the class of individuals described as the *mundi cordes, perfecte purgati*, who possess the *oculus mentis sanus*. The reiteration of such qualifiers each time the vision of the supreme truth in itself is affirmed must be deliberate. Who are these *perfecte purgati*? Grosseteste's *DeVeritate* does not identify them; yet it is imperative that we learn who they are, otherwise the charge of ontologism looks sustainable. Fortunately, the question need not be left a riddle, for two other texts offer reflections which we hope to show elucidate it, and provide besides a much less one-sided picture of Grosseteste's epistemology.

The first of these works is the *Commentary on the Celestial Hierarchy* which, in the passage defending the beatific vision, provides us with a definition of human life and development in terms of the vision of the supreme truth.[12] It is introduced here because it represents Grosseteste's main attempt to say what man is. When taken as the broadest interpretative background for his theory of knowledge it allows a unitary interpretation of many of the difficulties of both the *De Veritate* and the *Commentary on the Posterior Analytics*.

The Dionysian and the common Greek understanding of the scriptural statements, 'no man has seen God at any time', and 'no man shall see me and live',[13] forces Grosseteste to make a fundamental distinction. It is true, he admits, that in the conditions of the present life, where the intelligence is weighed down by the mass of corrupt flesh or by inordinate affections, the essence of God as he is in himself is not immediately visible. However, human nature understood in its integrity does possess the capacity for the direct vision of God, and will be able to exercise it when, with the help of the divine light, the body is no longer a hindrance and the affections are properly ordered. This will be the case when 'that which is

---

[12] That it is later than the *De Veritate* by a decade or so does not invalidate the comparison; the same doctrine is found, albeit less exhaustively stated, in other works, and there is no reason to suspect development or modification.

[13] John 1:18, 1 John 4:12, Exod. 33:20. See above, Part IV, ch. 2, i (pp. 296-9).

perfect shall have come and that which is in part be done away with' (I Cor. 13:12). We cannot normally see God directly in this life, but only in heaven. However, a small number of mystics has been favoured with the vision of God. At this point Grosseteste becomes cautious. He emphasizes the fleeting duration of the experience and the fewness of the privileged (Moses, St. Paul, Mary), and remarks that in the moment of vision the normal conditions of human life (namely, those derived from the fall) were so transcended that such persons were living above human nature and sharing the life and contemplation of the angels.

Nothing could be added to Grosseteste's words to render them more clear than they are. The small number of mystics of the Judaeo-Christian tradition do, he declares, share momentarily and by anticipation in the beatific vision and God reveals in them, as he did most signally in Christ, the ultimate possibility of human redemption, the vision, namely, of the Light no longer seen as in a mirror darkly, but face to face. The normal human condition is not that of illumination. We risk nothing in identifying the *mundicordes* and *perfecte purgati* of *De Veritate*, those who see the *veritas suprema* as he is himself, with the handful of mystics of the Christian tradition and with the blessed in heaven. In the normal human life, purgation is gradual and is not completed till after death. The ontologism of which Grosseteste has been accused, that excess which makes the difference between the state of the blessed and the *viatores* as regards illumination a matter merely of degree, could not be further from his mind; he is totally innocent of it.[14]

II THE *COMMENTARY ON THE POSTERIOR ANALYTICS*

If it be granted that the form of illumination-theory which Grosseteste propounded was the one universally accepted in the schools of his time and not some idiosyncratic aberration, it still remains for us to determine the degree of welcome he

[14] Gilson, art. cit. 97–8: 'Une théorie de la connaissance qui ne met pas de différence entre la lumière par laquelle les bienheureux voient Dieu dans le ciel, celle par laquelle Adam, ou une âme libre du corps comme l'était la sienne avant la chute, connaissent l'intelligible, et celle enfin par laquelle nous le connaissons nous-mêmes, mais qui n'admet entre ces modes de connaissance que des différences de degré fondées sur la liberté relative de l'âme à l'égard du corps dans chacun de ses divers états, c'est bien là ce que Grosseteste nous enseigne.' Gilson does not employ the term 'ontologism' to describe the extreme illuminationism he attributes to Grosseteste.

accorded to the Aristotelean theory of scientific knowledge. We shall find the distinctions acquired above to be of vital help in interpreting the *Commentary on the Posterior Analytics*; however, a danger exists in this undertaking too. If we remain fixated upon the distinction of agent and passive intellects (terms not used by Grosseteste) and the opposition between illumination and abstraction, then the conclusion that Grosseteste's thought 'moves on a level that is entirely Augustinian and totally foreign to Aristoteleanism', imposes itself necessarily,[15] but blinds us at the same time to the real preoccupations of the commentator and to his openness towards the Philosopher. The texts must be allowed to speak fully, and must not be interrogated from a single point of view.

Since the commentary is extremely long, I have been forced to choose out six main passages in which the kernel themes of epistemology are treated.[16] I have attempted to interpret these in the light of the whole work, and indeed of the other works of Grosseteste.

a) 'Omnis demonstratio est de incorruptibilibus'

Grosseteste raises a question concerning the crucial condition for scientific knowledge: how are the objects of demonstration incorruptible, when they are found as universals in singulars which are themselves corruptible? In his answer he distinguishes four kinds of universal, which are variously perceived by the

---

[15] Ibid., p. 98: 'Chaque fois en effet que le problème de la connaissance des intelligibles se trouve soulevé, Grosseteste recourt à sa métaphysique de l'illumination pour le résoudre, sans faire la moindre allusion à la distinction aristotélicienne entre l'intellect possible et l'intellect agent.' p. 95: 'La seule solution qu'il apporte au problème de la connaissance humaine des formes intelligibles, c'est précisément cette infusion de la lumière divine qui se déverse sur l'intellect humain.' p. 98: 'En droit, nous sommes aptes à recevoir de Dieu l'intelligible sans le secours du sensible; en fait, la sensation n'a d'autre fonction que de raviver la lumière qui, partie de Dieu, traverse l'Intelligence angélique et parvient jusqu'à la partie la plus haute de notre propre intelligence; c'est donc finalement par elle seule que nous connaissons.' Apropos of this last quotation, with its strongly Avicennian resonances: is it not strange for an author to find in a commentary on the *Posterior Analytics* a determinant influence of Augustine and Avicenna and none whatsoever of Aristotle himself? For a discussion which lends support to Gilson's conclusions, cf. L. Lynch, *Med. Stud.* 3 (1941), esp. pp. 172-3, where it is roundly denied that Aristotle exerted any appreciable influence on Grosseteste, with a sole exception made for the physical corpus.

[16] These are, in the Venice 1514 edition (to which all references are given), fols. 8c-d, 16d-17b, 20c, 22b-c, 25a-d and 39d-40a. See the edition of the commentary by P. Rossi (Florence, 1981), pp. 139-42; 212-6; 240-1; 255-7; 278-86; 403-5.

created mind, according to their different modes of existence. The first are the uncreated reasons in the divine mind, the causes of both being and knowledge for all things below them; they are identified with Plato's ideas in the archetypal world. They are principles of knowledge as well as of being, because a created intellect which is purified and separated from phantasms can contemplate the first light and can see all of creation most truly and manifestly in the eternal *rationes*.[17]

The second mode of existence of universals is in the mind of the angel (*intelligentia*), whose complete knowledge of the rest of creation is derived directly from contemplation of the exemplars. Such knowledge can be communicated to a human intellect, '*qui non est ad purum defaecatus ita ut possit lucem primam immediate contueri*', and in the mind thus illuminated it forms the principles of understanding. Clearly, the universals thus far mentioned are incorruptible.

Again, the causes of all that occurs in the lower world are to be found in the heavenly bodies, and can be discovered through natural inquiry by a mind which is not capable of the direct sight of the divine or angelic light. These universals also are perpetual, though the products of their causal action are subject to generation and corruption.

A fourth source of universal knowledge is form, which, as the intelligible element in the whole composite, can be classified according to genus and species (of the Aristotelean, not the Platonic kind, Grosseteste is careful to point out), thus giving rise to true definition and demonstration. A mind not capable of reaching scientific knowledge according to this method can know only the accidents accompanying the true essences of things. Grosseteste finds it unclear how the principles of knowledge in these last two cases are incorruptible. It must be, he suggests, either because form does not become corrupted in itself, even though matter does, or for the reason that the permanence of species is preserved by the temporally-unbroken succession of individuals within them.

When it is realized that this passage is underlain by the same dualism, of the present condition of human nature and its future state, as is to be found in the *Commentary on the Celestial*

[17] '. . . cum intellectus purus potest in his defigere intuitum, in istis verissime et manifestissime cognoscit res creatas, et non solum res creatas sed ipsam lucem primam in qua cognoscit cetera.' (fol. 8^c).

Hierarchy—and so much is apparent from the shared language —there is little that is mysterious in it. The opposition is not between Augustinian illumination on the one hand and Aristotelean abstraction on the other, but between human knowledge in its present limited condition and the direct vision of God. No hesitation is necessary concerning the *intellectus purus* which can contemplate the first light.[18] It is instanced in the first place in the angels, and by extension in the blessed and in mystics at the moment of vision. On the other hand, human nature as it actually is has only lesser sources of knowledge available to it. The first of these is inspiration by the mediation of an Intelligence (or angel—the Avicennian language should not blind us to the Christian content of the thought). At a lower level still there is scientific inquiry concerning the unchanging principles of nature. Here the methodology of Aristotle is the guide, and that is what the *Posterior Analytics* is all about. Gilson finds it disappointing that Grosseteste did not choose to expatiate in this connection upon the active intellect, and hastens to supply his silence by referring to Augustinian illumination. What Grosseteste, on the contrary, offers his reader is a hierarchy of human knowledge on the way to perfect knowledge, *gradus ad lucem primam*. All the possibilities of acquiring the principles of knowledge are hierarchized, from pre-scientific opinion to the beatific vision; they are affirmed as capacities of human nature, here and hereafter. Within these grades Aristotle's science has a lowly place, but it does occupy a definable area of competence within which it is valid.

b) *'Occasione defectus alicuius sensus est defectus alicuius scientiae'*[19]

Grosseteste's summary of Aristotle's argument for the above proposition reveals at its highest perfection that quality of clarity which assured his commentary constant use for three hundred years after his death. It deserves to be quoted, as a measure of his grasp of Aristotelean principle.

The proof of this is as follows. Sense apprehends singulars, wherefore the lack of a given sense means the incapacity to apprehend certain singulars.

[18] *Pace* Gilson, p. 93.
[19] Ibid., fols. 16^d-17^b. The greater part of Grosseteste's comment is printed by Gilson, art. cit. pp. 96-7 n.

Now since induction is made from singulars, the defect of a sense necessitates the impossibility of induction from the singulars which are objects of the missing sense. When, in turn, induction from such singulars is ruled out, so also is the intellect's universal knowledge of them, since the universal can only be arrived at by way of induction. The absence of the universal in the understanding removes the possibility of demonstration, which can only begin from universals; and if there can be no demonstration, it follows that there can be no scientific knowledge. *Ergo*, the basic lack of a given sense means the absence of the corresponding science.

The thesis thus enuntiated and proven constitutes a condensed version of the Aristotelean theory of the relationship of sensation to intellectual knowledge. Once again Grosseteste chooses to give his personal view on the question in independence of the Philosopher's text: '*Dico tamen. . .*'.

Is it the case that all knowledge comes through the senses ('*cum sensus adminiculo*')? Grosseteste enlarges the horizons within which the question is posed and spreads before us the entire Christian metaphysics of knowledge and being. All knowledge is found in the divine mind from eternity, not of universals only but of singulars as well, the latter being known abstractly and without material accidents. How different this knowledge is from ours, which cannot know individuals in the purity of their essence, but only as mingled with accidents! The angels too see all singulars and universals in the illumination they receive from the first light, and by reflection upon themselves they know all those things which come after them in the hierarchy of being. They have no need of sense and abstraction. The highest part of the human soul, the *intelligentia*, which functions without a corporeal instrument, would have complete knowledge by illumination and without recourse to the senses' aid, were it not encumbered and darkened by the conditions that are normal for fallen man.[20] Indeed, some men do perhaps enjoy such illumination as the souls of the blessed receive, but then only to the extent that they are absolved from the ordinary human lot. In the normal case, the higher human powers are lulled to sleep by the weight of the flesh.

Granted these conditions, the awakening of reason can take place only through sense experience. The senses' repeated

---

[20] The example given by Grosseteste, of the form of humanity individualized by multiplication in matter, is fairly commonplace, but may have been derived from Avicenna, *De Anima* II, 2 (ed. Van Riet, pp. 115-16).

meeting with their objects stirs the reason and ferries it down to meet the *sensibilia*. Reason has the capacity to discriminate and resolve the objects perceived only in a confused way by the senses. It differentiates colour, size, shape, and body, which are perceived as a global unity by sight, and by a process of division and abstraction it arrives at the knowledge of essence and accidents. Abstraction is made from the repeated experience of singulars, one and the same thing being judged by reason to be present in many instances. Thus, both the abstraction of simple universal ideas (*universale incomplexum*) and the perception of universal relations of cause and effect arrived at by, for instance, medical science (*universale experimentale*),[21] can only be attained by the aid of the senses, granted the actual fallen condition of human nature. In such circumstances as we are now submitted to, the lack of a sense implies the lack of both kinds of universals that are won from the singulars which that sense alone is capable of perceiving, and therefore of demonstration and of scientific knowledge.

Can the reason be assigned as to why the soul is thus clouded over by the corruption of the flesh? Grosseteste thinks that it can and proceeds to apply one of his deepest convictions to the point at issue. The soul's capacity to understand (*aspectus*) is inseparable from its loves (*affectus*) and cannot transcend them. When these are turned to the body and the seductions of matter that surround us they entice the capacity for truth to dally with them and distract it from its true light, the sun of the intelligible world, leaving the mind in a darkness and idleness which only begin to be relieved when it issues through the external senses into a light which is a reminder (*vestigium*) of that other Light, its birthright. The stimulation of natural light becomes then the initial moment of a quest for the true Light. The measure of the mind's success in finding the pathway of light is the degree to which the soul's love can transcend ephemeral objects.[22]

There is no need to underline the fact that, once again in this passage, the central emphasis is placed, not upon a contrast or opposition of some kind between Augustinian and

[21] Grosseteste here inserts an example of how the medical thesis that scammony attracts red bile can be demonstrated from experience. It is of some interest for his scientific methodology, but not of relevance here.

[22] Cf. *Comm. in Hier. Cael.*, ed. McQuade, pp. 275-7.

Aristotelean theories of knowledge, but upon the different conditions of knowledge here in this present life and in beatitude and the mystical union. Grosseteste effectively reaffirms Aristotle's thesis, 'one less sense, one less science', and with it the necessity of sense-knowledge as the foundation of all higher knowing-dimensions; but he puts all this into a Christian perspective, which makes it a partial truth rather than the whole.

First, he reaffirms Aristotle's conclusion and makes it fully his own, without intrinsic restriction of any kind. Granted that there are no innate ideas[23] and that only a tiny number of human beings is privileged with the direct vision, sense-knowledge cannot but represent the foundational possibility of all knowledge. Grosseteste, of course, could not agree with Aristotle that the senses are purely passive, or that matter can act directly upon the soul; in both questions regarding the actual operation of sensation he remained firmly Augustinian. That, however, makes no difference in the present context of discussion,[24] for however sensation actually functions and manages to communicate its messages to the spiritual soul, without it the abstraction process, whereby the universal common to many individuals is grasped by the intellect within the material of sensation, would be impossible. Apart from the work of the senses there would be no beginning of knowledge, no abstraction nor demonstration from universals, no science. Aristotle is right on all these points. His doctrine of the genesis and construction of scientific knowledge is not placed in opposition to, nor relativized by contrast with, other available models of the elaboration of demonstrative science, and the commentator is satisfied that Aristotle's theory corresponds exactly to the conditions of knowing imposed upon human nature as we actually experience it. If it were admitted that these conditions spring from the nature of things, that they have never been nor could ever be otherwise, then there would be nothing left to say; the authority would have spoken.

How different and how much more complex is the awareness of human nature and its possibilities that the Christian commentator brings to his pagan text! It is not as though he had at his disposal a Christian methodology of medicine, physics, and the other sciences different from that of Aristotle;

[23] This thesis is defended in a later passage (fol. 39ᵈ) and examined below; see pp. 340–45.   [24] *Pace* Gilson, art. cit., p. 95.

as far as Grosseteste is concerned, there is no other that stands
as rival to Aristotle's. But his own vision of man includes an
awareness of historical occurrence and supernatural call which
relativizes Aristotle's consciousness of the human condition,
not by denying its intrinsic validity in the area of methodology,
but by outflanking the claim to absolutely inclusive truth implicitly laid by the Philosopher and revealing its limitations.
When measured against all that the Christian revelation brings
as a light upon man's nature and destiny, Aristotle's admirable
theory of knowing receives a set of brackets and is governed
by coefficients of which he could not possibly have had even
the suspicion of knowledge. Here we are at the central meaning of Grosseteste's *Commentary on the Posterior Analytics*,
and have found perhaps the key to the interpretation of his
attitude to Aristotle and to pagan wisdom as such. Grosseteste
shows no signs of feeling threatened by Aristotle's view of
human knowledge, because he does not feel himself to be a
member of a rival philosophical sect (for instance, of Avicen-
nians or Augustinians). His commentary is rather a celebration
of the superiority of Christian and revealed truth over the
sapientia huius mundi. From the advantageous position granted
by the Christian perspective on God, the angels, and man fallen,
redeemed, and glorified, he can see and adopt Aristotle's wisdom as reflecting one level of truth, but no more than that.
The Christian revelation, giving absolute truth and certitude,
opens up perspectives upon human knowledge the very existence of which remained unsuspected on the part of the Philosopher, living as he did in pagan society. Aristotle could not
have known that the human situation which he regarded as
natural and given was no more than the result of a primordial
historical lapse. He could not have known that human nature
has been invited to the direct enjoyment of the Light of God's
face, that the capacity for this vision underlies the whole course
of human life and development, and that its realization even
in the present life is a possibility, which has in fact been instanced in certain privileged cases. Grosseteste, therefore, while
commenting upon Aristotle, returns time and again to the
contrast of *sapientia nostra Christiana* and the highest attain-
ments of unaided human reason, with a purpose which must
have been evident and unmistakable to his contemporaries.
The fact that he did not feel any need to explain the obvious

necessarily renders the work somewhat opaque and difficult to modern readers.

This interpretation of Grosseteste's attitude with regard to Aristotle may claim support in his division of the rational powers of the soul, which we have already discussed. He adopted from Christian sources a schema consisting of *intelligentia, intellectus,* and *ratio*, which made a place for Aristotle's division of powers (understanding and scientific reason), while at the same time revealing its limitations; these are part of the human capacities only, not their entirety.

Intermezzo: The body's drag upon the soul

The most saliently non-Aristotelean element in the texts explored is, of course, the dualism of body and soul and the location of the effects of original sin more prominently in the body, in the negative sense of 'the flesh'. So typically Neoplatonic is the negative language employed by Grosseteste that it is tempting to think that he equated the Christian doctrine of original sin with the Neoplatonic doctrine of the evil body. That such was not the case could not easily be shown from the *Commentary on the Posterior Analytics* alone.

The language of bodily corruption was present in some degree in all of Grosseteste's sources, save Aristotle, so that even a single word can often bear multiple resonances. Grosseteste's most typical expression, for example, *'anima per corpus corruptum obnubilata et aggravata est'*[25] (and its variant formulations: *'intelligentia mole corporis obnubilata et aggravata'*,[26] *'aggravamur mole corporis corrupti'*,[27] *'mole carnis praegravati et nube peccati caliginosi'*[28]), is founded upon Wisd. 9:15: *'corpus quod corrumpitur aggravat animam'*; but the image is repeated not only by Christian sources (*'anima praegravatur corporis societate . . . amat carcerem suam'*—Pseudo-Augustine)[29] but by Arabic ones as well-Avicenna, for example: *'cum autem aufertur de nostra anima ipsa aggravatio et impedimentum . . .'.*[30] The soul is described by Grosseteste as *'obnubilatus*

[25] *Comm. in Anal. Post.*, fol. 17[b].
[26] Ibid., fol. 17[a].   [27] Ibid., fol. 22[c].
[28] *Comm. in Hier. Cael.*, ed. McQuade, p. 252.
[29] *De Spiritu et Anima*, ch. 48; *P.L.* 40, 789.
[30] *De Anima*, V, 5, ed. Van Riet, p. 132.15. That the dead weight here referred to is the body there can be no doubt: 'anima impedita est in corpore et ex corpore' (ibid., p. 131.10). Cf. p. 80.68: 'est [anima] apta ad conservandum se ab impedimentis sibi accidentibus ex consortio . . .'.

corporis contagione',[31] and in Avicenna it is *'inquinata ab his quibus [deprimitur] deorsum'*.[32] The higher faculties are lulled to sleep (*sopitae*),[33] as in Pseudo-Augustine the soul is *'corporeis passionibus consopitus [animus]'*.[34]

To adjust the perspective somewhat it must be pointed out that the soul is not spared either. Already in the *Commentary on the Posterior Analytics* we have seen that the mind is clouded by the weight of the soul's loves, so that the openness to light through the body's senses becomes a first moment of truth for the soul.[35] In the *Commentary on the Celestial Hierarchy*, likewise, the 'obstacle impeding the soul' is that of flesh and inordinate affections combined.[36] In the *Hexaemeron* it is the whole man who is fallen and redeemed, so that the *imago restaurata* reaches all the way from the *intelligentia* to the body.[37] In Grosseteste's fullest treatment of redemption (written after 1240) his doctrine of original sin is fully orthodox, and the effects of the fall are said to be found in the disorder of all the faculties, from the knowledge of God and freedom, through the affections, to sexuality.[38] There the extreme Neoplatonic language of the corrupt body is no longer to be found. Grosseteste did not look upon the restoration of the capacity for the beatific union as a matter of the soul's increasing freedom from the body[39] but as a liberation from the totality of the conditions which have unhinged human nature.

c) 'Lux spiritualis quae superfunditur rebus intelligibilibus'[40]

Certainty is predicated of objects in their relation to the mind of the knower. Now there is a spiritual light which illuminates intelligible objects and the mind, just as the sun does visible objects and the eye. The intelligibility of the objects of understanding is in proportion to their receptivity of the spiritual

---

[31] *Comm. in Anal. Post.*, fol. 40ª.
[32] *De Anima* IV. 2, ed. Van Riet, p. 29.91.
[33] *Comm. in Anal. Post.*, fol. 17ª.　　　　[34] *De Spiritu et Anima, P.L.* 40, 781.
[35] *Comm. in Anal. Post.*, fol. 17ᵇ.
[36] *Comm. in Hier. Cael.*, ed. McQuade, pp. 193-4.
[37] See ed. Muckle, *Med. Stud.* 6 (1944), pp. 164 ff.
[38] *Ex Rer. Init.*, ed. Gieben, *Coll. Franc.* 37 (1967), pp. 123-6.
[39] Cf. Gilson, art. cit., p. 98. This doctrine is found in a crude form in the *De Spiritu et Anima*, ch. 48, *P.L.* 40, 815.
[40] *Comm. in Anal. Post.*, fol. 20ᶜ; quoted by Gilson, art. cit., pp. 91-2 n.

light, in other words to their more or less close natural resemblance to the Light. Such objects as are more receptive of the light that is shone upon them are more visible by the mind (*oculus interior, acies mentis*), which itself is a created spiritual light; since they are more perfectly penetrated by light they give greater certitude to knowledge.

This brief but very explicit illumination-text calls for a number of remarks.

The light-metaphor used has an added element over and beyond the teaching of *De Veritate*. There the comparison of the spiritual light to the sun was restricted to the illumination of the objects of knowledge and of sight, whereas here the sun acts upon bodies and upon the eye.[41] How the sun acts upon the eye (as is explained in the *Hexaemeron* and *De Operationibus Solis*)[42] has been outlined in the course of our treatment of sensation and need only be recalled here. Grosseteste adapted the Arabic visible spirit to the Augustinian theory of the activity of the sense-organ. The spirit, the instrument of the soul in sensation, shares the nature of the sun's light, so that the action of the soul upon it releases its energy in the form of rays which stream out from the eye to objects. The activity of light is, therefore, present both in the perceiving eye and in the colours it activates. (We reserve till later a fuller discussion of the importance of the analogy between sight and intellection for Grosseteste's theory of mind.)[43]

This text reaffirms the Augustinian light-metaphysics, according to which natures closer to the divine are more light-filled and intelligible, and hence capable of giving rise to a higher degree of certitude in the mind of the knower. It is in the nature of things that the more spiritual a being is, the more it is knowable, in contrast to material beings.

It is still not defined, however, in what state the *acies mentis* is adapted to attain the degrees of certitude in knowing which correspond simply to the natural hierarchy of light. No doubt, in the natural and best state of man (before the fall; in mystical knowing; in heaven), illuminative knowledge confers certitude upon the mind, in a degree corresponding with the objective

---

[41] 'Dico ergo quod est lux spiritualis que superfunditur rebus intelligibilibus, et oculus mentis qui se habet ad oculum interiorem [et] ad res intelligibiles sicut se habet sol corporalis ad oculum corporalem et ad res corporales visibiles.' (ibid.).

[42] *Hexaem.*, fol. 203b, and *De Oper. Sol.*, § 7, ed. cit., pp. 70-1 and nn. See above, ch. 4, ii a (pp. 296-9).   [43] See the following section iii (pp. 346-51).

intelligibility of each grade of reality, so that the principles of knowledge and of being are placed in the direct equation of Platonic idealism. That, however, is not the given state.

If this interpretation should seem tendentious, it must be admitted that it receives striking confirmation in the passage next discussed.

d) 'Scientia certior est quae prior'[44]

Aristotle does not demonstrate his twenty-fourth conclusion, Grosseteste complains, to the effect that the science whose objects are prior in the order of nature is more certain; but it follows from the notion of certitude as he has defined it (fol. 20[c]). The science of the incorporeal separate substances is more certain than that of incorporeal substances linked to bodies, which in turn provides greater certainty than the science of bodily substances. Aristotle remarks that psychology is more certain than those other natural sciences which handle natural bodies in process of change.[45] This assertion does not run counter either to the doctrine we accept concerning the clarity of mathematics, claims Grosseteste, or to the assertion of Ptolemy, that certainty is greater in mathmatics than in metaphysics,[46] because we maintain that things divine are more visible to a mind that is in a wholesome state and not overclouded by phantasms, just as those material bodies which receive the strongest light are more visible to the healthy eye accustomed to bright objects. But for as long as our minds are weighted down by the mass of flesh and the love of material things, they are like an 'affected' eye, which functions better in the presence of dark and shadowy objects than in that of brightly-lit ones. In the actual condition of our intellect ('*humano igitur intellectui qualis est adhuc in nobis*'), mathematical objects have the highest certitude and visual aids can be usefully employed in their understanding; whereas if the intellect were in the state of excellence appropriate to it, divine things would be the most certain of all and the relative priority and nobility of the natures of things would determine the degree of certitude with which they can be known.[47]

[44] *Comm. in Anal. Post.*, fol. 22[b–c] (tr. Crombie, *Grosseteste and Experimental Science*, p. 129).

[45] *De Anima* I. 1, 402a, 2–5.

[46] The reference to the *Almagest* is given in Crombie, p. 129 n.

[47] '. . . sed intellectui tali qualis oportet esse secundum statum suum optimum

One could not wish for a more lucid statement of the distinction between the natural hierarchical order of intelligibility that is based upon the creative outpouring of the *suprema lux*, and the actual epistemological order which, being the reverse of the former, begins in shadows and only ends with the vision of the objectively first light. The order of knowledge remains the inverse of the order of being for as long as the conditions of the present life last and the distortion introduced by the fall continues to prevail. The spirit possesses a natural desire to reverse these conditions, a desire which is not different from its innate orientation to the vision of God; but the first can only be first and the last, last (the principles of our knowledge will only coincide with the principles of reality), when the *status optimus* is attained and the *imago* is fully restored by the visible presence of the divine glory. The theory of illumination applies to the state of perfection, towards which we are growing, but Aristotle is the philosopher of the darkened mind. Nowhere is it made more clear than in this passage, that the difference between the epistemological and spiritual conditions of fallen and glorified man is something more profound than the mere matter of degree that has been alleged.

e) 'Est igitur intellectus sicut visus colorati'[48]

The interest of this lengthy discussion is limited for our present purposes to the comparison between the habits of *intellectus*, *scientia*, and *opinio*, on the one hand, and, on the other, vision —a comparison Grosseteste thinks useful in order to illustrate the specificity of each of the terms involved.

Intellect, the habit of first principles, is like the sight of a coloured object, whereas scientific knowledge resembles the sight of a coloured body of a certain shape and size and in a certain state of motion or rest—all of these differences being perceptible only through colour. Both differ from *opinio*, analogously as the act of seeing performed in optimal light-conditions and by a healthy organ differs from a perception distorted by an infected pupil and an unfavourable medium. In other words, opinion struggles to see the same objects as *intellectus* and *scientia*, but cannot grasp their essential

sunt res divine certissime, et quanto res sunt priores et natura sublimiores, tanto certiores.' (*Comm. in Anal. Post.*, fol. 22ᶜ).

[48] See fols. 25ᵃ⁻ᵈ.

purity and is limited to the phantasms of changeable material things.

To return to the comparison between *intellectus* and *visus*: the intelligence and the mind are both illuminated by a light which brings about the act of vision, as the sun does the act of sight.[49] Now, strictly speaking, that spiritual light is, considered in itself, the first and brightest visible light (as is the sun in nature);[50] yet just as we call the coloured *object* the essentially first visible object (meaning the first to *receive* the solar light), we can call the object which first receives the spiritual light, the first visible object.

A complicated way of making a simple point, perhaps, but a passage of importance from the point of view adopted here. It would be possible, no doubt, to interpret the statement concerning the illumination of the mind and the intelligibles in a sense consistent with the principles which have guided our exegesis so far and to claim that the model of illumination does not necessarily apply within the present, fallen order. However, such an interpretation should be renounced in favour of the more obvious one, that Grosseteste is here restating the traditional Augustinian illumination-theory. The context and the language indicate as much. Grosseteste uses the technical Aristotelean vocabulary of the objects of intellect (*intelligibilia*, *prima*) and of demonstrative science (*scibilia*), which are grasped by the *visus mentalis*, whose components are *intellectus* and *scientia* (*ratio*). Now, it is clearly affirmed that both categories of objects, separately and technically named, and in addition the *visus* itself, are illuminated by the *lumen spirituale*.

If this interpretation is the correct one, then this is the first passage we have come across in which the model of illumination is adopted with reference to the objects of scientific

---

[49] 'Et iterum dico quod est visus mentalis apprehensivus intelligibilium, et sunt res visibiles ab illo visu quas dicimus intelligibiles et scibiles; et est lumen quod superfusum visui et visibili facit visionem in actu, sicut facit in visu exteriori lux solis.' (fol. 25b).

[50] 'Et licet istud lumen spirituale sit maxime et per se primum visibile, sicut est lumen solis ab oculo corporali, tamen eo modo quo dicimus coloratum, eo quod est proximum receptivum luminis solis, esse per se et primum visibile, dicemus apud visum interiorem illud quod proximo et per se recipit lumen spirituale esse per se et primum visibile, et visum interiorem primi et per se visibilis interioris dicemus intellectum in transitu isto . . . Est igitur intellectus sicut visus colorati.' (ibid.).

intuition and knowledge, as these are known in the actual given order; it is the only one of its kind that I have turned up in the whole book. It is brief and is introduced in an unobtrusive way. Indeed, its presence can be taken as confirming what we have maintained, namely, that Grosseteste nowhere in this work (nor in the others we have examined) sets himself to oppose Aristotle's theory of knowledge to St. Augustine's, or, more precisely, illumination to abstraction, as though these represented mutually exclusive choices or were the possessions of conflicting schools. He created no scenes in epistemological discussion, as he did over the denial of creation by Aristotle; he rather pursued his own objectives in undisturbed tranquillity. There is, of course, interest in asking how he reconciled illumination with the activity of the intellect in abstraction; his own writings, however, provide no conclusive answer. It is a matter for speculation, which we postpone for the present.[51]

f) 'De principiis, qualiter fiant cognita'[52]

The last important text from the commentary to merit our examination is also Grosseteste's most concise statement concerning the genesis of universal ideas (*principia*).[53]

It is apparent that neither do we possess universal ideas from the beginning in act, nor are we totally ignorant of them; they are present in us potentially and are drawn out into actuality. Sense knowledge would seem to constitute our potential possession of them; we acquire them in act as follows.

A particular sense apprehends its determinate aspect of reality always in a unique object individualized in space and time, conditions which likewise apply to the sense itself and prevent its grasping more than one object at a given moment.[54] This indissoluble bond with the uniquely individual leads us to describe sense as being rather the occasion of science than

[51] See section iii, p. 350.

[52] *Comm. in Anal. Post.*, fols. 39^d–40^a.

[53] The ambiguity of the word *principia* has been noted. Here it certainly means 'universal concepts' and not '*dignitates*'. The following paraphrase of the passage is expanded by inclusion of complementary material from other works of Grosseteste, the supplementary sources being indicated *toties quoties*.

[54] Ibid., fol. 23^c: '. . . sensus talis est quod ipse est apprehensivus rei alicuius signate, et non simul apprehensivus alterius rei, quia necesse est scire rem signatam in loco signato et tempore signato, quare non sentit nisi rem unam signatam.'

its cause.[55] On the other hand, sensation represents already a first degree of abstraction: since the organ apprehends the *species sensibilis* without taking in the matter of the object,[56] its operation is already immaterial or spiritual in a qualified sense.[57] The primitive nature of the abstraction from matter which the senses effect is revealed by their confused, approximate and relatively external grasp of reality. Like the matter from which they are drawn, the likenesses present in sense are fluid, imperfect and indeterminate.[58] Compared with reason, sense grasps reality in an undifferentiated and incomplete way, without either the distance or the grasp of wholeness and essence which characterize reason's superior knowledge. When its limitations are admitted, however, it still remains true that sense gives real knowledge of the real. The meeting of material and spiritual in which all knowledge consists is found already in sensation, where the sensible form actively generates a likeness (*species*) in the surrounding medium[59] and in the organ, and the intentionality of the knower can compare the *species* and the form of the real object. The act of apprehension knows no difference between the *species* and the form of the real,[60] and sensation is therefore an active process in which both object and subject are involved.[61]

The activity of the *species* engendered in the sense-organ bears it through the spirit-filled nerves to the brain, where the information provided by the various senses is first actively synthesized by the *sensus communis*, which is still the agent

---

[55] Ibid., fol. 23C: 'Similiter non contingit scire sensibilia, neque sensus est causa scientie, sed occasio . . . Huius autem ratio est quod sensus talis est quod ipse est apprehensivus rei alicuius signate, et non simul apprehensivus alterius rei . . . cum sola universalia et demonstrabilia sciantur, manifestum est quod non contingit sentire per sensum.'

[56] *Hexaem.*, MS cit. fol. 223ᵃ.

[57] *De Lineis*, ed. Baur, p. 60.25: 'In sensu enim ista virtus recepta facit operationem spiritualem quodammodo et nobiliorem.'

[58] 'Racio vero diiudicat integritatem atque veras prosequitur differencias, sed sensus invenit quedam proxima et confusa veritati, accipit vero racio integritatem. Racio . . . accipit vero a sensu confusam ac veri proximam similitudinem.' (*Comm. Phys.*, ed. Dales, pp. 4-5). The details of the contrast *sensus–ratio* and the rather disparaging attitude to undependable matter are taken from Boethius, *De Musica*, I, 1, *P.L.* 63, 1167d.

[59] Except of course in the case of touch, where the *species* is directly impressed upon the most material and least active of the senses.

[60] *Hexaem.*, ed. Muckle, p. 163.

[61] Both, of course, share in the active nature of light.

of the intention of the soul and therefore represents the sense-reality, as it is and as it was present in the *species* of the particular senses, but more adequately than in them, because as common sense it contains the reintegrated product of all five sense-operations.[62] The *species* united in the *sensus communis* are passed into the memory for retention, again without further change in their nature. Memory receives not only the forms of sense, but also the *intentiones* detected by the *vis aestimativa*.[63]

All of these powers are shared by man and brute alike, although, of course, human reason and freedom have a presidency over their operations such as animals cannot have. In man, the store of memories built up arouses reason, so that experience in its higher, properly human dimensions can result.[64]

The knowledge first acquired through the external senses and prepared by the internal senses provides the material from which our understanding (*intellectus*) abstracts the universal idea. Universals are not ideas existing apart from the particulars and intuited by the mind without reference to the senses[65] (as Plato thought),[66] but are beyond the particulars without being apart from them. The formal cause of any material thing (which in union with matter produces the composite), is not an abstract reality when taken in its relation to the whole composite, but it is never the less the intelligible element in the whole and the cause of the knowledge we have of it, as well as being a cause of the thing's being.[67] Form is potentially

---

[62] *Hexaem.*, loc. cit.

[63] Grosseteste scrupulously notes that he is using *memoria* to cover both the *imaginatio*, which retains forms which have been sensed, and the *memoria proprie dicta*, which retains the *intentiones aestimatae*; *Comm. in Anal. Post.*, fol. 39[d].

[64] 'Sed in rationabilibus iam contingit ex multis memoriis, excitata ratione, fieri experientiam; in brutis vero non est hoc.' (*Comm. in Anal. Post.*, fol. 39[d]).

[65] 'Ex sensu igitur fit memoria, ex memoria multiplicata experimentum, et ex experimento universale, quod est praeter particularia, non quasi separatum a particularibus . . .' (ibid.).

[66] 'Ponebant enim platonici rationes quasdam in mente divina, aeternas, per se subsistentes, divinas, intellectuales, ad quas dicebant omnia esse et fieri, quas et species et ideas vocabant, et tota et universalia, separatas autem a creatis omnibus, et ideo dictas universalia et tota, quoniam unaquaequae illarum una existens, habet multa ex illa et secundum illam facta, separabiliter ab illis existens.' (*Comm. in Hier. Cael.*, ed. McEvoy, pp. 81-2).

[67] *Comm. in Anal. Post.*, fol. 8[d]: 'Quarto modo cognoscitur res in sua causa formali que est in ipsa, a qua est hoc quod est, et secundum quod in ista forma, que est pars rei, videtur ipsa eadem forma . . . vel secundum quod in ipsa videtur

abstract in reality, but really abstract only in the mind, where alone it is predicable of many: *res inventa in multitudine.* In the abstract way in which it exists in the mind, it is freed from the individualizing conditions of space and time, which govern all individuals and sensations; it is *semper et ubique.*[68] The universal is not in place, and it has abstract existence only in the mind; yet because the mind somehow becomes all the objects it assimilates and is spiritually there wherever its object of understanding is (just as love dwells with the beloved), the universal, which is its instrument in knowing, is everywhere.[69] The universal concept is neither one nor many, but both unity and multiplicity can be attributed to it.[70] It is a single act of understanding, whose unifying power over particulars is like that of the generative source of light over against the light as received by many objects: just as between the sun's light and its multiple rays in the air there is a play of identity and difference, of unity and multiplicity, so it is with the concept and its instances. The immateriality of the concept removes it from change and variability.[71]

The process by which the universal concept is found in the particulars is described by Grosseteste both as *'inductio'*, which refers to its derivation from many singulars experienced by the senses, and as *'abstractio'*, which connotes the attainment of the essence without its individuating accidents.[72]

materia que similiter est pars rei, ipsa forma non est genus vel species; sed secundum quod ipsa forma est sicut totius compositi, et secundum quod ipsa est principium cognoscendi totum compositum, sic est genus vel species et principium essendi et predicabile in quod . . .'. The concluding phrase, 'et hoc est sententia Aristotelis de generibus et speciebus', is not to be interpreted as though Grosseteste were distancing himself from the idea, but as a counterpoint to 'et hee sunt quas vocavit Plato ideas et mundum archetypum'. Grosseteste can accept both theories of universals by identifying the Platonic exemplars with the divine ideas. His criticism of them was that they are separate from things (here with Aristotle) and not placed in the divine mind (with Augustine).

[68] 'Universale autem [i.e., as opposed to sense-knowledge] cum sit res inventa in multitudine, non est possibile sentire, quia quod reperitur in multis non est in loco signato et tempore signato, quia si esset in loco signato aut tempore signato non esset universale inventum in omnibus; universale namque est semper et ubique.' (ibid., fol. 39d).

[69] Ibid., fol. 23c; a reference to Aristotle's famous remark in *De Anima* III 8, 431b, 20.

[70] Ibid., fol. 20d.

[71] Ibid.; cf., however, the end of this paragraph (fol. 20d), where, with untroubled Platonism, he puts the universals close to the first being because they are immaterial.

[72] In fol. 17a, twice; the second mention associates abstraction with division

In one respect, when taken in their actual employment, sense and intellectual knowledge resemble each other, for both are governed by intentionality. The immaterial *species apprehensibilis* begets a likeness in the understanding which the *intentio animae* joins and compares with the real, in order to effect judgement—for it is the real that we know, not a fiction of the mind.[73]

It is manifest from this description of the genesis of universals that they are neither innate nor derived from existing knowledge; sensation is their foundation, and in its absence there could be no abstract knowledge.[74] If sense is called the occasion rather than the cause of demonstrative knowledge, this is said only by way of contrast with the universal concept, not in order to play down its foundational role. Sensation cannot give science, because it is limited to particulars individuated in time and place; concepts can, because they are not so limited.[75] The paradox involved in the study of knowledge is, that if the battle for scientific knowledge is to be won, then sense is the first rank that must make a stand, even though it is also the weakest. Only its success can inspire the massed ranks of imagination, memory and, finally, understanding, *'qui est nobilissima virtutum apprehensivarum, apprehensiva universalium primorum incorruptibilium'*.[76]

From this composite image of Grosseteste's theory of abstraction certain conclusions may be drawn.

The theory presents an eclectic appearance, as an awareness of the sources lying behind it and the variability of the vocabulary quickly reveal. The Augustinian emphasis on the active involvement of the sense-organ in perception is promoted at the expense of Aristotle's view (reinforced by Avicenna) that the object alone is active in the impression of its *species*.

Augustine's division of the soul into *intelligentia, memoria,* and *amor* provides the structure of the text which states the theory of intentionality most clearly. The distinction of sense and reason is Boethian in origin, and carries with it Neoplatonic

of the accidents from the substance and the grasp of 'unum et idem secundum iudicium suum in multis'.

[73] *Hexaem.*, ed. Muckle, p. 163.
[74] *Comm. in Anal. Post.*, fol. 39d.
[75] Ibid., fol. 23c. Sensation is not an occasion in the extrinsic sense which Gilson accords it (art. cit., p. 98).
[76] Ibid., fol. 39d.

attitudes to the instability of matter, which are quite non-Aristotelean. The structure of the interior senses and the view taken of their contribution to the elaboration of perception stems, of course, from Avicenna. When all these subtractions (or better, bracketings) are duly made, however, not only is what remains very close to Aristotle's thought, but the sweep of thought in the theory as a whole aims at fidelity to Aristotelean abstractionism, and achieves it without any fundamental distortion.

The conclusion that Grosseteste accepted in essence the Aristotelean doctrine of the abstraction of universals from sense-knowledge is fully justified, even when allowance is made for the foreign elements included by him in his presentation. None of these strikes at the basic principle of Aristotle, namely, that concepts are not innate but are acquired from sensation by a process of abstraction which has nothing to do with occasionalism, or pre-established harmony, or the theory that the mind is able to elicit knowledge from within itself, the opportune stimulus being provided by an otherwise dispensable sensation-process. Grosseteste's attitude to sensation was definitely positive.

His idea of the universal concept corresponded closely to that of Aristotle. Grosseteste was able to make a firm distinction between the Aristotelean universal and the Platonic idea. He considered that Plato—whom he almost always treated with sympathy and even indulgence—had a profound intuition, but that the idea failed to do justice either to the intrinsic intelligibility of the real or to the divine mind; the ideas were in neither place where they should have been. Aristotle, on the other hand, had no suspicion that the universals which are *principia cognoscendi* for us in the present state are the created product of other universals, in which *principia cognoscendi* and *essendi* coincide, namely, the eternal ideas. But Aristotle was right at the given level, that of the foundations of demonstrative science in the present condition of mankind.

Grosseteste was to make full and even creative use of the Aristotelean methodology in his own scientific tractates. About its validity he had no doubt, about its comprehensiveness, no illusions: it uncovered, after all, only the veil of truth, and offered access only to the literal sense of the book of creatures.

III THE *INTELLECTUS AGENS*

What significance, if any, is to be attributed to the silence of the *Commentary on the Posterior Analytics* concerning the Aristotelean distinction of the active and passive intellects? For Étienne Gilson, the silence of Grosseteste is eloquent. In a work in which illumination is a main theme, the absence of the Aristotelean distinction can only mean that Grosseteste remains firmly anchored in the ideas of Augustine and Avicenna.[77] His only answer to the problem of the knowledge of intelligibles is a reassertion of the traditional illumination-theory.[78] Sense-knowledge has the very limited role of awakening the soul to receive illumination (pp. 95-6). The Aristotelean doctrine of knowledge is not admitted, and the chief distinction upon which it turns, that of active and passive intellects, is unknown. A doctrine in which illumination is so much stressed can only allow the mind a purely passive role in acquiring concepts.[79] If Grosseteste had been forced to employ the language of Aristotle, his only possible option would have been to identify the active intellect with God, as Roger Bacon attests he did; hence he belongs to the school of *augustinisme avicennisant* (pp. 97-8).

We find none of these conclusions fully acceptable and some flatly mistaken. It is not true that Grosseteste was unaware of the language of passivity and activity with regard to the intellect, even though he admittedly does not use the distinction in the commentary and his '*intellectus agens*' (as we have seen) and '*intellectus materialis*' (as we hope to show)[80] would not have been recognized by Aristotle. Sensation is in Grosseteste's view something more than a mere occasion for intellectual knowledge. Far from spreading the Augustinian illumination-theory all over his commentary, Grosseteste makes only one demonstrably Augustinian use of it, a use not qualified by his customary distinction between the conditions

---

[77] Gilson, art. cit., pp. 91-2.

[78] Ibid., p. 98 (see n. 15 in the present chapter, p. 327).

[79] 'L'opposition d'un S. Bonaventure à la doctrine de Grosseteste ... nous fait ... apercevoir le caractère purement passif que la doctrine de Grosseteste reconnaît implicitement à l'intellect humain. Une lumière intégralement reçue, obscurcie par le péché, ravivée par la sensation et s'en libérant progressivement jusqu'à pouvoir s'en passer de nouveau, c'est tout ce que Grosseteste nous attribue en fait d'intellect.' (p. 98).

[80] See Part IV, ch. 4, iii, pp. 306-7, and ch. 5, v, p. 357.

of knowledge in this life and in the illuminative state of glorification. Grosseteste did not oppose the two theories of knowledge (Augustinian and Aristotelean) upon the same level and find them mutually exclusive, rather did he feel free to accept the Aristotelean doctrine, without restriction, at the appropriate level. His theory of illumination (as known to us from the *De Veritate*) is not ontologistic nor otherwise so extreme as to merit the opposition of St. Bonaventure. The human intellect is not merely passive but has an active part to play in abstraction. Finally, the argument from silence must always be used circumspectly: if Grosseteste did not mention the distinction of *intellectus agens* and *passivus* in the course of commenting the *Posterior Analytics*, neither had Aristotle himself done so in the work itself. Indeed, Grosseteste expressly remarks that other powers of the soul not mentioned in the *Analytics* are handled in the *De Anima* and the *Nicomachean Ethics*. Gilson's case for claiming that he identified the active intellect with God must rest in the final analysis with the twice-repeated testimony of Bacon, to which we now turn.

Bacon affirms in the *Opus Maius* that he twice heard William the Bishop of Paris determine (*sententiare*) before the full convocation of the University that the active intellect cannot be a part of the soul. The Lord Robert of Lincoln, Friar Adam Marsh and other such *maiores* held the same doctrine.[81]

The agent intellect is not a power of the soul, contends Bacon in the *Opus Tertium*, and all the wise of ancient times and up to the present identified it with God. He twice heard the Venerable William of Auvergne dispute with and reproach opponents and prove them wrong on this point. The Lord Robert and Brother Adam, '*maiores clerici de mundo, et perfecti in sapientia divina et humana*', held the same doctrine. When asked by certain friars what the *intellectus agens* is, Adam responded, 'The crow of Elias'. He meant by this that it is either God or an angel, but did not want to say so because the question was a taunt and was not asked for the sake of the truth.[82]

Evidently, Bacon is better informed about the teaching of Auvergne and Marsh than about Grosseteste's, concerning which he offers no supporting evidence, only the assertion:

[81] *Op. Mai.* II 5 (ed. Bridges III, p. 47).
[82] *Op. Tert.* 23 (ed. Brewer, pp. 74-5).

Grosseteste taught that the active intellect is not a part of the soul, but either God or an angel. How far is he a reliable witness to Grosseteste's doctrine? The question can indeed be generalized: how much did Bacon know of Grosseteste at all? He was not his pupil, nor is there any evidence that he knew him personally.[83] He did have access to the books Grosseteste bequeathed to the Franciscans in Oxford, but it is arguable that he actually read only some of the works on natural science and sampled his translation of Pseudo-Dionysius. There is no indication that he had read the exegetical works (he certainly did not know the *Hexaemeron*),[84] though he may have used the brief *De Operationibus Solis*. His references to Grosseteste, with the single exception of the testimony under discussion, consist of generalized praise of his method as a natural scientist and as a translator; but the metaphysical doctrines of Grosseteste he does not discuss, and he had little sympathy with the light-metaphysics as such. The readiness with which Bacon appealed to the example of Grosseteste and Marsh in the two areas of philology and natural philosophy is no warrant of a broad and unselective reading of their less spectacular works.

Besides, Bacon's testimony implies more than the simple equation of the active intellect with God: '*Deus, sive angelus*', is what he states Marsh held. Now Bacon never mentions Marsh but that he associates him with the Lord Robert; and the Dionysian commentaries of the latter, which he had read (or at least sampled), have illumination by God through the angelic hierarchy as a central theme.[85] It may have seemed to Bacon writing in 1268—as indeed it has done to others since— that such emphasis on illumination justified placing Grosseteste on the side of the angels, and was inconsistent with the attribution of an illuminating function to the human intellect. It is plausible to suggest that he was drawing a conclusion

---

[83] Bacon was in France till around 1247. If he had been associated with Grosseteste after that time, is it likely that he would have praised his translations of Pseudo-Dionysius and Damascene, yet never have referred to the rendering of the *Nicomachean Ethics*, by far the most celebrated translation Grosseteste achieved— and finished, *c*.1247?

[84] Otherwise he would have known that Grosseteste did not hold the theory on the origin of the soul which Bacon claimed to have been universal among English theologians of the previous generation, see Ch. 4, iv, p. 314.

[85] Compare the passage in the *Comm. in Anal. Post.* (fol. 8[c]), where the *intelligentia creata* is the illuminator of the impure intellect unable to see the first light in itself.

from Grosseteste's known teaching, rather than basing his evidence on a clear written statement or even upon oral tradition among the English friars, as he did in the case of Marsh.

If Bacon's testimony is not allowed to settle the matter summarily (and we consider that it is not qualified to do so), we are thrown back upon the evidence available from Grosseteste's own known doctrines. The matter is speculative, of course, and no certainty can be attained, but I believe that a degree of probability can be reached.

The first principle from which we begin is that Grosseteste accepted the Aristotelean epistemology as true in all its essentials, while continuing to hold the Augustinian theory of illumination. He both approved the doctrine of abstraction and was convinced at the same time that absolutely true and necessary knowledge is available to the human mind as it is at present, only if both object and subject be illuminated by God.

Now if he held abstraction it could only be the case that he allowed the intellect an active nature, because in the abstraction-process the essence has to be discovered behind the individualizing accidents, the shared nature hunted out from the range of instances embodying it. That Grosseteste did in fact regard the intellect as active is something easily shown from his vocabulary. The strange dualism whereby, according to Gilson's interpretation, the mind is active in sensation and purely passive in intellection, was totally foreign to him, not only in fact but in principle.[86] The height and depth of Grosseteste's personal thought was the metaphysics of light, and within this system the mind was considered as a created spiritual light (*irradiatio spiritualis*) engendered by the *lux suprema*. All light has an innate expansive tendency to multiply and diffuse itself in imitation of its source. The soul, then, is of its very nature spiritual energy and in it essence and activity are indissociable. It radiates. The language used of it is energetic and outgoing: *aspectus, intentio, extensio, protensio, appetitus; venatur universale, dividit, inspicit* (etc.).

Is the doctrine of the activity of the intellect different from that of the active intellect? It is, and for two reasons. Grosseteste shied away from making real distinctions within the simplicity of the soul. So much is already apparent from his

[86] A point excellently brought out by Fr Gieben in his unpublished thesis. The conclusions he arrives at there agree in general with those we defend here.

enumeration of its powers, which presupposes indeed the existence of different objects of knowledge, and hence of correspondingly differentiated powers of cognition in the soul, but does not erect these powers into distinct faculties. His readiness to conceive of the soul as a simple light made him hesitate to divide the intellect into active and passive 'parts'; and it was a hesitation he was not alone in experiencing,[87] for in some measure the thinkers of the entire first half of the thirteenth century felt disinclined to sacrifice the integrity of the soul to a rigid adherence to the strict division implied in Aristotle's terminology.

There is another and more profound reason why Grosseteste should not have used the Aristotelean distinction in an orthodox manner. It is plausible to suggest that he would have regarded unqualified passivity and activity as attributes inapplicable to the soul and its faculties. The two extreme poles of passivity and activity are matter and God; the soul is neither the one nor the other.[88] Its nature is spiritual, and hence active, at a much higher level than that of physical light, but if it were to be considered purely active, the transcendence of the first light would thereby be denied. It must, therefore, be active *lux*, dependent in its very activity upon the *actus purus*. In a metaphysics of light, illumination is the most natural way of describing the relation of the rational creature to the Creator. The dependence involved is one of being; but all being, in the degree that it is active, is light. Both the nature and the activity of the created spiritual light must, therefore, depend upon God.

Regarded from within this perspective, Grosseteste's doctrine of illumination, as we find it developed in the *De Veritate*, need not be thought to exclude (much less *intentionally* to exclude) the activity of the human mind in acquiring true knowledge, but should be understood as an assertion that the mind's activity is dependent upon a higher activity still. Ultimately, the attainment of absolute and necessary knowledge is not the unaided achievement of the creature. But how are we to conceive the causal relationship between the supreme

[87] Neither William of Auvergne nor St. Bonaventure, for example, admitted a real distinction between the soul and its faculties.

[88] Which is the reason why Bonaventure also refused to make a real distinction between active and passive intellects.

and the dependent activity, the interplay of which results in created knowledge?

Here the optical illustration used by Grosseteste himself comes to rescue us from mere speculation. Vision and knowledge are paralleled time out of number, and always the same constants appear: the intelligible light activates both the mind's object and the intellect, as the sun does the colour *and the eye*. Now, granted the physiological theory which holds that the eye actively emits rays to the object as well as receiving them, and the conviction on the part of Grosseteste that the spirits of the eye are of the same nature as the *lumen solare*, Grosseteste can only conceive of the sun's influence upon the eye as the activation of a matter itself intrinsically energetic, by that supreme generative source of energy, which stands to all its emanations as cause of their being and activity. When the analogy is transferred to knowing it reinforces the conjecture made above concerning the relation of the intellect to the *lux suprema*: the mind is of active nature, like the spirits, yet requires the influence of the *actus purus* to render it capable of exercising its natural activity of knowing the necessarily and absolutely true.

The non-correspondence between this vision and the Aristotelean distinction of agent and passive intellects is apparent. On the one side, activity (created) implies activity (uncreated) as its correlate; on the other, activity is the complement of passivity. If Grosseteste had been forced to use this unsympathetic language, he could only have said that both God and the intellect are *intellectus agentes*, God chiefly and the intellect dependently,[89] but such a statement would have represented an obscuring and not a clarification of his thought, so he wisely avoided it.

IV ILLUMINATION BY A SEPARATE INTELLIGENCE

In a passage from the *Commentary on the Posterior Analytics* already examined we read as follows:[90]

Iterum in luce creata quae est intelligentia est cognitio et descriptio rerum creatarum sequentium ipsam, et intellectus humanus qui non est ad purum defaecatus ita ut possit lucem primam immediate intueri, multoties recipit irradiationem a luce creata quae est intelligentia, et in ipsis

---

[89] As Peckham did (Gilson, art. cit., pp. 99-100).
[90] See Passage a (fol. 8^C), in the preceding section (pp. 327-9).

descriptionibus quae sunt in intelligentia cognoscit res posteriores quarum formae exemplares sunt illae descriptiones. Cognitiones enim rerum sub-sequentium, quae cognitiones sunt in ipsa mente intelligentiae, sunt formae exemplares et etiam rationes causales creatae rerum posterius fiendarum; mediante enim ministerio intelligentiarum virtute causae primae processerunt in esse species corporales. Hae igitur ideae creatae sunt principia cognoscendi apud intellectum ab eis irradiatum . . .

This tantalizing remark is preceded by the assertion that the pure intellect can contemplate the exemplars *in luce prima* and followed by a declaration of the possibility of scientific knowledge of natural causes and forms.

What makes it tantalizing is its Arabic ring.[91] The separate intelligence, which contemplates the first light and passes on its irradiation to the human intellect, resembles all the more the Avicennian *dator formarum* in that Grosseteste attributes to it a role in the production of things in the sublunary world. One wonders whether he has strayed from his purpose, at least as we have interpreted it, namely, to recommend the epistem-ology of Aristotle and yet to celebrate the transcendence of the Christian view of man over the pagan and worldly one.

No other passage in the same work nor any text from the same period of Grosseteste's life casts any direct light on the riddle. However, the *Commentary on the Celestial Hierarchy*, which is posterior by a decade or more, suggests an inter-pretation which we consider has everything to recommend it.

The text we have in mind deals likewise with a communi-cation of light from the *lux suprema*, through the agency of the *intellectus*, to man,[92] but the atmosphere and language differ widely from the statement in the *Commentary on the Posterior Analytics*. Dionysius's manner is sacral, his repeti-tive expressions almost incantatory, and the distinction of sacred and profane knowledge is, of course, passed over in complete silence. Grosseteste imitates him in all this as best he can, and for the most part with real sympathy. Like his

---

[91] Pierre Duhem (*Le Système du monde*, t. V, p. 348) quotes the same passage and comments: 'Elle est bien surprenante, cette page tracée par la plume de l'Evêque de Lincoln et qui pourrait être signée du nom d'Avicenne, cette page où se formule, avec une si précise clarté, l'enseignement des Néo-platoniciens hellènes et arabes au sujet du *Nous*, de l'intelligence du Premier créé. Nous avons quelque peine à croire qu'elle représente la véritable pensée de Robert Grosse-Teste; nous soupçonnons qu'il a seulement voulu résumer la doctrine des philosophes sans lui donner son adhésion, et, bientôt, nous en aurons l'assurance'.

[92] See ed. McQuade, 187-9.

author, Grosseteste speaks of illumination always in the context of hierarchy and salvation. However, we are told much more of the formal metaphysical structure of communication, from God through the angelic hierarchy to the human, than of its content. It is the rare glimpse of what it is that the angels actually reveal which makes our passage of almost unique interest. The chief ministry of the angels was to give the law to Moses, Dionysius insists, but God used his messengers to man both before and after the Law, to show our fathers the path of truth, to reveal visions of heaven, and to deliver prophecies. Grosseteste, as usual, is less sparing and more generous in his exposition that the Areopagite in his text, and he adds something.

What can Dionysius mean by the 'sacred orders' which the angels reveal to men at God's behest? Grosseteste has no hesitation: he must mean the angelic and human (or ecclesiastical) hierarchy and, beyond these, the grades (*ordines*) of the whole hierarchy of being; for, of course, order as such leads the mind to God. 'Order' here may also, he thinks, refer to the archetypal world, which is the creative exemplar of all number, weight and measure and which the angelic hierarchy contemplates.

What, then, of the 'visions of transcendent mysteries' which are referred to? Aided by a scholion of Maximus, Grosseteste has no trouble in identifying these. The mysteries are the hidden things of the divine life, the Trinity and Unity of God, the Incarnation and the time of its occurrence—such things, in short, as lie beyond the investigation of any creature, as were revealed to Daniel and to Paul rapt in the third heaven, and as fill the Apocalypse of St. John. And, of course, angels were sent to reveal the divine will concerning future events, as at the Annunciation.[93]

The *intellectus* of Grosseteste's translation and commentary on the *Celestial Hierarchy* is recognizably biblical (despite the metaphysics), whereas the *intelligentia* of his commentary on Aristotle is not so, at first glance; yet I suggest that they are the same creature in different guises. The latter gives the impression that his revealed illuminations would take the form of mathematical, scientific, and in general thoroughly profane information about the nature and causes of things; but the

[93] Ibid., pp. 211-12.

impression is misleading, for he is an angel in academic dress. His message would likewise be salvific, would concern the 'order of things' (including the historical order), through which the mind is re-raised and the image of God restored. Grosseteste's was an age when learned men delighted in giving to the world of religious meaning as scientific and metaphysical a statement as might be—a way they had of affirming the universality and catholicity of its truth. But one cannot disguise an angel; the brightness will shine through.

V ILLUMINATION AND SYMBOLIC KNOWLEDGE IN THE DIONYSIAN COMMENTARIES

'*Dans toute sa culture, le moyen âge est l'âge du symbole, autant et plus que celui de la dialectique.*'[94] This remark of Père Chenu is verified in the later thought of Grosseteste. Though his development is not marked by abrupt transitions and brusque volte-faces, in the wake of which hitherto accepted opinions are overthrown and totally replaced, it is never the less possible to discern the path of a thought which proves itself receptive of new influences and seeks to expand its capital of ideas by wide investment in promising areas. So it was with the enthusiasm for Pseudo-Dionysius, which gave him occupation for around four years (1239–43) and left a lasting imprint on his mind. His meeting with a theology in which symbolism played a very fundamental role was surely no accident, in the case of a theologian who was increasingly ill at ease with the growingly conceptual and dialectical theology of the university faculties.

That Grosseteste's conversion to Dionysian theology was not without some intrinsic relationship to his Aristoteleanism can be seen from his preoccupation with epistemological problems, as evidenced in the commentaries on Pseudo-Dionysius, and, more precisely still, with a single theological issue of unique importance, then as now: if our knowledge is acquired by abstraction from sense-experience and our thinking is always accompanied by phantasms (and in our present state it cannot be otherwise), how then do we manage to know and name the purely spiritual, and especially the transcendent God? The answer he reached contained a strong element of the Dionysian 'pathos of the symbol', as we may term it: while

[94] Chenu, *La Théologie au XIIe siècle*, p. 161.

the invisible cannot be exhaustively conceptualized, all that is spiritual can be darkly surmised by an intentionality which dwells within the congenital energy of material and historical symbols, particularly those canonized by sacred use.

The traditional optimism concerning the ability of the mind as spirit to know the spiritual,[95] the product of centuries of uncritical saturation in Neo-Platonism and its easy recourse to the language of illumination and intuition, fell victim to the intellectual ascesis to which the Aristotelean scientific methodology summoned its readers. We have seen Grosseteste commit himself to the proposition that no direct intuition of the spiritual and transcendent is available to fallen man, and this was an opinion which he continued to maintain with undeviating firmness. His reflections on the place of the phantasm in intellectual knowledge constitute a detailed elaboration of it.

The image of God which Adam bore brought with it the direct knowledge of God as he is in himself, of the spiritual world and of the order of the entire creation, as revealed in the divine mind.[96] The disruption of this wisdom and of its connate knowledge by sin[97] leaves us the senses as our only natural access to truth. Grosseteste gave early adhesion to the doctrine that all knowledge, even the highest, is accompanied by a phantasm or material image, which is the highest product of the sense-faculties,[98] but the phantasm as he regarded it was loaded with the ambivalence inherent in all human operations after the fall, being at once our openness to the world and the foundation of our abstract knowledge, and closure or limitation, as compared with the natural freedom of the intellect in its best state to know the totality of being. The phantasm, which retains the *appendicia materialia*, can have grave consequences already for the knowledge of natural essences and causes, as when its interference induces the intellect to take accident for essence and in consequence to reach only opinion instead of the demonstrative knowledge of a whole genus and its species. Natural science, of course, can always

[95] The *De Spiritu et Anima* is a case in point.
[96] *Ex Rerum Init.*, ed. Gieben, p. 121.
[97] Referred to by medieval theologians globally as 'humanum genus condemnatum et sauciatum in naturalibus virtutibus' (ibid., 126), a phrase whose formulation preceded the distinction of nature and grace and managed to survive it to bedevil later theology, even long after the rise of *Dogmengeschichte*.
[98] *De Statu Causarum*, ed. Baur, p. 125.

hope to progress from opinion to epistemic knowledge through further effort at definition. When, however, the object of inquiry is not given in experience, the attainment of abstract knowledge is threatened by the unavoidable superimposition of the phantasm, in a way that permits no definitive purification of the object; in such a case the phantasm can only be regarded as a direct impediment to knowledge. It is for this reason that Grosseteste attributes to the phantasm all the negative qualities of the faculty of imagination that were a commonplace of the Augustinian tradition. The phantasm threatens the very possibility of knowing that which transcends our experience, and infiltrates something of the spatio-temporal conditions of our earthly existence into all our attempts to know the purely spiritual.

Grosseteste has left us a refutation of the errors of Aristotle which provides an exemplary illustration of the interference of the material phantasm with the mind's processes.[99]

Human reason even as it now is can reason to the existence of God from mutable and composite things, and can form a notional idea of the attributes of God. Aristotle and other philosophers prove God's immortality and intemporality; but no one should be deceived by his conclusion, for the discourse of reason can convince us of the truth of many things, without always being able to give us an understanding of their essence. There have been men who have demonstrated the spirituality of the intelligences and the reality of God, without understanding the content of either notion. They see reality under the influence of corporeal phantasms—as it were, the sun through a cloud—and continue to speak and think the very opposite of what the discourse of reason has revealed to them.[100] Aristotle reasoned to God's eternity, yet affirmed of him much that is unfitting, for through following the image of temporal extendedness he failed to grasp the essential simplicity of eternity. Having failed in that, it was only natural that he should err concerning the perpetuity and infinity of time and motion, imagining the existence of a time before all time and a motion before all motion—just as people imagine a space beyond space, *ad infinitum*. The *falsa imaginatio* of the infinity of time led on to the further images of perpetual

[99] *De Finitate Motus et Temporis*, ed. Baur, pp. 105–6; ed. Dales, *Comm. Phys.*, pp. 147, 153–4.
[100] *De Finitate*, ed. Baur, p. 105.12 ff.

motion and of a creation coeval with God.[101] Time and space became the measure of all reality and resulted in Aristotle's denial of the transcendence and eternity of God.

That sensation with its highest product, the phantasm, and more generally the body, are in Grosseteste's view not alone in bearing the wounds of original sin, has already been noted. The soul is not a pure spirit struggling with a corrupt body which alienates it from its true perfection, as in Avicenna;[102] its own distortion is equally as real as that of the body and it clouds the knowledge of God with anthropomorphism, as happens when men think that God is cruel, or evil, or uncaring, like themselves.[103] They refuse to raise their loves beyond the coloured world and cannot expect to know that which they do not care for. While the effects of original sin are thus diffused over all the actions of man (and indeed over the whole of creation), they are apprehended with particular force in the *adminiculum materialitatis* which accompanies all the knowing of which we are now capable. This is the main emphasis of the contrasts between angelic and human knowledge of God (so much a feature of Grosseteste's *Commentary on the Celestial Hierarchy*—where, however, 'angelic' represents at the same time the highest capacity of human nature).[104] Grosseteste finds it necessary to institute a language of the understanding, one that is properly modulated to express the unconditional prevalence of spatio-temporal conditions over it, and to highlight its present distance from the ideal. He repeatedly speaks of the *intellectus materialis*; *admixtus*, in opposition to the *intellectus immaterialis*; *abstractus*; *nudus*; *purus*; *impassibilis*. Such terms bear no relation to the thought of Avicenna, still less to that of Aristotle, for it is not separate faculties of the

---

[101] Grosseteste concludes on every occasion with an explanation of the philosophers' error in terms of *affectus/aspectus*.

[102] *De Anima* V, 5, ed. cit., pp. 131-2.

[103] '. . . quidam, ex distortione animae suae Deum iudicantes, reputant Deum aut crudelem cum iuste punit, aut remissum vel non curantem mundum . . . aut iniquum . . . Noli sic de Deo iudicare: non sunt haec distortiones in Deo, sed in animae tuae distorto speculo.'—*Dictum 60*, ed. Gieben, *Franc. Stud.* (1964), p. 157. Cf. *Comm. in Hier. Cael.*, ed. McQuade, pp. 252, 275-7, 290.

[104] *Comm. in Hier. Cael.*, ed. McQuade, p. 180: 'Ipsae [the angelic hierarchies] enim *intelligibiliter*, hoc est simplicis intelligentiae et liberi arbitrii conatu, et nullo materialitatis adminiculo, *reformantes*, hoc est reformare enitentes seipsas ad deiimitativum pro sua possibilitate, et *aspicientes supermundane*, hoc est nullo admixto materiali fantasmate, *ad thearchicam similitudinem*, hoc est ad ipsam divinitatem . . .'

mind which are in question, but differing states of the same human soul, now and in its glorified condition.[105]

Such is the darkness into which illuminatory symbols are introduced and hence the insistence throughout the Dionysian commentaries that we can know God and the heavenly hierarchies only in so far as we are illuminated by the thearchic Light, whose rays are reflected down through the mirrors of the angelic orders to reach mankind. Illumination is a synonym for revelation, and even for the redemptive knowledge which is the beginning of divinization.[106]

When we try to come closer to the nature of illumination we find that it divides itself into two forms, according to its two mediate sources, namely, the sensible realities of the world and the Scriptures. There is no divorce between these, because the former is opened to us by the latter: without the guidance of the *liber scripturae* we could not read aright the *vestigia Dei* which are written into the nature of things. Now both material things and the Scriptures which interpret them take the form of symbols of the invisible realities and offer to faith a light in the darkness to which reason is reduced. Symbols alone and their interpretation provide the possibility of signifying beyond the limits of spatio-temporal experience. Revelation (illumination) is proportionate to our *analogia* or mode of being and divine providence adapts it to our condition. Even though our knowledge (like our love) finds those things objectively farthest from simplicity and unity to be clearest,[107] its weakness can be turned into strength and the dark things around us lit up to form a pathway of light:

. . . subiungit causam quare eloquia per symbola et imagines sunt

---

[105] The influence at work is of course that of Pseudo-Dionysius and the terminology is found in Grosseteste's works only in the commentaries, e.g., *Comm. in De Div. Nom.*, ed. Ruello, p. 137 § 12: 'Intellectus humanus phantasmatibus admixtus et non ab omni materiali phantasia abstractus non potest comprehendere incorporalium puram informitatem abstractam ab omni formatione et figuratione et situ materiali'; see ibid. p. 139, § 16 and 17. The *intellectus purus et separatus a phantasmatibus* is mentioned already in the *Comm. in Anal. Post.*

[106] *Comm. in Hier. Cael.*, ed. McQuade, pp. 138, 181, 294, 360. The same holds true of the illumination of the angelic hierarchies; illumination is always associated with the notion of hierarchy and the re-raising of the mind and will, and is therefore salvific. Augustinian illumination does not appear much in the commentary.

[107] Ibid., ed. McEvoy, p. 2: '. . . nostro namque intellectui sunt manifestiora, quae plus a prima simplicitate et unitate sunt elongata et compositiora, et quae econverso se habent occultiora.'

nobis tradita, quae causa est quod *thearchicus radius nobis* adhuc infirmis *supersplendere non potest, nisi circumvelatus varietate sacrorum velaminum*, hoc est imaginum et figurarum, ut sic sursum ducamur, et nisi sit praeparatus *paterna providentia connaturaliter et familiariter* nobis *his*, hoc est . . . quae sunt apud nos, nobis cognita, quemadmodum neque sol corporeus oculo corporeo adhuc infirmo visibilis est si immediate ei contiguetur. Sed neque forte si distet a sole, solo aetheris interstitio, ipso sole aethere intermedio circumvelato, et radio ipsius sua penetratione per aethera aliquantulum connaturalius visui praeparato, adhuc poterit solis refulgentiam sustinere. Forte etiamsi elongetur a sole tam aetheris quam aeris intermedio, aere existente sereno, et radio ipsius utriusque medii per transitum magis adhuc connaturaliter visui praeparato et contemperato, nec adhuc sic poterit in solem directe figere contuitum et rectam radii solaris incidentiam sustinere, sed oportebit simul cum his nobis rore vel nebulae interpositione solem circumvelari, et ab his omnibus mediis radium ipsius praeparari et connaturari et contemperari infirmo visui.[108]

Nunc enim ad illa [i.e., the divine lights] pertingimus per sacra velamina symbolorum et subfigurationum tam in eloquiis quam in sanctorum patrum traditionibus contentorum, tunc autem, cum erimus docibiles Dei, sine velaminibus omnia nude conspiciemus.[109]

Grosseteste would not have been a man of his own times had he not reflected upon symbolism long before Dionysius caught a grip upon him. History and nature, Augustine had taught, are a double road of access to the knowledge of God and his designs.[110] But already in the *De Cessatione Legalium* of Grosseteste the common medieval symbiosis is apparent: the book of nature is read in the Book as such. Scripture teaches us sacred history, but it also teaches us to read the mysteries of redemption as these are signified within the natural order.[111] *Res et verbum*, thing and interpretation, together constitute a sacramental order. Traditional ideas on the four senses of

---

[108] Ibid., ed. McQuade, p. 16; cf. also pp. 10, 11, 252, 330, and the important note of Maximus, p. 12: 'Quoniam sine typis et symbolis non possibile nos in carne existentes conspicere immaterialia et incorporalia . . .'.

[109] *Comm. in De Div. Nom.*, ed. Ruello, p. 153, § 51.

[110] *Enarr. in Psal.* 45, 7 (*P.L.* 36, 518): 'Liber sit tibi pagina divina, ut haec audias; liber sit tibi orbis terrarum, ut haec videas . . .'; cf. Chenu, op. cit., pp. 169-70.

[111] 'Et quia actus hominum voluntarii non possunt sciri nisi per sensum aut per historiam . . . oportuit ut illius populi vita et conversacio prophetalis nostre salvacionis, in auctenticam redigeretur scripturam. Similiter quia signacio nostrae reparacionis per creaturarum species non est patenter nota humano generi, oportuit ut creature huius mundi nostram reparationem significantes per auctenticam scripturam manifestarentur nostre reparacionis misterium designare, ut videlicet adiutorio illuminationis scripture agnosci posset que creature quibus suis proprietatibus et qualiter signarent misterium salvacionis nostre.' (*De Cess. Legal.*, MS Bodleian lat. th. c. 17, fol. 164ᵃ).

Scripture provided Grosseteste, as we shall see, with a first definition of the symbol.

It is scarcely an exaggeration to say, however, that the world of Dionysian symbolism differed from the traditional Western ideas almost more than it agreed with them. Though based on Scripture, it avoided historical or typological symbolism in favour of cosmic. While it dwelt with the symbols used in the sacred tradition, it held to an objectivism which resisted the idea of conferred meanings, so that *res* encroached largely upon *verbum*.[112] All reality is a theophany in circular motion from and to the Good, and anagogy shades off into mystagogy when the symbol is grasped as initiation into the mystery which it makes present. This apophatic climate of thought, governed as it was by the dialectic of divine immanence and transcendence, was at a far remove from strict conceptualization and allegorization. In the eyes of Pseudo-Dionysius the symbol is irreducible to thought precisely because the mind must pass through matter towards an unknowable good.[113]

Grosseteste's success in following the mystagogue was not complete, but it was none the less substantial. Symbols, he insisted, are more than signs. A sign is used either naturally or conventionally to impress what it signifies on the mind, or to reawaken a memory of the signified. It can be multiplied to reinforce the impression, but it remains essentially technical.[114] Yet even signs have a possibility of ambiguity, one which increases as the move is made from the natural towards the conventional. To take two examples: a picture of a lion may be seen as a likeness of a lion or as a purely formal structure; depending upon the perception of the observer, it reveals or conceals another reality.[115] A book can be perceived by an unlettered layman only as a series of strokes, whereas a reader perceives the strokes as letters which represent elements of

---

[112] Chenu, *La Théologie au XIIe siècle*, pp. 176-7.

[113] *Comm. in Hier. Cael.*, ed., McQuade, p. 23.

[114] *De Cess. Legal.*, MS cit., fol. 158ª.

[115] *Comm. in Hier. Cael.*, ed. McQuade, p. 50: 'Numquid sicut pictura leonis respicienti eam ut imaginem et similitudinem animalis seu leonis est manifestatio et rememoratio animalis seu leonis, respicienti vero eandem picturam solum ut linearum quandam divaricatam protractionem et ut nullius similitudinem aut imaginem, nullius alterius est manifestatio seu rememoratio, sed magis visu et intentione ipsius detentis et attentis in solam ipsam linearum figurationem ipsa pictura est eis ipsius rei quae depingitur absconsio.'

sound.[116] Thus a sign leads the intention to the apprehension of something else only if it is perceived as a sign and not as a reality in itself.[117]

The signs taken from history and nature and used in the Scriptures have a character other than the purely conventional, for what they signify is a mystery. The word used signifies in the first place a reality, but the intention is led beyond the literal referent to a secondary meaning, pertaining to the order of human salvation.[118] The symbol enters into its own, however, when it is employed to reveal something that would otherwise be totally hidden from intellectual or discursive thought, such as purely spiritual being. Grosseteste grasped very clearly the essence of Pseudo-Dionysius's theory of the symbol: a material reality, or a reality of our total experience, which has its own density and yet an innate energy sufficient to lead the mind into a dialectic of continuity and rupture, of affirmation and denial, in which the invisible is brought close without being possessed and dominated.

This is illustrated very clearly in the first chapter of his commentary on the first *Hierarchy*.

[116] 'Item laicus omnino ignorans etiam quid vel qualis est liber et quid vel qualis est littera, inveniens librum et inspiciens in eo in solas litterarum varias protractiones intendit, nihil per ipsas protractiones aliud ab ipsis protractionibus imaginans aut intelligens; unde nullius rei alterius sunt ipsae litterarum protractiones ei manifestatio, sed magis intentio in illas solas est ei rerum aliarum occultatio. Sciens autem quoniam litterae sunt protractiones quas videt, scit quod aliud significant, agnoscens vero uniuscuiusque litterae figuram propriam et cuius vocis simplicis signum est, per figuram visibilem imaginatur elementum audibile, et est figura visibilis eidem manifestatio audibilis elementi.'(ibid.).

[117] 'Sic omne signum consideratum in quantum signum ducit intentionem considerantis in aliud apprehendendum per ipsum; consideratum autem in quantum est res sola et non signum non ducit in aliud . . . sed magis suae apprehensioni alia interim occultat.' (ibid.).

[118] '. . . significant verba primo loco res, sed ibi non sistit signantis intentio, quia intendit ultra per res primo verbis signatas signare secundo aliquid [fol. 164d] quod ad humani generis pertineat salvacionem.' (*De Cess. Legal.*, MS cit., fol. 164c-d). The context of discussion is the relationship of the literal sense to the other three, but that Grosseteste intends these statements to cover symbolism in a wider sense is confirmed when some ten lines further on he takes up the scriptural symbol of fire as representing the divinity. The *Comm. in Hier. Cael.* (ed. McQuade, pp. 23-4) refers to the sacred disciplines (themselves symbols of celestial contemplation), 'quae per res materiales immaterialia demonstrant, quas *discursivas*, seu decursivas seu digressivas seu exitivas dicit, quia in illis per litteram primo significantur res, et non ibi sistit nostra comprehensio, sicut in disciplinis liberalibus, sed ultra tendit ut per res primo per litteram significatas alias res intelligat, et has frequenter multimodas, quia eisdem rebus et anagogicum aliquid et allegoricum et morale plerumque comprehendit'.

It is only in and through material symbols, especially those used in the Bible, that we can begin to know (contemplate would be too strong an expression for our earthly state) the divine and spiritual.[119] The symbols are variously named by Grosseteste: *figurae*; *formae*; *plasmationes*; *velamina*; *notae*; *signa*. The only apparent difference he makes among these designations is that some (*figurae*, *signa*) refer more often to *res*, others (*plasmationes*, *velamina*) to elements of language. If we take *formae*, *figurae*: form, the intelligible reality and principle of existence of the things of experience, gives us access to the being of the thing, and allows us to say what species it belongs to, what acts are typical of members of that species, and what their proper functions are.[120] Such forms, taken in their concrete reality (visible, material, three-dimensional), are made figures, signs, or notes of realities that lie beyond matter and dimensions, in such a way that the knowledge we have of the former serves to signify what is less known, the spiritual. One example is the sun, which, taken in the concreteness of its being and functioning in the world, is made a symbol of God, the Father of lights.

The clarity of this elucidation of symbolic functions leaves little to be desired. When, however, Grosseteste introduces the notion of analogy in the immediate sequel, the reader may feel some concern: is he about to evacuate symbolism of its specific endowment by reducing it to the analogy of proportions, something more or less conceptualizible? It is with relief that his true intention is grasped. Preoccupied still by the relation of the mind's present capacities to its future state, he states the equation: as the material symbol is to the spiritual reality, so is the capacity of our present knowing, forced as it is to aim at the spiritual through matter, to the purified understanding which shall be capable of direct contemplation.

[119] Ibid., p. 20.

[120] Ibid., p. 21: Grosseteste has posited an identity between *forma* and *figura*: 'Et quia per figuram rei corpoream plurimum cognoscimus, et cuius speciei est res, et quod individuum in illa specie, et per proprias figuras singulorum membrorum qui sunt actus proprii eorundem, et ex integra figura totius qui sit actus proprius huiusmodi speciei conveniens, et sic figurae rerum corporee visibiles et materiales et dimensionem habentes sint nobis plurimum note et signa invisibilium et dimensionem in se ipsis non recipientium omnem formam corpoream in quantum est nobis nota, et signum rei minus note, seu intelligibilis et incorporee, solemus figuram et typum ipsius rei vocare, sic dicimus solem visibilem, qui corporaliter omnia illustrat mundana, esse figuram et typum dei, qui spiritaliter omnia illustrat intellectualia'.

Unreflectingly perhaps, Grosseteste brings out a fundamental requirement of symbolism in his discussion of natural forms as invitations to a metaphorical leap, for whether he was fully aware of it or not he puts Aristotelean science at the service of symbolism. As a result, the stress he places upon the 'real' natures of objects assures them of a firmness and stability, without which they could not successfully function as symbols. Confirmation of this is provided by the experience of earlier medieval generations.

An object which is chosen as the material support for a symbolic operation should not be etherealized but must be allowed to retain its full ontological and epistemological density. The symbolic value sought in the world's contents by the earlier Middle Ages managed to short-circuit symbolization by inducing a premature deflection of interest from the nature of the thing to the moral or spiritual reality it was held to signify. This was a natural consequence of the scientific ignorance of that early period, and equally, by rebound, a cause of its prolongation. It led inevitably to a disintegration of the very foundation of symbolic reference. When the object's natural consistency was dissolved and it was denied its place within an interlocking system of natures, causes and effects, it ceded its concrete dimensions and became etherealized. No longer possessing a reality of its own, bereft of the opacity of matter, the object was reduced to being a pure transparency of the sacred. It sacrificed its terrestrial being to Reality and its intelligible structure to the miraculous mentality. When the *res* was no longer dwelt with for its own sake but looked through, it lost, simultaneously with its material individuation and epistemological autonomy, the very capacity to symbolize, *donner à penser*, since the dynamism inherent in its intrinsic properties was volatilized by the effective denial of any retaining substrate. The energy which could have relayed a true aim at the invisible was thus bypassed, and the short-circuit left the mind an illusory freedom to find God and the spiritual world everywhere in nature. Such a freedom, however, fell little short of caprice, in proportion as nature was seen no longer as sacramental but as quasi-miraculous. The natural result of the encroachment of *verbum* upon *res* was artifice, ingeniously and at times eruditely set to work in typology, allegory, and moralization; but loss of objectivity and of conviction preyed upon the outcome. The spontaneous power of the symbol was evacuated and replaced by largely conventional associations. The symbol became denatured and yielded its ground to the mere sign.[121]

Aristotelean naturalism came to rescue symbolism from its confinement in convention and the progress begun already in the twelfth century continued during the thirteenth. It is not without an element of paradox that experimentalism should be regarded as having restored an elementary realism, which profited and revitalized the symbolic consciousness, so commonly assumed to be the connatural possession of Platonism in all its forms. The paradox is diminished when once it is seen that the cosmos of natures regulated by the laws of Aristotelean physics was placed before minds whose cultural noetic was not limited to natural knowledge but was opened upon the transcendent. The search for the beyond came in consequence to be regulated by the natures and properties of things as those were scientifically aimed at; with that, a solid support for the transcendental intention of the search was provided. We have seen Grosseteste insist that it is as members of species and genera, as endowed with consistent properties, and in their material reality, that the things of this world lend themselves as *notae et signa* to theological symbolization. This was the programme to which he himself was won over.

The nature of the symbol is that it both reveals and hides what it symbolizes. The hiatus between visible and invisible reality is not bridged by the symbol, as it would be by an established logical continuity; it is confirmed by the rupture which makes every resemblance to be unlike—in Dionysius's paradoxical expression, a *dissimilis similitudo*. Grosseteste follows him, it is true, in distinguishing *similes* from *dissimiles similitudines*, but he emphasizes the inadequacy of either kind to provide strictly conceptual understanding.

The distinction of like and unlike resemblances is not understood by Grosseteste as referring in the first instance to material and spiritual being, for it is, after all, the knowledge of the latter which is the whole object of symbolic signification.[122] He

---

[121] See McEvoy, *Rech. Théol. anc. méd.* 41 (1974), pp. 52-3.

[122] *Comm. in Hier. Cael.*, ed. McQuade, p. 39, opposes 'dissimiles similitudines, viles, terrestres, et omnino materiales', to 'similes similitudines, nobiles, caelestes, aliquatenus immateriales'.

contents himself with the contrast of *terrestria* and *caelestia*, the latter referring to the nobility of the physical heavens and their luminaries. No doubt this distinction is aided by the light-metaphysics, which regards pure physical light as attaining to the borders of spiritual nature; but at least it is not unsympathetic to the thought of Dionysius.[123] With Dionysius, then, he asks whether the opinion is justified which would hold gross and terrestrial things too vile to be made symbols of the spiritual hierarchies. Should not what is objectively closer to the celestial intellects at once bring our minds nearer to them and obviate the risk of a too-gross understanding of their attributes? Against this is to be counted in the first place the biblical symbolism, which holds nothing to be in principle too mean to bear some vestige of its origin. Dionysius flatly denies that like resemblances provide more effective symbols of the spiritual and that grosser things lead the mind astray. On the contrary, provided that the material element is seen as symbolizing and not positively representing spiritual being, like and unlike resemblances form a single complementary system of referential values. As a good interpreter, Grosseteste allows Dionysius his view, and agrees that our minds can use all the help they can get from whatever is connatural and familiar to us. Nevertheless, a quiet but insistent superimposition of the Augustinian distinction of image and vestige runs through his commentary. In the *Commentary on the Celestial Hierarchy*, as in the *Hexaemeron* and elsewhere, *imago* denotes a certain imitative likeness of God found in rational creatures, whereas the vestige is a distant and partial representation. While no name signifies what God is in himself, and no univocal predication of attributes to creatures and their Creator is possible, words like life and light come nearer to resembling him, as we know him from his externalizations, than do less worthy material properties, and constitute the middle sequence of a triple hierarchy, the upper reaches of which are spirit and its attributes.[124] Thus the language of negation ('invisible', 'infinite', 'unknowable'), which asserts God's transcendence, is rejoined by another modality of negation, in which God is more praised by the denial of the higher than of the lower power, and a

---

[123] *Comm. in Hier. Cael.*, ed. McQuade, p. 41 (quoted in Part II, ch. 2, p. 109, n. 124).

[124] *Comm. in Hier. Cael.*, ed. McQuade, pp. 59 ff; *Comm. in De Div. Nom.*, ed. Ruello, p. 165, § 80.

positive dimension of naming emerges, purified by the fire of negation; naming, that is, not what he is in his unapproachable mystery, but what he has made and done, the *benefici processus ad extra.*[125]

The development of a symbolic theology presupposes a theory of the interpretation of the symbol. In this area Grosseteste has little new to say; he contents himself with a general reference to the illumination which reflection on Scripture brings to our minds. The Bible teaches us by offering symbols both like and unlike, various modes of doctrinal teaching and the four kinds of meaning. Theology is consequently an analytic science which invokes all operations of the mind: translation of words from one language to another, a differentiated understanding of the realities referred to, and an interpretation on several levels for each word; in short, all the resources at our command for progressing from the known to the unknown. If we think of the unknown, for which we search, as being linked to the known, from which our search sets out, the method in its various forms is seen to be analytic, Grosseteste argues, since it 'releases' and manifests new meaning.[126]

The chief inadequacy in this theory, so far as it touches symbols, is its tendency to assimilate symbolical reference to a quasi-pictorial model of representation. The elements of the symbol are to be analysed, Grosseteste proposes, and coupled with the spiritual (angelic) properties they most aptly signify.[127] For all their love of the symbol the medievals never quite escaped from the temptation to tame it at a theoretical level and make it the servant of conceptual theology. If this fault was typical of the Scholastics' attempt to extend reason ever further into experience, Pseudo-Dionysius himself must carry part of the blame, for a metaphysics which laid claim to a conceptual knowledge of the properties and organization of

[125] See Ruello's very fine introductory study to his edition of *Comm. in De Div. Nom.* ch. 1, *Arch. Hist. Doctr. Litt. M.A.* (1959), pp. 101 ff., and his *Les 'noms divins' et leurs 'raisons'.* . . (Paris, 1963), pp. 155–73.

[126] See Part II, ch. 2, n. 124.

[127] *Comm. in Hier. Cael.*, ed. McEvoy, p. 124: 'Reflectamur inquam resolutorie, hoc est ipsorum symbolorum resolutionem faciendo, dividendo ea in suas partes et proprietates, omnia et omnes definiendo, et quid spiritale quo corporali convenientius significari possit investigando, et omnem materialitatem ab immaterialibus significatis abiciendo, ipsa spiritalia tandem intellectu abstracto speculando . . .'. *'Intellectus abstractus'* is to be understood in terms of its use elsewhere in the Dionysian commentaries, as already defined.

the heavenly hierarchies exposed itself to the temptation to invoke symbolism only to give expression to the concepts with which the symbolic elements were paired.

That the theory was in fact more inadequate than the practice is suggested by Grosseteste's comments on the symbolism of fire in Chapter Fifteen of the *Celestial Hierarchy*, whose main elements we shall retail in summary form.[128]

Fire is the primary symbol of spiritual being in the Old Testament, particularly in Daniel and the Major Prophets. It is applied both to God and to the angels, especially the seraphim, whose name means 'burning'.

Elemental fire is disseminated in all compounds without losing its nature in their complexion, just as God is at once immanent and transcendent. The nobility of celestial fire reveals the divine eminence raised beyond finitude, as fire is above the sublunary world. The symbolism of fire is cognate to that of light. The brightness of fire signifies the divine light in whom no darkness is to be found, but as fire too can hide itself in layers of flint before bursting into open activity in tinder, so God is manifest in his acts but hidden in his nature. The clarity of fire flames out and reveals its surroundings, as God's illumination reveals truth to the mind. Fire converts matter to its own activity, and is lord of all it touches, just as God shares his nature with us, being as prodigal of his grace as fire is of its warmth and lustre.

There is a sort of circumincession which promotes symbiosis among primary symbols, and which leads Grosseteste from fire, through light, to life. Life is born and renewed from warmth. All its functions (growth, nutrition, fructification and health), depend on vital heat, which is sustained ultimately by the transcendent fire of the heavens. Warmth is a powerful symbol of God's gift of life and of its renewal by grace.

The divine essence cannot be grasped, any more than fire can be confined within a material container; it devours the bonds which seek to restrain it. Fire purifies, rises from the earth's centre to the empyrean, is regular in its motion both on earth and in heaven, and is the active agent among the elements; in all these ways it symbolizes God's nature. It absorbs without being absorbed, even as God divinises his rational creatures without sacrificing his own nature. Fire needs no

[128] Ibid., ed. McEvoy, pp. 133-59.

other element to activate it, but is and shows immense power, to which no limits can be set. But if fire is mighty, it is not always dominating, for when the tiny spark is not caught it disappears as suddenly as it has leaped—like the divine spark we have failed to catch in the tinder of the virtues. For all its generosity with light and warmth, fire remains undiminished in its energy.

Enough has been said to show that Grosseteste's understanding of symbolism is real enough. Even though he is always liable to allegorize and moralize, a genuine sense of the power of symbolic realities is undeniably present. He has an Aristotelean feeling for the consistency and regularity of natural events, but over all knowledge stands the mind opened upon the transcendent, drawing wisdom from the sacred tradition and stirring the embers to incandescence.

Chapter 6
The Place of Man in the Cosmos

THE history of ideas is still written as though the parallel of macrocosm and microcosm was submerged after its vigorous airing during the twelfth century[1] and only surfaced again in Nicolas of Cusa, to become a favourite theme of the fifteenth-century Renaissance.[2] This view can, no doubt, be explained by the silence of the historians of thirteenth-century philosophy with regard to microcosmism.[3] The impression could be gained that the twelfth century gloried in microcosmism because it was a predominantly Platonist and humanistic age, while the thirteenth, being Aristotelean and rationalistic, treated the theme with reserve or even disapproval. The truth which this impression approaches is that microcosmic parallels were derived from Plato, not from Aristotle, and that every reinfusion of Platonism into Latin Christianity witnessed a quickening of interest in them: Scottus Eriugena, the School of Chartres, and Florentine Platonism all bear this out. The error which preys upon the notion advanced becomes serious in proportion as the legitimate contrast between the two centuries is pushed to an opposition which, today, few historians would care to defend. We can say at once that if microcosmism lived on into the thirteenth century it did so because Aristoteleanism was at no time in sole possession of the stage; there were side-shows in plenty.

---

[1] Studies on twelfth-century microcosmic themes are relatively numerous and will be referred to in the course of this chapter.

[2] For the relevant texts of Cusa see E. Zellinger, *Cusanus-Konkordanz* (Munich, 1960), pp. 244-5, and for a commentary, E. Colomer, 'Individuo y cosmos en Nicolás de Cusa', in *Nicolás de Cusa en el V Centenario de su morte (1464-1964)*, vol. 1, pp. 67-88; E. Cassirer, *Individuum und Kosmos in der Philosophie der Renaissance* (Berlin-Leipzig, 1927), pp. 41, 68-70, 97, 115-8.

[3] The only studies I have found which are expressly devoted to microcosmism in the thirteenth century are E. Stadter, 'Die Seele als *minor mundus* und als *regnum*. Ein Beitrag zur Psychologie der mittleren Franziskanerschule', in *Miscell. Med.* 5 (1968) 56-72; J. McEvoy, 'Microcosm and macrocosm in the writings of St. Bonaventure', in *S. Bonaventura 1274-1974*, vol. 2, pp. 309-43; G. Verbeke, 'Man as a "frontier" according to Aquinas', in *Aquinas and Problems of his Time*, ed. by G. Verbeke and D. Verhelst (Mediaevalia Lovaniensia. Series 1, Studia V) (Leuven-The Hague, 1976), 229 pp.; pp. 195-223.

One possible reason why the microcosmic ideas of the thirteenth century have not yet found a place in histories of philosophy is that the majority of the texts which develop them are not philosophical but theological: commentaries on Genesis and on Mark 16:15, and questions on the creation of the body in commentaries on the *Sentences* and the *summa*-literature.

A difficulty of a more intrinsic character is the bewildering profusion of microcosmic themes in both ancient and medieval periods. The kernel notion is that the same order can be affirmed of two terms, a great and a small, where the small is human nature or one of its constituent parts or aspects. However, the protean nature of medieval microcosmic speculation and the astounding degree of amplification and associative richness of which the seminal idea proved itself capable, defy treatment under a single heading and make even the work of classification one of formidable complexity.[4] Add to that, that thirteenth-century sources do not provide systematic treatises on microcosmism, most references wearing the air of asides, illustrations, or passing and decorative hints at what everybody knows to be the case, and the challenge to the historian is evident. He is not often so conscious of his subjective intervention, of putting order upon the formless and multiple, as when he attempts to present, in a unified and synthetic way, thoughts so generally diffused and fragmentarily expressed as to denote the presence almost of a mentality, rather than a doctrine, in his sources. Yet when the whole series of microcosmic *leitmotive* is assembled from the writings of St. Bonaventure, for example, it is undoubtedly something like a central key to the organicity of his system that is recovered, an architectonic element of unquestionable revelatory power for his thought, taken as a whole. Something similar holds true of Grosseteste.

There were good *a priori* grounds for believing that Grosseteste might have something worthwhile to say concerning microcosmism. When he attended the schools as a young man the glory of Chartres was still a recent memory; indeed, one

---

[4] Cf. the two main attempts at classification, both well meditated and yet very different in the result, by R. Allers, 'Microcosmus from Anaximandros to Paracelsus', *Traditio* 2 (1944), 319–407, and M. Kurdzialek, 'Der Mensch als Abbild des Kosmos', in *Miscell. Med.* 8 (1971), 35–75.

can with plausibility find a direct influence of its thought upon his scientific interests. Moreover, he received his first education in an age which played incessant variations on the microcosmic theme. His later-acquired Aristoteleanism was never doctrinaire, and despite his admiration for the Philosopher he was prepared not only to complete him, but to modify such basic notions as, for instance, the nature of the quintessence. He was avid of Neoplatonic ideas. He produced the only original cosmogony of the later Middle Ages, and the double concern which invigorated this speculative enterprise was sympathetic to microcosmic speculation: the passion to grasp the whole of the physical cosmos in its moment of generation and the desire to translate the creation-narrative into a philosophical prose, and one of which the Fathers would have approved. Lastly, we may mention his preoccupation during his later years with the assimilation of the Greek patristic tradition, which, from Gregory of Nyssa through Maximus the Confessor to John of Damascus, preserved a rich seam of microcosmic speculation. One might expect all these interests to catalyse in a single reaction; it is with a sense of excitement that one discovers an achievement which transcends the expectation.

I *MINOR MUNDUS*: A SYMBOLIST EXEGESIS

Published among the philosophical *opuscula* of Grosseteste is a fragment of fourteen lines, entitled, '*Quod homo sit minor mundus*'.[5] If it really is from his hand and is not simply an excerpt from something he had read,[6] it gives us a point of departure for the microcosmic speculations of his maturity. I render it as follows:

God, in himself great, made man for himself. The human body consists of flesh and bone. Now it is divided into the four elements, for it contains a portion of fire, air, water, and earth. The nature of earth is in its flesh, that of water in its blood, of air in its breath, and of fire in its vital heat. Moreover, the fourfold division of the body represents the forms of the

[5] Baur, *Werke*, p. 59.

[6] In one of the three MSS listed by Baur (ibid., p. 77*) it is made into a prologue to the *De Statu Causarum*. Such a positioning is as inherently unlikely as Baur's own suggestion, that it is the final paragraph of the *De Luce*; but it may well indicate the embarrassment of a copyist as to know what to do with it. Perhaps it is best regarded as one of the *cedulae* with which Grosseteste worked, jotting down ideas as they occurred to him and waiting to use them on a good occasion.

four elements, for the head is borne towards the heavens and has two lights, as it were sun and moon, while the breast is linked with the air, the one sending forth its breath as the other the breath of its winds. The belly is likened to the sea, the collection of its humours being like the 'gathering of the waters', and the extremities, finally, are compared to the earth, being, like it, arid and dry.

The paragraph speaks with the genuine accent of the early twelfth century. Perhaps Grosseteste lifted it bodily from a source. Certainly each phrase of it could be paralleled from the literature before 1150, as the following example illustrates:

Grosseteste: Ratio terrae in carne est, aquae in sanguine, aeris in spiritu, ignis in caliditate vitali.	*Hon. Augustodunensis*: Habet namque de terra carnem, ex aqua sanguinem, ex aere flatum, ex igne calorem. *Elucidarium*, I, 11 (*P.L.* 172, 1116).

The form of microcosmism implied is elementaristic (the same elements constitute man as make up the world), with no evidence of a structural analogy (the same arrangement, order, law to be found in both). The scientific content is, however, so meagre, and the lack of discrimination between causality and signification so complete, that the physical is quickly dissolved into allegorical symbolism, to produce an air of wonder. It is the body, not the person, which is a microcosm of the world, in conformity with the twelfth-century tendency to build a microcosmism of either body or soul, but not both. The pictorial *naïveté* of the symbolism, whereby the human frame is divided into head, breast, belly, and extremities, recalls the favourite theme of quadripartite man and its graphic illustrations: four parts of man, or humours, or members, symbolize the four corners of the earth, the four elements, the four winds . . . There is nothing in this poor baggage that the meagre resources of the Latin tradition could not have provided[7]—no trace of Greek physics, Arabic medicine, or Cappadocian theology, the three transforming influences on later microcosmic speculation.

A later and more critical period would make use only of such microcosmic ideas as survived the winnowing wind of philosophical criticism. The exuberances of the twelfth-century imagination, which were largely untrammelled by scientific

[7] See, e.g., Isidore, *De Natura Rerum*, c. 11, 'De mundo' (*P.L.* 83, 977); *Etymologiae*, IV, 5, 3; Bede, *De Temporum Ratione*, c. 35 (*P.L.* 90, 457 ff.; ed. Jones, Cambridge Mass., 1943, p. 246).

knowledge, were not to survive attempts at transplantation into the different climate of the following age. In many ways, Grosseteste is the figure in whom the transition from the twelfth to the thirteenth century is most visible, and it is fascinating to follow him as he hesitates, decides, criticizes, and constructs. If by 1240 or 1250 a consensus had formed that many of the microcosmic speculations of the previous century had been naïve and hasty, the credit goes largely to him for his pruning of the more exaggerated symbolic outgrowths and for the reserve he increasingly felt about the hypothesis of an animated world and heavens.[8]

II THE WORLD-SOUL

Though the conception of a World-Soul had always been known to the Latin theologians through Chalcidius, Macrobius, Seneca, and Augustine, it gained few adherents until about 1120. The reading of the *Timaeus* fertilized minds already rendered receptive by a twofold aim: to produce a philosophical *hexaemeron*, and to express the partnership of a regular and autonomous nature with man, whom she enfolds. The cosmic religion of the *Timaeus* fed both initiatives for a period of a half-century.[9] Almost as soon as we meet the hypothesis of the World-Soul, in Abelard, it is at the centre of theological controversy. Seduced by the perennial Christian temptation to harmonize biblical faith and Neoplatonic spiritualism, the 'stormy petrel' identified the *anima mundi* with the Holy Spirit, who indwells all creation[10]— a speculative *démarche* which was promptly condemned by the Council of Soissons.[11] William of Conches,[12] Thierry of Chartres,[13] and

[8] Since the idea of a World-Soul is reached by extrapolation from the human constitution to the nature of the visible cosmos, it represents a form of microcosmism which we may call anthropocentric. On its historical development, see Cornford, *Plato's Cosmology* (London, 1937); Allers, art. cit., pp. 323, 351-67; Kurdzialek, art. cit., pp. 37-43, 62-5; Chenu, *La Théologie au XIIᵉ siècle*, pp. 34-43; Kranz, *Kosmos* (Bonn, 1958), pp. 167-74; D'Alverny, 'Le cosmos symbolique du XIIᵉ siècle', in *Arch. Hist. Doctr. Litt. M.A.* (1953), pp. 69-81; Tullio Gregory, *Anima Mundi. La filosofia di Guglielmo di Conches e la scuola di Chartres* (Florence, 1955).

[9] Chenu, *La Théologie*, pp. 21-34.

[10] *Theologia Summi Boni*, ed. H. Ostlender, *BGPM* (Münster, 1939), Bd. 35/2, p. 13; *Theologia Christ.* I, 4 (*P.L.* 178, 1156).

[11] Allers, art. cit., p. 359.

[12] *Glosae super Platonem*, ed. Jeauneau (Paris, 1965), pp. 415 ff.

[13] *De Septem Diebus*, ed. N. Häring, *Arch. Hist. Doctr. Litt. M. A.* 22 (1955), p. 193.

Bernardus Silvestris[14] all believed in the existence of a World-Soul, which brings order everywhere in the visible cosmos, as the soul does in the body. At the end of the century, long after the sharp condemnations of the pagan idea by Cîteaux and St. Victor, Alain de Lille was careful to disguise the same Platonic notion in the dress of *Natura*, personified as the unifying moving-principle of a living universe.[15] The attraction of the hypothesis lingered in the early years of the thirteenth century, given new force by the Avicennian idea of the animation of the spheres. Thus William of Auvergne, whose cosmology manifests a strong influence of the *Timaeus*, made a place for the World-Soul.[16]

I have found eight references to the hypothesis in Grosseteste, testimonies to an evolution of considerable moment in his thought. Putting them in an order which has a strong chronological plausibility, we can begin with the *De Sphaera*, which identifies the *anima mundi* as the efficient cause of the diurnal motion of the heavens.[17] In the *De Motu Corporali* and *De Motu Supercaelesti* (which must be taken as a pair), Grosseteste reveals an awareness of the historical and philosophical similarity between the moving soul of Plato and the first Unmoved Mover of Aristotle.[18] The First Mover of *Metaphysics XII* moves as final cause, *'desideratum et intellectum'*; what then of the efficient cause of celestial motion? The heavenly bodies, Grosseteste responds, are animate, but not, of course, sensate. Their souls possess a single faculty of knowledge and desire of the good.[19] The moving substances or souls of the

[14] *De Mundi Universitate*, ed. Barach-Wrobel (Innsbruck, 1876), 31, v. 68; see Gregory, *Anima Mundi*, pp. 80 ff.

[15] Chenu, *La Théologie*, pp. 30 ff.

[16] Gilson, *A History of Christian Philosophy in the Middle Ages*, p. 256.

[17] 'Super hos duos polos, ut diximus, circumvolvitur coelum cum omnibus stellis et planetis qui sunt in eo motu aequali et uniformi per diem et noctem semel, cuius motus causa efficiens est anima mundi.' (Baur, *Werke*, p. 13. 32-5). Since Grosseteste would not have committed himself to this hypothesis as late as 1236, Sarton's attempt to make the *De Sphaera* depend upon Sacrobosco's work of around that date fails; with Baur, we would postulate a much earlier date of composition, *c.*1215-20. For the references to Sarton and Baur, see Thomson, *The Writings*, p. 115.

[18] *De Motu Corporali*, ed. Baur, p. 91.12.

[19] 'Ex hoc manifestum est quod haec corpora sunt animata, et non habent de virtutibus animae nisi virtutem intellectivam et desiderativam, quae movet in loco, et haec est quae immediate facit motionem in ipsis. Motor enim qui movet per modum desiderati et intellecti, sicut dictum est, est virtus separata.' (ibid., p. 94. 27).

heavenly bodies act as conjoined movers, for only the First Mover is separate or abstract.[20] They are only formally distinct from their bodies.[21] The number of these souls is to be calculated by the number of the bodies and of the motions which they produce. The relative speed of their motion depends upon the proportion between each soul and its body.

It is not easy to say whether Grosseteste is merely reporting Aristotle's views or making them his own. To establish the evolution of thought of which we have spoken, however, we need only acknowledge two evident facts. He speaks no longer of a single World-Soul but of a plurality of celestial souls. In fact, in his later works the *anima mundi* in the strictly Platonic sense is mentioned only once more, in a dialectical passage concerning the nature of time (in the notes on the *Physics*), and it is treated purely as an hypothesis.[22] In the second place, Grosseteste feels (*c*.1230) no strong philosophical unease with the animation-theory, and its attractive force is sufficiently strong to form a category of his interpretation of Aristotle. He finds it natural to think of the Aristotelean deities as animating souls, as it were, the Platonic soul multiplied and individuated to realize a new distribution of functions, such as Avicenna had imagined to be appropriate to the souls which emanated from the Intelligences.

We come lastly to a group of three texts, all certainly written in the decade between 1230 and 1240, in which Grosseteste's attitude to the animation-theory is more reserved, even sceptical, than could have been foreseen from his writings up to that date. In the sermon, *Exiit Edictum*, the human soul is represented as being a microcosm of all spiritual being, sharing in intelligence with the angel; and, he adds, if the heavens are animate with one soul or more, 'as certain philosophers maintained', then the human soul will share in its (or their) nature also. Such a soul, or such souls, would assure the regularity of celestial motions and would be personally united with the bodies they animate, just as the human soul is with the body. Grosseteste makes the hypothesis deliberately wide, to include

[20] Ibid., p. 100.
[21] Ibid., p. 93. 15 ff: 'Sed sicut ostensum est in VIII [*Phys.*, VIII, 4] intentio eius quod est motum a se est quod motus componatur ex motore et moto, motore distincto a moto secundum designationem et non secundum esse; non est abstractus sed coniunctus.'
[22] See ed. Dales, pp. 95. 25-96. 5.

the speculations of Plato, Aristotle, and Avicenna, the ideas, therefore, with which he had been able successively to identify himself; but now he keeps his distance.[23] His procedure in the *De Operationibus Solis* is identical: certain philosophers, he remarks, postulated a living principle of heavenly motion, which might be a soul of the heavens, or again might not . . . but beyond subjunctives he is not prepared to go.[24]

The motivation for his reserve is made clear in a passage of the *Hexaemeron*:[25] he felt the weight of patristic opposition to the idea. Damascene, he points out, knew very well how much effort philosophers had expended in order to prove that the heavens have a soul or a plurality of souls, united to the celestial bodies or separate from them, as moving intelligences; yet he says that the heavens are inanimate and insensate and that no one should think of them otherwise.[26] There is no support for the animation-hypothesis in Scripture. If the heavens are said to rejoice, exult, bless the Lord, etc., this is merely by way of prosopopoeia, and has no philosophical implications. The case of Augustine he finds singularly instructive, for having been tempted by the Platonic hypothesis in the *Enchiridion*, he retracted what he had written as *temere dictum*, not affirming its falsity, but rather finding no rational or scriptural support for its truth.[27] Jerome's voice is loudest of all; he includes the animation of the heavens in his enumeration of the errors of Origen. Having established the clash of opinion between the philosophers and '*auctores tanti*', Grosseteste, surprisingly, still hesitates. He bewails his ignorance and refers to the limits of human knowledge concerning the universe; but still he does not definitively reject the philosophers' notion. It is not

---

[23] See ed. Unger, p. 22.

[24] *De Operationibus Solis*, § 4, ed. McEvoy, p. 67: 'Si vero secundum intentionem quorundam philosophorum caelum primum haberet intelligentiam et vitam rationalem, et moveret primum caelum intelligendo moventem se (sive solum praesidentem, sive unitam ei in unitatem individualem); et moveret caelum desiderando primam causam, cui appetens assimilari moveret caelum ad participationem omnis situs sibi possibilis, satis convenienter hic posset dici species caeli vita illa rationalis movens caelum per formam caeli corporalem.'

[25] Oxford, *MS Bodl. lat. th. c. 17*, fol. 214^d.

[26] Grosseteste quotes literally from the *De Fide Orthodoxa* (ed. Buytaert, p. 83): 'Nullus autem animatos caelos vel luminaria existimet; inanimati enim sunt et insensibiles.' Cf. Basil, *Homilia III in Hexaemeron*, 9 (*P.G.* 29, 76 AB).

[27] In the *Retractationes* (I, 2) Augustine refers, not to the *Enchiridion*, but to the *De Immortalitate Animae* 15, 24 (*P.L.* 32, 1033). Cf. also *De Musica*, VI, 14, 43.

unlikely that Augustine's attitude weighed more with him than that of the others: where Scripture is silent no complete certainty is attainable, and the arguments of the wordly philosophers do not justify a definite conclusion in the present case. The implication is strongly present throughout the passage that divine omnipotence cuts all *a priori* dogmatism away by the roots and leaves no basis for deductive arguments concerning the nature of heavenly motion.[28]

What was probably Grosseteste's final word on the whole matter is found in the question, raised in his commentary on the *Celestial Hierarchy*, as to whether the angels differ in species.[29] One possible *differentia* could be instanced in the case of the intellects which move the spheres. If they are conjoined movers, united to their spheres as souls to bodies, it would follow that such movers would be differentiated by their bodies, each of which differs in species from the other. However, this is not the case, Grosseteste affirms (*pace* Avicenna, the reader wants to prompt).[30] The movers of the spheres are not conjoined to them but are separate substances, and therefore require no specifically different motive-power.

This conclusion brings us to the end of a long process of development, in which Grosseteste personally made the transition from twelfth-century Platonism to thirteenth-century cosmology. He never showed signs of being moved by that quasi-mystical veneration of the World-Soul which some authors had professed, but seems none the less to have accepted the reality of the World-Soul at an early period, to have felt the attraction of the animation-hypothesis consistently, and to have distanced himself from both only after a struggle. His treatment of the question lacks the firmness commonly found in writers of the generation after him, but gains in interest by being the record of a thought-itinerary which those thinkers were spared. He may, besides, have influenced later writers, for Bonaventure quotes the same texts of Damascene, Augustine, and Jerome in refutation of the World-Soul[31] as Grosseteste had done, and adopts the same conclusion concerning the

[28] See Part III, ch. 1, iii c (pp. 188–200).

[29] See above, Part II, ch. 3, ii, pp. 140–45; ed McQuade, p. 240.

[30] Cf. *De Caelo et Mundo, Opera Omnia* (Venice, 1508), fol. 39d; Gilson, *A History of Christian Philosophy in the Middle Ages*, pp. 196–7. For Avicenna, each soul is a species of its own, differentiated by the body which it moves.

[31] *In II Sent.*, d. 14, a. 3, q. 2 (II, 347–50).

motion of the heavens, one which generalizes itself in his own generation to include all the major Scholastics.

III THE EXEGETICAL TRADITION: IMAGE AND MICROCOSM

A casual-seeming remark of St. Gregory came to constitute the *locus classicus* for microcosmism in the Latin tradition (which otherwise had few opportunities to introduce the theme), and was appealed to by exegetes and theologians down to the thirteenth century. Commenting on Christ's instruction to 'preach the Gospel to every creature' (Mark 16:15), Gregory asks whether we are meant to believe that creatures lacking human dignity could hear the word of truth. Evidently not, he answers; it is man who is called 'every creature', since he shares in every level of created reality. He has existence like stones, life like trees, sensation like the animals, and understanding in common with the angels. He is accordingly 'in a sense all things'.[32] St. Gregory further associates this exegesis with the place accorded man in the creation-narrative: all of creation is found in man, since it was on his account that all things on earth were made.

The temptation to amplify the exegesis of Gen. 1:26 by associating it with microcosmic motifs chosen from Platonism was much older than Gregory, indeed, older than Christianity. An anthropocentric conflation of the Genesis text affirming man's lordship over the earth with the idea that he is a summary of all that is, had been effected already by Philo, in whom the tendency had been, however, towards the allegorization of the relations between man and the created world.[33] Origen introduced the *rapprochement* of the two ideas into Christian exegesis, constructing out of their union an initiation into the more central theme of man made in the image of God.[34] In the tradition which was founded upon his exegesis, microcosmism was attuned to emphasize the finality of a creation all the aspects of which point to the coming of man as the perfection and lord of the universe, the image of God.

The close connection between man and the visible cosmos was further reinforced by the context of salvation-history, into

[32] St Gregory, *Homiliae in Evangelium* 29 (*P.L.* 76, 1214).

[33] Philo Judaeus, *Opera*, ed. K. Cohn and J. Wentland (Berlin, 1896 ff.); *De Opificio*, p. 69.

[34] Origen, *Homiliae in Genesis* I, 11, tr. J. Doutreleau, *Sources chrét.* 7 (Paris, 1943), p. 78.

which St. Paul introduced his vivid image of the creation groaning in parturition, while awaiting the revelation of the sons of God (Rom. 8:18-22). Here once again the words *'omnis creatura'* occur; for many medieval writers this coincidence only reinforced the temptation to link the exegesis of all three passages to a common basis and inspiration. Man, it was repeated, was created on the sixth and last day, because all things were made for him and placed under his dominion. He contains all other creatures at some level of his being, and they were made to serve him so long as he remained obedient to God; with his fall the entire creation lost its wholeness and finality.

Almost every mention of *minor mundus* in Grosseteste's writings reveals him as heir to this long Christian exegetical tradition. It was second nature for him to associate 'man for whom all other things were made', with *omnis creatura*, the Gospel text,[35] and similarly with the idea of the disorientation introduced by the fall into the whole cosmos—a major theme of *De Cessatione Legalium*. There can be little doubt that he thought of the relationship between man and 'all creatures', with the fully-deliberate ambivalence given that expression in the Gregorian tradition, as being a matter of revelation, so clearly did he find it affirmed in Mark, Genesis, and Romans. In his case the microcosmic development of the Genesis narrative was enriched and deepened by his avid reading of the Greek Fathers. Basil, Gregory of Nyssa, the *De Structura Hominis* of Pseudo-Basil, and Damascene brought him a range of themes which the narrow and somewhat formalized resources of the native Latin tradition could not equal. The Western tradition, on the other hand, had its own wealth, for although Augustine had paid no attention to microcosmic themes as such he had bequeathed to his successors a rich theology of the image of God in man. The central interest of Grosseteste's hexaemeral contribution lies in the conscious and thorough-going effort he made to bring the essentially Greek theme of microcosm and macrocosm into the mainstream of Latin thought and thereby to unite the Eastern and Western

---

[35] *Hexaemeron*, MS cit., fol. 233ᵈ: 'Secundum autem quod habetur in littera die septimo intelligitur allegorice quod deus die septimo complet opus suum, id est hominem, propter quem facta sunt cetera, et qui etiam in evangelio dicitur omnis creatura, quando transfert eum de labore huius corruptibilis vitae in septimam etatem quiescentium.' Cf. fol. 227ᶜ, a comment on Gen. 1: 26.

theologies of man's place in creation. In his later years, Grosse-teste never spoke of man as the microcosm without returning to the great themes of the creation-narrative: man, the image of God and ruler of the earth, is man in his vicarious nature, *minor mundus*. Though he used a variety of microcosmic themes he had no difficulty in holding them in harmonious unity, indeed, the difficulty is for the reader to hold them apart and to explicitate the invisible transitions within the motifs which interweave in his exposition.

The discussion that follows takes these themes one by one, roughly as they occur in Part VIII of the *Hexaemeron*, and draws remarks made elsewhere by Grosseteste into this frame-work. It is forced to make as little separation of philosophy and theology as Grosseteste himself did. No more than St. Bonaventure, the great synthesizer of the medieval micro-cosmic themes, did Grosseteste have a philosophy of *minor mundus*; the philosophical elements contained in his doctrine —such as the body's representation of the whole visible cos-mos, and man's mediate station between sensible and intelli-gible orders—were never meant to stand alone, and his vision must be recaptured in its wholeness.

a) *Imago Dei*

'*Faciamus hominem ad imaginem et similitudinem nostram.*'[36] Grosseteste wishes to base his whole theological anthropology upon the theology of image, but he has no fear of exhausting his theme. Image is a brief word, but if its significance could be teased out in all its implications, 'the world could not hold the books which would be written'. 'Let *us* make', implies that the triune God is speaking, wherefore his 'image' contains the most secret thing in God's life and the most sacred in man's, his highest dignity. Of course, it is by imitation and not by equality that man is like to God. Still, to explain what that likeness is would demand unfolding all that the triune God is, and finding God's qualified imitation in man. God is 'all in all': the life of all living things, the beauty and form of all that is lovely; his image will likewise be in some way all things.[37] To explain the image would at the limit demand an absolute

---

[36] See ed. Muckle, p. 157. All references to Part VIII of the *Hexaemeron* are to Muckle's edition. The section printed covers fols. 224a-227a in the Bodleian MS.

[37] 'Quapropter et homo in hoc quod ipse est imago Dei, est quodammodo omnia', Muckle, p. 158.

knowledge of God, creation, and man, not only in their natures, but in the relation of each term to the other two, but such knowledge is not given to man. It is with a profound sense of mystery, therefore, that the writer will give a grain of the sand, a drop of the rain, an atom of the world's machine, in place of the total knowledge which only God possesses.

Grosseteste has manifested his purpose in the first lines of his treatise. He has forged a direct link between man as image, 'representing' by imitation all that is in God, and man's as being the representative nature that includes 'somehow' all things. In what follows he can pass from the one form of representation to the other with perfect logic, by invoking the coincidence of both in a single term.

Exempla, reflections of the Trinity,[38] can be found everywhere in creation, but the light they bear to the mind varies in intensity, according to the grades of creaturely being. A structural analysis of Grosseteste's text reveals that he is working according to a carefully-constructed plan, though, typically enough, he does not announce it but leads his reader's attention to the matter rather than the form. It is human nature which is made the axis of the ascent towards the Trinity, the pivot between matter and spirit. The sensible world exhibits vestiges of the Trinity in a series of triads that are to be found in each thing: matter, form, and composition; magnitude, species, and order; number, weight, and measure,[39] reveal the attributes of power, wisdom, and goodness, which are appropriated to the three Persons. However, it is light that provides the most manifest vestige among material creatures, as it begets splendour and warmth in its indivisible substance.[40]

There is a point where matter meets the immaterial. In sensation, a material form generates a species in a sense-organ, and both form and species are united in a judgement by the purely immaterial intentionality of the soul. In one of his major texts on the powers of knowing Grosseteste follows the triadic vestiges of the Trinity through sensation to the imagination, and with that to the threshold of purely spiritual life (*Hexaemeron*, Part VIII, Chapter Four).

[38] Grosseteste refers to them as proofs (*Hexaemeron*, ed. Muckle, p. 162, and *Dictum 60*, ed. Gieben, *Franc. Stud.* 24 (1964), p. 149).

[39] The latter two triads are not, of course, restricted to sensible being, but are transcendentals of the created order; cf. *Dictum 60*, ibid., p. 153.

[40] *Hexaemeron*, ed. Muckle, p. 163.

It is in the powers of the soul, *intelligentia, memoria,* and *amor,* that the highest exemplification of the Trinity is to be found, for this *suprema facies animae* requires no material organ, as the lower faculties of cognition do.[41] As a natural endowment it makes man a natural image of God, even before there is mention of the grace which assimilates him to the divine nature. Following Augustine, Grosseteste distinguishes *imago naturalis, reformata* and *deformata.*[42] The natural image consists in rationality and liberty, and is never lost; the renewed is deformed by sin and restored by the Holy Spirit.[43]

This account of the ascent from material vestige to spiritual image calls for a word of comment.

In passing from matter to spirit, the one a vestige, the other a full image of the Trinity, Grosseteste situates man at the junction of material and spiritual creation. So much is made clear at the transition-points from material things to the conjunction of bodily and spiritual being that is effected in sensation, and thence to the intellectual acts, performed by the soul alone. The microcosmic dimension, latent in the passage examined, is explicitated later in the same section,[44] when he quotes from the *De Hominis Opificio* of Gregory of Nyssa: man is situated between spiritual and material being, and, by the association of matter with spirit, a fitting blend of nature is formed in the last being created.[45] In Grosseteste's own words, man is the most admirable creation produced by the divine wisdom, for in him are joined in personal unity the highest creation, namely, free intelligence, and the dust of the earth. The union of things so distant in nature arouses in us wonder at the power of God.[46]

[41] See ed. Muckle, p. 164: 'Et ita secundum hanc supremam virtutem unam et simplicem dicto modo memorantem, intelligentem et diligentem est homo summa similitudo et per hoc imago unius Dei Trinitatis.'

[42] *De Trinitate* XIV, 14, 18 (*CCSL* LA, pp. 445-6).

[43] *Hexaemeron,* ed. Muckle, p. 166.   [44] Chs. 10-11, ed. Muckle, pp. 172-3.

[45] *P.G.* 44, 145B ff. Grosseteste quotes from the translation of Dionysius Exiguus.

[46] *Hexaemeron,* ed. cit., p. 172: Man is 'maxime inter cetera opera admirabilis, coniuncta est enim in homine in unitatem personae suprema creatura, rationalis videlicet et arbitrio libera intelligentia, cum creatura infima, videlicet . . . cum pulvere sumpto de terra . . . Et quid tam distancium coniunctione artificialius aut mirabilius potest excogitari?' See the same idea in St. Bonaventure, *In II Sent.* d. 1, p. 2, a. 1, q. 2, f. 1 (II, 41b), where he gives *rationes congruentiae* for the intermediary place of man, as manifestation of the power, wisdom, and goodness of God.

The idea thus expressed was probably the most widespread microcosmic theme in the later Middle Ages. Its authorities were good. The *De Natura Hominis* of Nemesius (attributed by the Latins to Gregory of Nyssa or Basil) gave it perhaps its clearest formulation,[47] but the doctrine that man stands at the confines of the intelligible and sensible worlds was found also in St. Gregory Nazianzen and Maximus the Confessor, and was finally canonized by inclusion in the *De Fide Orthodoxa*.[48] Scottus Eriugena was apparently the first to make use of the idea in the Latin world, and his rich language of microcosmism reveals the central place it held in his thought: man is *medius*, *medietas atque adunatio*, and *copula*, as well as *officina omnium*.[49] The idea of man as the link of creation was reinforced for the Scholastics of the twelfth and thirteenth centuries by the concurrence of almost all the Neoplatonic sources they encountered, including (in addition to the patristic ones mentioned) the *Liber de Causis* and Asclepius. The Latin vocabulary, expanding under these influences, appealed to such descriptions as '*horizon et confinium spiritualis et corporalis naturae*'.[50] The terms *confluxus*, *concursus*, *compositio*, *copula*, and *connectio* also assumed a cosmic dimension, to render the unification of the created order which human nature effects.[51]

It is also worth remarking that the step by step ascent from matter through the senses to the mind is something Grosseteste shared with St. Bonaventure's *Itinerarium*, written some thirty years later. In itself, of course, this is not surprising, for both were disciples of Augustine, schooled by him in finding their way from the *regio dissimilitudinis* to the domain of

---

[47] Trans. in A. Schaefer, 'The position and function of man in the created world according to St. Bonaventure', in *Franc. Stud.* 21 (1961), p. 270, from the text printed in *P.G.* 40, 512B.

[48] *De Fide Orthodoxa* II, 2 and 12; ed. Buytaert, pp. 68, 112. Man is 'quaedam copula visibilis et invisibilis naturae.'

[49] *De Divisione Naturae* II, 4 and III, 7 (*P.L.* 122, 530D, 733B). Cf. *Homélie sur le prologue de Jean*, ed. Jeauneau, *Sources chrét.* 151, pp. 336–8, and the texts of the sermon and the *De Divisione Naturae* there examined. This homily is frequently quoted by Grosseteste as a work of Chrysostom. The *Ambigua* of Maximus was Scottus's chief source for the microcosmic theme, as his language and frequent quotations attest (cf. *P.G.* 91, 1093D–1096A). See the fine observations of T. Gregory in *Giovanni Scoto Erigena* (Florence, 1963), pp. 34–42; and on Maximus's ideas, L. Thunberg, *Microcosm and Mediator. The Theological Anthropology of Maximus the Confessor*, (Lund, 1965), pp. 140–51.

[50] St Thomas Aquinas, *In III Sent., Prologus.*

interiority, where God may be directly encountered. It is interesting that St. Bonaventure's first three chapters are explicitly microcosmic in orientation,[52] yet differ from Grosseteste's exploitation of similar motifs. The genius of Bonaventure was to produce a form of noetic microcosmism which, while it owed something to both Aristotle and Augustine, was his own creation. The sensible world, or macrocosm, enters the microcosm through the gates of the five senses and does so moreover in its totality. The identity of physical and sensible qualities postulated by Aristotle as the foundation of sense-knowledge is made to rejoin the Augustinian parallelism of the five elements and the five senses.[53] The active emotional accompaniment of sensation, *oblectatio*, the affective response to colour, sound etc., is likewise a macrocosmic referent. In the *maior mundus* thus microcosmically assimilated and interiorized as a totality, man can read the first vestiges of God's presence. By reflecting upon the conditions of *scientia certitudinalis* the mind finds within itself the image of the Knower, the morning light promising the noonday brightness. If, physically speaking, man is a product of the macrocosm, this causal relationship takes a dynamic turn in the openness of that natural light, which man is. All the elements of this noetic development of microcosmism were, of course, already present in Grosseteste's work, but were never drawn together there into the single constellation they present in Bonaventure's little masterpiece.

Since the natural image was located in the spiritual faculties of man, the symbolic motifs which had decorated microcosmism in Platonism and Stoicism remained the natural expression of the Christian exegetes. Reason alone places man above the animal, which is often so much better endowed than him at the level of sensation.[54] Yet the transcendence of man is already apparent in his bodily perfection, which manifests his superiority over the beasts. All living creatures are inclined to the earth, whereas man's upright stature signifies his

---

[51] Cf. the quotation from Pierre de Tarantaise reprinted in A. Schaefer, 'The Position and Function . . ., (Washington, 1965), p. 45 n.

[52] *Itinerarium* I, 5; II, 2–3; III, 1.

[53] For the references, see our study of St. Bonaventure, mentioned in n. 3 above.

[54] *Comm. in Hier. Cael.*, ed. McEvoy, p. 162: 'Habet quoque homo respectu irrationalium animalium minimum secundum sensum, quia sunt plura de illis quae homine acutius vident, clarius audiunt, melius olfaciunt et sapores discernunt et subtilius tangunt . . .'.

godliness and his eternal spiritual destiny.[55] The eyes, directing their gaze upon the heavens, are a symbol of contemplation of the eternal.[56] Symbolism merges imperceptibly into moralization, as the stature of man, indicating his place between beast and angel, is made to symbolize the call to turn from the lower world to contemplation and become an angel rather than a beast.[57] The effect of microcosmic considerations, here as elsewhere, was to highlight reason and nobility, the attributes of the interior man, and to focus all of creation's value upon his liberty, on the use of which the fulfilment of the cosmos hangs.

The Genesis narrative made man the last thing created, and offered Christian exegetes a further opportunity to celebrate his dignity as the image of God. Grosseteste remained in a long tradition when he repeatedly associated man as image with his late formation (*ultimus*), his vicarious nature (*omnis creatura*, *creatura mundi*), his lordship over the earth (*dominus et rector*), and his centrality within creation (*finis omnium*). All of these attributes are predicated in terms of a stricter logic than appears at first sight, for each of them is possessed by man precisely as the highest likeness of God, the *summa similitudo* of the divine nature. They are imitations of the divine being on the part of a creature, as Grosseteste declared, in a daring statement, to the effect that all that we can attribute to God befits man also, as the *summa et propinquissima similitudo*[58]—a claim which did not risk pantheistic confusion, because it remained always qualified by the logic of analogy: nothing can be predicated univocally of God and the creature.

[55] *Hexaemeron*, ed. Muckle, p. 166. The oldest expression of the idea in philosophical literature is found in *Timaeus* 90A, but it owed its classical formulation to Ovid, *Metamorphoses* I, 84-6.

[56] *Comm. in Hier. Cael.*, p. 160: 'Habet etiam homo virtutes visivas corporeas ad superius, id est ad superna conspicienda apta in corpore sitas. Habet quoque homo figuram corporis rectam et directam . . . in quo gerit typum angelorum . . .'.

[57] *Hexaemeron*, fol. 223[b]: 'Erecta enim statura soli homini servatur. . . . Quapropter homo, si corporis voluptate fedatur obediendo luxui ventris et inferioribus eius partibus, comparatus est iumentis irrationalibus et simul factus est illis . . .' Cf. *Comm. in Hier. Cael.*, ed. McEvoy, p. 162: 'in quo gerit homo typum angelorum, secundum quod ipsi sensibus carent corporeis, qui in hoc debet secundum sibi possibile convenire cum angelis, videlicet quod aversus a sensibus et sensibilibus, totus convertatur ad intelligibilia.'

[58] *Hexaemeron*, ed. Muckle, p. 168: '. . . quaecumque dicuntur de Deo aliquo modo imitatorio etiam homini congruunt . . . Unde etsi in aliis creaturis luceat aliqua Dei similitudo, nec tamen elucet in illis Dei imago, quia imago est summa et propinquissima similitudo.'

The ideas referred to, of man as the last creature made, as 'all things' and as lord and centre of the universe, generate a strong mutual attraction which renders their separate analysis difficult; the circumincession of each idea within the others is meant to lead the mind back again and again to an enriched idea of the image of God. We are forced for the sake of clarity to dismember the group and to treat the ideas one by one.

b) Ultimus

It was because of his dignity that man was made last, for he has lordship over creation.[59] If his creation was reserved for the sixth day, that can only have been because the work of all the preceding days was required to constitute human nature, for all other things are found at some level of man's composite being. It was fitting that he should appear last, as lord and master, but the natural order of things assured this in any case and it could not have been otherwise. Human nature presupposes all other creatures. It is as though God made individual editions of his wisdom and afterwards edited them to produce a *summa* containing the contents of them all.[60] The first day's work saw the creation of spiritual light, the angels, and the suceeding days the production of the material universe through all its kinds, down to the dust of the earth. From these two opposites, spirit and matter, a last being could be made who contained highest and lowest in creation,[61] and who, therefore, had a unique manner of being. By being a union of extremes, man's was the median being, his nature a summary of all the species of natural things, with all of which he communicated at some level of himself.

Such a nature is a task, as Grosseteste insists when drawing out the moral and spiritual senses found in the letter. Man is made of highest and lowest, of noblest and vilest natures, that the consciousness of his worthier part may forbid contempt for what is lowly in him and despairing abandonment to the vices of the flesh. From the corresponding sins of the spirit,

[59] Ibid., p. 173. Cf. *Comm. in De Div. Nom.*, ed. Ruello, p. 152, § 48, where Grosseteste attempts to find the same idea in Pseudo-Dionysius.

[60] *De Confessione*, ed. Wenzel, p. 240. Grosseteste's expressions, 'man needs all things', and 'is taken from all', correspond to Eriugena's term *officina omnium*, a happy metaphor which, however, seems to have gone out of currency by the thirteenth century. On its meaning, see Ph. Delhaye, *Le Microcosmus de Godefroy de Saint-Victor. Étude théologique* (Lille, 1951), p. 160.

[61] *Hexaemeron*, ed. Muckle, p. 172.

self-elation, ambition and presumption, he is in turn dissuaded by the consciousness of his humbler parts, wherefore a realistic self-awareness guides and equilibrates him in the middle ground between self-elation in pride and pusillanimous self-abasement.[62] Moreover, the ascending sequence of natures that is present in his median being, from the organic through the sensitive levels of his nature, invites him (as the Bible does too, by calling him 'the last of creatures') to philosophize on the state of his soul,[63] his noblest part. Once again we find a microcosmic theme leading towards interiority, the natural atmosphere of an Augustinian mind: *'homo namque simpliciter est homo interior'*.

c) *Quodammodo omnia*

It is because he is the image of God that man is in some way all things, and therefore the last, the summary of creation.[64] From the Augustinian theology of image Grosseteste derived that central form of microcosmism which we may call elementaristic. There can be no doubt that he considered this to be the basic content and meaning of the *minor mundus* theme, for it is developed at length on a number of occasions and merits references throughout his works.[65] His treatment of it is lent vigorous interest by the wide philosophical background it presupposes, and it repays closer study.

The briefest statement of elementaristic microcosmism is, that man shares the natures of all other creatures; his being is a representative image of creation.[66] As spirit, he shares the nature of the angels; the soul's only specific difference from

---

[62] *Hexaemeron*, MS cit., fol. 238ᵃ. (There are two fols. numbered 238; see the first of these).

[63] Ibid., ed. Muckle, p. 174: 'Quod igitur novissimum post omnia factum hominem scriptura commemorat nihil aliud quam de statu animae philosophari nos debere latenter informat, necessaria quadam rerum consequencia, id quod perfectum est in postremis insinuans. Natura etiam rationalis cetera quoque continet, et incrementa germinis, et sensus animantis . . . consequentia igitur naturae veluti per gradus quosdam dico proprietatum ab inferioribus ad perfectiora conscendit.'

[64] Ibid., p. 158: 'Quapropter homo in hoc quod ipse est imago dei est quodammodo omnia.'

[65] The *De Cessatione Legalium*, the sermon *Exiit Edictum*, *De Confessione*, and *Hexaemeron* are the main *loci*.

[66] 'Homo communicat in natura cum omni creatura', *De Cessatione Legalium*, ed. Unger, p. 14; 'homo, in cuius natura aliquo modo continetur omnis alia', *Ex Rerum Initiatarum*, ed. Gieben in *Coll. Franc.* 37 (1967), p. 128; 'Homo minor mundus dicitur, quia participat per modum aliquem omnibus naturis mundi maioris.' *Exiit Edictum*, ed. Unger, p. 20.

them is found in its capacity and natural desire for union with the body.[67] The powers of the soul, however, extend to the animal and vegetable functions of life, and its incarnation roots it in the world of growth and perception. It shares all the functions of plant and animal life and is a microcosm of cosmic and spiritual life, in all their forms. The human body is the highest product of cosmic nature and shares the nature of all elementated things.[68]

The sources used by Grosseteste to develop his theme are multiple. The idea that the elements of nature are also those of man can be traced back in the history of ideas to Heraclitus, who declared that man is made of the cosmic elements, fire, water, and earth—fire representing consciousness and being identified with the one wisdom which pervades all things. Elementaristic microcosmism is a highly speculative achievement, proclaiming as it does that all natures or elements found in the cosmos universally turn up also in the many-levelled being which man is.[69] In its historical development at the hands of Greek and Christian philosophers it received many specifications[70] and restrictions.[71] Later Western references to elementaristic microcosmism tend in general to restrict the range of its application to the body, paralleling the four elements with the four humours,[72] and by symbolic extension the four

[67] *Exiit Edictum*, p. 22; *De Cessatione Legalium*, p. 14; *De Confessione*, p. 240; *Comm. in Hier. Cael.*, ed. McQuade, p. 241; *Hexaemeron*, ed. Muckle, p. 173.

[68] 'Corpus autem humanum habet communionem in natura cum omnibus naturis corporalibus . . .' (*De Cessatione Legalium*, p. 14).

[69] One of the few faults in Allers's fine article is his tendency to underrate this form of thought. I find it difficult to accept that it has 'no philosophical implications' (art. cit., p. 347), or that it is accorded 'no prominent role in its noteworthiest proponents'. In Greek philosophy, for example, it was found to be hostile to materialism, entering as it did quite naturally into a constellation of ideas already present in Heraclitus: soul–fire–heavens–divinity. If structural microcosmism has more philosophical profundity, as Allers argues, it should not be forgotten that the distinction of elementaristic and structural microcosmism is a historian's analytical device and not an historical datum. It would be difficult to find ancient or medieval texts in which the presence of the elements in man did not also imply in some degree a working of universal cosmic law, or some such structural element.

[70] For example, Empedocles's fixing of the number of elements at four; Aristotle's theory of mixture, with its medical antecedents and centuries-long influence.

[71] Neither the Neoplatonic One nor the Judaeo-Christian God is in any sense an element of reality or a level within human nature.

[72] *Isidori Hispalensis Episcopi Etymologiarum . . . Libri XX*, ed Lindsay, vol. 1 (Oxon., 1911): IV, v, 3; *De Natura Rerum*, ch. 11; *De Mundo*, (*P.L.* 83, 977).

seasons, winds, and corners of the earth.[73] Under the influence
of the Latin Encyclopaedists, the doctrine became something
of a *locus communis* in the twelfth century, although the treatment it received, while progressively reinforced by borrowings
from Greek patristic theology,[74] always exhibited a strong
tendency towards symbolism and allegorization,[75] something
which threatened to erode the cosmological basis of the idea.
The needed corrective was supplied by the recovery of Arabic
medical literature, whose influence is visible already in Adelard
of Bath and Daniel of Morley. In their thought a new consistency and scientific depth extended to the doctrine of the elements and humours, and the assimilation of medical theory
shaded off latterly into the recovery of Aristotelean natural
philosophy.[76] The effects of the developments thus inaugur-
ated made themselves felt in the formation of a more sober,
factual, and rational doctrine of microcosmism, in which the
scientific kernal studied by the *physicus* and *medicus* was distinguished from the looser symbolical associations of the native
Latin tradition.

This evolution can be observed in its completion in Grosseteste's later thought. It can be claimed that his ideas on the
functioning of the human body aimed at respecting the scientific data, while endeavouring to structure them within a
framework of elementaristic microcosmism, a cadre which
enabled him to integrate the resulting theory into a conception
of the whole man, body and spirit, and to put the entirety of
the anthropological doctrine so articulated at the service of
the Christian vision of man.

In a passage whose beguiling simplicity conceals the art
with which it is constructed, Grosseteste presents the physiological side of human nature as being the highest product of
the cosmos, fully dependent upon it, yet representing its unification.[77] The heavenly bodies, he begins, are light and are in

---

[73] Bede, *De Temporum Ratione* ch. 35, (*P.L.* 80, 457 ff.; ed. Jones (1943),
p. 246).

[74] See, e.g., *De Fide Orthodoxa* II, 12, (*P.G.* 94, 925).

[75] Honorius Augustodunensis, *Elucidarium* I, 11 (*P.L.* 172, 1116); Bernardus
Silvestris, *De Mundi Universitate*, ed. cit., p. 53; Hildegaard of Bingen, *Liber
Compositae Medicinae de Aegritudinum Causis, Signis atque Curis*, ed. Pitra,
Analecta Sacra, t. 8 (1882), p. 469.

[76] H. Schipperges, 'Einflüsse arabischer Medizin auf die Mikrokosmosliteratur
des 12. Jhdts.', in *Miscell. Med.*, Bd. 1, p. 140.

[77] The passage turns up in three works in almost identical form: *De Cessatione*

continuity with elemental fire. Fire shares the quality of warmth with air, air, in turn, that of humidity with water, and water that of frigidity with earth. The matter of all the mixtures, from which result stones, metals, plants, and animals, is drawn from the four elements; so the human body, which, being composed of the elements, shares the natures of all bodily things. These natures are continuous from the centre of the earth to the height of the luminous heavens. The entirety of the great world is therefore reproduced in the body.

It requires only a little reflection in order to perceive that an entire cosmology is neatly summarized in these few lines. Aristotle's sublunary world is present in the theory of elementary qualities and their chemical mixtures; but that something more than Aristotelean theory is present, is readily apparent. For one thing, the Philosopher would simply not have admitted a continuity ranging upwards from earth (element or planet, it makes no difference), to the quintessence. Grosseteste can do so only on the basis of his own cosmological theory, with its clear dependence on patristic and Neoplatonic ideas.

The ease with which Grosseteste could claim a natural share for the body in all material reality was something that disappeared with his own generation. So much emerges from the contrast between himself and St. Bonaventure. In his treatise on the human body in the *Commentary on the Sentences* the latter proposed a curious question for discussion: whether Adam's body should have been produced from the purely celestial nature of light (and so conformed to his soul), or constituted by a mixture of the sublunary elements.[78] The double question derives its significance from microcosmism. Preoccupied with the centrality of man in creation, Bonaventure must hold that the body is fully representative of the universe, something it could not be if its components mirrored either the lower or the supralunary world, but not both at once. When he passes to his second question, as to whether the body consists of the four elements, a whole dimension of conflicting loyalties is revealed. Some kind of Neoplatonic communication between the elements and light, in the body as well as in the world, must be established. Yet the *fundamenta* marshalled

Legalium, ed. Unger, p. 14; Sermon, *Exiit Edictum* (ibid., p. 24); and *Hexaemeron*, MS cit., fol. 223^c.

[78] *In II Sent.*, d. 17, a. 2, qq. 1–3.

by St. Bonaventure all presuppose the Aristotelean physics, according to which the quintessence is fundamentally different in nature and activity from the elements and cannot enter into composition with them, having no contrary quality.[79] A mixture is rarefactible and corruptible, and therefore also generable, whereas the fifth element is incorruptible and ungenerable. It cannot form part of human nature in its substance, therefore, but only in its power and through a kind of conformity.

He considers that to deny the quintessence its transcendence runs counter to the received philosophical understanding of the universe, and he finds himself forced to regard the denial as being unworthy of the authority of St. Augustine, which it alleges in its support; Augustine, he opines, both can and should be interpreted otherwise. Bonaventure likewise rejects the weakened form of the same view, according to which light links and conciliates the elements, without actually mixing with them. In discussing one of the key cosmological disagreements between Neo-Platonism and Aristoteleanism, therefore, he opts resolutely for the latter's solution and follows it to a point which threatens to preclude any real elementaristic microcosmism, for the chasm of natures posited by Aristotle between heavens and earth was hostile to such speculations. St. Bonaventure is satisfied that it is sufficient if there is something in the body which resembles the heavens; this he finds, of course, in the spirits, the intermediaries between soul and body.

The differences between Grosseteste's and Bonaventure's conclusions should not, of course, be exaggerated. After all, the body can validly be regarded as the product and mirror of the universe even if it is thought to be composed only of the lower elements, for it is the influence of the celestial bodies which stabilizes and perfects all the chemical mixtures of the sublunary region. The significant difference is that, in the eyes of St. Bonaventure, Aristotle's authority in physics is supreme, whereas Grosseteste experiences no profound conflict in subordinating one of the vital *differentiae* of peripatetic physical theory to a patristic-Neoplatonic theme, which attracted him for a variety of reasons. He understands the Aristotelean doctrine, but feels perfectly free to depart from it.

Had Grosseteste felt the need to answer the Aristotelean

[79] Ibid., a. 2, q. 2 (II, 421b–422a).

objection to his teaching, the means were to hand. The body results from a mixture of the four elements and shares also in the nature of light, which serves as *medium* of the soul's action upon it. Just as surely as its *calor vitalis* derives from a predominance of elemental fire within the mixture, so also do the spirits share in the nature of light, as found in the sun;[80] this is clear from their properties of subtlety, clarity or luminosity, and sublimity. They conform to the definition of light as 'a self-generating bodily substance, highly subtle and close to the immaterial'.[81] The spirits are active, harmonious (that is to say, remote from the contrariety of opposed qualities), and, in the case of the visual spirits at least, luminous and transparent. Grosseteste calls them *lumen*, and even on one occasion *lux pura*. Their situation relative to the body is like that of celestial light with respect to the lower elements: their activity mediates immaterial states of the soul to the nerves and muscles, and finally to the gross members in which the element of earth predominates.[82]

This doctrine of the spirits, heavily indebted as it is to Augustine and Avicenna, finds a natural place within the light-metaphysics. Light is found to be the first form of corporeity in all material things, the cause of their activity and the source of their energy. Viewed in this perspective, all parts of the material universe and all bodies possess a fundamental kinship on the physical level, a dimension that is missing from Aristotelean physics. The increasing grades of activity present in vegetable and animal bodies, which accompany greater complexity of organization, are the work of more concentrated light-activity within progressively subtler and more penetrable matter, in which light dominates ever more easily over the contrariety of the elemental qualities. The highest grade of material organization, the delicate human nervous system of spirits, is a concentrated physical light generated and influenced by the heavenly bodies in virtue of a fundamental

---

[80] *De Operationibus Solis*, § 7, ed. McEvoy in *Rech. Théol. anc. méd.* 41 (1974), pp. 70-1.

[81] *Hexaemeron*, MS cit., fol. 203ᶜ: 'significat enim [i.e., light] substantiam corpoream subtilissimam et incorporalitati proximam, naturaliter sui ipsius generativam . . .' The quality of subtlety is defined in the *Comm. in Post. Anal.*, fol. 25ᶜ, as follows: '. . . subtile est corpus quod contiguatum aliis dividit se in partes graciles acutas et ingerit se per porros minutissimos ipsius donec totum penetraverit; sicut acetum et alia penetrativa non quiescunt donec penetraverint totum cui contiguantur, et ideo dicuntur subtilia.'   [82] See ch. 3, ii (pp. 278-89).

identity of nature, which is only overlain (and not overcome) by their mutual differences in perfection and quality of operation. It can be seen from this upon what a strong philosophical foundation Grosseteste's elementaristic microcosmism rests. We must now consider its broader implications for his view of man's place in creation.

d) *Quodammodo ipse finis omnium*

If man is the last creature, the summary and microcosm of all things made in the first to fifth days, that is because his creation was first in the order of divine intention, which means in turn that he is last, in the plenary sense of 'end'.[83] Paradoxically, this implies that his nature is not only taken from all others but is the exemplar of all others, since they were made in view of his coming;[84] and if that is so, then human nature provides the unity of all creation.[85] There are three aspects of the emanation of things from God, Grosseteste tells us; all three refer to man, and through him are referred back to the Creator. First come the stages or grades which make the universe an order, each of which presupposes and resumes all the preceding ones. Man is the summary of them all. Secondly, one final cause: all things aim at man's coming. And thirdly, the reflection of all things in one is to be considered.[86]

One brief formula employed by Grosseteste to express the final concentration of all things upon man is of special interest. In the course of one of his sermons he refers to '*ipse homo qui creatura mundi dicitur, quia quodammodo ipse finis omnium*', and he adds, somewhat obscurely, '*non ut habetur 2° Physicorum*'.[87] The force of the reserve thus elliptically

---

[83] *De Confessione*, p. 241: '. . . una finalis causa, cum propter hominem facta sunt singula.'

[84] 'Ultimo namque facturam quandam, hominem scilicet, statuit altissimus quasi praedictorum omnium exemplar et ex omnibus acceptam . . . (ibid.).

[85] '. . . ad modum facientis singulas editiones suae sapientiae et in summam unam redigentis.' 'Omnium creaturarum in uno relucentia . . .' (ibid.).

[86] 'Tria ergo considerantur in fluxu rerum: gradus in creatione, quia prioribus omnibus indigent posteriora; una finalis causa, cum propter hominem facta sunt singula; omniumque creaturarum in uno relucentia. Quod enim gradatim processerunt omnia, manifestatio est quod ab uno exierunt; et quod in uno cuncta relucent, probatio est quod per illum a quo sunt stant; quod enim propter unum sunt cetera signum est quod in illud per quod stant tendere debent universa.' (*De Confessione*, p. 241).

[87] The sentence continues, '. . . vel creatura dicitur mundi, quia in eo quodam-

expressed is to underline the distance between the Christian idea, that man is the end of creation universally, and Aristotle's perfectly limited assertion that all artifacts derive their meaning and purpose in different ways from human need. It is not unlikely that Grosseteste was among the first to adopt and amplify the happy formula. Its later currency in the Franciscan School at Paris did not lead to forgetfulness of its Peripatetic origin, but did on the other hand remove it from its original context. St. Bonaventure several times attributes the fully amplified notion of the finality of man to the Philosopher.[88] St. Thomas felt it necessary, perhaps in consequence, to sound a warning note and to recall attention to the limited extension of the original remark.[89]

The centrality of man within creation is a theme of frequent occurrence in Grosseteste's later works.[90] Man shares a single spiritual destiny with the angel, but his way of attaining it is unique and other creatures are means which he can freely use on his way, for they were made by God, not immediately for himself as the rational creature was, but only to return to him mediately.[91] Once again, it is human freedom that is both exalted and challenged by man's central position, for if the whole world serves man's needs, even the gyrations of the heavens[92] and the uninhabited tracts of the earth,[93] his physical requirements and his body itself are subordinated to the needs and the destiny of his spiritual part.[94]

modo totus reservatur mundus, ut philosophi vocaverunt illum minorem mundum; haec enim creatura mundi scilicet homo per ea quae facta sunt, id est per creaturas, potest conspicere invisibilia dei . . .' (London, *Brit. Library, MS Royal VII. D. 15*, fol. 54r). I owe this reference to the unfailing kindness of Fr Gieben. Grosseteste's reference is to *Physics* II, 2, 194a, 34–5.

[88] *In II Sent.*, d. 1, p. 2, a. 2, q. 2 (II, 45b); cf. d. 15, a. 2, q. 2.

[89] St Thomas Aquinas, *Comm. in VIII Libros Physicorum II*, 2, lect. 4, 8 (Leonine ed., *Opera Omnia*, vol. II, p. 66a).

[90] See especially *De Cessatione Legalium*, MS cit., fol. 164b: '. . . omnis creatura huius mundi sensibilis propter hominem facta est', a very typical formulation. See also ibid., fols. 165d and 179b, and ed. Unger, pp. 16, 28; *De Finitate Motus et Temporis*, ed. Baur, *Werke*, p. 265; and *Hexaemeron*, MS cit., fol. 199d.

[91] *Hexaemeron*, ed. Muckle, p. 170. Cf. *De Confessione*, p. 241: 'Manebunt igitur secundum se in praesentia salvatoris extrema suae creationis, scilicet homo et angelus, et medium in extremis.'

[92] *De Finitate Motus et Temporis*, ed. Baur, *Werke*, p. 265; *Hexaemeron*, MS cit., fol. 199d.

[93] *Hexaemeron*, fol 228c–d; by maintaining the overall climatic balance which allows favourable conditions for life.

[94] *Notulae in Ethicam*, Oxford, *MS All Souls 84*, fol. 31b: 'Nam res exteriores

To express the unity which human nature brings to creation Grosseteste uses the image of the circle, with its immemorial connotations of perfection, completion and order. The grades of being as they flow out suggest the image of a chain, each successive link of which is farther from the first cause.[95] Since each level of being is found in human nature's all-sidedness, it is as though the whole catena of created natures from angel to matter were turned back from the straight line of *exitus* into a circle whose decisive link is man's nature.[96] Grosseteste is at pains to point out in both central unity-texts that it is only the created order 'from first to last' (i.e., from angel to man) that is thus unified, 'as it were in a principal member', such as every genus of unity seeks. The gap between the creation thus unified and God remains infinite, and can only be bridged by the Incarnation of the Word. Microcosmism serves his theology of the Incarnation beautifully.[97]

e) *Dominus et rector*

The themes of finality, centrality and unity intertwine to lend strength to an idea which can in turn be seen as subtending them all: man is lord of creation and has dominion over the entire visible universe, a lordship which makes him to be the image of the Lord.[98] Since things are subject to him as their master, he can make use of them by natural right, as aids provided for his nature; but under the law of the author of all.

Man's dominion over the cosmos is natural in the being who alone among visible things has intellect and free will. All the knowledge which he acquires, even on the purely natural or scientific level, gives him power over the natural order.[99]

sunt propter hominem, corpus autem propter animam, sicut materia propter formam et imperfectum [?] propter agens principale.'

[95] See *De Confessione, Prologus*, p. 239.

[96] *Exiit Edictum*, ed. Unger, p. 22: 'Homo praeterea habet in suo corpore ... communicationem cum omnibus naturis corporeis; in anima quoque communicat cum angelis intelligentiae natura, et per hunc modum est quaedam circularis concatenatio adinvicem omnium naturarum.' Cf. pp. 13-4.

[97] See McEvoy, 'The absolute predestination of Christ in the theology of Robert Grosseteste', in *Mélanges Bascour*, pp. 212-230.

[98] *De Confessione*, p. 241: 'Et hoc est quod dixit Dominus: Faciamus hominem ad imaginem et similitudinem nostram. Dedit ei Dominus cunctorum dominium, quorum fuit factus exemplum.' Cf. James 3:7, 'Omnia domantur et subiecta sunt naturae humanae', a reminiscence of which is associated again in the *Hexaemeron* with the concept of image (ed. Muckle, p. 167).

[99] *Comm. in Hier. Cael.*, ed. McEvoy, p. 162: '... habet inquam superativum

Freedom, however, has its history, and the universe which is the setting for its use is not indifferent to its exercise, for the lordship of man means that the destiny of the visible cosmos is placed by the Creator in his hands, and the history of his relationship to the cosmos is a history of the use and abuse of his liberty to serve God. The macrocosm being instituted for the microcosm, a change in the latter cannot leave the former, and the relations between the two, unaffected. The metaphysics of macrocosm and microcosm inevitably turns into history.

The universe was made for man, but for man in his best state, namely, that of free obedience to God.[100] All creatures by nature seek to possess whatever share they can have in the supreme Good. Material things manifest some mute vestige of God, but the inner movement of desire within all things becomes conscious in man, who can read the book of creation and utter the praise of its Lord and his. The universe is like a republic in which each citizen does what lies in his power to preserve the safety and peace of the community, but a special role is allotted to the ruler and governor.[101] His decisions affect the state in a unique way and he is answerable for the good ordering of the whole.[102] The fall of its leader had catastrophic results for the state, no part of which was spared the damage that came about through the sin of its Lord.[103] After man's sin the mute praise of creation no longer found its destined voice. The feudal oath between lord and serf was broken by the former's infidelity to his suzerain; obedience was no longer owed him by his servants. Instead, what had been man's solace and joy became his tormentor; fruit turned to thorn in his hands.[104]

omnium scilicet corporalium virtute et potentia intellectus, data ei secundum abundantiam . . . et praepotens est et dominatur, et quasi potestative tenet omnia post celestes substantias. Est etiam homo superativus secundum scientiam rationalem ratiocinatione acquisitam.'

[100] What follows is based upon *Ex Rerum Initiatarum* (ed. Gieben, pp. 122 ff.). reference being made to the *De Cessatione Legalium*.

[101] The image of creation as a republic is a Stoic theme retained by St. Augustine, for whom providence is found everywhere active 'in ista totius creaturae amplissima quadam immensaque republica', (*De Trinitate III*, 4; *CCSL* L. p. 136). Cf. Grosseteste's *De Libero Arbitrio* (ed. Baur, *Werke*, p. 155. 17).

[102] *Ex Rerum Initiatarum*, ed. Gieben, p. 123.

[103] *Exiit Edictum*, ed. Unger: 'Omnis quoque creatura corporalis deteriorationem passa est in lapsu hominis.'

[104] *Ex Rerum Initiatarum*, pp. 123, 125; 'Sicut enim eum redarguit terra per tribulorum et spinarum pro frugibus germinationem, sic omne elementum et etiam

Man's loss of dominion over the universe is nothing more than a consequence of his continuing inability to master himself, for if he cannot dominate the appetites of his own body, neither can he hope for due order in the macrocosm, which is like the body's extension.[105] He is like the father of a large family who, when in health, had respect enough to keep harmony in his household, but who in illness has lost control.

The Redemption won in Christ likewise has cosmic effect. It is the initiation of a gradual reconciliation between mankind and God the Father, and by that very fact a restoration of the finality of creation and the well-being of its entire order.[106] Man once again becomes the leader and representative of all things. It is in his cosmic nature that the other creatures will finally be saved,[107] for they have no destiny in themselves save to serve the preparation of the Church triumphant, whose head, Christ, is the end of all creation.[108] When the numbers of the heavenly courts are completed the world's machine will cease to work, the heavens, which turn only to cause the generation of man and all that ministers to him, will cease their motions, and with the quiet of the spheres will come the end of time.[109] Man, the end and unity of creation in the order of *exitus*, is in the *reditus* the means of the return of all things to the one source.

If the destiny of the macrocosm is ordained to that of man, it is not surprising that there is no creature that lacks some

caelestia corpora, per innumeras punitiones, de quibus nunc non est sermo interserendus.'

[105] *Hexaemeron*, fol. 227b.

[106] *De Cessatione Legalium*, fol. 182b: '... mors ipsius fuit reconciliatio humani generis cum deo patre et omnium creaturarum restitutio in esse suum optimum et finale, quod amiserant peccante homine . . .' Cf. *Ex Rerum Initiatarum*, p. 128: 'Si igitur divina providentia non evacuatur, necessario consequitur ut homo, cuius natura non continetur in aliqua alia, et in cuius natura aliquo modo continetur omnis alia, salvetur.'

[107] *De Confessione*, p. 241: 'Cum ergo non propter se sed propter alia facta sunt, non in se sed in illis pro quibus facta sunt salvari necesse est. A quo enim aliquorum pendet perfectio in eodem erit eorundem salvatio.'

[108] *De Cessatione Legalium*, ed. Unger, p. 16. 28: '... quapropter finis omnium factorum in hoc mundo sensibili est ecclesia triumphans, et maxime finis omnium esset illius ecclesiae caput unicum.'

[109] *Hexaemeron*, fol. 199d: 'Cum enim omnia propter hominem sint, ut compleatur videlicet humana generatio usque ad complementum corporis Christi quod est ecclesia, motus celorum non erit nisi propter generacionem hominum, et eorum que hic inferius ministrant homini.' (cf. *De Finitate Motus et Temporis*, ed. Baur, p. 265. 7–13).

similitude of the mystery of salvation. Their natural forms are like letters, whose meaning designates the benefit the Incarnation and the Redemption have brought to man.[110] Nor is it surprising that the history of the world and the six stages of the individual human life are comparable, for the same Orderer is at work here who made all things in six days.[111] This type of parallel or syntagmism between the cosmic-hierarchical gradation of the world and the historical ages of salvation-history was to become a prominent element in St. Bonaventure's theology of history.[112]

f) Exemplar omnium

Man's microcosmic nature does not make him merely the natural lord of the visible world, it accords him rather a unique status among creatures, one to which not even pure spirits can attain. Philosophically considered, of course, his uniqueness can be found implied in all that has been said up until now: the union of spirit and matter occurs only in man; only he is *'in medio collocatus'*, and only in him are all other natures of things to be found, as in their final cause. Grosseteste has still something of importance in reserve for us, however, and in an intrepid *tour de force* he crowns the special place of man by an explicit reversion to the idea of image, conceived of this time directly in terms of exemplarism.

Already in the *De Intelligentiis* of *c*.1228 Grosseteste had drawn upon Ambrose and Augustine in order to liken the soul to God, and indeed not so much the soul taken in itself as spiritual image, but in its relation to the body. Ambrose drew a parallel between the soul's government of the body as everywhere unifying, ruling and moving it, and God's causal ubiquity in creation.[113] St. Augustine reached a formula for the divine omnipresence which remained classic: God is *'sine situ praesens, sine loco ubique totus'*, in creation; the soul is *'sine situ praesens'* (etc.) in the body, and is thus a microcosm

---

[110] *De Cessatione Legalium*, fol. 164^c.

[111] *Hexaemeron*, fols. 229^d-230^b.

[112] For example, in the *Breviloquium*, prol. 2 (v, 204a). Cf. J. Ratzinger, *Die Geschichtstheologie des hl. Bonaventura* (Munich, 1957), pp. 44 ff.

[113] St. Ambrose, *De Dignitate Conditionis Humanae*, c. 1, (*P.L.* 17, 1135A): . . . et haec est imago unitatis omnipotentis Dei quam anima in se habet.' Similar parallels between God and the soul are also found in the Greek patristic tradition, in Gregory of Nyssa, for example; see Gilson, *A History of Christian Philosophy in the Middle Ages*, pp. 57-8.

of the relations of spirit to matter universally. The idea seemed to be capable of further development and received it in due course. In the *De Confessione* a series of parallels are drawn between God and his image. Of all things that are, man is the best, because he is the equal of all the others but is unequal to any one (*De Confessione*, p. 241). Taken in his body he is equal to the lowest of things, in his soul to the most noble; but taken in his whole being, he is the most worthy creature of all, because the most like to the Creator: as in God all things stand as in their cause, so the whole universe of creatures is mirrored only in the *minor mundus*, as their effect. The analogy is audacious. In only two beings are all things found, in the divine exemplar and in the last of all natures. Creation flows out from the unity of the Word in a graduated and differentiated plurality of natures, only to flow together again into the polyphonous unity of human nature. Man and God, infinitely apart, are thus daringly paralleled across the *universitas*, which each in his way contains.

In other ways also man is the fullest created representation of God's nature. The grades of created reality, which are all found in him, point to the single source of their orderly diversity. The mirror he is for all the universe is a reminder of the divine exemplarity. The finality of a creation guided towards a single product and summary, human being, is a sign that all things made have but one ultimate end. Hence man's uniquely representative nature and his central function in the universe are the best symbols of God's creative action, of his provident conservation of creation universally, and of his recalling all things to unity; the first, middle, and last moments in which the perfect order of things consists find full expression in human nature.[114] We have here a symbolic vision which is measured by basic metaphysical principles and is no longer under the sway of untempered imagination.

An interesting aspect of the connection between image and exemplarity forged by Grosseteste is the extension of the concept of image to the whole person, which is of considerable significance for his view of human nature. In the earlier works we examined it is always in the soul that the likeness of God

---

[114] *Hexaemeron*, MS cit., fol. 228a: Man as image shares at an infinite distance all that is in God, even creativity, for he must actively co-operate with God in the renovation of the creation, the order of which has been upset by sin.

is to be sought (albeit sometimes the soul taken in its relationship to the body), whereas in the later it is in the totality of
man that the fullest creaturely representation of God is
achieved. We are here in the presence of yet another of those
small but meaningful displacements of accent, which in their
sum manifest a noteworthy—though hitherto unnoticed—evolution in Grosseteste's thought. Turning back from the *De
Confessione* to the *Hexaemeron*, we are able to discern within
the theory of image presented there an anticipation of the
same idea.

Grosseteste quotes St. Augustine and (Pseudo-) Basil to the
effect that only the human reason, or consciousness, the 'interior man', is the image of God in the highest and fullest
sense. He is himself prepared, however, to go beyond both
authorities[115] and declare that the image as renewed by grace
extends its transforming effects to the depths of human nature,
and that it includes the whole person. The highest part of the
soul draws the lower powers into its likeness, and conforms
their acts, and even the organic body, to its nature. It impresses
and seals the whole man, whom it is destined to govern, with
the vestige of the Trinity, which it has received by its direct
ordination to God and which it in turn mediates to the whole
hierarchy of powers beneath it. On those more proximately
subordinated to its activity it confers a more express likeness
and on those further off a more derivative one, but no depth
of nature is too far removed from it to miss altogether the
transforming action of its powers. If the reformed image is to
be found already perfect and entire in the mind, still its natural
tendency is to extend itself to the whole person. The mind or
intelligence acts like the heavenly ether with regard to the
lower elements. Made luminous by the immediate gift of the
sun's light, it transmits the splendour it has received to the air
and further downwards still to the water and the thick, bodily
earth. Each element absorbs what it can of brightness, but
receives it only through the influence of that highest power,
which alone is a pure reflection of the sun's glory. Thanks to
the mediation of the ether, the whole receptive medium from
the earth to the sun's sphere glows with a single illumination;

[115] And to interpret them in his own sense. The texts are from *De Genesi ad
Litteram* III, 20, 30 (*CSEL* 28 3/2, p. 86), and *De Hominis Structura*, oratio I,
6-7 (*P.G.* 30, 16-17); see *Hexaemeron*, ed. Muckle, pp. 165-6.

so too the intelligence passes on the likeness of God which it bears in itself, till the entire human being becomes an image of the Trinity—a lovely macrocosmic parable of the effects of grace.[116]

IV MICROCOSMISM AND HUMANISM

In affirming the cosmic nature of man as the summary and focus of creation universally, microcosmism exalted human nature's grandeur and became at once the basis and the expression of a humanistic outlook. In a number of medieval writers, Grosseteste and Bonaventure in particular, microcosmic speculations were among the instruments used to counteract a kind of conspiracy, hatched within their Neoplatonic sources. Neo-Platonism, in almost all its prominent Greek and Arabic representatives from Proclus through Pseudo-Dionysius to Avicenna, was in its central focus of interest an angelology rather than an anthropology: the intelligences emanate from the principle more immediately than does man, who is often represented as in some way their product and always as lower and further removed from God than they. The Latin theologians had to struggle at various crucial moments against the negative consequences which this subordinationism threatened to produce in the theory of man; one of the weapons they used in the conflict was microcosmism. If man is the unique microcosm of creation, even in a sense of the divine nature, he can no longer be completely in the shadow of the angels' wings: in essence he is the equal of the pure spirit, and in one respect its superior. Through this means, the contempt for the body and the world, that widely-diffused product of unhealthily spiritualizing currents of thought, began at least to be counteracted. In illustration of this thesis three sets of passages in Grosseteste's works can bear examination, in all of which the being of man and angel are compared. In one of these the angel emerges as in certain respects man's superior in the hierarchy, in another as his equal and in the last as his inferior. How nuanced Grosseteste's conclusions were can be observed from the progression thus brought into focus.

What Pseudo-Dionysius affirms in the *Celestial Hierarchy* purely and simply of the angels, Grosseteste, with his much stronger anthropological interest, frequently turns into an

[116] *Hexaemeron*, ed. Muckle, pp. 164–5.

explicit contrast with human nature, though oftentimes without changing the wavelength of the ideas. The angels receive the divine enlightenments directly from God, without intermediary, and therefore more intensely than lower creatures.[117] From their direct vision of God comes a greater simplicity of nature and activity than is possible to man as he now is. Their minds are fixed in knowledge and love because they lack materiality and all that follows from it, mutability and the consequences of the fall. Their activity of adoration is 'undeclining', 'irremissible', 'inflexible', and their purification not gradual, but established by preservation from sin. Their contemplation of the divine essence reveals to them the totality of creation in a single glance, without the need to reason from known to unknown. In all these respects man is the inferior of the angel, working out his salvation in the lowest degree of the spiritual hierarchy, in matter, time, and darkness. He is subject to the angels, needing their illumination as willed by God to bring the light of knowledge and the warmth of love into the sphere of shadows he presently inhabits.

If this hierarchical emphasis could not but surbordinate man to his superior, its influence was never allowed to preponderate over the more egalitarian Augustinian thesis that all rational creatures are equally possessed of the image of God.[118] It is only the soul's natural desire for the body which differentiates it from the angel, not the possession of intellectual powers of lesser capacity. With the patristic tradition Grosseteste rejects the rabbinical belief that the angels had a share in creation, and discountenances more firmly still any suggestion of man's having been made somehow in the image of the angelic spirit (*Hexaemeron*, MS. cit., fol. 201^b). He considers it axiomatic that a creature made in the image of God and bearing the closest created likeness of the Trinity cannot have another creature as its superior in nature.[119] His Augustinian conviction

---

[117] *Comm. in Hier. Cael.*, ed. McQuade, p. 179; ibid., pp. 141, 154, 288, 315, 341, 350; ibid., p. 219: '. . . infirmus homo necesse habet formationibus per angelos subici et edoceri per angelos medios in symbolicis formationibus divinas visiones.'

[118] Ibid., p. 123: 'Intelligimus autem hic omnem rationalem creaturam ad imaginem dei conditam, licet de homine specialiter dicatur quod conditus sit ad imaginem et similitudinem dei, et licet a subtilius perscrutantibus assignetur ratio quare homo solus et non ita expresse angelus in scriptura dicatur ad imaginem dei conditus.'

[119] Ibid., ed. Muckle, p. 159: 'Sed secundum quod homo est imago et summa

that 'only God is greater than the soul',[120] made no fundamental concessions to his reading of Dionysius. All rational
creatures are destined for the vision of God, man not less than
the pure spirit, and the direct and personal nature of that
vision does not admit of weakening by the insertion of created
intermediaries and the introduction of subordinationism.[121]
Grosseteste seems to imply that it is only the fallen state of
man which calls the ministering aid of the angelic hierarchies
to his support.

The final doctrinal strand in the comparison between angel
and man can be introduced by a remark of Grosseteste's to
the effect that the Scriptures nowhere expressly refer to the
angel as the image of God.[122] We are to understand that every
rational creature was made in the image of God, but that this
designation is applied to man in a special degree. Grosseteste
leaves it to *'subtilius perscrutantibus'* to enquire how this is
so.[123] We shall not go far wrong in forging a direct link between
this hint and the explicitly microcosmic passages in the *Hexaemeron* and other works.[124] Man is 'more especially' the image
of God, as *minor mundus*. It is his privilege to mirror the
integrality of creation, and hence of the divine ideas that generated it. Only through man's free co-operation can the total
divine plan be effectuated, so that all other things are somehow
saved in his restoration.[125] Equal to the angel in the nobility
of his spiritual destiny yet on a level with the vilest of creatures
through his body, man's being, taken all in all, is without exception the most wonderful and worthy that exists. His God-
given dominion over all creation can be extended to include

similitudo dei non habet creaturam se superiorem, quia si esset superior, esset etiam
illa dei similitudo maior, creans autem esset maior a se creato; esset igitur illa
concreans creatura homine maior et non maior.'

[120] Ibid., MS Cit., fol. 215[b], in the context of his sharp attack on astrology;
cf. *Dictum de Gratia*, ed. Brown, *Fasciculus* ... p. 299: '... cum ipse [God] solus
sit animae natura sublimior.' *Comm. in Hier. Cael.*, ed. McQuade, pp. 187 ff., the
passage on the beatific vision, rejects *any* created intermediaries—theophanies in
the first instance, but also, in the context, angels.

[121] See Part II, ch. 2, pp. 94–5.

[122] Something already noticed by the Fathers; see St. Ambrose, *Expos. in
Psalmos*, n. 108 (*CSEL* 32, 211): man is made *ad imaginem* and the angel *ad mini-
sterium*.

[123] Ibid., pp. 123–4, quoted in n. 118 above.

[124] Grosseteste avoids invoking microcosmic themes in the Dionysian commentaries. It is not unlikely that he was aware they would be unsympathetic to
the interpretation of his author.

[125] *De Confessione*, p. 241.

even the angels, since they are ordered to his service and 'he who sits at table is greater than him who ministers'.[126] The seal was set upon man's uniquely privileged nature by the In-carnation, through which the Word redemptively assumed the entire universe as represented by human nature.

V ANCIENT AND MEDIEVAL ANTHROPOCENTRISM

That the Greek world-view which was partly taken over by medieval Christianity was permeated by religious feeling, is not to be doubted, but it could be claimed with justification that its religious elements were largely in contrast to Christian faith, or at least that the comparison between the expectations raised by the two views manifests a strong disparity of at-mosphere between them, especially with regard to the question of man's place in the cosmos.[127] I am strongly tempted to think that the combination of microcosmism and image-theory found in Grosseteste was aimed consciously at reducing that disharmony. In the Greek conception canonized by Aristotle, man is set on the lowest sphere, that area farthest from God, from perfect being and love, amid the turmoil of elements and the war of generation with decay. He is central only within the sublunary world, as its highest animal product; but this world itself lies as it were beneath the city-wall of the great universe. Man is a marginal creature within the whole, for-bidden its royal court by the decree which proclaims all hybris. What love there is in this cosmos rises towards the First Mover, by whose desire the intelligences turn the spheres; but it re-mains for ever unrequited. In the Judaeo-Christian tradition, on the other hand, 'love of God' loses its unidimensional and transitive character ('love *for* God'): that God has first loved his creature lies at the centre of faith. God is the centre, not the outer circumference of the universe of being, and his love must descend to earth before it can rise again. According to a perceptual shift inevitable within the ambiguity of the medi-eval Christian *Weltbild*, light and warmth must spread from the central creative point of brightness concentrically to the

[126] *Hexaemeron*, fol. 227ᶜ: 'Sed quid sibi vult quod ait, universeque creature? Numquid non creatura est angelus? Aut homo preest angelo? . . . an forte ideo dictum est homini quod presit universe creature, et ita etiam quod presit etiam angelis, quia maior est qui recumbit quam qui ministrat.'

[127] See the fine chapter on the heavens in C. S. Lewis's *The Discarded Image*, pp. 92-121.

outermost sphere of the light's diffusion.[128] Thus the intelligible universe of Christian metaphysics somersaulted the world of sense, much as *sapientia* imposed a noetic *metanoia* on its Greek philosophical convert. Centre and circumference manifested a tendency to change roles according to the frame of discourse chosen, cosmological or metaphysical.

This profoundly different image of God and his 'place' cannot leave mankind made in his likeness without a share in the divine patrimony. Ways will be found of celebrating man's centrality among creatures. He will be emancipated from the soil to gain the freedom of the *respublica universitatis*, in which he has citizen's rights.[129] The cosmos will become a stage, on which man plays the dramatic parts before a breathless auditorium, or a stadium, in which man wears the colours of his polity and runs to gain the prize for the honour of all. Indeed, the very language of man's microcosmic centrality in creation is breathless with the paradox it announces, and in which a Greek could only have seen his world stood on its head. Man has the lowest place in the world, but at the same time the highest function; he is the last of creatures, but only because he was first in the creative intention. Greek metaphysics met its *coup de grâce* and moment of truth all in one, when it was turned to psalmody and chanted antiphonally with the gospel of deliverance in a cosmic liturgy of thanksgiving: the final is the cause of causes—I am Alpha and Omega; the first in intention is the last in the order of execution—he has raised up the humble and made the last to be the first. In this new world, between height and depth, empyrean and element, there is span without distance or remove, since he who is Lord emptied himself and took the form of a servant, dwelling among us. Earth is still the physical centre, lying farthest from the circumference of the universe, but for all that it is the centre in a new and altogether more dynamic sense, for as the humblest are those who receive the most, so the point of convergence of all heavenly 'influence', light, and illumination,

---

[128] Thus Dante, *Paradiso* xxviii, 25 ff., quoted by Lewis, p. 116.

[129] Cf. the use of political metaphors for man's relationship to the cosmos in *Ex Rerum Initiatarum*, pp. 126 ff., a microcosmic type of metaphor which projects the order of the city upon the whole of creation. The earliest occurrence of this theme in Christian literature is met with in the *Apostolic Constitutions*, where man is called *cosmopolitēs* and the perfection of the *dēmiourgia* of creation; see Roques, *L'Univers dionysien*, p. 53.

bears the Incarnate Word, the centre of the universe, the locus of cosmic hope, 'the diadem in the golden ring of creation'.

VI ELEMENTS OF NOETIC MICROCOSMISM

We have traced the interlocking by Grosseteste of microcosmic notions of various provenance with the theology of image and its associated biblical themes. The conclusion that emerges is that this whole complex as well as all its strands (which can only with difficulty be unravelled) was focused upon the nobility of that in the soul which is the supreme natural image of God, namely, reason and freedom. The whole became a triumphant recovery of the grand Augustinian theme of interiority, as the road of the mind towards that presence which is more intimate than the most intimate self within us. At this point we must note a lacuna in Grosseteste's microcosmic theory, in the somewhat surprising failure on his part to develop fully a further systematic possibility of microcosmism, what we have referred to as noetic microcosmism.

This lacuna is surprising, because all the elements required for the development are present in Grosseteste's thought—only their systematic linking is wanting. From Augustine comes the parallel of the five senses with the five elements of the universe.[130] Each sense assimilates an objective quality by which a segment of the world's reality is made available to the mind. The identity of the physical with the sensible qualities, widely accepted in the ancient and medieval periods (and defended explicitly by Aristotle),[131] is taken to imply that there is no element nor quality which is not sensed, nor any sense wanting; wherefore the totality of the cosmos has access to the mind through the five doors of sense. From Augustine also originated the view that the hierarchy both of sense objects and of the senses themselves is due to the mixture of active light with the lower and relatively passive elements, which incorporate or receive it into mixtures. Thus it is the finest particles of matter, atoms of light, which vibrate within objects struck to produce sound-waves, and in the sense-organ it is likewise the spirits, luminous in varying degrees within a hierarchy reaching from sight to touch, that actively receive the

---

[130] See ch. 4, ii, a (pp. 296–7) for details.
[131] *De Anima III*, 1, 424b 20 ff: that there cannot be more than five senses.

stimulus and transmit it to the brain. A single order, therefore, is found to govern the macrocosm and the microcosm at the level of sense-knowledge, and in the openness of the natural light which man is, a dynamic assimilation of macrocosm by microcosm overtakes the causal relationship of the cosmos to its highest physical product, the organic body.

'The human soul is in some way all things'. We have seen how Grosseteste drew Aristotle's tantalizing adage into the orbit of Alcher's division of the psychological powers, and ultimately into the Augustinian dynamic of the desire to know and possess all things in the vision of God.[132] The mind has (or rather it is) intelligence, to know God, intellect, to recognize the first principles and know purely spiritual being, and *ratio*, to construct demonstrative science of the world and its contents. It has access in principle, therefore, to all dimensions of the macrocosm, its only restriction being of a temporary character, namely, that it does not have actual knowledge of all things in this life, and it has knowledge of God's inner life only through his revelation of himself.

The elements of noetic microcosmism are all present, but they never spring into the single, inevitable synthesis which they were to form in the *Itinerarium*. Several times in Grosseteste's texts it seems that the implicit attraction among the various ideas is becoming magnetized and is on the point of forming a unifed field. Man as the image of God is all things; his body is a product of all the elements; his senses assimilate all the qualities; his mind can know all regions of being and is destined to know all in God. If Grosseteste did not arrive at seeing the latent dynamism of these themes, he prepared the way richly for Bonaventure, who did; and, after all, had that translucent mind not written as he did after Alverno, perhaps no one ever would have done.

We have now said enough about Grosseteste's use of microcosmic themes to let a single rich text speak for itself. An interesting reflection, one to which we shall have to return, is suggested by the fact that the passage we translate here is a prologue to a treatise on the sacrament of penance, written by a bishop for the eminently practical purpose of instructing his priests. What was it, in the man and in his world, which made him begin with a sketch of all creation (angel, heavenly

---

[132] See above, ch. 4, iii (pp. 299-312).

bodies, mixtures of elements, vegetable life, minerals), and only thus make his way to man? Why did so many authors of his time (for in this, of course, Grosseteste was no exception) delight in reporting every fact, idea and institution to what Lewis calls 'the Model' as a whole?

It is a delight to enumerate the themes which Grosseteste manages to cram into a passage that seems leisurely and casual. All the associations of image which we have analysed are there, the transitions between them made with an art that conceals itself in innocence. One after another they follow: last creature, exemplar, summary; equal to all but unequalled by any; the union of extremes, the unique exemplar of God, the best of creatures, the lord of the world, the image of God.

In the last place the All-high established a product, man, who would be at once the exemplar of all [the grades] mentioned and drawn from them all, as one might do who wrote individual works containing his wisdom and then edited them into a *summa*. For man is on the same level as the angel in his soul, his sensibility relates him to the animals and he shares his lowest organic level with all growing things, while certain parts of his body bear a likeness to other material things. In his physical aspect, therefore, he resembles the most lowly of things and so is imperfect, but his soul is the equal of the highest creature and hence most noble. Taken in all of what he is, however, he is the most worthy creature that exists. For I maintain that man resembles the Creator more than does any other thing made, for as all things stand in God as their cause, so too all shine forth in man as their effect, which is why he is called a tiny world. And since he is the best of all, being equal to all together yet equalled by none, they commonly owe him natural obedience; so he is the image of God. The Lord said, 'Let us make man in our image and likeness'. He gave him dominion over all things , for man had been conceived as the model of the whole universe.

(*De Confessione*, pp. 240-41)

A paean to man as he was instituted, this passage was at one and the same time a prelude to forty pages of deadly sins (and almost equally mortal penances) and a sublime exaltation of essential human nature, which made it possible to look at what man is and does, yet not despair at heart.

VII ORDER, HIERARCHY, AND ORGANIZATION

From its origins in Plato microcosmic speculation was an attempt to situate an ambiguous and threatened human nature within the widest context of meaning available; the totality of being, seen as sacred order, was made a home for man, supporting him in his struggle and hope. Something akin to this microcosmic feeling had pervaded the deepest dimension of myth, as man had enacted the analogies between the macrocosm and the microcosm in rituals, which had had the effect of opening the celebrating community to the world and enabling it to share in the sacrality of cosmic order.[133] It was in the guise of myth that Plato presented his mature doctrine, in the great unfinished trilogy of *Timaeus, Critias,* and *Hermocrates.*[134] The *Republic* had imagined a city-state whose institutions were to be based upon the unalterable characteristics of human nature. Plato, however, at no stage considered man's nature to be self-grounding. In the *Critias* he aimed at transferring this state from the ideal realm to real existence, but placed it in the remote past of Athens, nine thousand years before. As a prologue to this entire endeavour, the *Timaeus* was to recount the myth of creation and develop a cosmology linking human nature, with its moral and political task, to the very structure of the world, and through that to the forms on which the latter was modelled by the Demiurge. Plato hoped in this way to remove morality definitively from the region of convention and arbitrary enactment, by locating it in the harmony of the soul, itself an imitation by human wisdom of the harmony of the World-Soul's heavenly thoughts and revolutions; the latter in their turn are only a reproduction, the best that could be realized by the divine Reason working in intractible matter, of the order prevailing in the world of forms. The soul is supported in its struggle with its lower parts by the knowledge that even universal reason cannot forcefully subjugate matter, but must persuade necessity's errant causality to imitate the harmony of mind. Plato's late vision embraces, therefore, four related *kosmoi*: the human soul, the classes of the city, the visible universe and the forms, so that microcosmic parallels are multiform and the idea of order develops a uniquely rich polysemy.

Ethico-political and religious considerations prompted many medieval thinkers to attempt to locate man within a total framework of order, which reached upward through social

[133] This is what Eliade (*Myth and Reality*, pp. 5–6) calls 'the deepest dimension of myth'.

[134] I am indebted to Cornford's commentary on the *Timaeus* (*Plato's Cosmology*) for the essentials of what immediately follows.

organization, the material universe, and the spiritual world, to the *rationes aeternae*. The end-product of this search was the theory of natural law perfected by Aquinas, but historically the process towards it is no less interesting than the result itself. Grosseteste has a great deal of light to cast upon that evolution, as reflection on his ideas of order, hierarchy, authority and responsibility, and Church and State, amply reveals.

We can begin inductively by examining his ideas on the Church, the foundation of all social organization in the Middle Ages and the matrix and model for all those institutions in the creation of which medieval Christendom was so prolific. It is clear for a start that Grosseteste did not regard the structure of the Church and the functions exercised within it as casual empirical forms that had evolved through history and by the exercise of free human initiative. When he asserted that the Church is an order or hierarchy he was already by that fact identifying it as the visible, earthward projection of the angelic choirs; the structural parallel between celestial and ecclesiastical hierarchies was a theme which yielded to no other in importance in his late writings on the Church.

Grosseteste's exposition of the concept of hierarchy in his commentary on Pseudo-Dionysius is both lucid and personal: lucid in its substantial recovery of the meaning of the Greek text, and personal in the additional depth he lends tò the notion.[135] He takes his point of departure in St. Augustine's definition of order, which makes explicit reference to grades and establishes a solid traditional basis for the assimilation of the more specifically Greek elements in the notion of hierarchy: 'Order is the arrangement of graded elements in which each is given its due place'—the definition recurs frequently in Grosseteste's works.

The general notion of order is specified with the intention of reaching the very precise concept of that hierarchy which is the secret of the life of the angelic community. The first note to be added, 'sacred', is given a teleological character: it signifies direction to God, as the last and best end of the *reditus*. A sacred action is one which is aimed at the end; similarly, a sacred thing is so denominated from the teleological reference it bears. To order a diversity into hierarchical form

---

[135] See Part II, ch. 3, i: 'The nature of hierarchy'. There is no need to repeat what was said there.

is, then, to assign to each member its place, so that it is established neither higher nor lower nor otherwise than is simply its due, in such a way that the whole arrangement leads (or is brought back) to the enjoyment of God. A further note is the knowledge granted to all members of the hierarchy, and which itself shares in the sacrality of order, since it reveals God's presence in all things. Such knowledge can be by direct vision or by faith; it always includes consciousness of the order of the Church militant and triumphant and of one's place within it, and with that frees one from the temptation to imagine oneself higher or lower than one actually is. This knowledge governs the conduct of human life and the activity which it informs is the third note of 'hierarchy'. Sacred operation is an expression of love, whose gravitational pull draws the whole being of a rational creature towards the objects of its desire. These in turn are loved individually and in their totality according to their place within the whole order of things. In this ordered love the whole arrangement (*dispositio*) of being becomes determinant of the affective life of the microcosm, for each thing is loved in a way that corresponds to its place in the real order, that is, it is accorded its own *pondus* or weight, as that was determined by him who arranged (*disposuit*) all things in number, weight, and measure. Ordered love loves the things below it with measure, in so far as they profit it towards the knowledge, praise, and love of their Creator. It loves its equals, men and angels, in the same measure as itself, but God immeasurably, as he alone deserves to be loved. Within the hierarchy of rational creatures, the law of their natures is likewise that of loving activity. Each order is responsible for the good of that below it and each individual for the needs of the following one, for handing down to him what he has received from the one above him, and leading him up to knowledge and love, just as he for his part is led back by his superior. Hierarchy thus defined in its formal elements applies with equal force to angels and men.

Before considering the implications of hierarchical order for human organization it is necessary to reflect on the other great manifestation of divine order, namely, the physical cosmos, in which also the number, weight, and measure universally present within creation have produced an ordered hierarchy of grades.

The sphere of the empyrean is the first body, the most simple and least corpulent in its matter.[136] Subtle and solid in nature, it is light-filled and hence of the highest energy. It is the greatest of things in respect of quantity, the highest in place, the absolute measure of location and time in the universe which it contains and bounds. It is of the greatest beauty; the full range of meaning of the Greek word 'cosmos' was not lost on Grosseteste here, any more than when he commented upon Pseudo-Dionysius. Since each material thing receives activity from God in the measure of its own aptitude and receptivity, and the heavens contain the supreme physical manifestation of the divine operation, God is fittingly said to have placed his throne in them, and they are called his dwelling place.

This firmament imparts its continuous, simple, and uniform motion to the whole universe, first to the other celestial spheres (those of the fixed stars and the planets), and through them to the sublunary world. The law of the transmission of motion and energy is that the higher, being closer to the primal source of both, possesses them in higher degree than its inferiors, which are more multiple and have less capacity for activity. Each succeeding sphere receives motion from its superior, but its rotation is slower and its capacity for transmission more limited. This is most manifest in the spheres of the elements: fire revolves more quickly than air, and motion is weaker still in water. The degree of dependability and regularity likewise decreases the lower we come in this universe, in favour of variability and chance.

Earth is the central and lowest sphere, least of all in nobility, quantity, motion and activity. Its substance is vile by comparison with the pure dignity of the heavens; all that it has, it has received from their influence. Its very lowliness at the centre, however, makes it the focus of the congregation of light from every part of the universe. The celestial bodies, especially the sun and the moon, are the chief agents of generation, corruption and growth on the earth's surface,[137] since they precontain

[136] The following exposition is based on the lovely chapter on 'the heavens and the earth' found in the *Hexaemeron*, MS cit., fols. 206ᶜ to 207ᶜ, and later remarks found on fols. 210ᶜ and 214ᶜ.

[137] *De Fluxu et Refluxu Maris*, ed. Dales, pp. 72 ff. Early in his career, while still under the influence of Arabic astrology, Grosseteste privileged the moon, 'quae virtutes caelestes mundo coniungit inferiori.'—*De Artibus Liberalibus*, ed.

as causal agents all the figures and species of terrestrial forms.[138]
The sun, in particular, is the cause of activity, growth, regularity and beauty in the earth; most physical and meteorological phenomena can be traced back to its causality, for it
is like the central organ of the body of the universe, giving
unity, light, and beauty to the whole.

We have sketched two worlds, the invisible order of the
angelic hierarchy and the visible, material universe. They
achieve an astonishing degree of likeness to one another. Each
is hierarchical, continuous, and centralized. Both exhibit order,
the due arrangement of higher and lower in a graded series.
The hierarchical features of the angelic world find a physical
analogue at the cosmic level, where the lower material potencies are impressed by the higher concentrations of energy
and activity and each superior thing stands as agent to its
patient inferior, polarizing its causality and transmitting to it
what it has received from those higher still. The being and activity of each member of the ordered ensemble is apportioned
in the measure of its own analogy, in the Dionysian sense,
that is to say, according to the degree of receptivity allotted
to it by its place within the entire system. Justice rules the
whole, giving each thing its due. Both orders, the material
scarcely less than the spiritual, have a sacral character, deriving their form from divine decree, and being directed to God
as their end. Teleology rules over all. Even in the physical
universe there is a parallel to that love which is the operation
of the spiritual hierarchy, for all things desire to participate
in the supreme good in so far as their nature allows: the mute
but never the less dynamic expression of the good by the universe is a form of admiration, praise and love for the Creator
and an invitation to man to take up the praise in words.[139]
Even the metaphor used by Grosseteste when defining the
love of the celestial and ecclesiastical hierarchies is significant: *amor est pondus et collocatio rationalis naturae*, where
pondus, in addition to its Augustinian and biblical resonances
('*amor meus, pondus meus*'; '*numerus, pondus et mensura*'),

Baur, p. 5. 26. In the *De Fluxu* the sun's rays are the predominant influence on
the quality of the air and the moon's on the water.

[138] *Comm. Phys.*, ed. Dales, p. 52. 29; a theme of common occurrence.
[139] *Ex Rerum Initiatarum*, ed. Gieben, p. 123; cf. *Comm. in De Div. Nom.*,
ed. Ruello, I, § 71-2.

has as its primary referent the 'kindly stede' allocated to each nature in the Aristotelean universe, the place towards which its gravity brings it within the entire order and where it comes naturally to rest. Finally, both orders are ruled by the Neo-platonic law of continuity of natures[140] and of participation, whereby any genus presupposes a first member in whom the perfection shared is centralized, united and focused.[141]

There is abundant evidence that Grosseteste was fully alive to what we may call the syntagmism of the material and spiritual *kosmoi*, which situate human nature at their coincidence. Three instances will be given, the last of which leads us directly back to anthropological considerations.

In the prologue to the *De Confessione* the basic dynamism of *exitus* is made to imply that some things are directly subject to God in their essence and operation (understand the highest spiritual order of seraphim and the firmament), others only *secundum mediationem*, in such a way that lower are subordinated to and dependent upon higher. As Grosseteste himself says, 'It is the order of creation that things farther removed draw less from the springs of true existence because they are farther from what is truest.'[142]

In the second place, the ground of the analogy that holds between the laws of the material and spiritual universes is given in the metaphysics of light; the laws of operation of being dynamically conceived apply universally and without restriction. This is the ontological basis for the light-metaphors that abound in Grosseteste's later works, by means of which the operation of physical light is made to bear reference to the life of the spirit.[143]

A third piece of evidence is found in a marginal *notula* on the *Nicomachean Ethics*, which I reproduce:[144]

Quia autem a substanciis separatis aliquid detur hominibus evidens sit ex ipsa communicatio [ne] hominis cum substanciis separatis secundum

---

[140] Interesting sidelights on the idea of continuity are provided in the *Hexae-meron*, fol. 223ᵃ, and in the prologue to the *De Confessione*.

[141] This is the principle to which Grosseteste appeals in *De Operationibus Solis* § 6, to ground the centrality of the sun in the universe.

[142] *De Confessione*, ed. Wenzel, p. 240.

[143] *Comm. in Hier. Cael.*, ed. McQuade, pp. 4–5, 11, 409, 463, etc.

[144] The note is authenticated by the abbreviation-sign linking it to the text. I give the relevant section of the text, the words placed in parentheses being inter-linear additions of Grosseteste:
'*Siquidem igitur et aliud aliquid deorum* (i.e. substanciarum separatarum) *est*

intellectualem virtutem. Sicut enim corpora inferiora recip[iunt] suas perfectiones a corporibus superioribus, ita intellectus inferiores ab intellectibus superioribus.

Something like a world-view becomes visible here, behind an apparently innocent Christianization of the Aristotelean conception, that man may share the happiness of the gods 'or something close to it'. The gods or separate substances become angels, the intermediaries through whom the One God sends his gifts to mankind: *'omne datum bonum et omne donum perfectum'*. In the ambivalence of the Latin word *communicatio*, two central ideas are present: man *shares in* the angel's nature, and the angel *communicates* to him what it has received from above. By the law of creation and order, lower spirits receive their perfections from their superiors; within the vertical strength of the Aristotelean and Arabic world, the lower spheres and elements are likewise impressed by the higher. The tiny *notula* could easily be overlooked, but it is in effect something of a key-text. Aristotelean cosmology and Dionysian angelology are drawn by it into unity under the idea of hierarchy, and are focused upon man, who stands as a union of opposites at the confines of spiritual and material being, receptive to higher influences in both dimensions of his composite nature.

One of the characteristics of the medieval model of the universe which lends it its unique architectonic quality is the ease with which analogies could be discovered between its parts, as severally explored by minds which remained at the same time intent upon the totality. Within this model of the world, every region becomes something of a microcosm of the whole, so pervasive is the force of parallels drawn between different areas. It is in anthropology that the habit of cross-reference is most enriching; we may say that the mental inclination to refer across the diagram of the world subtends all of medieval microcosmism. The medieval political writer, say,

---

donum (scil. datum) *hominibus, rationabile et felicitatem dei* (scil. supremi) *datum esse et maxime* (i.e. quia ipsa est optimum inter dona humana) *humanorum quantum optimum.*' Interestingly, the interlinear method of annotation used here is probably a clue to how Grosseteste began to write a literal commentary. Interlinear insertions explain the words of the text, longer marginal notes accumulate as he reflects, and the whole would be worked into a flowing text in a final stage of redaction.
The margin has been trimmed by the binder.

or the theologian, can draw upon the models of organization exhibited in the arrangement of the celestial spheres or of the angelic choirs and can find analogies as between the models themselves. The results of this widespread mental mechanism in terms of social thinking have yet to be explored. Perhaps they can only be revealed in their full dimensions when the law of thought which governs them has been explicitly formulated. A discussion of microcosmism cannot be dispensed from defining the meaning of 'cosmos', in its highest and most abstract sense.

VIII COSMOS AND THE ETERNAL PLAN

Though the speculations which take their origin from the couple macrocosm and microcosm are multiform, the examination of their ancient and medieval forms suggests the existence of a lowest common denominator. In every speculation of this type two terms are posited, a great and a small, of which a single (or at any rate analogous) constitution and order, cosmos, is affirmed. Again, as far as the ancient and medieval periods are concerned, we may hazard the generalization that the microcosm is always human nature—the body, the soul, or some of their parts, the whole composite human being, or the social nature of man and the order this subtends. That being said, further generalizations should be avoided. However, another general contextual factor may be claimed as a constant within the Christian period. All Christian microcosmologists construct their parallels in the awareness that a creative infinite being is the source of the *universitas rerum* and its orderer in both nature and history. Thus the order-concept, which is the first presupposition of any paralleling of microcosm and macrocosm, receives a basic metaphysical qualification which brackets the empirical order of universe, hierarchy or polis.

Taken in themselves, many of the parallels which Grosseteste (and most other medieval writers) drew between macrocosm and microcosm must seem to the modern reader whimsical and arbitrary, unworthy even. The following are a few examples.

The spiritual and moral sense of the six days of creation can fittingly be found in the six natural ages of individual human life. On the sixth day man was made in God's image, and in the sixth age, *senectus*, despite his weakness of body, he reaches the age of wisdom and often of mental vigour, passion being stilled. Of like congruency is the unfolding of the entirety of time and the history of mankind in six stages that parallel the development of human life from *infantia* to *senectus*[145] and reaching their consummation at the end of the sixth age; the unfolding of time should correspond to the order of creation and parallel the ages of the microcosm, for whom the universe was made.[146]

The six ages of man and of the world, like the six days of creation, await their fulfilment in the seventh, in which perfection is attained. Seven is the number of universality and betokens a great mystery; it is found in both macrocosm and microcosm. Man is a septenary as God left him on the seventh day, for in him are to be found the four elements of the world and the three powers of the soul. He is perfected by the seven virtues (the four cardinal and three theological)[147] and the seven gifts of the Holy Spirit, and comforted by the seven beatitudes.[148] The *septenarium* likewise embraces many natural phenomena, such as the seven planets and the seven metals they govern, the phases of the moon changing at regular intervals of seven weeks, the stages of foetal development, as distinguished by medical art, and the kinds of motion. There is so much more to say '*in admirationem et investigationem perfectionis septenarii*',[149] even without going into the sevens mentioned in the Bible—which Grosseteste, writing at leisure, cannot resist doing; and at a length which leaves the reader in no doubt of the importance he attached to the idea and the delight its pursuit afforded him. Thus, like St. Bonaventure a generation later, he manages to introduce the number-symbolism which was associated with microcosmic speculation from earliest times. Indeed, Bonaventure goes further and claims that seven as an element of

---

[145] A microcosmic parallel between individual development and the history of humanity which has not ceased to find adherents even in modern times: much of Comte's philosophy depends on it.

[146] *Hexaemeron*, MS cit., fol. 229^d; cf. St. Bonaventure, *Breviloquium*, prol. 2 (V, 204a).

[147] *De Confessione*, pp. 247-9, an extended allegory in which the seven elements of human nature are paralleled with the seven virtues and vices.

[148] *Hexaemeron*, fol. 233^d.

[149] *Hexaemeron*, fols. 234^d-235^a.

order in microcosm and macrocosm takes its origin from the archetypal world.[150]

As in the universe all movers are governed by a first, so in the Church's order all powers should be regulated by a single principal mover, wherefore all Christians owe obedience to the Lord Pope.[151] Just as in the celestial hierarchy each order is subordinated to its superior and is responsible for its inferior, so it must be in the visible Church, *'omne unum unitur in aliquo principali uno'*.[152]

The same type of thought-mechanism may be applied to the individual. The universal order of justice requires that there be a first in every kind as a measure of the others: a first mover, a first place, a first light-giver; in the human body, a first organ (*membrum principale*, the heart),[153] and in man, a single power that rules all others, freedom.[154] In every multitude that is ordained towards one end there must be a single governing-power to rule the rest, Aquinas will claim in the course of his theological defence of monarchy; for every multitude is derived from unity.[155]

What weight can we give to these parallels drawn between things at first sight different? It must be said at once that there is no contemporary analogy for the spiritual perception or even argument that these examples represent. It would be misleading to regard them, in the lack of an evident hermeneutical key, as being rather piquant and at best ingenious *rapprochements* between human nature and the broad universe of spirit and matter inhabited by the bookish mind of the Middle Ages, happy accidents of symmetry interpreted by a poetic imagination and accentuated by a pious or moralizing intent. In fact, the basis of these considerations is one with the profoundest source of Christian microcosmism as such, even of that kind of reflection which, like elementaristic microcosmism, for

---

[150] *Hexaemeron*, coll. 16, 7 (V, 404b); cf. Allers, art. cit., p. 342, and Ratzinger, op. cit., pp. 16–21.

[151] A prominent theme in the Lyons dossier; cf. St. Bonaventure, *De Perfectione Evangelica* q. 4, a. 3, arg. 21 (V, 192a).

[152] *Epistolae*, ed. Luard, no. 36, p. 127.

[153] *Comm. Myst, Theol.*, ed. Gamba, p. 52. 7; cf. *De Cessatione Legalium*, ed. Unger, p. 13. One of the sources for the idea may well have been Avicenna's *De Anima* V, 8, ed. Van Riet, pp. 176–7.

[154] A universally-accepted notion, often made to coincide with the analogy between the soul and God's providence in the universe; cf. St. Bonaventure, *De Perfectione Evangelica*, concl. V, 194a.

[155] *De Regno* I, 1, 19; tr. Phelan-Eschmann, pp. 6, 12.

example, seems in retrospect more scientific and respectable. Something like a strong expectation of order surrounds and infiltrates Grosseteste's view of things and the order he suspects he will find is always in origin the same, namely, the eternal plan, which is the expression of God's orderly, interior life. Regarded thus, microcosmism becomes a product of a mentality which, in an effort to give content to the idea of human nature, searches out relationships between man and creation universally and believes that the resulting analogies are informative concerning the typical traits of God's action *ad extra* and his nature *ad intra*. The most diverse things will be found to be in relation—the heavens and the head—or will manifest similar laws of action—physical light descending as a celestial influence on the body, spiritual light as the divine action in the soul. The likenesses thus exhibited are not to be thought of as being mere empirical contingencies, rather has every analogy the character of a new confirmation, at once of the homogeneity of God's plan and of the resemblance of creatures to him, and therefore in a measure to each other. This belief in the integrity and consistency of the eternal law inevitably reflects itself in anthropology, where its ethical implications are evident in the theory of natural law.

IX THE DIVINE NATURAL LAW

The fact that Grosseteste made frequent reference to the natural law without ever defining it set him apart from the Parisian Scholastics of his day, who initiated a philosophical analysis of the meaning of law.[156] Grosseteste remained within the older tradition, for which the idea of natural or divine law was so all-embracing that one could only work towards it, without ever being able to descend upon it from above, in a deductive fashion. Grosseteste always presented the natural law as the divine law, an obligation placed immediately by God upon the entire spiritual world. Each category of creature (men, angels) obeys it in function of its specific nature and conditions. It is known by reason, is acquired, therefore, rather than innate; it comprehends the entirety of precepts that are just in themselves and are not simply imposed by decree. His best remarks on natural law (found in the *De*

---

[156] P. Michaud-Quantin, 'La notion de loi naturelle chez Robert Grosseteste', in *Actes du XIe Congrès International de Philosophie*, vol. XII, p. 168.

Cessatione Legalium) relate it directly to man's historical growth under divine providence, rather than to nature conceived in a static, atemporal sense, so that it is said to include all that God eternally foresees and ordains in the order of things, including the Incarnation and the Church, together with all that follows from them for humanity. That is not to say that it changes; though man grows in stature before it and through it in the course of time, it is an ethico-religious ordination, grounded upon the divine exemplar ideas and defining for the rational creature the conditions of its conformity to its creative origin, to that which is most profoundly the truth of its nature and destiny: deification. If Grosseteste had had to define it further he would have done so by opposing it in the first place to divine positive law and then to all legal enactments and arrangements of purely human institution, such as civil and ecclesiastical law, traditions, mandates, and precepts of man, which ideally should be adjusted to the natural order of things.[157] Awareness of the subordinate role of human law demands, at the very least, that it should never contradict nor interfere with the divine natural order.[158] Finally, while the natural law is ordained for both man and angel, it is learned and obeyed by man in a world which is itself governed by the divine law and created in number, weight, and measure according to the same divine exemplar, so that in the foreordained harmony of things the nature of each material creature instructs man concerning some aspect of his duty.[159]

We can give a somewhat more exact idea of how the divine and natural order of things is revealed to the human mind. In a homily given to his priests on the subject of pastoral guidance, Grosseteste instructs them how to place the divine plan before their people.[160] The good shepherd must pasture his flock in the fields of Scripture and creation and instruct them in the life of Christ. In the field of Scripture a treasure lies hidden, for under the surface of the letter is the spiritual understanding, containing the seed of moral precept.[161] The second pasture is the broad field of creation, where the beauty of every fruit

[157] Lyons Dossier, ed. Gieben, p. 381.

[158] Ibid., pp. 380, 385, and *Epistola 24*, ed. Luard, p. 76.

[159] Brown, *Fasciculus*, p. 260.

[160] *Ego sum pastor bonus*, ed. Brown, ibid., pp. 260–4.

[161] Ibid., '. . . quia in latitudine scripturae sub sensus historici superficie latet thesaurus desiderabilis spiritalis intelligentiae.'

and flower is an example of something good in human actions, even a sign of divinity. Christ used both the moral precepts of Scripture and instruction based on the lessons and parables of creation in the course of preaching the Kingdom, but for us it is his work and sufferings that are the supreme revelation of the divine plan and object of preaching. As Grosseteste recalls elsewhere, *'Omnis Christi actio nostra est lectio / Ut in eius omnino actione discamus / quid nos ad eius imitationem faciamus.'*[162]

The method, by which the order revealed in the books of Scripture and creation and in the person of Christ was converted into an infallible guide for the Christian, was the spiritual exegesis of the Bible. This technique was used by Grosseteste in imitation of the patristic tradition, in order to give concrete expression to the demands of order, as revealed by God. A charter of the rights of the homilist, it established two basic freedoms. In the first place it allowed the preacher to approach the teaching of the Bible with a mind opened upon the totality of the divine plan. Each incident, person, and thing in the sacred narrative was made into a transparency, through which the actions of God and their effects as a whole, in the Old and New Testaments and in creation, were revealed. The preacher was thus enabled to forge a link between the Word of God in its dimension of universality and his present situation or preoccupation. The matter in hand—a duty or abuse, a state in life, or a moment of conflict—was elevated from the isolating factors of time and place, and regarded *sub specie aeternitatis*, in the light of the whole divine plan; it was opened up to the abiding presence of grace, through the authoritative application to it of the Word of God, and in this way the vertical dimension of eternity was introduced as support and foundation into the fragile impermanence of time.

The habits of thought which Grosseteste acquired through the pedagogy of spiritual interpretation underwent an imperceptible transposition to become the hermeneutic that guided his reading of the book of creatures. The transition is easily comprehensible. Both books have the same author and must bear the same message for man. No less than the figures of Moses and Melchisedech will the sun, moon, and stars, the seasons, elements, and angels, bear typological or tropic reference to human being and salvation. This project for the

[162] *De Cessatione Legalium*, MS cit., fol. 159[a].

interpretation of the world imposed upon him a consistent
effort to explore the dramatic and symbolic properties of the
cosmos, in order to place them at the service of man's search
for the meaning of the total ofder of things. It was an effort
the fate of which was indissolubly tied to the fortunes of the
spiritual sense whose method it prolonged, and one to which
the medieval model of the cosmos was, perhaps uniquely,
sympathetic. In Grosseteste's case, at least, it acquired a degree
of firmness and originality by being based upon a coherent
philosophical view of the orderly functioning of secondary
causes in nature. His application of the method made repeated
references to the two great transcendental prototypes of social
organization, drawn from the physical universe and the celestial hierarchy. The order of each was regarded as an expression of the divine law and was made paradigmatic for the
social orders of Church and realm. The application of both
archetypes can be plentifully illustrated from Grosseteste's
late works.

X TRANSCENDENTAL PARADIGMS OF HUMAN ORDER

We could not wish for a clearer statement of the foundations
of ecclesiastical order than that found in Chapter Three of the
Commentary on the Celestial Hierarchy.[163] The concept of
hierarchy embraces three realities, the divine, the angelic, and
the human. It signifies in the first place the creative wisdom
and power 'reaching from one end of the earth to the other
and ordering all things for good' (Wisd. 8:1). The entire creation is an *ordo* expressing God's nature. Analogously (for
nothing can be univocally predicated of God and creatures),
it denotes the angelic and the human hierarchy.[164] '*Ecclesiastica est angelicae imitatrix*';[165] the good order of the Church,
which consists in the divine ordination of sacred ministers to
rule over the people, must struggle to reproduce within itself
the order and beauty of the angelic choirs, as far as is possible

---

[163] See ed. McQuade, p. 137. I have not read the whole of the *Commentary on
the Ecclesiastical Hierarchy*. The bibliographical tradition records a *Sermo in Hierarchiam Triplicem* of Grosseteste, which Thomson managed to identify (*The
Writings*, p. 181); this may well have the parallels between the celestial and ecclesiastical hierarchies for its theme. It has thus far remained unstudied.

[164] *Comm. in Hier. Cael.*, ed. McQuade, p. 137: '. . . cum nihil positive dictum
possit de creatore et creatura univoce dici, manifestum quod ierarchia univoce dicta
non potest universalius dici quam de angelica et humana.'

[165] *Ex Rerum Initiatarum*, ed. Gieben, p. 140.

in the conditions of time and space.[166] Order on earth can never, of course, be as strongly confirmed as among the angels, being a more fragile and precarious thing.

Everything in the earthly Church must be done according to the example of the Church triumphant, since the same teletarchic and thearchic law rules all sacred operations in both.[167] That same decree which founded hierarchy enjoins that every act be performed with the end of sacred order in mind, namely, the return of the rational creature to God. It forbids any agent's usurping the office of its superior in the hierarchy, commands that it carry out to the full its proper function for the good of all (the common good must always be preferred to the individual, needless to say),[168] and commands the superior to hand on to its inferior the enlightenments it has received, giving neither impossibly much nor unduly little but exactly in the measure of the recipient's capacity.[169]

The theology of holy orders in the Church is presented as a deduction from the hierarchical order manifested in the celestial choirs. As each heavenly order passes on something of its power to its inferior, it retains at the same time some aspect of power which is inappropriate to the lower grade. Similarly, no higher order in the Church militant passes on the fullness

[166] *Comm. in Hier. Cael.*, ed. McQuade, p. 430: 'Est namque bona ordinatio ecclesiasticae hierarchiae, quae consistit in sacro principatu et sancta sacrorum dispositione hominum a deo aliis ad regimen sacrum propositorum, assimulata secundum possibile bonae ornationi angelicae, quae consistit in bona ordinatione ornatum faciente, et in hac assimulatione habet hic et habebit amplius in futuro angelicam decentiam.'

[167] Lyons Dossier, ed. Gieben, *Coll. Franc.* 41 (1971), p. 360. It is scarcely necessary to remark that Grosseteste is not enunciating a new principle here; the conviction that the Church on earth is the visible reflection of the heavenly courts was omnipresent in the sphere of influence of Pseudo-Dionysius in the high Middle Ages; it became a commonplace among Grosseteste's own generation, even something of an ideology in the following one, when it came to play a role in the conflict between seculars and mendicants. Pseudo-Dionysius gave formal accreditation and systematic form to this idea, but it anteceded him, being traceable also to Pseudo-Jerome in the fifth century. Most valuable on its history and place in thirteenth-century ecclesiology is Congar, 'Aspects ecclésiologiques de la querelle entre mendiants et séculiers dans la seconde moitié du XIIIe siècle et le début du XIVe', *Arch. Hist. Doctr. Litt. M.A.* 28 (1961), 35-151; pp. 114-38.

[168] *De Cessatione Legalium*, MS cit., fol. 179^C: 'Commune bonum privato bono semper est praeferendum.'

[169] *Comm. in Hier. Cael.*, ed. McQuade, pp. 126-7; 'Lex igitur teletarchica et thearchica est, ut quicquid hierarchice in teletis agitur, fine revocationis et reductionis in deum agatur, nec usurpat agens quod supra se est, nec omittat quod sibi conveniens est, nec tradat alii quod est supra suam receptibilitatem, nec deneget ei quod suae congruit susceptibilitati.' cf. Lyons Dossier, pp. 360-1.

of its power to the lower grade, but retains some incommunicable aspect which the inferior could not appropriately receive; the priest cannot ordain or confirm as can his hierarch, the bishop, nor can the deacon consecrate the body of Christ.[170] As well exemption as usurpation is forbidden, however, for both are *ordinis perturbationes*.[171] If a coadjutor is deputed to a subordinate task he must attribute his power and function to the superior and be always answerable to him, but in no respect can he be exempted from the superior's jurisdiction, any more than can an angel from that of the seraphim.[172]

The spiritual interpretation of Moses as the type of the prelate comes to reinforce the principle derived from the nature of hierarchy. Moses received from the Lord the government of the whole people. Though he associated others as tribunes and judges with himself, he retained the *plenitudo potestatis* to correct, reform, and judge cases. While dividing the burden, he did not lessen his authority but retained it entire. In all hierarchical works, the ministry of assistants is effected through the power received, and the superior performs their actions more truly than they do themselves, because more ultimately.[173]

The Dionysian principle of the centralization of power, that is, its progressive concentration at each higher level of the hierarchy, was Grosseteste's main theological defence of the *plenitudo potestatis* of the papacy, an idea to which he gave unreserved adherence. One is tempted to conclude that Pseudo-Dionysius's arrival was opportune for Scholasticism, and helped to furnish it with a cosmic theology which appeared to support the ecclesiastical centralization produced by the Gregorian Reform and its sequels. No doubt the lawyers' theories of papal jurisdiction and of *plenitudo potestatis* were

---

[170] Ibid., p. 127: Grosseteste's most formal treatment of orders is a lengthy application of the principles of the *Ecclesiastical Hierarchy*, in which purgation is appropriated to the deacon, illumination to the priest, and the fullness of thearchic government to the bishop: *Ex Rerum Initiatarum*, ed. Gieben, pp. 136–40. Cf. *Letter to the Dean and Chapter of Lincoln* (1239), ed. Luard, p. 361: 'Hoc namque modo concordat ecclesiasticae hierarchiae ordinatio cum caelesti, in qua quicquid potest inferior ordo potest superior, et non econverso.'

[171] Lyons Dossier, ed. Gieben, *Coll. Franc.* 41 (1971), p. 361.

[172] *Parabola*, ibid., pp. 270–2.

[173] Letter 127 dwells at inordinate length on this point, hoping to convince the chapter that they cannot be exempted from episcopal visitation (Luard, pp. 358–60).

put forward for the greater part in furtherance of an ideal of a united Christendom, set free at every level from parish Church to Rome itself from secular domination of the *Eigenkirche* type; but for the more speculative sort of mind a rationale of a purely legal or administrative kind no doubt appeared insufficient. Grosseteste assented to the theory of *plenitudo*, not on empirical, juridical, practical, historical of any other imaginable grounds, but quite simply because it appeared to him to be indissociably a part of the whole scheme of things, as known by reason and biblical faith. That is why he sometimes chose to advocate it in terms of analogies with the celestial hierarchy.[174] More frequently still, perhaps, he compared it to the ordering of the physical universe, thus invoking the second great transcendental paradigm of order we have mentioned. I can do no better than quote from the *Memorandum* which he read to the Pope at Lyons, during that dramatic audience on 13 May 1250:

This most Holy See is the throne of God and like to the sun of the world in His sight. Whence just as there exists causally in the sun the whole illumination of this world, its vegetation, nutrition of sensible life, augmentation, consummation, conservation, beauty, and grace; and just as the sun keeps producing these effects in this sensible world and so makes and keeps this sensible world perfect, so this most Holy See ought to have all these things, spiritually understood, within itself causally, and ought to cause all these things to flow unceasingly into that whole spiritual world of which it is the spiritual sun, and so save that spiritual world. Otherwise, just as, if the causal reasons in the visible sun should fail and the influences from it upon the world, straightway this whole sensible world would perish; even so if there should fail in this spiritual sun the spiritual causal reasons, which correspond to those of the sun, and from it the corresponding influences upon the spiritual world, this world of which it is the sun must needs perish, and it must be the cause and be guilty of this perdition, especially since these causal reasons and the influences therefrom are in its free power. But God forbid, God forbid that this sun, altogether shining in its intelligence and always straight in its justice, should at any time be turned into darkness and turn black like a piece of sackcloth, or retrogress like a planet.[175]

This parallel between the Pope and the sun recurs several times in Grosseteste's episcopal writings,[176] and the analogy

---

[174] Letter to Cardinal Gil de Torres, *Epistolae*, p. 126, transl. by Pantin, 'Grosseteste's relations with the Papacy and the Crown', in *Scholar and Bishop*, ed. Callus, pp. 184-5.

[175] Pantin, ibid., pp. 186-7. For the Latin text, see Gieben's edition of the complete Lyons dossier, *Coll. Franc.* 41 (1971), pp. 361-2.

[176] See, e.g., *Epistolae*, N. 36, p. 125: '. . . et sicut in mundo sensibili sol iste

between the physical macrocosm and the Church's organization shows itself capable of yet fuller exploitation. To give an instance: the prelate is like a star in the firmament. He reflects the sun's light to earth and so, without usurping the sun's glory, co-operates in directing kindly, warming light upon the earth. The prelate must have the radiance of the faith and of wisdom within him, and by his good example must enlighten the people under him. He must be constant like the fixed star, uniform in motion and removed from the conflicting gravities of the world, the ballast of concupiscence, the buoyancy of pride. Fixed in his order as in a firmament, he should no more ambition to rise in the world's eyes than a star does to ascend to the empyreum.[177]

I cannot resist paraphrasing a few lines of Grosseteste's *Expositio in Galatas*, so revealing are they for his views on the relationship between the natural world and the human moral order. St. Paul's outburst, '*O insensati Galatae*' (Gal. 3:1) may sound over-severe, comments Grosseteste; but the sin of the Galatians in falling away from the true Gospel was the basest of all—to begin in the better state and end in the worse, when every motion in the natural world tends on the contrary to acquire some perfection and every agent acts to reduce the unformed, imperfect, and material to form, perfection, and act. The Galatians ran counter to the order of the whole universe, in turning from spirit to flesh even while this corruptible lower world is tending towards renewal and the donning of incorruptibility, as our very bodies are to resurrected life. Growth to fullness of being is inscribed in the order of things. The world in its creation, progression, and consummation convicts the Galatians of stupidity, in reversing the progress they had made.[178]

The recourse had by medieval writers to transcendental paradigms of social organization is something that is likely to repel the modern reader. 'Platonism for the people', Nietzsche called it, and our instincts today prompt us to concur in the

conspicuus suo praeeminente lumine mundi tenebras purgat, singulariterque mundum illustrat, motuque suo ordinatissimo, ut opinantur mundi sapientes, caeteros motus corporales naturales ordinat et regulat; sic in orbe ecclesiae summus pontifex vicem solis obtinet . . .'

[177] *Letter to the Dean and Chapter*, *Epistolae*, p. 389; sermon to clergy, ed. Brown, *Fasciculus*, p. 268.
[178] Oxford, *MS Magdalen College 57*, fol. 10ᵃ.

dismissal. Preferences aside, however (and tastes in world-views are less fully accountable to the unbiased *ratio* than we generally incline to believe), we must try to appreciate this mentality for what it historically was. When accorded its full weight as an essential part of the medieval view of things it may help to rid us of a prevalent contemporary myth concerning the type of central position man is supposed to have arrogated to himself, until the successors of Copernicus dislodged him from his cosmic pedestal and taught him humility before the infinite silences of the heavens. In fact the representations of man's central position in the world formed a composite picture, thematically portraying his lived relations to his milieu of life, and all the shades and tints used in the painting cannot easily be reduced to primary colours; the basic composition may be relatively simple, yet the disposition of its elements taken separately and together, and the relation of foreground to background, difficult to assimilate at a glance. We have argued that the creationist view and its conjugation with the theme of order exalted the uniqueness and centrality of human nature within being. However, a corrective must be added, namely, that man was conceived of as unique only within an encompassing order which, as it situated and framed his nature, likewise set limits to his freedom and directed it.[179] Order there must be in human affairs, else chaos and disruption, *horror et perturbatio*, must prevail, Grosseteste proclaimed. Now, the general medieval conviction was that man is not reduced to imagining various possible arrangements of society and attempting to enact the object of his arbitrary choice. A celebrated line of the *Consolatio* remained active in the medieval mind: '*Cum mores nostros totiusque vitae rationem ad caelestis ordinis exemplar formares.*'[180] The universe, they felt, was there to provide prototypes of good order,

---

[179] Père Congar may have exaggerated somewhat in claiming that 'their dominant world-view was a cosmological and cosmocentric one, rather than an historical one centred on man and his designs.' ('Two factors in the sacralisation of Western society during the Middle Ages', in *Concilium* 5 (1969), p. 32). A cosmological view can be made to focus upon man just as surely as an historical one, although with somewhat different consequences for the idea of human nature.

[180] Boethius, *De Consolatione Philosophiae I, Prosa* 4 (43), (*CCSL* XCIV, p. 7; *P.L.* 63, 615). Quoted by Grosseteste when introducing the spiritual sense of the heavens, *Hexaemeron*, MS cit., fol. 217ª. Note the ambiguity of *caelestis ordo*: the regularity of astronomical phenomena, and the good order preserved by the angelic choirs can both be understood under it.

just as ancient authorities were at the disposal of the writer
to offer him models and themes of literary composition; hence
it would have been considered an acknowledgement of human
poverty rather than of wealth, to have to invent purely and
simply at such a late stage of history.[181] In art and human
affairs there was only one order that imposed itself as unques-
tionably worthy of the cult of imitation—the one actually
given, the real order of the nature of things, embracing the
world visible and invisible, grounded directly and infallibily
upon the divine archetypal ideas, and affirmed by the infinite
creative will.

'But the whole view is so laden with sacrality', the modern
reader wants to object. This is perfectly true. The disposition
of ecclesiastical functions within the Church derives from
God's covenant with Israel and its renovation in Christ; to
subvert it would be to rebel against his will and do him injury,
Grosseteste declares.[182] The high Middle Ages saw the full
development of a dimension of sacrality which had been initiated in some respects already in earliest patristic times,
namely, the use of Old Testament models of cultic priesthood
and theocratic kingship, hierocratic views of law and paternal-
istic theories of government.[183] The transcendental models of
order reinforced the sense of sacrality, their sacred and eternal
cosmos advocating fixity, conservatism, and paternalism in
social relations. No doubt medieval man did not possess any
sharp awareness of human historicity and did not regard social
structures as a cultural product proposed and disposed by
majorities. Time and change were regarded more generally as
signs of the fragility of human nature and of man's imperfec-
tion than as positive dimensions of his being; the tendency,
however, was to feel grateful that the assurance of good order
was not left quite at their mercy, but was founded upon something older and nobler and more dependable than chance, the
whimsy of human will, or the ambition of individual rulers.
The divine will was acknowledged to be the ultimate foundation of all order, and 'order' connoted good, and ultimately

---

[181] See Lewis's characteristically original remarks on the medieval literary
imagination and conception of originality, *The Discarded Image*, pp. 206 ff.

[182] *Letter to the Dean and Chapter, Epistolae*, p. 371.

[183] For references to Christian literature from post-Apostolic times (Clement
of Rome) through the Irish Penitentials and the Carolingian writers, see Congar,
art. cit., pp. 28-31.

sacred, order. However, the sacral forms with which order invested itself did not exclude a large measure of creativity, of a certain kind at least. Key medieval words such as *reformatio*, *renovatio*, and *legis emendatio*[184] made reference to the new vitality with which old ideas could be invested, and it cannot be forgotten that medieval society was prolific in the creation of social institutions. Indeed, almost all modern institutions with the exception of the technocratic state and the more repressive forms of democratic government trace their origins back to some medieval ancestor. The matrix of social organization was the Church, the central sacred institution, of which so many others became microcosms: monastery, hospital, school. It is too little realized today that a strong sense of the purpose of things, the presence and acknowledged reality of paradigmatic—that is to say, guaranteed—models of organization (the properties of a more or less sacralized world-view), can combine with the perception of a social need or duty (which is the property of humanity simply speaking), to produce a strongly-motivated and humanly creative service of others. Indeed, perhaps nothing else can do this in quite the same measure or degree.

An apt illustration of precisely this is found in Grosseteste's case. In the *Propositum* presented to Pope Innocent he describes a new institution which he has established in Lincoln Diocese.[185] Prompted by his pastoral responsibility for souls and by the thought of the strict account he must one day render for each one entrusted to his oversight, he has instituted a personal visitation of clergy and people in each rural deanery. He sketches in a few tantalizingly brief details of his practice on such occasions, before remarking that on the occasion of his very first circuit certain people approached him and remonstrated with him: '*Domine, novum facitis et inconsuetum.*' His answer: novelties adjudged constructive in the light of the Gospel can only be a blessing, for they build up the new man. Only such innovations as bind a new burden onto the backs of the oppressed and, by inflicting oppression, offend against the Gospel of Christ, are to be condemned and suppressed, as merely further evidence of the devil's astuteness. The sequel has relevance to our theme. Grosseteste confesses his present fear that the new institution could indeed turn into a burden

---

[184] Ibid., p. 33.   [185] Lyons Dossier, ed. Gieben, pp. 375-7.

for the people of Lincoln, if his successors make it an excuse
to extort procurations, something which they could legally do
and which, he claims, the Archbishop of Canterbury is doing,
putting the money of the poor into his treasure-house. Evidently, his profound consciousness of the sacredness of his
office and of the institution he served in no way reduced
Grosseteste's sharp awareness of the ambiguity of human situations and institutions. If the latter are not served altruistically
by idealists their sacred character will neither prevent nor
delay their degeneration.

A third typical contemporary reaction to the sacralized
order of the Middle Ages is disquiet at the lack of social mobility it entailed. Certainly texts are not wanting (any less than
are historians) which suggest that the hierarchical subordination, which was so much a feature of the macrocosmic order,
imposed social rigidity in the name of a sacralized conception
of order.

There is a fundamental truth about medieval society here,
for the feeling was widespread until the end of the Middle
Ages (indeed it remained so well beyond that), that, as in the
universe, so also in society providence has fixed an order of
ranks or estates (*ordines*), whose very name connotes higher
and lower without possibility of emancipation. Each individual
has a place assigned him within the system, has the corresponding rights and duties and belongs to a class firmly integrated
into the overall order. Even in the communes, which had no
theoretical place in the feudal order and, indeed, challenged
its basic presuppositions, there remained a great deal of the
class and rank system of any limited social order. The un-
attached individual was an anomaly, worthy of excommunication or exile. The place of the individual within the group
was covered with the mantle of sacrality, which promised
'salvation through obedience within one's rank',[186] forbade
self-exaltation through pride or ambition, condemned envy
or contempt for the station of others and advocated humility.

An over-schematized account of social relationships and

---

[186] Quoted by Congar from the *Hortus Deliciarum* (art. cit., p. 32). There is an
abundant literature on the estates; two of the best contributions are Ruth Mohl,
The Three Estates in Medieval and Renaissance Literature (especially pp. 277–83),
and Congar, 'Les laics et l'ecclésiologie des *ordines* chez les théologiens des XIe et
XIIe siècles', in *I laici nella Societas christiana dei secoli XI e XII* (Milan, 1968,
pp. 83–117.

the mentalities surrounding them has the disadvantage that it appears to tell us something of how those relationships were actually lived, but does not in fact do so. However, if the lived relationship is something that has left insufficient trace to allow us identify with it 'from inside', there do still remain elements within the abstract, schematic account in respect of which we can avoid some excesses of historical *naïveté*. In this connection a glance at some relevant remarks of Grosseteste provides a few important correctives.

Grosseteste was at once a thinker who had done his share of reflection on the nature of order and a moralist well able and accustomed to express his fundamental convictions; yet I cannot recall that any word of his directed people to love their rank in society. Speaking to his priests, of course, he admonished them to put away ambition: *'unusquisque maneat fixus in suo ordine'*.[187] But this was good sense, not ideology: he did not want the Church to be beset with place-seekers. He could treat of *superbia* without mentioning social rank, save to warn that it is liable to incite people to envy or contempt.[188] His *Dictum de Humilitate*[189] defined the virtue as 'the love of remaining in the rank befitting one according to all one's conditions', but the reader looking for a canonization or sacralizing reinforcement of the estates will be disappointed, provided he takes the trouble to read on, for the order of which Grosseteste speaks is not social rank but something altogether more metaphysical, namely, the position of man in the universe: man is subject to God alone, equal to the angel, and superior to all other things. The *conditiones* he refers to are the sinfulness in one's own human nature and in history, the virtues one possesses, and so on. The moral is, that the man who knows himself thoroughly can begin to love his *status*, or as we might

---

[187] See ed. Brown, *Fasciculus*, pp. 268 ff.

[188] *De Confessione*, pp. 264–6. '*Superbia*' is defined in the *De Cessatione Legalium* (fol. 161^C) as '*amor excellentiae propriae*'; it is the refusal of the rational creature to accept his creatureliness and his place in the order of creation.

[189] Brown, op. cit., pp. 190–1: 'Humilitas est amor persistendi in ordine sibi congruo secundum omnes conditiones suas. Verbi gratia, prius amat in universali persistere in ordine sibi congruo, non tamen cognoscit quis status sibi congruat secundum conditionem humanae dignitatis, neque quis sibi congruat secundum quod ipse est peccator, neque secundum quod patiens, vel justus. Cum autem cognoscit iam quod status debitus humanae dignitatis est subesse soli deo, parificari angelis et animabus sanctis, praeesse omni alii creaturae, incipit amare hunc statum quem prius non amavit nisi in universali.'

say, can affirm his existential reality. Probably Grosseteste himself made little enough of social distinctions; his opponents among the Lincoln canons did not allow him to forget that he was of humble origin. Certainly in judging the fitness of candidates for pastoral office he was guided by considerations quite other than those of birth and connections. The latter at least would, if anything, have prejudiced rather than enhanced a candidate's chances of promotion.[190]

It is even now still not totally impossible for us to conceive how social rank and its attendant duties were once the un-questioned guide to responsibility, understood largely in terms of devotion to the allotted task and obedience to natural superiors. What we, perhaps, find more difficult to recover imaginatively is the double sense of responsibility and security felt by men who lived their lives under the sacred canopy of an overarching cosmic and divine order, through which time and history, those bearers of the seeds of mutability and cor-ruption, were made to enter into symbiosis with eternity, to form a single firmly-integrated order of things. The elements of sacrality contained in this view did not, we have argued, outrule social and institutional creativity, but they did help to structure it and perhaps to set some kind of upper limit to its efforts.[191] Above all, however, the sense of objective order was felt to demand an adequate response from man, the one unfinished project needed to complete the whole. We are free to believe that the idea of hierarchy, for all that it may have been rigid, moved the finest souls to a profound sense of re-sponsibility, to something, therefore, that transcends historical conditioning to achieve the universality of the human as such. Grosseteste himself is an excellent example of this.

XI ORDER, RESPONSIBILITY, AND REFORM

Most of Grosseteste's ideas about the Church, the pastoral office, and the realm, were not original with him, but they were profoundly personal to him. His uniqueness lies in the thoroughness with which he brought them to their furthest

[190] Many of the candidates he rejected were well-connected people; see Strawley, 'Grosseteste's administration of the Diocese of Lincoln', in *Scholar and Bishop*, ed. Callus, pp. 159–61.

[191] Congar convincingly explains the absence of revolutionary movements, as distinct from uprisings, from the Middle Ages, as a result of the deep-rooted rever-ence for sacred order.

practical conclusions and turned them into a programme for the reform of the Church. We can only guess at what it was that prompted him to relinquish his teaching in order to undertake the duties of a bishop, but his sense of responsibility must have been an important factor. As a scholar-bishop, he brought to his task all the theological and philosophical learning of a lifetime, something that gave him a breadth of vision transcending petty politics and the comfortable clerical world of customs and rights. His intellectual interests remained alive and continued to evolve after his elevation; his public life was, as it were, their dramatization on the stage of history.

The idea which guided everything he did in those last years was *cura pastoralis*, the responsibility of the prelate or priest for his flock. To have a place in the Church's hierarchy is to be subject to the laws of hierarchy, namely, order, knowledge, and operation. The very substance of hierarchy is the Scriptures[192] and its sole end is the salvation of souls. Knowledge of revealed truth, therefore, and active involvement in ministry are the chief conditions for office in the Church. Power without involvement is a pretence, and without knowledge is like a weapon placed in the hands of a maniac.[193] No art is higher than the divine task laid upon the pastor, and all other human arts are subalternate to it.[194] Grosseteste allowed no function in the Church, not even the study and teaching of theology, to be preferred to direct involvement in the ministry. Other pursuits have their value and legitimate place within the entirety of human occupations and needs, but as the whole world exists for man, serving his temporal requirements, and the body exists in turn for the soul (as matter for the sake of form and the imperfect for the sake of the perfect), so the salvation of souls is in the nature of things the highest vocation and the most urgent task.

The abuses which Grosseteste attacked, first in his own diocese and later, inevitably, in the universal Church, were all abuses of what he considered to be the due order of things and which involved placing some human interest or institutional consideration before the Church's divine task. His *gravamina*

---

[192] Lyons Dossier, pp. 382-3.
[193] 'Potestas autem sine operatione cassa est et vana, et sine scientia, quasi gladius in manu furiosi.' (*Comm. in Hier. Cael.*, ed. McQuade, p. 139).
[194] Lyons Dossier, p. 365.

were set out lucidly in the dossier which he presented at Lyons in May 1250.[195] They are penetrated by the attention to underlying principle that was typical of him and came always to the fore in situations of conflict. His charges were chiefly five.

The stagnation of the Church's work has one sole cause, the appointment of bad pastors. Grosseteste's mode of argumentation may be characteristically academic,[196] but its logic is unassailable and its criticism clear-sighted. Pastors who neither preach nor live by the Gospel are stranglers of the sheep and make the Church a den of thieves. By destroying the spiritual life of those placed in their care they are worse than sodomites. They intensify the darkness of those who wait in shadow for the coming of the light. He can no longer keep silence. The cause and fontal origin of this evil is the Roman Curia. Its purpose and duty is to correct abuses, its actual occupation, however, appears to him to be to multiply them by provisions, dispensations, and the collation of benefices on unscrupulous place-seekers. It provides for the temporal ease of its favourites, regardless of the spiritual fate of souls, for whom Christ was crucified. Wolves dressed as shepherds, paralytic steersmen in the middle of a storm, impotent, negligent, ignorant sybarites, Herods with hands steeped in the blood of innocents, are given the places which demand other Christs at the service of the people. Benefices are conferred for venal, temporal motives. Nepotism, reward for services rendered, temporal advancement, and favour-currying with the influential, are all taken for granted.

The next two charges are laid against the constrictions placed upon the liberty of action of the Church. Grosseteste regarded with particular aversion the exemptions which removed whole areas of Church life (monasteries, cathedral chapters, and appropriated benefices) from the influence of episcopal visitation and correction. The bishop must be free to regulate the whole region for which he is answerable, just as the Pope is answerable for the whole Church. His consecration gives him that right, and hierarchical order demands that it be guaranteed. All things must be done in the earthly

---

[195] Pantin, art. cit., examines them in detail. Our only purpose here is to trace within them the presence of his view of order or the nature of things.

[196] See, e.g., Lyons Dossier, p. 353: 'Unius enim una est causa, et oppositorum causae oppositae. Et quia boni corruptio est mali oppositi generatio, malique pastores sunt causae corruptionis fidei et religionis christianae . . .'

Church as in the celestial hierarchy, and there is no such disruption of order there as exemptions by dispensation.[197]

Grosseteste had very definite ideas on the relation of Church to realm. Priesthood was instituted to govern for eternal peace, kingship for temporal, he reminds Henry III;[198] neither should interfere with the other's domain, both should co-operate. However, the Church receives its power immediately from God, the prince, on the other hand, from God through the Church, so that in the last analysis *'uterque gladius, tam materialis quam spiritualis, gladius est Petri'*.[199] Whereas pagan kings looked to the wisdom of philosophers for guidance, Christian princes have the Gospel for light, hence the virtues expected of them as rulers: tolerance and charity to their subjects, respect for the Church's freedom and its goods.[200] At his investiture the king receives a non-sacramental anointing conferring the gifts of the Holy Spirit, for his chief need is wisdom to guide his subjects. The 'knowledge, counsel, and fear of the Lord' conferred upon him refer to the sensible world, the intelligible world, and God, 'so that after the pattern of the order of the world and of the angelic orders, according to the laws written in the eternal reason of God, by which he rules the whole of creation, the king also may rule in orderly fashion (*ordinabiliter*) the commonwealth subjected to him.'[201] Typically, Grosseteste has less to say concerning the dignity of the king's position and more about the responsibility it imposes; he was no courtier. He complains to the Pope that the freedom of the Church is not being respected by the secular power, in England above all.[202] His liberty as a bishop to interrogate the laity of his diocese has been threatened by the king. Royal

---

[197] Ibid.: 'Quos permaxime condecet et perurget districtissimi mandati Dei summa necessitas diligentissime considerare et circumspicere, ut omnia in aedificatione Ecclesiae militantis faciant secundum exemplar Ecclesiae triumphantis ... In qua insuper nulla est inferioris a suo naturaliter et divina dispositione superiore exemptio, ac per hoc nulla ordinis perturbatio, nulla nisi sacre facta quorumcumque dispositio vel administratio.'   [198] *Epistolae*, ed. Luard, p. 348.

[199] Ibid., p. 90.   [200] *De Cessatione Legalium*, MS cit., fol. 171^d.

[201] Quoted by Pantin from *Epistolae*, p. 349. Cf. the classical statement of this theme in Aquinas's political philosophy, where the microcosmic dimension of human government, modelled on God's providence in the universe, is explicitly formulated: 'In things of nature there is both a universal and a particular government. The former is God's government, whose providence rules all things, the latter is found in man, and it is much like the divine government; hence man is called a microcosm.' (*De Regno* II, 1, tr. Phelan-Eschmann, pp. 53–4).

[202] Lyons Dossier, pp. 373–5.

jurisdiction is encroaching upon ecclesiastical, and while Christ divided the two sovereign powers in order to free the Church and her ministers from secular affairs and the judgement of blood, the king is not slow to employ clerics in wordly business, even as judges.[203] Finally, the law of the land is not in all respects in conformity with the Church's, nor indeed with the divine natural law.[204] In all these ways the natural freedom of the Church, which was guaranteed in the Great Charter and by the anointing oath, is being restricted, and people fear that it will disappear altogether.

Grosseteste charges the Curia with the worst kind of pragmatism. He finds as a typical excuse for the abuses that are being committed, that the times are evil and malign and much must be dissimulated and tolerated for the common utility of the Church.[205] He warns severely that no smallest evil may ever be done in order that good may come of it. The worse the times and the world's actions, the more principled must be the behaviour of those who are not of this world. Where the battle is most thickly joined it is manly struggle that is to be prescribed, not capitulation nor wholesale flight; the world's evil must be answered by an intensification of the good. In one point the charge of expedience is unmistakeably aimed through the Curia at the Pope himself, for the policy of war against the empire was the personal responsibility of Innocent IV. The command of Christ to Peter, 'Put up thy sword', has gone unheeded, Grosseteste protests, and the Church has taken to using the weapons of the world. People everywhere fear that the awful threat of Christ may come upon it and it may perish by the sword it has taken up (p. 367). Has the thought not occurred to his listeners, that the recent sufferings of the Church may be a well-merited penance? (p. 369).

The final charge is a summary of them all. Its occasion is the commission given to Grosseteste to complain on behalf of the English clergy against the procurations that are being exacted by the Archbishop of Canterbury at visitations.[206]

[203] See *Epistolae*, p. 92, and Pantin, art. cit., for the case involved.
[204] In the question of bastardy (Pantin, pp. 204-6).
[205] Lyons Dossier, pp. 358, 368: 'Nec proponat quis in excusationem talium quod dies mali sunt et mundus in maligno positus est, et ideo oportet multa dissimulare et sustinere pro communi Ecclesiae utilitate, quasi ob hoc esset vel minimum malum faciendum ut eveniat quantumcumque bonum . . .'
[206] *Conquaestio Cleri Anglici*, ibid., pp. 373-5. The documents which follow

Canonically these are legal, but they come from the Archbishop's greed, not need, and are an occasion of scandal to the
Gospel. It is a case where legality is limited by the overriding
demands of the divine and natural law. Canterbury is laying
new yokes upon backs that can bear no more. He is a tyrant
according to the strict definition found in Aristotle, for he is
seeking his own profit, not that of his subjects.[207] But if he is
found wanting already by pagan standards of what is due, how
much more so by those of the Gospel and the natural divine
law! Grosseteste invests the case with an importance that raises
it above its essential sordidness. The whole Church needs to
undertake a re-examination of the balanced compromise it has
struck in so many areas of its life between divine and human
law. Positive law with all its apparatus is no more than traditions and commands devised and imposed by men; these
may never be elevated to a level of equality with the Scriptures,
which are identically natural and divine law, nor usurp their
authority. People such as Canterbury, who claim the sanction
of common law for self-enrichment at the expense of others,
reverse the natural order of things, for natural justice sets an
absolute limit on human legal arrangements. The laws of men
can never be absolute. Hence if there is even the suspicion of
a conflict between common and natural law, there is an absolute duty to follow the law promulgated by the divine reason
and to disregard the human. Against the self-interest that
takes dishonest refuge behind the sanction of custom and traditional right, Grosseteste summons the prophetic tradition
of both Testaments to accuse and expose the workings of
legalism and phariseeism in the Church of his times. Needless
to say, his efforts were not aimed at the institution of ecclesiastical or civil law itself, for while he may not have had much
interest in law[208] he never attacked more than its abuses; his
main concern as philosopher and reformer was to establish its
limits as human law, and as a humane man to relate its application to the actual circumstances of life. Regarded from his
point of views, procurations which are legitimate in themselves

(nos. 4 to 7 in Gieben's edition) contain a justification of the position Grosseteste
has taken up in the *Conquaestio*.

[207] *De Rege et Tyrannide*, ibid., pp. 377–81.
[208] *Epistola* 24, p. 96: 'Ironice autem adnectis me scire omnes leges, cum sim
homo legum imperitus.'

can, according to circumstances, be unjust in virtue of the harm
their exaction creates and the disproportionate burden they
impose;[209] in this case the single, absolute measure of right
and wrong sets *a priori* limits to established rights. Again, no
purely human law is of universal application, because the
legislator cannot foresee the circumstances of each case that
will fall under it. To deny the need for jurisprudence and
epieikeia would be to divinize the law by equating it with the
natural law, which alone applies to all men always, since it was
devised by an all-foreseeing providence. The moderation of
positive law is the good judge's way of serving natural justice.[210] There was more than good sense and humanity in what
Grosseteste said, there was a critical sharpness which reflected
the increasing frustration with which he saw his reforming
efforts dissolve into litigation. Tutorial appeals by deposed
priests, endless sessions of legal quibbles, subtleties and empty
solemnities wear a pastor down, he complains, till life itself
becomes a burden.[211]

Grosseteste did not shrink from applying his ideas on human
and divine law to the actions of the Pope. The order of reason
and nature demands, he argued, that the *virtus influens* should
be able to do more than the *recipiens*,[212] wherefore all owe
ready obedience to the commands of the Holy See. However,
it is the same natural and divine law, that antithesis of legal
positivism, which places *a priori* limits on the *plenitudo potestatis*, as it does on all human power, even that which is exercised in the name of Christ.[213] At Lyons Grosseteste gave
warning of the tragedy which could overtake the Church if
papal power, which is power for constructive and evangelical
action, should forget its nature and be carried by hybris to
command wrongdoing. Any Christian obeying such a command
would incur no less a sin than the Pope who issued it;[214] he
fervently wishes that it may never come to that. Three years
later, however, it did, to cause him the most bitter moments

[209] Lyons Dossier, pp. 383-4.   [210] Ibid., p. 386.
[211] Ibid., p. 364-5.   [212] *Epistolae*, ed. Luard, p. 364.
[213] Lyons Dossier, p. 362.

[214] Ibid: '. . . praesidentibus huic sacrae sedi, in quantum indutis Christum et in
tantum vere praesidentibus, in omnibus est obtemperandum. Si autem quis eorum,
quod absit, superinduat amictum cognationis et carnis aut mundi aut alicuius
alterius praeterquam Christi et ex huiusmodi amore quicquam praecipit Christi
praeceptis et voluntati contrarium, obtemperans ei in huiusmodi manifeste se
separat a Christo et a corpore eius quod est ecclesia . . .'

of his life. In nominating his own nephew a Canon of Lincoln, Pope Innocent issued a challenge to the bishop's whole view of things. It remained only for the latter to draw the consequences of the highest loyalty he acknowledged, that of conscience to the natural order of things, the order from which the Pope had by this act separated himself, and, out of that loyalty, to rebel and disobey. The juridical limits of papal authority which he could have invoked have been set out by historians, but the profound meaning of what he did can only be seen in terms that are beyond law, for he acted out of a conviction that was rooted at its deepest point in a sense of values whose foundations were set in the whole cosmic order.

If there is a single principle that can be isolated as the ultimate underpinning of Grosseteste's sense of order it is that God has arranged the universe in the best way: *Deus melius facit*.[215] The equation once drawn by the Greeks between κόσμος and εὔκοσμος was reaffirmed on the basis of a metaphysical faith and made the premiss of a rational optimism that manifested itself in the scholastic renaissance, from the School of Chartres to Aquinas. This faith lent support to the idealist and reformer. The idea of order which it elaborated assumed a transcendental character, both as a category of thought and as a challenge to Christian action. It was from this direction that the medieval impulse was derived, to situate every domain of discourse, and even every social initiative, by referring it back to the totality—an emphatic assertion that all the realities of experience derive their meaning from an englobing divine plan. Collectivities were referred already by their names (*ordines, universitates*), through the higher levels of collection they served (the Church, mainly) to the ultimate community, the *universitas rerum*. Something like a material symbol for the pervasive sense of the wholeness of reality can perhaps be found in the mechanical clock, which was the invention of the later thirteenth century, and which, ill-content with the limited function of a timepiece, made itself into a cosmological device ambitioning to tell the time of the whole world-machine.

---

[215] 'Deus facit melius in ordine universitatis . . . minimaque ratio sit apud summam rationem efficax ut melius fiat.'—*De Cessatione Legalium*, MS cit., fol. 158[b].

In commenting Genesis Grosseteste declares that the goodness of a thing, namely, the activity for which it was specifically made, and the benefit accruing from it, can only be appreciated when its order to itself and to the other things in the universe is correctly understood.[216] The young Socrates, Plato tells us, proposed at first to investigate the good or purpose of each thing, but despaired of accomplishing the project and turned instead to human nature. Medieval Christians revived the Socratic ambition and believed that they could discern the good of all things, and of man as a part of what is. They accepted the principle which later ages found themselves forced to abandon, namely, that there can be no adequate anthropology that is not inscribed within a cosmology. It was the unashamed realism of their account of universal order that lent a characteristic density and richness of content to their idea of man. Order, they were convinced, pre-existed the coming of man and is objective, right down to the colours with which God (not the great architect of later ages, but a finished artist) painted the works of his hand, before any human eye was there to desire their sight. Man is in no degree the measure of this order. Though his nature is 'equal to all things and greater than any', he must take his place within it. The medievals found multiple ways of defining this place in virtue of a network of analogies taken from the megacosmic term of comparison and applied to the microcosm. And since they believed the given order of things to be a good order, an expression of the interior life of a God who is without envy, they felt that human initiative could not aspire higher than to reproduce the sacred order in its life and even in its cultural expressions; *ars imitatur naturam*, was the whole of political philosophy until after Aquinas.[217] Order, affirmed Grosseteste, is the link between beginning and end, between efficient and final causality. Order in thought and action can only be found in us if we never lose sight of our end, which is also our beginning.[218] The supreme and unquestioned ideal for man was to reproduce within himself the universal harmonious order.

---

[216] *Hexaemeron*, MS cit., fol. 202^C: 'Bonitas autem rei consistit in accione propter quam res specialiter facta est, et eiusdem accionis utilitate et in ordine eiusdem rei ad se et ad alia quaequae in universitate.'

[217] *De Regno II*, 1, ed. cit., p. 54.

[218] 'Est enim ordo colligatio principii cum fine. Qui enim in omnibus agendis tendit in suum principium, ordinate dicitur agere.'

Physics opened upon ethics; both were considered to be parts of a single wisdom.

Grosseteste's reforming activity cannot be dismissed as that of an unbending and doctrinaire academic unable to make the transition from schoolroom metaphysics to the world of practical affairs, with its very different rules. His ideal it was to inform practice by contemplation, to make faith programmatic for activity. The philosophical scope of his vision made him a sign of contradiction, for his idealism, grounded in metaphysics and religion, was unafraid of uncomfortable conclusions and intolerant of expedients that were not grounded in principle. His lofty views led to intransigence only when he found compromise with the comfortable ecclesiastical world of established custom and vested interest impossible. He recoiled from the sight of position and power being used without responsibility, from means being wrenched from their context of meaning and elevated into ends. What others, who did not share his intensity of conviction, regarded as self-justifying or tolerated as immemorial practice, he identified as abuse of order, and his fearless courage could not be silenced in its denunciations, nor his acquiescence commanded or bought. *Abeant studia in mores.* Metaphysics and political praxis coincided in him as in few other philosophers; there perhaps only Leibniz was his equal. Both failed to impose their metaphysically-grounded reforms on the Church, but the failure in action done out of idealism and the highest love is better than most success: '*Hic amor est res suavissima, quia nulla nisi per amorem sunt suavia, et sine amore omnia sunt amara.*'[219]

[219] *De X Mandatis*, London, *Brit. Library MS VII. f. 2*, fol. 184[a].

CONCLUSION

Conclusion

GROSSETESTE'S intellectual achievements mark him out as the most striking personality in the academic and ecclesiastical life of England during the second quarter of the thirteenth century. Of his contemporaries who achieved celebrity at Oxford, men like St. Edmund of Abingdon or John Blund, none emerges with the same clarity before the eyes of the historian. Of course, incomparably more of Grosseteste's actual writings have survived to perpetuate his memory and to attract increasing numbers of scholars to their study. However, the survival of such a large and diversified literary output cannot simply be regarded as a lucky accident; it bears its own eloquent testimony to the place which Grosseteste held in the memory of succeeding generations of scholars, and not in England alone. During the two or three generations after his death a number of patient and dedicated people, whose names have not come down to us, spent months and years of their lives in editing his notes, collecting his correspondence, and putting together and copying his philosophical writings and his sermons. Most of these men worked, no doubt, either at Lincoln or in the priests' library of the Franciscan convent at Oxford, to which Grosseteste bequeathed his books. It was in that library around 1310 that William of Alnwick studied the bishop's marginal notes on the *Physics* of Aristotle, and that Thomas Gascoigne spent happy and rewarding hours, in the course of repeated visits between the years 1433 and 1456, copying copious extracts from the glosses on St. Paul, which were not to survive the upheavals of the sixteenth century. Through his eyes we can imagine the room where the bishop's books, neatly catalogued (Thomas appears to have had a record of the catalogue-numbers), stood on shelves of their own, preserved, with the bishop's rush slippers in a display case, as relics of the man who was believed to have been a saint. Leland visited the library shortly before the suppression of the house in 1538 and reported in a letter to Thomas Cromwell that, much to his chagrin as a connoisseur of medieval

English books, the books of the bishop had vanished, 'stolen by the Franciscans themselves', and that the library itself was in a melancholy state, all cobwebs, cockroaches, and woodworm.

It is not surprising that the friars should have been tenacious of the memory of Grosseteste, nor that Gascoigne should have taken pride in being heir to him in the chancellorship and the same tradition of ecclesiastical learning. Grosseteste's intellectual achievement, matched as it was by his singular moral stature and his sincere devoutness, drew him inevitably into prominence within the institutions that were quickening with new life, first the new university, then the mendicant orders and the school at the Minorite Friary, before finally he was raised to the episcopate. More, perhaps, than we of today, the medievals were apt to weigh the significance of a human life in terms of the contribution made by it to institutions and, through these, to the common good of society. The University of Oxford kept the memory of Grosseteste's fame as chancellor of the time of its origins; and rightly so, for he had been the leading force in the formation of the intellectual tradition of Oxford and he had helped to shape the syllabus and launch the interests, including natural philosophy and mathematics, that gave the university an intellectual physiognomy which it carried down to the times of Bradwardine and Wyclif. Grosseteste put Oxford on the map in European terms by his championing of the Aristotelean movement there, at a time when its smooth development at Paris had been made impossible by Church intervention.

If we leave aside Grosseteste's outward, institutional achievements and concentrate upon those of the strictly intellectual order, the picture of him becomes somehow less focused. This is not because his contribution was not distinctive (he was for one thing the only original thinker among the Scholastics who was capable of translating from the Greek, and by his talent for philosophical scholarship he raised the interpretation of ancient texts to a new level), but rather because it is difficult to characterize his thought, even his philosophical thought, to which he owed most of his renown. He has been claimed for a variety of schools and philosophical currents: idealist, empiricist, Platonic or Neoplatonic, Avicennian and Aristotelean. Yet he belonged to none of them, not, at least, in any clearly-defined and exclusive way. For one thing, he never wrote a systematic work, either as a philosopher or as a theologian, with the result that even his most original thoughts were never fully synthesized and harmonized. It may be that he planned at one stage to write a summa of theology, as Fr Callus has suggested; if so, he does not appear to have got far with it, even though he was a man who liked to finish whatever he took up; that was a habit which corresponded to his need to feel in control of his actions and responsibilities. He was, indeed, methodical, as his working method, his sophisticated and persevering use of indexing-symbols for his library and, above all, his astonishing productivity, confirm. But it is one thing to be methodical in one's working habits, to assemble and animate a team of collaborators and helpers, to undertake tasks of inhibiting difficulty, magnitude and variety, to see these as the months and years go by take on form and shape in the reality of ink and parchment, always to have several projects going at the same time, only to look beyond them to yet more that remain to be tackled when the present lot is disposed of; it is quite another to work and think and write systematically, in a way, that is, that bends the multiplicity of problems and tasks into a unitary whole and that ties up the loose ends of inspiration into a single web of thought. Grosseteste did not achieve the latter kind of success, and it may be that he was not capable of it. His age, in a double sense, was against such an achievement.

First of all, Grosseteste, like all his contemporaries, experienced two different and contradictory intellectual impulses: the centripetal one of tradition and the centrifugal one of ancient and Muslim science. The theologians at Paris such as William of Auvergne, William of Auxerre, and Philip the Chancellor, like Grosseteste, remained faithful to the theological tradition in which they stood, and which had received its salient delineaments from the towering figure of St. Augustine. Yet they read and sought to assimilate books by the pagan greats, Aristotle and Ptolemy, and by the Muslim commentators who had inherited the classical tradition of science and philosophy. Eclecticism was the inevitable result of the struggle to fuse the two. Grosseteste had his own singular approach to the situation of his times, and it was an approach that becomes explicable if we take into account the age at which he launched himself into the new currents.

Roger Bacon stated, correctly and perceptively, that Grosseteste's mastery of the sciences was made possible only by the length of his life: *'propter longitudinem vitae suae'*. The remark is worth pondering. The unique length of Grosseteste's career allows us to follow, more clearly than in any of his contemporaries, the transition from the thought of the late twelfth century down to the period when Averroës was becoming known in the universities. The little work which he wrote on the liberal arts carries with it the atmosphere of the schools around 1200, in its blend of the traditional Latin Neo-Platonism of Augustine and Boethius and the new Arabic science. The musical harmony of the universe and the sympathy of its parts, the *anima mundi*, the relations of microcosm and macrocosm and other such Platonic and Stoic teachings, made up the stuff of his thought at this point. We can only take up the story again after a long interruption, when we find him back in Oxford as a theologian, *c.*1220. Now scholars are in agreement that Grosseteste's theology, for all his breadth of learning, was of a conservative kind that continued in the tradition of the biblico-moral school of Paris, and many have commented on the apparent paradox of the innovative natural philosopher and the old-fashioned theologian. I would argue that we have here not so much a paradox as a natural effect of age. At the time when Grosseteste launched himself into the research which issued in his scientific and Aristotelean works he was already over fifty years old. He appears to have wakened up only at this rather late stage to the realization that something new was happening and that Aristotle was at the centre of it. It was becoming possible to construct a new kind of natural explanation, indeed, to develop the world-picture anew. Grosseteste felt attracted by the thought and determined to play his part in the movement. It is not so strange, after all, that he continued in a theological mould while pioneering the philosophical movement. A new generation of theologians was growing up, at Oxford at least, who had read Aristotelean logic and natural philosophy already as youngsters in school and who found it natural to make theology more scientific by importing something of philosophical matter and distinctions into it. Grosseteste was fully aware of the new trend (at least by the time he became bishop) and he reacted to it with dismay, writing from Lincoln an open letter to the Oxford theologians to admonish them to return to the ways which had made Paris great in former times and to keep biblical study rather than systematic construction to the fore. In practice, of course, he himself was more relaxed, for even in his pastoral works the new style asserted itself in the commingling of philosophy and theology and of Greek and Latin ideas. He even sought to show that, in psychology at least, St. Augustine and Aristotle were not really so far apart, after all. The *rapprochements* which he made were not always justifiable (the issues were clouded by pseudonymous writings and by the Arabic commentators), but they gave him deep satisfaction and reassurance.

It must be remembered that Grosseteste, when he was nearly sixty, approached philosophy with a double interest, in nature and natural phenomena, and in the development of the theological heritage. He may have felt himself too old by then to undertake really systematic work in theology, but he did feel impelled to develop the traditional doctrine of creation by learning from the philosophical tradition. His little book on light and his great commentary on the Aristotelean theory of science are the best illustrations of this ambition and at the same time the crown of his philosophical work. The former took up the idea, launched in the hexaemeral works of Basil and Augustine, that matter is at bottom a form of light or energy and turned it into a philosophical idiom that was in step with the newly-discovered peripatetic cosmology. The *Commentary on the Posterior Analytics* likewise showed the Christian Aristotelean at work in expounding the relationship of the mind to the world of nature and placing it within the larger metaphysical and theological context of the finite spirit and the light of Being itself. I think we can trace the same theological impulse, oddly enough, in his explorations of the tides, thunder, climates, the rainbow, and so on; Grosseteste was sure that the correct understanding of the wonders of the natural world would lead to an enriched appreciation of the literal sense of the Bible and of the book of creatures.

I have placed the composition of all of these works after 1220 but before Grosseteste's election to Lincoln. However, we scarcely need Bacon's reminder to realize that the extraordinary literary fertility of these years reposed upon a preparation of three decades of wide reading, book-collecting, lecturing, and annotating. If the silence of the earlier part of his career stands in sharp contrast to his later productivity, it was in reality the indispensable foundation for the latter.

Grosseteste's most original contribution to philosophy lay in his metaphysical intuition that visible light is the reflection on the face of nature of the inner workings of a geometrically-functioning reality. It was doubtless this inspiration and its development that Roger Bacon had in mind when he remarked that Grosseteste forsook the Aristotelean sciences and their method. Not only did this thought, and the lines of speculative inquiry to which it gave rise, have no foothold within the Aristotelean explanation of nature, they escaped altogether from the framework provided by the division of speculative knowledge in the peripatetic tradition and opened it up in ways that anticipated the scientific theory of three or four centuries after Grosseteste's time. The spring which launched the idea itself was Grosseteste's faith in the Father of Lights, who 'dwells in light inaccessible', and whose works are all in differing degree *'luciformes a luce prima'*—light-formed, deriving from the First Light. It was what Grosseteste called the *'magna magni Augustini auctoritas'* which led him in speculative directions to frame the metaphysics of light: Baeumker's designation must, however, be used rather loosely of Grosseteste's thought, which was never systematic in the highest degree; never the less it is certain that the equation of being with light underlay a great deal of his thinking. Within that equation and the field of tension which it established were contained most of his characteristic themes and approaches: the triune nature of God, the Creator as Mathematician, the hierarchical order of creation containing the world of spirits, life, and matter, the immediate presence of first principles to the mind, the constitution of space, and the energy of material being. That the Creator was a mathematician was Grosseteste's own original idea; for the rest he drew his inspiration from Augustine, Basil, and the Pseudo-Dionysius.

The central idea of the metaphysics of light is that being, all being, is active and dynamic; to be is to exercise a fundamental act, to stand upon a foundation of energy, which in turn interlocks each thing with others in the totality and relates each thing immediately to the one source of all existence and manifestation. The anthropological emphasis corresponding to the metaphysics of light fell upon vision and located at the very core of human existence the transcendental tension of the subject towards the full manifestation of being. Only in the presence of the Good, the source of all light, can the many be grasped in the unity that is their common ground, the origin and end of all, the Father of Lights, the sun of the intelligible world, seen in and through all that is manifest to the eyes of the mind. Grosseteste could not have known how closely his development of the parallel between the visible sun, the source of both light and vision, and the sun of the intelligible world was to the *Republic* of Plato, for it was the development by St. Augustine of the Pauline and Johannine doctrines which he himself expounded, when he affirmed that the mind is made to enjoy the immediate vision of God, *'sine symbolo et parabola'*. In his writings on light, sight, vision, and illumination, Grosseteste notably anticipated the much more systematic doctrine of St. Bonaventure, whom he doubtless influenced in ways that are not yet fully clear to us. One aspect of his speculations concerning light, however, was not developed further until the age of Kepler, namely, the view that the sun has a functional centrality in nature and is the chief activating cause of all natural phenomena.

There is much that we do not know as yet concerning Grosseteste's influence upon later generations of thinkers. However, it is clear that no school of thought was created by his teaching; here, once again, Bacon has been proved right. Such influence as he had was diffused over many areas and was not continuous over time (aspects of his thought were being rediscovered at various times during the fourteenth and fifteenth centuries), but reflected his versatility as thinker, scholar, and writer, his richness in initiatives. Probably Adam Marsh was his only real disciple; unfortunately, his works have not been identified. Within the twenty or so years after Grosseteste's death Scholasticism developed in ways that differed widely from many of the impulses which he had contributed and his writings attracted the interest only of a small minority. To most they must have seemed somewhat archaic, diffuse, and unscholastic. In the year of his death (1253) both Bonaventure and Aquinas were presented as regents to the theological faculty at Paris; by then the phase which would see the full expansion of the *Sentences* commentaries and summas, of commentaries by way of questions and of *correctoria*, had been firmly initiated. Grosseteste would, if he knew of it, have been outraged by the disedifying behaviour of the secular masters in attempting to exclude the order men, but would have taken some measure of comfort in the thought that the new age of the schools would belong not to the seculars but to the mendicant friars, whom he had helped to get established. However, he would have been happy only if the new masters had been prepared to learn and teach both Greek and Hebrew and to accord priority to mathematics in the explanation of natural phenomena. If his example and his programmes had been followed the history of Scholasticism, in the Middle Ages and beyond, would have been different, and for the better. The Schoolmen honoured him for the part he played in the origins of the Aristotelean movement, but we today can acknowledge that he was greater, broader of vision, more versatile and, in the last part of his life, more thorough-going in scholarship than any figure in the Schools of Europe during the two centuries following his death.

APPENDIXES

Appendix A

*A catalogue of manuscript-discoveries, editions, and
translations of Robert Grosseteste's works for
the years 1940-1980*

S.H. THOMSON'S catalogue of Grosseteste's writings, which
was published in 1940, is an indispensable working-aid to
which all students of Grosseteste will be indebted for the fore-
seeable future.[1] Envisaged primarily as a list of MSS known
to contain works ascribed to the bishop, it yet managed to
transcend the genre by reason of the richness of information
which it offered to the researcher. The author visited more
than 140 libraries, and examined approximately 2,500 MSS
in which it was known or suspected that works of Grosseteste
were to be found. In the course of his travels he made many
valuable MS-discoveries, but the fortuitous nature of certain
of these led him to conclude that ample additions would in
due course be made to the list of MSS which he compiled and
that works of Grosseteste would be turned up of which no
trace existed at that time. The accuracy of his prediction has
been amply confirmed in the meantime.

The most important task which Thomson set himself was,
of course, that of determining the authenticity of the various
treatises attributed to Grosseteste by the bibliographical
tradition.[2] Naturally, Thomson placed the burden of proof
upon priority of MS-ascription, preferring to credit the earliest
available witnesses. Such procedure necessitates judgements
of date which are often delicate, founded as they are upon
purely paleographical evidence in all cases save those in which
the MSS are dated. However, other kinds of evidence were
used in a supplementary way towards determining authenticity,
so that arguments from content, style, and contemporary and

[1] *The Writings of Robert Grosseteste, Bishop of Lincoln 1235-1253* (Cambridge,
1940), xv+302 pp.

[2] A valuable assessment of the worth of this very uneven tradition, which
extends from Henry of Kirkestede (fl. 1400) to L. Baur (*Die Werke*, 1912), is to be
found in Thomson's introduction, pp. 4-9.

later testimony received their due place. Thomson did not attempt to exaggerate the accuracy of which paleographical methods of dating admit; he indicated that a suggested date allows for around twenty years latitude either way (p. 3).

A typical catalogue-entry in this book indicates the title by which a work of Grosseteste is generally known; its *incipit* and *explicit*; the MSS containing it; where and when it was published, if it has reached print; and supplementary information, the need for which varies according to the individual work, concerning chronology, authenticity, and the place of the work in literary history. Each writing is assigned a number and is placed in a category according to its content. The number and disparity of the categories suggested by the material reflect the extent of Grosseteste's interests: Translations, Commentaries (Biblical, Philosophical), Philosophical and Scientific Works, Pastoral and Devotional Works, Anglo-Norman Writings, Sermons, Letters, *Dicta*.[3] Taken altogether, *The Writings* represents a great mass of basic information on the works of Grosseteste, and a firm starting-point for further enquiry.

The purpose of the present catalogue is to update Thomson's book in some respects—and the qualification added is important, since these few pages are not intended as a complete revalorization of his work. I have tried to compile a list of MSS which were unknown to Thomson, but which have been discovered and studied by various scholars in the meantime. This list has been compiled from publications which have appeared between 1940 and 1980, so that there probably are some discoveries which it does not record, since they have not been made widely known. The second objective of this study is to list all the editions, complete or partial, and translations, of works of Grosseteste which have been published since 1940. This aim was governed by my own interest, which is doctrinal,

---

[3] Thomson regards the classification as an arbitrary one (ibid., p. 1), which, of course, it is not, though convenience must play a part in determining the category to which certain books are to be assigned. However, one feels that the convenient approaches the arbitrary when the *Hexaemeron* (which is, after all, a commentary on the opening chapter of Genesis), the *De Operationibus Solis* (a comentary on Ecclesiasticus 43:1-5), the *V Quaestiones Theologicae*, and the *De Ordine Emanandi Causatorum a Deo*, are placed in the category of 'Philosophical and Scientific Works'. It is remarkable that Thomson did not institute a category of 'Theological Writings' to accommodate, for example, the *Quaestiones*, and such treatises as the *De Cessatione Legalium* and the *De X Mandatis*.

and which has led me to examine the text of each edited work, in order to determine its degree of reliability as a basis for the study of the teachings of Grosseteste. Finally, some information has been added concerning the works edited. This has taken the form of indications bearing on their chronology. and, in a few cases, on their authenticity. I have not attempted to offer a general chronology of all the works listed by Thomson; such an undertaking would be premature, as regards both my ability to pursue it and the objective state of research on Grosseteste's writings. Also, while I have on occasion indicated that certain works listed by Thomson have been shown to be spurious, I have not undertaken to re-examine the credentials of all of the writings attributed to Grosseteste.

The division of material adopted is the following:

Section 1.   Manuscript-discoveries, 1940-1980
Section 2.   Editions and translations of texts of Grosseteste, 1940-1980
          A. Translations from the Greek
          B. Commentaries: Biblical, Philosophical
          C. Philosophical and Scientific Works
          D. Pastoral and Devotional Works
          E. Sermons and *Dicta*
Section 3.   Editions Planned

The following guidelines regulated the compilation of the catalogue:

1. MS-discoveries are listed in order according to the year of appearance of the publication which first drew attention to them. Immediately following the reference to the year I have put the title of the work of Grosseteste affected by the discovery.
2. The Latin titles of the works of Grosseteste are those found in Thomson's catalogue, as are the numbers and the order assigned to them. (For an exception to this rule, see No. 3 below).[4] The division of titles into categories is therefore also that of Thomson; I have followed it even where it is inadequate, purely in order to achieve ease of reference.
3. Where a commentary of Grosseteste is associated with one of his versions, as in the case of the Dionysian translations, the edition of the commentary is noticed and discussed with that of the translation which it accompanies, therefore in the section on translations.

[4] The other exceptions are, the *Quadratura Per Lunulas* (which Thomson classed as a work rather than a translation: *The Writings*, p. 113), and the version of Pseudo-Andronicus *De Passionibus* (catalogued by Thomson as a *dubium*, p. 233); both will be treated of in the section on 'Translations from the Greek'.

These pages would not have been so complete had they not profited greatly from Fr Gieben's bibliography.[5] This scrupulously exact work replaces all previous attempts at a bibliography of Grosseteste-materials. It is of particular merit in its commented survey of the early editions of Grosseteste's works.

1. MS-DISCOVERIES, 1940–1980

1942. *Opera Pseudo-Dionysii Areopagitae (iv) De Mystica Theologia* In his edition of Grosseteste's translation and commentary on the *Mystical Theology*, Gamba noted the following complete MS of the work which was unknown to Thomson:[6] Oxford, *Merton College, MS 69* (14th cent.)

Gamba furthermore names six MSS which contain alternating extracts from Grosseteste and Thomas Gallus on the *Mystical Theology* (p. 5):

Vienna, *Nationalbibliothek, MS 790* (14th cent.)
Vienna, *Schottenkloster, MS 396* (14th cent.)
Melk, *Stiftsbibliothek, MS 59* (B.24) (1456)
Melk, *Stiftsbibliothek, MS 427* (H.46) (1455)
Melk, *Stiftsbibliothek, MS 61* (B.26) (1476)
Munich, *Staatsbibliothek, Clm 18759* (15th cent.)

The following MS contains brief extracts chosen from all four Dionysian commentaries of Grosseteste:
Munich, *Staatsbibliothek, Clm 8827* (15th cent.)

1948. *Hexaemeron, De Cessatione Legalium* The attention of scholars was first drawn to the Oxford MS *Bodleian lat. th. c. 17*, in 1948, by Dr Hunt.[7] The MS contains William of Auvergne's *De Universo Spirituali et Corporali*, which is followed by Grosseteste's *De Cessatione Legalium* (fols. 158a–89d), and *Hexaemeron* (fols. 190a–243a). The former work is ascribed to Grosseteste in a fifteenth-century hand, but otherwise there is no rubric at the beginning or end of either treatise. There are, however, many corrections to the text inserted

---

[5] 'Bibliographia universa Roberti Grosseteste ab an. 1473 ad an 1969', in *Coll. Franc.* 39 (1969), 362–418. The works of Lacombe, Lohr, and Schneyer, mentioned in my general bibliography, are of particular importance for the MS-literature.

[6] U. Gamba, *Il Commento di Roberto Grossatesta al 'De Mystica Theologia' del Pseudo-Dionigi Areopagita* (Milan, 1942).

[7] 'Notable Accessions: Manuscripts', in *Bodl. Libr. Rec.* 2 (1941–1949) (No. 27), 226–7, 1 plate.

in the margins, and a number of subject-headings. The outstanding value of the MS derives from Dr Hunt's identification of the hand responsible for many of these additions with that which annotated *MS Bodley 198*, and which is without doubt Grosseteste's own. The corrections and additions referred to include several Greek words and phrases (fols. 159, 192^{r-v}, 193^r), also in Grosseteste's hand, which display an accuracy of accentuation and a fluency denoting a practised style of writing. The additions further include several of Grosseteste's concordantial symbols.

In view of the importance attaching to this codex it is unfortunate that its second part offers no clue as to its history in the later Middle Ages. The quire-numbers indicate clearly that the two parts of the present codex (namely, that containing the *De Universo*, and that consisting of the two works of Grosseteste) were originally separate. As we have it, the MS has lost its old initial fly-leaves, and is bound in nineteenth-century antique calf.

1950. *Aristotelis De Caelo et Mundo* First studied by D. J. Allan in 1950,[8] the *Balliol College MS 99* provided important new evidence concerning Grosseteste's activity as a translator. A full description of the codex is to be found in *Aristoteles Latinus I*, No. 343, p. 400; Allan quotes (p. 87) the relevant section from it concerning fols. 183^r-319^r, which contain Simplicius's commentary on the *De Caelo et Mundo* (see further under 'Translations', 10).

1955. *Quadratura Per Lunulas* Professor Clagett was the first to note that the *Quadratura*, which he identified as an extract from Simplicius's *Commentary on the Physics*, survives in two forms.[9] Version I is extant in a single MS, which was unknown to Thomson: Oxford, *Corpus Christi, MS 251*, fols. 83^v-4^r (13th cent.). Thomson (p. 113) listed eight MSS of the *Quadratura*, all of which represent the second version. To these Clagett adds one other witness, used by H. Suter in his

[8] 'Medieval versions of Aristotle, *De Caelo*, and of the Commentary of Simplicius', in *Med. Renaiss. Stud.* 2 (1950), 82–120.

[9] Marshall Clagett, 'The *Quadratura per Lunulas*. A thirteenth-century fragment of Simplicius' commentary on the *Physics* of Aristotle', in *Essays in Medieval Life and Thought* . . . ed. J. H. Mundy (New York, 1955), pp. 98–108. Clagett publishes both forms of the fragment (see under 'Translations').

edition of the fragment:[10] Bern, *Stadtbibliothek, MS A.50*, fols. 168r-9r.

1955. *Commentatores Graeci in Ethicam Nicomacheam Aristotelis* The Stockholm MS *Kungl. Bibl. V. a.3*, unknown to Thomson, was listed in *Aristoteles Latinus. Pars Posterior*, 1955, n. 1701. This thirteenth-century codex of French provenance contains important marginal notes of Grosseteste on the *Ethics* and the Greek Commentators.[11]

1958. *Opera Pseudo-Dionysii Areopagitae* In 1958 Miss Ruth Barbour reported a MS-discovery of fundamental importance for students of Grosseteste, MS *Canonici Graeci 97* of the Bodleian Library, Oxford.[12] This codex contains the Greek text of all four major Dionysian works, plus the ten letters, the *Prologus* and *Scholia* of Maximus, the epigrams and chapter-headings of each work, and other Greek material commonly found in MSS of Pseudo-Dionysius.[13] Further investigation revealed that this MS was undoubtedly the one used by Grosseteste as the basis of his Dionysian translations and commentaries. The identification of the MS was made on two grounds:
1. Paleographical evidence: a considerable number of Greek variants were added to the MS in a hand which is unmistakably that of Grosseteste, and two notes drawing attention to the subject-matter are his also.
2. A comparison effected between this MS and the Dionysian translations and commentaries of Grosseteste confirmed in a quite unambiguous way the conclusion arrived at on paleographical evidence.

1959. *De Sphaera, De Impressionibus Aeris* In an article published in 1959, Professor Thorndike described a number of MSS of scientific interest which are preserved in Italian libraries.[14] Three of the MSS which he mentioned contain

---

[10] *Zeitschrift für Mathematik und Physik* 29 (1884), 85-6.

[11] *The Greek Commentaries on the Nicomachean Ethics of Aristotle in the Latin translation of Robert Grosseteste*, vol. 1, ed. H. P. Mercken (Leiden, 1973), p. 50.

[12] 'A Manuscript of Ps.-Dionysius Areopagita copied for Robert Grosseteste', in *Bodl. Libr. Rec.* 6 (1958), 401-16.

[13] Ibid., p. 401, where a complete list of the materials is given. See also Part II, ch. 2, p. 74.

[14] 'Notes upon some medieval astronomical, astrological and mathematical MSS at Florence, Milan, Bologna, and Venice', in *Isis* 50 (1959), 33-50.

scientific treatises of Grosseteste, but only one of them was already known to Thomson (p. 116). In the Milan, *Bibl. Ambrosiana MS 35 Sup.*, fols. 27r–37r, which he dated to the fifteenth century, Thorndike found the *De Sphaera* of Grosseteste. The work is ascribed in the MS. The same codex contains Sacrobosco's treatise *De Sphaera.*

The Bologna *Bibl. Universitatis MS 154 (132)*, which represents a fourteenth-century astronomical collection, includes at fols. 28v–31r a *Liber de prenosticatione sive prescientia dispositionis temporum.*[15] This treatise appears from its *incipit* to be identical with the *De Impressionibus Aeris (De Prognosticatione)* published by Baur (*De Werke*, pp. 41–51), of which Thomson lists seventeen known copies. The *explicit* of the treatise as it occurs in the Bologna MS is, however, quite different from that given in Baur's edition and in Thomson's catalogue, which suggests that the Bologna codex may witness another redaction of the work.

The following are the *incipit* and *explicit* as found in the Bologna MS, and quoted by Thorndike:

Incipit. Ad praenotandam diversam aeris dispositionem futuram propter diversitatem motuum superiorum necesse est potestates signorum naturas planetarum qualitates quoque quartarum circuli descripti per revolutionem diurnam[16] perscrutari . . .

Explicit. Taurus Veneri quam Mars sequebatur Gemini Mercurio qui post Venerem fuerat deputati sunt. Explicit.

1962. *Extracts from the Fathers and from the Classical Writers* The number of MSS known to have been written or used by Grosseteste[17] was extended by a recent discovery of Thomson,[18] namely Vienna, *Nat. Bibl. MS lat. 1619*. The section of this codex which most concerns the student of Grosseteste runs from fols. 38a to 62b and contains *auctoritates* drawn from Jerome, Augustine, Chrysostom, Ambrose, and Solinus. Also included are fourteen letters of the supposed correspondence of St. Paul and Seneca. Short *notulae* of Grosseteste are to be found on fols. 45b, 46a, 53a, and 53b. Fols. 64a–6b are in a hand identified by Thomson as being

[15] Thorndike, art. cit. pp. 43–4.

[16] Thorndike read '*divinam*', which is certainly a mistake.

[17] See R. W. Hunt, 'The library of Robert Grosseteste', in *Scholar and Bishop*, pp. 121–45.

[18] 'An unnoticed autograph of Robert Grosseteste', in *Med. Human.* 14 (1962), 55–60.

Grosseteste's, and are entitled *Excerpta de primo libro Senece de beneficiis*. Thomson recalls that in the so-called *Concordantia Patrum* (to be found in the Lyons, *Bibl. Municipale MS 414*), a total of twenty-five references to Seneca are given, covering the first five books of the *De Beneficiis;*[19] since the Vienna MS quotes also from the last two books and from Seneca's *De Clementia*, it would appear that it was written by Grosseteste later than the *Concordantia* (i.e., after *c*.1225-30). The importance of the Vienna codex is that it reflects Grosseteste's wide reading, which extended to many of the authors of classical antiquity.

1963. *De Lineis, Angulis et Figuris* The *Bibliotheca Vaticana MS Ottoboni lat. 1870* contains on fols. 169-71v the text of the *De Lineis*, published by Baur among the philosophical *opuscula* of Grosseteste. The work is apparently unascribed in the MS.[20]

1967. *Hexaemeron* A further MS-discovery affecting the *Hexaemeron* was that of Fr Gieben, who turned up a hitherto-unnoticed copy of the work in the *British Library MS Harley 3858*, fols. 258c-334b. It is written in a fifteenth-century English bookhand and gives an excellent text.[21]

1967. *Stans Puer* To the MSS listed by Thomson (p. 150) as containing the poem should be added the following, discovered by Gieben:[22]

Cambridge, *University Library, Add. 6865*, fol. 1r (13th cent.)
Cambridge, *Gonville and Caius College, MS 417*, pp. 103-4 (15th cent.)
Cambridge, *Trinity College 0.5.4*, fol. 18ra (15th cent.)
Oxford, *Bodleian Library, lat. misc. b.3*, fol. 107v-r (15th cent.)
Aberystwyth, *Nation. Libr. Wales, Peniarth 356*, fols. 143-4 (15th cent.)

[19] Thomson, *The Writings*, pp. 122-4.
[20] The contents of the MS are listed by S. Collin-Roset, 'Le *Liber Thesauri Occulti* de Paschalis Romanus', in *Arch. Hist. Doctr. Litt. M.A.* 30 (1963), 111-98; see p. 121.
[21] R. C. Dales and S. Gieben, 'The *Prooemium* to Robert Grosseteste's *Hexaemeron*', in *Speculum* 43 (1968), 451-61; p. 453.
[22] See S. Gieben, 'Robert Grosseteste and Medieval Courtesy-Books', in *Vivarium* 5 (1967), 47-74, for a discussion and edition of the poem.

1967. *Liber Curialis* Thomson refers (p. 148) to a single MS of this work; Gieben notes two new witnesses (ibid., p. 69): Oxford, *Bodleian Library, Bodley 310*, fols 147va–8ra (Partial, containing two-thirds of the poem; late 14th cent.) Oxford, *Bodleian Library, Rawlinson C.552*, fols. 22va–3rb (Unascribed, late 12th-early 13th cent.) The ascription of the poem to Grosseteste must now be considered problematic.[23]

1970. *De Confessione II* Thomson (p. 176) listed seven MSS of the *De Confessione II* (Sermon No. 32). In the critical edition prepared by Dr Wenzel three further MSS are taken into consideration, and a fourth (fragmentary) one reported.[24] A description of all the MSS used is furnished in the introduction to Wenzel's excellent edition. The following are the signatures of the three new MSS.

Oxford, *Bodleian Libary, MS Rawlinson A.446*, fols. 3r–20r (mid-13th cent. unascr.) Oxford, *St. John's College, MS 190*, fols. 127r–42r (13th cent. unascr.) Hereford, *Cathedral Library, MS P.3*.xii, fols. 177r–86r (13th cent.). The fact that this MS contains an ascribed text of the *De X Mandatis* (fols. 186r–206r) probably explains the title attributed to the sermon on confession in it: *De Decem Praeceptis*. The scribe who copied the sermon did not ascribe it to Grosseteste.

1972. *Aristotelis Ethica Nicomachea* While preparing the critical edition of Grosseteste's translation of the *Ethics*, Père Gauthier discovered a MS which gives a good text of the *Recensio Pura*, and which he used in establishing the text: Dublin, *Trinity College C.2.8*, fols. 1r–73v, late thirteenth century.

1973. *Summa in VIII Libros Physicorum Aristotelis* Lohr draws attention to some MSS containing the *Summa* which were unknown to Thomson; see ' Medieval Latin Aristotle Commentaries, Authors: Robertus-Wilgelmus', in *Traditio* 29

---

[23] See Section 2 under no. 109, *Liber Curialis*.
[24] S. Wenzel, 'Robert Grosseteste's treatise on Confession, *Deus Est*', in *Franc. Stud.* 28 (1970), pp. 224–8. The fragmentary witness is MS *Lambert Palace Library* 523, fols. 122v–3r.

(1973), 93–197, p. 105. The work is probably inauthentic (see no. 26 *infra*, p. 483).

1973. *Commentarius in Libros Analyticorum Posteriorum Aristotelis* Lohr adds the following items to Thomson's list:
Berlin, *Staatsbibliothek lat. fol. 565*, fols. 31^r–76^v (15th cent.)
Salamanca, *Biblioteca de la Universidad 2028*, fols. 90^r–126^r (15th cent.)
Vatican City, *Bibliotheca Vaticana vat. lat. 760*, fols. 44^r–74^r (15th cent.)
Venice, *Biblioteca Nazionale Marciana Z. lat. 241*, fols. 52^r–97^r (AD 1442)
Vienna, *Dominikanerkloster 192/158*, fols. 110^r–45^r (13th/14th cents.)
Printed: Venice, *c*.1473–8 (GW 2390)
P. Rossi further extends this list in *Riv. Fil. Neo-scol.* 67 (1975), p. 506:
Padua, *Biblioteca Civica C.M. 187*, fols. 1^r–73^v (15 cent.)
Printed: Padua 1497 (Hain 10106)

1973. *Commentatores Graeci in Ethicam Nichomacheam Aristotelis* Lacombe (*Aristoteles Latinus II* n. 1701, 1939) listed the following MS, which Thomson appears to have overlooked:
Stockholm, *Kungl. Biblioteket V, a.3*
Mercken employed this MS in constructing the critical edition (no. 13 *infra*, p. 471).

1973. *Regulae Libri Priorum Analyticorum Aristotelis* Lohr (p. 103) notes the following MS:
Chicago, *University of Chicago UL 968*, fols. 44 ff. (AD 1471), unascr.

1973. *Summa in Ethicam Nichomacheam* Lohr (p. 105) reports that the Oxford MS, *Merton College 14*, contains on fols. 321–6 an incomplete text of this work.

1974. *Sermons* Schneyer reported in his monumental *Repertorium*[25] that the following MSS contain one or more sermons

---

[25] J. B. Schneyer, *Repertorium der lat. Sermones . . ., BGPM* 43, H.5, p. 191.

attributed to Grosseteste:
Cambridge, *Gonville and Caius Coll. MS 233* fol. 81vb
MS 351 fol. 86^r
Cambridge, *Trinity College MS B.15.38* fol. 42bis^r-3^v (Also contains chapter-headings of the *Test. XII Pat.* and what appears to be an extract from Grosseteste's translation of Suda, attributed).
London, *Brit. Libr., MS Royal 2. D. xxx* fols. 144, 133
Oxford, *Bodleian Libr., MS Bodley 857*
Prague, *Univ. Bibl., MS IV.G.31* fols. 79-87
MS VIII.F.3 fols. 65-79
MS XII.F.21 fols. 152-9

1979. *De Confessione III* In his study, 'Robert Grosseteste and the Pastoral Care', *Med. Renaiss. Stud.* 8 (1979), 3-51, Fr Boyle refers to a new MS of this brief work (which refers to itself as '*Speculum Confessionis*'):
Brit. Libr. MS Harley 5441, fol. 147^v.

1979. De Modo Confitendi (Canones Poenitentiales) Boyle has discovered a new copy in the
Nat. Libr. of Scotland MS 18.3.6. fols. 132^r-4^r.

1979. *Templum Domini* The following new MSS are reported by Boyle:
Kues, *Hospital MS 233*
London, *Brit. Libr., MS Arundel 507; MS Cotton Vespasian D.V.; MS Egerton 665; MS Harley 209*
Longleat House MS
Metz, *Bibl. de la Ville MS 521*
Oxford, *Bodleian Libr. MS Bodley 440; MS Tanner 110;*
Oxford, *Balliol Coll. MS 228*
Oxford, *Magdalen Coll. MS 109*
Oxford, *St. John's Coll. MS 93*
Paris, *Bibl. nat. MS lat. 543*
Wisbech, *Town Libr. MS 5*

1979. *Letter 128* Boyle reports a copy in the Red Book of the Exchequer,
London, *Public Record Office MS E. 164/2*, fols. 196^v-7^r

1979. *Sermon 32* Boyle has uncovered a copy of this treatise (going under the name of '*De Virtutibus et Vitiis*'):
Oxford, *Bodleian Libr. MS Rawl. A 446*

2. EDITIONS AND TRANSLATIONS OF WORKS OF
GROSSETESTE 1940-1980 [26]

A. *Translations from the Greek*

2. *Opera Johannis Damasceni* (i) *De Logica*   The *De Logica*
or *Dialectica* was intended by St. John Damascene to be an
introduction to the *De Orthodoxa Fide*. The first scholar to
attribute the *Dialectica* version to Grosseteste was Thomson,
who pointed out that the Latin corpus of the five works of
Damascene must be treated as a unit since all or several of
these works are persistently associated in the manuscripts
which he consulted (p. 45). Two of these five are authenticated by Grosseteste himself, namely, the *De Fide* and the *De
Hymno Trisagion* (cited by Grosseteste in his *Commentary
on the Celestial Hierarchy*).[27] The fact that in many cases one
or more of the three remaining works of the corpus appear in
codices accompanied by either of the authenticated works is
evidence in favour of Grosseteste's authorship of the whole
corpus of translations. Besides, the rendering of all five treatises is marked by the exteme literalness so characteristic of
Grosseteste's method as to make it almost unmistakable, even
in an age in which such word-for-word rendering was the rule.

In 1953 Owen A. Colligan edited the *Dialectica* from three
of the thirteen MSS of the work listed byThomson.[28] This was
the first element of the Damascene corpus to appear in print
in the version of Grosseteste. The edition is not critical; no
reason is given as to why only three MSS were chosen for its
construction. Of the three, none is singled out as a basis for
the edition, so that the printed text is arrived at by a process
of compilation from all three witnesses.

An examination of the text reveals that elements of a philological commentary, such as Grosseteste was later to provide

[26] I have reviewed and discussed in some detail the editions of works by
Grosseteste since 1940: 'Questions of authenticity and chronology concerning
works of Robert Grosseteste edited 1940-1980', in *Bulletin de philosophie méd.*
23 (1981) and 24 (1982). What follows here is a condensed summary of my
findings.

[27] The relevant passage from ch. vii of the *Commentary* is quoted by Franceschini in *Roberto Grossatesta, vescovo di Lincoln*, p. 42. Franceschini, however,
knew of no MS of the *De Hymno*.

[28] *St. John Damascene, Dialectica. Version of Robert Grosseteste*, edited by
Owen A. Colligan (New York, 1953), vii+63 pp.

in a much more extensive fashion for the Dionysian corpus, are already present in the *Dialectica* version. The notes prefigure the expansive style of the philological commentaries realized during the years 1239–43, when Grosseteste assumed the enormous task of editing, translating, and commenting on the Dionysian corpus; but the rather fragmentary indications of meaning and etymology contained in the version of the *Dialectica*, while being more than a novice's effort, represent an initial stage of the application of his method, and support the date of *c*.1238-9 suggested for the work by Callus.[29]

It should be noted that the *Prologus* which precedes the *Dialectica* in four MSS is not, as Thomson thought (p. 81), for the greater part a composition of Grosseteste, but a translation of the prologue to Damascene's writings published in *P.G.* 94, cols. 489-98. The translation is probably due to Grosseteste.

Dr Meridel Holland has edited Grosseteste's versions of *De Hymno Trisagio, Introductio Dogmatum Elementaris*, and the *Disputatio Christiani et Saraceni* (which appears in the MSS at the end of *De Centum Haeresibus*, as an appendix to the Hundredth Heresy). (See the reference to her Harvard 1980 doctoral thesis in bibliography).

5. *Opera Pseudo-Dionysii Areopagitae* In 1937 a team of Benedictine scholars working under the chairmanship of Dom Ph. Chevallier published their first volume of the translations of the *Corpus Areopagiticum*; in 1950 they brought their massive work to a close.[30] The main body of their two volumes consists of a synoptic edition of the Latin versions of Hilduin, Eriugena, Sarrazen, Grosseteste, Ambrose Traversari, Marsilio Ficino, and Joachim Périon, plus the French versions, all printed under the headline of the Greek text as found in the Paris MS, *Bibl. Nat. gr. 437.*

Grosseteste's complete translation of the *Divine Names* and the *Mystical Theology* is found in the first volume, that of the two *Hierarchies* in the second. The editors copied the text from a single MS (Paris, *Bibl. Nat. lat. 1620*), without correction. The witness is a reliable one, and indeed the

---

[29] Callus, 'The Date of Grosseteste's Translation and Commentaries on Pseudo-Dionysius and the *Nicomachean Ethics*', in *Rech. Théol. anc. méd.* 14 (1947), 186-210.

[30] *Dionysiaca. Recueil donnant l'ensemble des traductions latines des ouvrages attribués à Denys de l'Aéropage* . . . 2 vols. (Paris-Bruges, 1937-50).

tradition of Grosseteste's Dionysian translations seem to have been singularly free from major textual distortions, but clearly this edition does not abate the need for a critical edition of all four versions, a need that is even now not fully supplied.

5. (i) *De Caelesti Hierarchia* Grosseteste's preface to his translation of the *Celestial Hierarchy* was first edited in 1955, by I. Ceccherelli.[31] The text provided (which is based on three MSS) is quite reliable.

3. *Prologus Maximi Confessoris in Opera Pseudo-Dionysii*
4. *Scholia Maximi Confessoris in Opera Pseudo-Dionysii*
5(i) *De Caelesti Hierarchia* (translation)
20(i) *De Caelesti Hierarchia* (commentary)
21. *Notulae in Opera Pseudo-Dionysii*

One of the most useful contributions to Grosseteste-scholarship made in recent years was J. S. McQuade's edition of chs. 1–9 of the *Celestial Hierarchy*, in the translation of Grosseteste, and accompanied by Grosseteste's extensive commentary.[32] This work has unfortunately remained till now unpublished.

The text of the translation and commentary is critically edited from the eleven MSS known to contain the whole work. The MSS are described and the *stemma codicum* carefully drawn up (pp. 11-20). Furthermore, the Greek MS used by Grosseteste as the basis for his translation is edited by McQuade.[33]

The full content of this edition includes the *Prologue* attributed to Maximus the Confessor, with notes on the same by Grosseteste; the *Epigramma*, with a commentary by Grosseteste; the translator's *Prologue*, in which he discusses the rendering of Greek words into Latin; a summary of contents, translated and annotated by Grosseteste; and finally, the text itself, accompanied by the commentary.

---

[31] I. Ceccherelli, 'Roberto Grossatesta studioso di greco e una cosiddetta sua introduzione grammaticale allo studio della lingua greca', in *Studi Franc.* 52 (1955), 426-44.

[32] J. S. McQuade, *Robert Grosseteste's Commentary on the 'Celestial Hierarchy' of Pseudo-Dionysius the Areopagite: an edition, translation, and introduction of his text and commentary. A thesis presented for the degree of Doctor of Philosophy to the Queen's University of Belfast* (1961), 141+22+482 pp.

[33] *Bodleian MS Canonici Graeci 97*; see the note on this MS in the preceding section.

One of the contributions made by McQuade was the identification of Grosseteste as the translator of the *Prologue* of Maximus. Grabmann had assigned the Latin version of the *Prologue* to the twelfth century:[34] Franceschini suggested in 1933 that Grosseteste was the translator,[35] on the grounds that the *Prologue* is found to precede the text of Dionysius in most of the MSS of Grosseteste's version. Thomson accepted this suggestion (p. 52). However, no conclusive evidence for its authenticity was available until McQuade pointed out that the *Prologue* as it exists in the MSS of Grosseteste's version is incomplete, breaking off at the same point as the Greek text of *MS Canonici graeci 97*. This MS is followed by the Latin version of the *Prologue* in every detail, even in the case of a mistaken reading.

In his introduction McQuade provides the best discussion to date of Grosseteste's title to authorship of the version of the *Scholia* attributed to Maximus, which accompany the text of Dionysius.[36] His conclusion from all the evidence considered is that it is most reasonable to regard Grosseteste as the translator of the *Scholia*.

McQuade's work covered only the first nine chapters of the *Celestial Hierarchy*, together with the commentary on these. The remaining six were edited by the present writer.[37] The same principles of edition were applied and the same presentation of materials adopted as were decided upon by McQuade and employed in his edition.

5(i) *De Caelesti Hierarchia* (translation)
20(i)*De Caelesti Hierarchia* (commentary)   In a masterly article of 1952, Fr Dondaine reviewed the problem of the object and *medium* of the beatific vision in the writings of some thirteenth-century theologians.[38] He offered a rich

---

[34] *Mittelalterliches Geistesleben*, I (Munich, 1926), p. 462.

[35] 'Grosseteste's translation of the Πρόλογος and Σχόλια of Maximus to the Writings of the Pseudo-Dionysius Areopagita', in *Journ. Theol. Stud.* 34 (1933), pp. 362 ff.

[36] Numbered 4 in Thomson's catalogue.

[37] J. McEvoy, *Robert Grosseteste on the Celestial Hierarchy of Pseudo-Dionysius. An edition and translation of his Commentary, chapters 10 to 15*. A thesis presented for the Degree of Master of Arts to the Queen's University of Belfast, 1967, 124+240 pp.

[38] 'L'Objet et le "*medium*" de la vision béatifique chez les théologiens du XIIIe siècle', in *Rech. Théol. anc. méd.* 19 (1952), 60–130.

selection of unedited texts relating to the problem, the ninth of these being a substantial extract from ch. 4 of Grosseteste's commentary (pp. 124-5).

The text which Fr Dondaine edited relied upon three MSS: Paris, *Mazarine 787*; Paris, *Bibl. Nat. lat. 1620; Vatican Chigi A.V. 129.* It is quite accurate for purposes of doctrinal analysis.

5(iii)   *De Divinis Nominibus* (translation)
20(iii)*De Divinis Nominibus* (commentary)
While the translation and commentary on the *Divine Names* remains as a whole unpublished, its first chapter has been edited by F. Ruello, and appeared in print in 1959.[39] Of the ten known MSS of the work, the editor has used three in establishing the text, taking the *Bibl. Nat. MS lat. 1620* as the basis of his edition. A useful doctrinal study is appended, in which Grosseteste's interpretation of Dionysius is contrasted with St. Albert's, as found in his *Commentary on the Sentences* and in his *Commentary on the Divine Names.*

In *Appendice 3*, pp. 177-8, Ruello edits an important passage taken from ch. 5 of the *Divine Names*. The text again presents no problem, being also based upon the same Paris MS.

In *Appendice 8*, pp. 194-7, Ruello gives an important passage taken from ch. 2 of the *Commentary on the Celestial Hierarchy*, also dealing with the problem of the divine names.

In a wide-ranging article of 1946, Dom H. Pouillon surveyed the scholastic discussions of the transcendentality of beauty, between the years 1220 and 1270.[40] In an appendix he edited a number of extracts from Grosseteste's *Commentary on the Celestial Hierarchy*, especially ch. 4 of that work. The MS which he used was Paris, *Bibl. Mazarine 787*, which contains the commentaries of Grosseteste, but omits the text of Dionysius and the *Scholia* of Maximus. However, the text which this MS offers is reasonably good and its mistakes have been deftly corrected by Dom Pouillon. He also prints a few brief extracts from the *Hexaemeron*.

---

[39] F. Ruello, 'La *Divinorum Nominum Reseratio* selon Robert Grosseteste et Albert le Grand', in *Arch. Hist. Doctr. Litt. M.A.* 34 (1959), 99-197. For the text, see *Appendix I*, pp. 134-71.

[40] 'La Beauté, propriété transcendentale, chez les Scholastiques (1220-1270)', in *Arch. Hist. Doctr. Litt. M.A.* 15 (1946), 263-329.

5(iv)   *De Mystica Theologia* (translation)
20(iv)   *De Mystica Theologia* (commentary)
    In 1942 appeared the edition of Grosseteste's translation
and commentary on the *Mystical Theology*, the briefest of
the four Dionysian works.[41] Of the 17 MSS which contain
the work in its entirety the editor consulted nine, and used in
addition to these the Strasbourg edition of 1503, which represents an original MS-tradition. The Florence MS *Laur. Plut.
xiii dextr. ii* is used as the basis of the printed text.

9.   *Aristotelis Ethica Nicomachea*
**13.   *Commentatores Graeci in Ethicam Nicomacheam
    Aristotelis***
28.   *Notulae in Ethicam Nicomacheam* The role played by
Grosseteste in the rendering of Aristotle's *Ethics* and of the
Greek commentaries on it has received more discussion than
any other aspect of his work as translator from the Greek.[42] It
is now accepted that Grosseteste was responsible for the first
complete extant Latin version of the *Nicomachean Ethics*, and
that he translated in addition the Greek Commentators on the
Ethics, namely Eustratius, Aspasius, Michael of Ephesus, and
the Anonymous. Yet until 1963 none of Grosseteste's ethical
translations had been published, despite their enormous
importance as a basis for the study of the ethical literature of
the Middle Ages from 1250 onwards.[43] The past decade, however, has witnessed a dramatic improvement in the position,
with the appearance of three partial editions of the *corpus
ethicorum*, and very recently of the critical edition of the
Nicomachean Ethics in Grosseteste's translation, and the first
volume of a planned complete edition of his rendering of the
Greek commentators and of his *Notulae*. We will give a comment on each edition.
    W. Kübel published the text of the *Ethics* in Grosseteste's

---

[41] U. Gamba, *Il Commento di Roberto Grossatesta al 'De Mystica Theologia'
del Pseudo-Dionigi Areopagita* (Milan, 1942), 13+69 pp.

[42] For a complete bibliography on the subject, see S. Gieben, 'Bibliographia
Universa Roberti Grosseteste', in *Coll. Franc.* 39 (1969), nos. 74-9, 248-87.

[43] Thomson (*The Writings*, p. 65) erred in stating that Grosseteste's version
of the *Ethics* has been frequently published with the commentaries of Aquinas,
etc. In fact, the text published till now has been that of a revision of Grosseteste's
version, done probably by Moerbeke.

version as an accompaniment to his edition of the commentary of St. Albert on the *Ethics*.[44] He did not aim at providing a critical edition of the Grosseteste version, but used only three MSS. The translation of Grosseteste is given in full in the first of the three apparatus which accompany the edition; those elements of translation which appear in Albert's *Commentary* are italicized. Variant readings of the Grosseteste version are not given. The philological notes of Grosseteste concerning the translation of Greek words are separated by brackets from the text of the Greek commentators, within which they are inserted.

The medieval Latin translations of the *Nicomachean Ethics* have recently been published in a critical edition as vol. XXVI of *Aristoteles Latinus*, the product of years of dedicated scholarly work by Père Gauthier.[45] The editor had to face the daunting task of dividing up close on three hundred MSS containing the *Ethics* in whole or part, first into the six translations known to have been made, and then into families for each of them. The difficulty of his task was increased by the contamination frequently effected by one translation upon another; in the case of some MSS it is difficult to say which version predominates and which is supplementary.

The edited volume comprises four fascicules, the first containing the *Antiquissima Translatio (Ethica Vetus*, extending to Bks. II and III) and the three fragmentary components collectively described as the *Antiquior: Ethica Nova, Hoferiana*, and *Borghesiana*. The second prints the text of Grosseteste's version, the *Recensio Pura* as the editor calls it, and is the only one that directly concerns us here. The third has the *Translatio Recognita* attributed by Franceschini, Mansion, Grabmann and others to William of Moerbeke, but which Gauthier prefers to regard as anonymous, despite the by no means negligible external and internal evidence in favour of the Flemish Dominican's authorship.

[44] *Alberti Magni Opera Omnia*, t. *XIV, Pars 1, Fasc. 1. Super Ethica Commentum et Quaestiones. Tres libros priores primum edidit W. Kübel* (Münster, 1968), pp. xiv+219.

[45] *Aristoteles Latinus* XXVI 1-3. Fasc. primus, *Ethica Nicomachea*. Praefatio quam conscripsit Renatus Antonius Gauthier. Fasc. secundus, *Translatio Antiquissima* libr. II–III sive 'Ethica Vetus' et translationis Antiquioris quae supersunt sive 'Ethica Nova', 'Hoferiana', 'Borghesiana', ed. R. A. Gauthier. Fasc. tertius, *Translatio Roberti Grosseteste Lincolniensis* sive 'Liber Ethicorum'. A. *Recensio Pura*. Fasc. quartus. *Translatio Recognita*. (Leiden-Bruxelles, 1973) (see under MSS, 1972, above).

The *Recensio Pura* (L¹, also referred to by Gauthier as *editio maior*) is found in the MSS as the nucleus of either a double or a triple work, the other components of which are the Greek commentaries and Grosseteste's *Notulae*, or else by itself; in the former state it is known in twenty-two MSS, in the latter only in eleven, in its pure form, but in many more in a contaminated text.

Having examined the textual tradition of L¹ (*Recensio Pura*), Père Gauthier turns to the complicated question of the corrections and revisions Grosseteste's translation received. These were two in number. What Gauthier calls the *Editio Minor* (L²) of the *translatio Lincolniensis* is a text of the *Ethics* without the commentators and the *Notulae*. Gauthier refers to it as a *recensio*, and suggests that it probably derived from a single exemplar. Certain of its variants turn out to be, not scholarly corrections made from a Greek MS, as had been supposed, but notes of Grosseteste himself incorporated into the text. While being unable to establish the date and place of its origin, the editor has shown that it goes back to the negligence of a copyist, whose careless transcription of L¹ introduced a multiplicity of blunders into his exemplar—a very influential one, as it turned out.

The *Recensio Parisiaca* was the only real revision Grosseteste's translation received, and it lies beyond our present purpose.

Grosseteste's version of the commentaries of Aspasius on Bk. VIII and Michael of Ephesus on Bk. IX of the *Ethics* was published by Dr Stinissen in 1963.[46] The edition of Bk. VIII cannot be considered fully critical: it is presented by the editor as a translation, whereas due to the mutilated and fragmentary state of Aspasius's commentary on the latter portion of the book Grosseteste was forced to compose extensively, reworking what Greek material he had before him, but adding much that is purely and simply his own.[47]

Stinissen's edition is due to be replaced by the complete edition of the Greek commentators in their Latin translation undertaken by H. P. Mercken in co-operation with J. P. Reilly.

[46] *Aristoteles over de vriendschap. Boeken VIII en IX van de Nicomachische Ethiek met de commentaren van Aspasius en Michaël in de Latijnse vertaling van Grosseteste*, door W. Stinissen (Brussels, 1963), 14+183 pp.

[47] e.g., pp. 59–61 (on political constitutions), 66.00–18 (on kingship and the natural law); 126. 6 ff.

Mercken's first volume has already replaced his own critical edition of the commentaries on Bks. I and II;[48] vols. II and III will complete the edition of the commentaries and of Grosseteste's *Notulae*, and will provide a study of their influence on the scholastic exegetes of Aristotle.

Mercken's discussion of the date of the ethical translations brings up a new piece of evidence: the closing formula of the *Commentary on the Celestial Hierarchy* (1239–40) contains a substantial literal quotation from his own translation of Michael of Ephesus's commentary on Bk. X, so that the work of translating the Greek commentaries must have been at least begun by 1240 (p. 42). We know (since Pelster) that it must have been finished before St. Albert used it in his lectures on the *Ethics* (1248–52) and on the *Sentences*, Bk. IV (1249); Callus has argued convincingly that it was completed by 1246–7. We can add a tint to Mercken's picture: if the body of Greek ethical works translated by Grosseteste be taken as a whole, as the *Peterhouse MS 116* reflects it and he seems to have planned it, then there is good evidence that its collection and translation occupied the bishop and his assistants for close on ten years. He very likely began by translating the *De Passionibus* and the *De Laudabilibus Bonis*, small separable elements in the vast corpus, almost as soon as he was capable of independent translation, and scarcely earlier than 1237.[49] In 1239–40 he quoted from his own translations of the *De Passionibus* and from Michael's closing formula to the commentary on Bk. X; and in annotating the commentary of Eustratius on Bk. I he again quoted from the *De Passionibus*. This relationship does not prove that the translation of the entire corpus of Greek commentators on the *Ethics* was completed by 1240, for Grosseteste could have translated the final page of Michael's commentary out of curiosity and before settling down to the body of the text (as Mercken says), and he could have inserted the quotation from *De Passionibus* into the text of Eustratius on Bk. I at a later stage of redaction.[50] But when

---

[48] See above, n. 11, and also Mercken, *Aristoteles over de menselijke volkomenheid. Boeken I en II van de Nicomachische Ethiek met de commentaren van Eustratius en een Anonymus in de Latijnse vertaling van Grosseteste*, H. P. Mercken (Brussels, 1964), 72+209 pp.

[49] See n. 62.

[50] The quotation is the same as that found in the *Comm. in Hier. Cael.*, but is slightly longer; 'Est autem passio irrationalis animae motus et praeter naturam

this evidence is put together with Grosseteste's interest in ethical matters, as manifested repeatedly in his *Commentary on the Celestial Hierarchy*, it is enough to justify a slightly bolder conclusion than Mercken reaches;[51] and we may hope for further evidence that the work of translating and commenting on Pseudo-Dionysius and the ethical corpus proceeded simultaneously, the preparation and execution of the former occupying Grosseteste from *c*.1236-7 until 1242-3, the latter until 1246 or so.

Mercken establishes that Grosseteste possessed only one Greek exemplar of the commentators; he nowhere refers to variant readings. His version was original; the Greek compilation was itself a recent work, and was not translated before Grosseteste, nor after him.

The editor gives an illuminating discussion of the *Notulae* of Grosseteste. He annotated both the Greek commentators and the text of the *Ethics*. The notes fall into two classes, those inserted into the translated text, and marginal and interlinear comments. Mercken is able to prove that the distinction of marginal and inserted notes goes back to Grosseteste himself, who selected from among a large amount of explanatory material what he wanted to introduce into the text of the commentaries and decided what should be left in the margin. There is no variation in the MS tradition as regards the inserted notes, but none of the twelve MSS in which *marginalia* have been found contains all the *Notulae*. Mercken advances the pioneering work of Thomson[52] by his discovery that the Florence *MS Naz. Centrale Conv. Soppr. I.V.21* gives information on the notes it does *not* have. The MS contains the standard symbols used by Grosseteste to refer from the text to the *marginalia*, but numbers them throughout each book of the *Ethics*, even when the corresponding marginal note does not appear. The notes referred to can often be verified in other MSS at the same place. Some of the notes we possess are

impetui approximans, cuius sunt generaliores species quattuor: tristitia, timor, concupiscentia, voluptas.' (ed. Mercken p. 44.46-8). The editor has treated it as a *Notula* of Grosseteste.

[51] 'It is likely that Grosseteste by that time was at least contemplating a translation of the *Nicomachean Ethics* with the Greek commentaries. He was probably already organizing the execution of this project . . .', p. 42.

[52] 'The *Notule* of Grosseteste on the *Nicomachean Ethics*', in *Proceedings of the British Academy* 19 (1933), 195-218.

obviously incomplete, as scribes despaired of crowding them
into the margins of their copies, but by far the major part of
the 134 notes on the *Ethics* and its Greek Commentators still
exists in the surviving MSS.

Grosseteste did not compose the questions on the *Ethics*
which Bale alone of all the bibliographers attributed to him,
but Mercken demonstrates that portions of his *Notulae* form
already an impressive commentary, far from continous in form
and diverse in content, but adequate to its essentially subsidiary task of complementing the Greek commentaries, which
were of very uneven value.

BK. VIII is a case apart. The exposition of Aspasius is
fragmentary as well as inadequate, and the last part of the
commentary on the book is a mixture of translation, paraphrase, and purely Latin addition. Mercken brings good evidence from language and style to the effect that Grosseteste
was the author of these extensive additions, and I am convinced
that further study of their content would render certain his
'strongly-supported hypothesis' (Mercken, p. 63) to that
effect. Thus the references to non-utilitarian arts which exist
for the sake of *delectatio*, with the example of *citharizatio*
that immediately follows it, has a direct parallel in the *Commentary on the Celestial Hierarchy*.[53] The belief that music
can cure illnesses, and especially nervous disorders, is one
Grosseteste seems to have adhered to all his life (ibid.). The
remarks on friendship (p. 126) can sometimes be matched
from the unedited *De Cessatione Legalium*, and the treatment
of tyranny and kingship (the latter in a little *quaestio* raised
on the text: '*Numquid de lege naturali est regnum*'?) rejoins
a dominant theme of Grosseteste's later life, the nature and
exercise of authority.[54] There can be no doubt that he drew
on this part of his commentary when preparing to confront
Innocent IV.[55]

Mercken's edition is a model of scientific work. He marks
the gaps that existed in Grosseteste's single Greek MS and

[53] 'Sed sciendum quod aliqua est operatio seu actio quae est propter aliud
operandum, ut aedificatio et consimiles, aliqua vero est quae est finis sui ipsius,
ut citharizatio et casta delectatio'. ed. McEvoy, p. 153; cf. Stinissen's ed., *Aristoteles over de vriendschap*, p. 58.56 ff.

[54] Ibid., p. 66.00-18; cf. pp. 59-61.

[55] Compare Stinissen, ed. cit. pp. 59-61 with the Lyons dossier, ed. S. Gieben,
Coll. Franc. 41 (1971), 340-93, and especially with the *De Regno et Tyrannide*,
ibid. pp. 377-80.

records in an apparatus every divergence between the translation and the Greek original. In the principal critical apparatus he prints the brief marginal and interlinear notes bearing on translation and grammar. Grosseteste's chapter-divisions are retained.

10. *Aristotelis De Caelo et Mundo*
Simplicii Commentarius in De Caelo et Mundo Aristotelis
Until 1950 Grosseteste's version of the *De Caelo* of Aristotle was known only in a very fragmentary form from the *marginalia* of the *Vatican MS. Pal. lat. 2088,*[56] and the existence of a translation of Simplicius's commentary made by him was entirely unsuspected. D. J. Allan's examination of the Oxford, *Balliol College MS. 99* led to several important discoveries, which can only be mentioned here very briefly.[57]

The text of the *De Caelo*, Bk. II, is given in its entirety in the Balliol MS, in contrast to the greater portion of the other three books, which are quoted within the *Commentary* of Simplicius in the form of abbreviated *lemmata*. The reason for this has proved to be, that Bk. II is given in Grosseteste's version (identified by comparison with the authenticated *marginalia* of the Vatican MS referred to above), whereas the other books are drawn from the common version (probably due to William of Moerbeke). The two versions, furthermore, follow different Greek originals. Since the version of the Aristotelean text is inseparable from that of the accompanying commentary of Simplicius, it must be concluded that Grosseteste rendered both. Stylistic evidence supports this inference.

Allan illustrates the contrast between the two versions of the *De Caelo* and the commentary by publishing in double columns a number of parallel texts taken from each (pp. 93-104). The Grosseteste version comes, of course, from the Balliol MS. The edited extracts support Allan's conclusion that William of Moerbeke worked in ignorance of his predecessor's version (p. 106).

What was the full extent of Grosseteste's project of

[56] For comments upon these marginal extracts, see *Aristoteles Latinus. Praefatio*, p. 53, and *Specimina*, No. 18(a), p. 129; Franceschini, *Roberto Grossatesta, vescovo di Lincoln*, pp. 57-60; Thomson, 'The *De Anima* of Robert Grosseteste', in *New Schol.* 7 (1933), 201-21; *idem., The Writings*, 66-7.

[57] 'Medieval versions of Aristotle's *De Caelo* and of the commentary of Simplicius', in *Med. Renaiss. Stud.* 2 (1950), 82-120. See our note on the MS (1950).

translation? The extracts contained in the Vatican MS cover the *De Caelo I-III.1*, ending with the marginal remark, '*Hucusque d. R*[obertus] '. As regards the commentary of Simplicius, the position is less clearly in view. Allan rightly thought it reasonable to conclude that if Grosseteste rendered Bk. II he must also have translated Bk. I, and he adduced some slight positive evidence that this was the case. L. Minio-Paluello confirmed this conclusion when he discovered that MS *Vatican Lat. 2088* contains a number of variants written between the lines of Simplicius's commentary in the *Translatio Nova*, Bk. I; these derive from Grosseteste's version.[58] The evidence has recently been submitted to a thorough re-examination by Dr Bossier, who has shown that the variants are much more numerous than Minio-Paluello had thought, occurring in almost every line of the text, which must be considered a *translatio contaminata*, preserving many of the characteristic elements of Grosseteste's version of Simplicius's commentary, Bk. I.[59]

12. *Aristoteles de Virtute (De Laudabilibus Bonis)*
       Pseudo-Andronici de Passionibus Thomson had no hesitation in including what he called *Aristoteles de Virtute* among the authentic translations of Grosseteste, but placed the *De Passionibus* among the *dubia* (p. 233). However, Franceschini had already remarked that Grosseteste quotes a definition of *eros* from the *De Passionibus* in the second chapter of his *Commentary on the Celestial Hierarchy;*[60] relying upon this indication, L. Tropia compared the translation with Grosseteste's authentic versions, with results that place its genuineness beyond doubt.[61]

       Mme Glibert-Thierry has recognized that the Latin translation of the *De Passionibus* appears in two MSS (Cambridge,

---

[58] 'Note sull'Aristotele latino medievale I-III', *Riv. Fil. Neo-scol.* 42 (1950), 222-37; p. 225, n. 1.

[59] I am indebted to Dr Bossier for permission to refer to his researches. He has prepared a critical edition of the Latin translations of Simplicius's commentary on the *De Caelo* as a doctoral thesis at the *Katholieke Universiteit te Leuven.* It is to be hoped that publication of his edition will not be long delayed.

[60] E. Franceschini, *Roberto Grossatesta*, p. 115. The quotation is brief but quite direct (p. 102).

[61] L. Tropia, 'La versione latina medievale del PERI PATHON dello Pseudo-Andronico', in *Aevum* 26 (1952), 97-112. So literal is the translation that Tropia was able to employ it in reconstructing the faulty Greek text published by Kreuttner (ibid. pp. 103-4).

Peterhouse 116 and Klosterneuburg, *Stiftsbibliothek 748*) together with *De Laudabilibus Bonis*, whose translation is certainly due to Grosseteste, and has demonstrated from the Greek MS-tradition that the two are in fact parts of a single *opusculum*.[62] We can safely assume that Grosseteste rendered both parts of the *opusculum* together, and there is no great difficulty in dating them; the quotation in the *Commentary on the Celestial Hierarchy* places the work of translation before 1239, but not long before then, since Grosseteste can scarcely have begun to translate independently before *c*.1237. This opuscule is therefore one of the earliest of his translations. It preludes his great project of rendering a whole corpus of Greek ethics into Latin;[63] we may suppose that he decided to make a trial run by translating first the smallest unit in the great codices that awaited his attention.

Mme Glibert-Thierry has replaced Tropia's edition of the Latin version of the *De Passionibus* (made from the two MSS listed by Thomson) by a critical edition of the entire opuscule (Greek text and medieval Latin version). This is based on the two MSS containing the whole text of the translation, the four of *De Passionibus* alone,[64] and the fifteen in which the *De Laudibilibus* is preserved.[65] The little work was consulted by Albert the Great and Thomas Aquinas, among others, and the Latin translation was not replaced until the Renaissance.

Quadratura per Lunulas: a fragment of Simplicius's *Commentary on the Physics* Wrongly listed by Thomson among the scientific works of Grosseteste,[66] the *Quadratura* has been identified by Professor Clagett as being the version of a brief section of Simplicius's *Commentary on the Physics*.[67] It has survived in two forms, which Clagett distinguished and

---

[62] *Pseudo-Andronicus de Rhodes*, 'Περὶ παθῶν'. Édition critique du texte grec et de la traduction latine médiévale par A. Glibert-Thirry (Leiden, 1977), vi+360 pp.

[63] This is inherently more likely than Thomson's suggestion (*The Writings*, p. 68) that the *De Laudabilibus Bonis* was translated in 1245 (one of his favourite dates), after the Greek commentators on the *Ethics*.

[64] See Thomson, *The Writings*, p. 234, and *Aristoteles Latinus*. Codices descripsit G. Lacombe, I, notes 47 and 668.

[65] Twelve of these were listed in *Aristoteles Latinus*, and the remaining three represent discoveries of Père Gauthier which are announced in Mme Glibert's edition.

[66] *The Writings*, p. 113, where the title is given as *De Quadratura Circuli*.

[67] Marshall Clagett, 'The *Quadratura Per Lunulas*. A thirteenth-century fragment of Simplicius' Commentary on the *Physics* of Aristotle'.

published. Version I is a literal translation from the Greek, and is extant in only one MS, Oxford, *Corpus Christi 251*, fols. 83ᵛ-84ʳ (13th cent.); it has never been published before.[68] Version II is described as being 'a paraphrase of that translation done into more natural Latin', and survives in nine MSS, five of which are employed by Clagett in his edition.[69]

The unique witness to the text of Version I points to Grosseteste's responsibility for it, being ascribed as follows: '*Hanc demonstrationem inveni Oxonie in quadam cedula Domini*(?) *Linco[lniensis]*'.[70] It is by no means as certain that Grosseteste was responsible for Version II of the *Quadratura*; he may well have made the translation without having written the paraphrase. All we can say is that the MS-tradition of the second version associates it with his name. One of the nine MSS containing it is ascribed, '*Lincolniensis de quadratura circuli*', and three others contain one or more of Grosseteste's treatises on natural philosophy.[71] As Clagett points out, however (p. 101), the above ascription to Grosseteste may indicate merely that the paraphraser is being honest about his source, namely Version I.

B. Commentaries

(a) Biblical[72]

15. *Prohemium et Glosae in Libros Sapientiae et Ecclesiastici*
These glosses, considered by Thomson (p. 72) to be autographs of Grosseteste and ascribed by him to the bishop on purely paleographical grounds, can no longer be considered authentic. In his article on Grosseteste's library,[73] Dr Hunt suggests that they were probably composed by Alexander of Stavensby, Bishop of Lichfield (d. 1238), to whom the codex containing them, now in the possession of Shrewsbury School, originally belonged.

---

[68] It was, however, published in the same year as Clagett's edition in *Arist. Lat.* Codices descripsit G. Lacombe, Pars Posterior (Cambridge, 1955), pp. 98-9.

[69] Clagett, art. cit., p. 100. For the list of MSS, see ibid., p. 102; cf. also the note in Section 1, under 1955.

[70] Clagett, art. cit., p. 100.

[71] These are: Oxford, *Bodleian, MS Digby 190*; London, *Brit Libr. MS Royal 12.E. xxv*; Florence, *Bibl. Nazionale, MS I.V.18.*

[72] The *Hexaemeron* (*The Writings*, no. 49) should be classified under this heading, instead of under 'Philosophical and Scientific Works'; the same applies to the *De Operationibus Solis* (no. 62).

[73] 'The library of Robert Grosseteste', in *Scholar and Bishop*, p. 141.

17. *Commentarius in Epistolam Pauli ad Romanos v-xvi* A marginal ascription to Grosseteste, which Thomson (p. 74) took to refer to the whole text of MS *Gonville and Caius College 439*, has been shown by Miss Smalley to refer only to a quotation from Grosseteste's commentary on Romans, which is now lost, but was presumably among the works known to Gascoigne in the early fifteenth century.[74] The Gonville and Caius MS is a compilation of texts from authors whose names are noted in the margins, the whole being arranged so as to form a commentary on chs. v-xvi of the Epistle.

19. *Notulae in Psalterium seu in Commentarium Petri Lombardi in Psalmos* The marginal comments in *Lincoln Cathedral MS 144*, thought by Thomson (pp. 77-8) to be in Grosseteste's hand, are not so in fact, and the MS offers no other indication which would connect it with Grosseteste (Hunt, art. cit., p. 78).

(b) Philosophical

22. *Notula super Epistolam Johannis Damasceni De Trisagion* The theological note which Grosseteste devoted to the chief doctrinal issue dividing the Greek and Latin Churches, namely, the procession of the Holy Spirit, was appended by its author to his version of Damascene's little work on the *Sanctus*.[75] Its central contention is that the opposing formulations of trinitarian faith do not entail a real difference in belief. We may remark in passing that the *Notula* is the best authenticated of Grosseteste's writings, for, besides Scotus's witness to its genuineness,[76] all five surviving MSS containing the note attribute it to him; such unanimity of ascription is unparalleled in the whole corpus of Grosseteste's works.

The Council of Florence saw the triumph of Grosseteste's idea, whose origin on the Latin side was by then probably long forgotten; its decree of 8 June 1439 acknowledged the orthodoxy of both Eastern and Western formulations and decided that the same faith in the eternal procession of the spirit underlies them.[77]

[74] Smalley, 'The Biblical Scholar', in *Scholar and Bishop*, p. 70 n.

[75] For the MSS of the *Notula* see Thomson, *The Writings*, p. 80. See further J. McEvoy, 'Robert Grosseteste and the reunion of the Church', in *Coll. Franc.* 45 (1975), 39-84. See Part I, ch. 1, p. 26-7, for a translation.

[76] Duns Scotus, *Opera Omnia*, vol. V (Vatican City, 1959), p. 1.

[77] Mansi, *Concilia*, t. XXXI, cols. 1030-31.

23. *Prologus in Librum Johannis Damasceni De Logica*
Thomson considered (p. 81) that most of this *Prologus* was
Grosseteste's own. Callus, however, noticed that the Latin
prologue is no more than a translation from the Greek, the
text of which is to be found in *P.G.* 94, cols. 489-98.[78]

25. *Commentarius in VIII Libros Physicorum Aristotelis*
This commentary was compiled from *marginalia* of Grosseteste by an unknown (Franciscan?) editor, working before
the time of Scotus and almost certainly at Oxford. There can
be no doubt that the substance of the work comes from
Grosseteste (all three extant MSS ascribe it to him), but their
form is due to the editor. Dr Dales, the modern editor of the
work, found great difficulty with the text of all but the last
pages (Bk. VIII), where the five independent MSS of the treatise, *De Finitate Motus et Temporis*, came to the rescue.[79] The
De Finitate grew out of the same project of commenting on
the *Physics* as the notes, and it was at first associated with
these before being excerpted to form a separate treatise. The
notes were probably written over a period of years, the majority of them between 1228-32, although the notes on the
later books and the *De Finitate Motus* may have been written
a few years later still.

The text presents great difficulties to the reader and would
need to be corrected. It may well be that Grosseteste post-
poned the writing-up of his notes, rather than abandoned all
notion of writing a commentary, until such time as he would
have acquired sufficient knowledge of Greek to read the
Physics in the original.[80] It looks as though he never got back
to his notes.[81]

[78] 'Robert Grosseteste as Scholar', in *Scholar and Bishop*, p. 54.
[79] *Roberti Grosseteste, Episcopi Lincolniensis, Commentarius in VIII libros
Physicorum Aristotelis . . . edidit R. C. Dales* (1963), xxxii+192 pp.
[80] His sense of bondage and servility to the Latin versions of his time is well
expressed in a famous passage of the *Hexaemeron*, where he voices a complaint
which was to find a recurrent echo in Roger Bacon: 'contra quosdam modernos',
who try to interpret Aristotle in a Catholic sense, 'mira cecitate et presumptione
putantes se limpidius intelligere et verius interpretari Aristotelem ex littera Latina
corrupta, quam philosophos tam gentiles quam Catholicos, qui eius litteram incor-
ruptam, originalem [i.e., 'complete', 'unabridged'], Grecam, plenissime noverunt'.—
Oxford, *Bodleian MS lat. th. c. 17*, fol. 197.
[81] I have attempted to supply the references and to examine the extent to
which the Latin version helped or hindered Grosseteste's understanding of the
Physics; see 'The correspondence between Grosseteste's quotations from the Latin

26. *Summa in VIII Libros Physicorum Aristotelis* This abbreviated paraphrase of the *Physics* has recently been reprinted from the Venice 1552 edition of the commentary of Aquinas on the *Physics*, but without any reference to the seven extant MSS.[82] Six of these ascribe the work to Grosseteste, but the results of internal criticism point in the other direction.[83] There seems to be nothing save the attribution in the MSS to connect this summary with his name.

29. *Quaestiones in De Caelo et Mundo Aristotelis* Thomson published these brief questions as *authentica* of Grosseteste,[84] but the problem of their genuineness is inextricably bound up with that of the *De Anima* (see No. 33), which Thomson is alone in defending. Both works are known only in MS *Bodleian, Digby 104*, in which the copyist ascribed the former work to Grosseteste and transcribed the *Quaestiones* after them with a break of only one line. Thomson claims that the questions 'pursue the same method as the *De Anima* and exhibit the same viewpoint we find Grosseteste adhering to in his other physical and cosmological works.'[85] This seems exaggerated and difficult to sustain. Grosseteste had, indeed, very personal ideas on cosmology, but I cannot find any echoes of them in this fragment, and can see no reason to link it with his name.

C *Philosophical and Scientific Works*

32. *De Accessu et Recessu Maris (De Fluxu et Refluxu Maris)* The *Assisi MS Bibl. Comm. 138* ascribes this work, but the vital letters are fatally ambiguous and can be read either as '*a magistro R. Oxon*' or as '*A Oxon*', and taken to refer to Grosseteste (Pelster, Thomson) or Adam Marsh (Henquinet, Callus). Dales at first defended the attribution to Grosseteste but has recently rallied to the other camp.[86] However, the

translation of the *Physics* and the Greek text', in *Bulletin de philosophie médiévale* 21 (1979), pp. 52–62.

[82] Celina A. Lértora Mendoza and J. E. Bolzán, 'La *Summa Physicorum* atribuida a Robert Grosseteste', *Sapientia* 26 (1971), 19–74; p. 74 (republished in book form: *Robert Grosseteste. Suma de los ocho libros de la 'Fisica' de Aristóteles* ('*Summa Physicorum*'): texto latino (Buenos Aires, 1972), pp. 150).

[83] Dales, 'The authorship of the *Summa in Physica* attributed to Robert Grosseteste', in *Isis* 55 (1964), 70–4; Lértora Mendoza, 'La *Summa Physicorum* y la filosofía natural de Grosseteste', in *Sapientia* 26 (1971), pp. 199–216; p. 216.

[84] *New Schol.* 7 (1933), 218–21.

[85] *The Writings*, p. 86.

[86] R. C. Dales, 'Robert Grosseteste's scientific works', *Isis* 52 (1961), p. 391;

new arguments which he brings forward do not appear to me very forceful ones, and I find that, in style, method and content, the *De Fluxu* has its natural place within the authentic works on natural philosophy. Moreover, two of the three MSS brought to light by Thomson attribute its composition to Grosseteste, and in the third it is accompanied by an authentic work.

The question was written while the author was '*in scolis suis*', so the Assisi MS informs us; in Grosseteste's case, that was before 1229-30. It refers to Al-Bitrūji on the tides, which was translated *c*.1220 by Michael Scot, and it was probably written between 1225 and 1229. It has been edited twice. Franceschini published a text based on three of the known MSS.[87] Dales has provided the critical edition.[88]

33. *De Anima* Thomson's grounds for accepting the authenticity of this treatise included both external and internal evidence.[89] The external is strong; the single MS in which the work is known to exist is early and ascribed. The internal evidence was, however, weakened considerably when Keeler noted a strong similarity between the doctrine of the *De Anima* and the early teaching of Philip the Chancellor.[90] Daniel Callus succeeded in establishing the extensive dependence of the treatise on the *Summa de Bono* of Philip, which was not edited till about 1230; he thus outruled Keeler's suggestion that the *De Anima* might be Grosseteste's report of Philip's lectures, taken while the former was studying at Paris,[91] and concluded to its inauthenticity on the grounds that Grosseteste was little likely to put together such a derivative work *c*.1230, the period of his greatest originality. The dependence

'Adam Marsh, Robert Grosseteste, and the treatise on the tides', in *Speculum* 52 (1977), 900-01.

[87] 'Un inedito di Roberto Grossatesta: la *Questio de Accessu et Recessu Maris*', in *Riv. Fil. Neo-scol.* 44 (1952), 11-21.

[88] 'The text of Robert Grosseteste's *Questio de Fluxu et Refluxu Maris* with an English Translation', in *Isis* 57 (1966), 455-74.

[89] 'The *De Anima* of Robert Grosseteste', in *New Schol.* 7 (1933), 201-21; cf. *The Writings*, pp. 89-90.

[90] 'The dependence of Robert Grosseteste's *De Anima* on the *Summa* of Philip the Chancellor', in *New Schol.* 11 (1937), 197-219.

[91] 'Philip the Chancellor and the *De Anima* ascribed to Robert Grosseteste', in *Med. Renaiss. Stud.* 1 (1941), 105-27.

in turn of the *Summa de Bono* upon the *Summa Duacensis* does not affect his argument.[92]

37. *De Calore Quaestio* This brief *quaestio* is known in only one MS, namely, Madrid *Bibl. Nac. 3314*, which contains six *opuscula* of Grosseteste, identified by Thomson in 1935.[93] Thomson printed it in 1957.[94] The terse manner of the *quaestio* and its lack of anything like literary style may indicate that it is a *reportatio*; in any case, it is not a finished work. The only authority cited is Aristotle. The date of *c.*1230 is suggested by Thomson, but since the *quaestio* is a school-exercise, one assumes that it must be earlier than 1230, by which time Grosseteste had certainly begun teaching theology. A date *c.*1220-25 might be more appropriate, though any suggestion must be made tentatively.

38. *De Calore Solis* A. C. Crombie published a translation of this *opusculum* as an appendix to his essay on Grosseteste's scientific method.[95] He considers it to represent an outstanding example of Grosseteste's method at work in the solution of a physical problem. The text which he used was that published by Baur, which has not been re-edited since (*Die Werke*, pp. 79-86).

40. *De Cometis et Causis Ipsarum* Baur published two texts of the *De Cometis*.[96] Following a MS-discovery in 1933, Thomson edited a fuller text of the same work.[97] The *De Cometis* appears also in the Madrid MS referred to above, and in the Florentine codex, *Bibl. Riccardiana 885*, fol. 214.[98] These give a more complete text than any yet known to exist. They were utilized by Thomson in order to provide an addition

[92] 'The *Summa Duacensis* and the Pseudo-Grosseteste's *De Anima*', in *Rech. Théol. anc. méd.* 13 (1946), 225-9. A comparison of the doctrine of the *De Anima* with the authentic works tends to confirm its inauthenticity; cf. E. Bettoni, 'Intorno all'autenticità del *De Anima* attribuito a Roberto Grossatesta', in *Pier Lombardo* 5 (1961), 3-27.

[93] *The Writings*, pp. 92 ff, 110 ff.

[94] 'Grosseteste's *Quaestio de Calore, De Cometis*, and *De Operacionibus Solis*', in *Med. Human.* 11 (1957), pp. 34-5.

[95] 'Grosseteste's Position in the History of Science', in *Scholar and Bishop*, pp. 98-120; for the translation, see pp. 116-70.

[96] Baur, *Die Werke*, pp. 36-41.

[97] 'The text of Grosseteste's *De Cometis*', in *Isis* 19 (1933), 19-25.

[98] See n. 93.

to the text of the *De Cometis* which he had already published in 1933.[99] The two MSS employed do not differ much; the variants are noted. The *De Cometis* was probably composed shortly after the year 1222.

45. *De Finitate Motus et Temporis*   Baur's edition of this short treatise was based upon two MSS containing Grosseteste's *Commentary on the Physics* and one independent MS of the *De Finitate*. The discovery by Thomson of four additional witnesses to the text of the *opusculum* necessitated a new edition, especially since two redactions of it can be clearly distinguished.[100]

The text of the *De Finitate* published by Dales in his edition of the *Commentary on the Physics* employed four MSS, the three containing the entire *Commentary*, and the Prague MS witnessing the longer redaction of the *De Finitate*.[101] Dales himself provided the critical edition of the *De Finitate* based upon all the known MSS, and representing therefore both redactions (which differ in fact only in length).[102] The date of composition suggested by the editor (c.1235) is very acceptable.

49. *Hexaemeron*   The first publication in the present century of an extract from the *Hexaemeron* falls outside the chronological limits of this survey, but deserves none the less to be mentioned here.[103] The edition of the *Hexaemeron* projected by Muckle bore fruit in two articles, the first of which was printed in 1944,[104] before being abandoned by him.

In the second of his two articles Muckle published chs. 24 and 25 of Part Two of the *Hexaemeron*, in order to illustrate the use made by Grosseteste of Greek sources, especially patristic ones, in the compositon of his major theological work.[105] The importance of the text is not confined to the

[99] 'Grosseteste's *Quaestio de Calore, De Cometis, and De Operacionibus Solis*'.
[100] Thomson, *The Writings*, p. 98.
[101] *Commentarius in VIII libros Physicorum Aristotelis*, ed. Dales, pp. 144-55.
[102] 'Robert Grosseteste's treatise *De Finitate Motus et Temporis*', in *Traditio* 19 (1963), 245-66.
[103] G. B. Phelan, 'An unedited text of Robert Grosseteste on the subject-matter of theology', in *Rev. Néoscol. Phil.*, 36 (1934), 172-9; for the text, see pp. 176-9.
[104] J. T. Muckle, 'The *Hexameron* of Robert Grosseteste. The first twelve chapters of Part Seven', in *Med. Stud.* 6 (1944), 151-74. What Muckle in fact published comes not from Part Seven, but from Part Eight of the *Hexaemeron*.
[105] 'Robert Grosseteste's use of Greek sources in his *Hexameron*', in *Med. Human.* 3 (1945), 33-48.

interest of its sources, but lies in its development of the theme of light.

In his article on beauty as a transcendental in the Scholastic tradition,[106] Dom Pouillon edited a few brief extracts from the *Hexaemeron*, using the London, British Museum MS *Royal VI.E.v.* The passages are important and the text is reliable.

A short text from the *Hexaemeron* was included by Unger in his treatment of the reasons for the Incarnation in Robert Grosseteste.[107] The text is based on the *Bodleian MS lat. th. c.17*, and the *Brit. Mus. MS Royal VI.E.v.*

In 1964 Dales and Gieben decided to undertake the critical edition of the *Hexaemeron*. They have offered as the first-fruit of their labours an edition and study of the *Prooemium* to the work.[108] This preface takes the form of a gloss on two of the letters of St. Jerome, nos. LIII *Ad Paulinum Presbyterum*, and XXVIII *Ad Desiderium*, which commonly preceded the text of Genesis in medieval MSS of the Bible. The glosses of Grosseteste on the two epistles display a certain knowledge of Greek, in illustration of which three passages from the *Prooemium* are edited and discussed.

50. *Grammatica* Thomson had reservations about the authenticity of this work but gave it the benefit of the doubt (p. 101). Its editor, Dr Reichl, has (rightly in my view) ruled out its ascription to Grosseteste.[109] There is no reason on grounds of style or content to connect it with his name, and the external evidence is weak, to say the least of it. Roger Bacon knew it, despised it, and made no connection between it and his hero.[110]

The work possesses interest in its own right, as a precursor of the genre of speculative treatises on grammar which began to appear in Paris in the second half of the thirteenth century.

[106] 'La Beauté, propriété transcendentale, chez les Scolastiques (1220-1270)', in *Arch. Hist. Doctr. Litt. M.A.* 15 (1946), 263-329; cf. also the note under 20 (iii), p. 470.

[107] 'Robert Grosseteste, Bishop of Lincoln (1235-1253), on the Reasons for the Incarnation', in *Franc. Stud.* 16 (1956), pp. 23-25.

[108] 'The *Prooemium* to Robert Grosseteste's *Hexaemeron*', in *Speculum*, 43 (1968), 451-61.

[109] K. Reichl, *Tractatus de Grammatica. Ein fälschlich Robert Grosseteste zugeschr. spekulative Grammatik. Edition und Kommentar* (Munich, 1976), 224 pp.

[110] See *The Greek Grammar of Roger Bacon*, ed. E. Nolan and S. A. Hirsh (Cambridge, 1902), pp. 57 ff.

Its anonymous author may have been an English Master of Arts, well schooled in the traditional Latin treatises of grammar and with a comprehensive if perhaps somewhat external knowledge of Aristotle and the newer philosophical literature, writing towards 1250, scarcely much later than that in any event, and ready to attempt something new.

58. *De Luce (De Inchoatione Formarum)* This little treatise, which represents Grosseteste's most central contribution to cosmology, was edited by Baur among the *opuscula*,[111] but the edition as a whole sadly lacks the critical status claimed for it, and the text of the *De Luce* is vitiated by serious errors. Its critical edition is a priority of Grosseteste studies. An English translation based upon that edition was published in 1942 by Riedl.[112] The *De Luce* was almost certainly composed between 1225 and 1230.

62. *De Operationibus Solis* In 1928 Martin Grabmann published his notes on MS 3314 of the *Biblioteca Nacional*, Madrid, drawing attention to its great importance for the history of thirteenth-century philosophy.[113] The thorough investigations made by Thomson between 1935 and 1940 failed to uncover any further witnesses to the text, which he subsequently published in 1957, but with many errors.[114]

Though the *De Operationibus* is not ascribed to Grosseteste in its single MS witness, the external evidence for its authenticity is quite good. The commentary is preceded by four certainly authentic works, none of which is ascribed, and followed by a treatise which is ascribed to Bacon, which may be taken as an indication that the copyist was aware of a difference of authorship as between this and what went before. I have argued in my edition of the work for a date of composition after *c.*1232, when Grosseteste had as much scientific information and interest as he ever acquired, and before 1235,

[111] Baur, *Die Werke*, pp. 51–9.

[112] *Robert Grosseteste on Light (De Luce)*. Translation from the Latin, with an introduction by Clare C. Riedl (Milwaukee, 1942), 17 pp.

[113] 'Mittelalterliche lateinische Aristotelesübersetzungen u. Aristoteleskommentare in Handschriften spanischer Bibliotheken', in *Sitzungsb. d. Bayer. Akad. der Wiss.*, philosophisch-philologische u. historische Kl., 1928, 5. Abh., pp. 120; 63–70. cf. *Mittelalterliches Geistesleben* II (Munich, 1936), pp. 114, 126, 139, 191, 197, 291.

[114] *Med. Human.* 11 (1957), 34–43. See notes on nos. 37 and 40 above.

when his knowledge of Greek was already considerable.[115] Although the little work was not composed for his scholars in the Oxford friary, it may have resulted from his biblical lectures there and we can be confident that, like the *Hexaemeron* and certain of his other biblical commentaries and treatises, it is a precious witness to the character of his theological lectures, and consequently to the nature of his influence upon the first generation of English Franciscans.

66. *De Quadratura Circuli*: see *Quadratura per Lunulas* (under 'Translations').

67. *V Quaestiones Theologicae* Callus, who edited these questions from the single MS in which they are known, is a forceful and persuasive defender of their authenticity, drawing attention to the doctrinal parallels which they exhibit in relation to other works of Grosseteste.[116] These similarities do not seem to me to warrant the firm authentication of the questions. I regard the external evidence of authenticity as being stronger than the internal, namely, their association with authentic treatises in the MS. Grosseteste's authorship of them cannot be established with certainty until further evidence turns up.

D. Pastoral and Devotional Works

77. *De Cessatione Legalium* The date of composition of this biblical treatise has not yet been fixed with certitude. Thomson ascribed it to *c*. 1231, and connected it with a project of converting the Jews.[117] Smalley, however, correctly observed that the work has no missionary aim, but that it was written for students of theology; she considered that it would fit very well into the end of Grosseteste's lectorship.[118]

Grosseteste himself offers us one hint as to the date of composition of the treatise. In the course of a vigorous defence of the truth of Christian faith he protests that if what has been

[115] J. McEvoy, 'The sun as *res* and *signum*: Grosseteste's Commentary on Ecclesiasticus ch. 43, vv. 1-5', in *Rech. Théol. anc. méd.* 41 (1974), 38-91. The Latin text occupies pp. 62-91.

[116] 'The *Summa Theologiae* of Robert Grosseteste', in *Studies . . . Presented to F. M. Powicke* (Oxford, 1948), pp. 180-208.

[117] *The Writings*, p. 121. Thomson relied on Stevenson's judgement in this matter, *Robert Grosseteste, Bishop of Lincoln*, pp. 99 ff., as also did L. M. Friedmann, *Robert Grosseteste and the Jews* (Cambridge, Mass., 1934), pp. 21 ff.

[118] 'The Biblical Scholar', in *Scholar and Bishop*, pp. 81-2.

preached concerning the Lord for more than one thousand two hundred and thirty years were no more than a tissue of lies of human invention, it could never have won acceptance, since falsehood is darkness, and of its very nature ephemeral.[119] The *terminus ante quem*, as Miss Smalley points out, is fixed by a quotation from the *De Ecclesiastica Hierarchia* of Pseudo-Dionysius, which corresponds verbally with the translation of Scottus Eriugena; Grosseteste would have quoted his own version had it been available. He translated the *Ecclesiastical Hierarchy* second of the Dionysian works, probably in the year 1239-40.

The LXX translation is quoted at least seven times,[120] reference is made explicitly to the versions of Aquila and 'Sinaicus', and more generally to '*alia translatio*'.[121] Grosseteste's manner of contrasting these versions with each other and with the Vulgate proves that he is no newcomer to Greek. If Grosseteste began to learn Greek around 1232 (and there seems to be no reason for positing an earlier date), we must conclude that the *De Cessatione* cannot have been composed before *c.*1234, and perhaps not until several years after that date. It is in any case posterior to his *Commentary on Galatians* (which dates from 1231-5), to which it twice refers,[122] and is roughly contemporaneous with the *Hexaemeron*.

The only part of the *De Cessatione* to have received a contemporary edition[123] is the latter part of the second section, which is headed in the MS *Cur Deus Homo*.[124] The editor based his edition on the *Bodleian MS lat. th. c. 17*. The published text is, however, unreliable.

78. *Concordantia Patrum*   One of Thomson's many valuable discoveries was a topical concordance of the Bible and Fathers,

[119] Oxford, *Bodleian MS lat. th. c. 17*, fol. 181[C]: 'Si enim esset falsitas humanae inventionis quod de fide in Dominum Jesum iam predicatum et receptum est plus quam per mille et ducentos et triginta annos, quomodo sic invaluisset apud sapientes et insipientes, pauperes et divites, humiles et potentes . . . presertim cum falsitas tenebra sit et sui natura evanescens?'

[120] MS cit., fols. 172[a] (twice), 172[b], 172[c] (twice), 172[d], and 173[a].

[121] Ibid., fol. 172[b], 172[a], 175[a]. The Greek New Testament is also quoted, fol. 177[a] (Gal. 3:28).

[122] MS cit., fols. 167[d] and 185[c].

[123] Part 1 was published in London in 1658; see Thomson, *The Writings*, p. 121.

[124] 'Robert Grosseteste, Bishop of Lincoln (1235-1253), on the Reasons for the Incarnation', D. Unger in *Franc. Stud.* 16 (1956), 1-36; for the text, see pp. 3-18.

found in the Lyons MS, *Bibl. Municipale 414*, fols. 17ᵃ–32ᵃ.¹²⁵ Its title reads, '*Incipit Tabula Magistri Roberti Lincolniensis episcopi cum addicione fratris Ade de Marisco, et sunt distincciones ix quarum hec prima est de deo*'. The first four pages of the *Concordance* are occupied by around four hundred symbols, each of which is assigned to a theological topic. In the *Concordance*, the same signs are repeated, and under each one are grouped references to Scripture and to patristic writings; pagan authors are often cited in the right-hand margins. The *Concordance* is incomplete, some blank spaces being left.

Thomson has returned to the study of the work, with an edition of the first pages of the Lyons MS, containing the list of symbols and topics.¹²⁶ As typographical reproduction of the symbols proved unfeasible, Thomson chose instead to publish a photographical reproduction of the folios containing the indexing-symbols and to provide a transcription of the *tituli* to which the symbols refer. The utility of this edition is clear: it will enable librarians and researchers to add to the list of MSS known to contain these symbols,¹²⁷ and so to provide more abundant evidence of the literary activities and influence of Grosseteste's circle.

Thomson, in the course of the same article, edited fol. 20ᵃ of the *Concordance*, as an example of the whole work, giving the reader the opportunity to judge for himself the extent and richness of Grosseteste's and Marsh's reading.¹²⁸ Finally, Thomson provided a list of all the works cited in the *Concordance*, including those of pagan authors. This has the advantage of revealing the predilections of Grosseteste and Marsh for the Latin patristic tradition: Augustine, Jerome, Gregory, and Ambrose predominate.

As Thomson remarks, the omission from the list of certain works, which we would have expected to find present, renders it quite certain that the Lyons MS in no way provides a complete record of its compilers' reading. It seems possible that the MS as we have it represents a stage in the production of a

---

¹²⁵ 'Grosseteste's Topical Concordance of the Bible and the Fathers', in *Speculum* 9 (1934), 139–44; also *The Writings*, pp. 122–4.
¹²⁶ Thomson, 'Grosseteste's Concordantial Signs', in *Med. Human.* 9 (1955), 39–53.
¹²⁷ For the complete list cf. Hunt, *Bodl. Quarterly Record* 4 (1953), 241–55.
¹²⁸ Thomson, 'Grosseteste's Concordantial-Signs', p. 46.

much fuller work. The paucity of references to the works of Damascene and Pseudo-Dionysius, as well as to the *Ethics* of Aristotle, would seem to date the project attested to by the Lyons codex to a period somewhat before *c*.1237. Thomson's suggestion,[129] that the period of relative leisure enjoyed by Grosseteste, between his illness of 1232 and his consecration in 1235, may have permitted him to begin work on the *Concordance*, is interesting, but perhaps it is better to bring the compilation back some years, to 1225-30.

80. *De Confessione II* (Sermon No. 32, *'Deus Est'*) Recent MS-discoveries have revealed that this work first existed as a separate and anonymous treatise, and was not incorporated into the sermon-collection before the middle of the fourteenth century.[130] This finding is confirmed by the style of composition employed, for the work is relatively long, compact, and unrhetorical. The *De Confessione* is a handbook for priests, a representative, therefore, of the genre which flourished for many years after the Fourth Lateran Council. Dr Wenzel, who has recently edited the *Deus Est*, has suggested for it a date of composition somewhat before 1250, but the evidence he offers in support is purely paleographical, being based upon the estimated age of the oldest known MS. The only evidence I can find which would permit the acceptance of a solid *terminus post quem* consists in the character of the prologue, in which Grosseteste uses Dionysian language and ideas to give expression to the themes of creation, hierarchy, and participation. Since Grosseteste does not seem actually to quote from the Dionysian works it is not possible to say whether his translation and commentary were already in existence at the time of composition of the *Deus Est*;never the less,it is strongly indicated that the treatise should be assigned to the period accompanying and following his absorption in the works of the Pseudo-Areopagite. The year 1239 would seem on these grounds to be the earliest possible *terminus post quem* for the *Deus Est*. Further precisions must, however, await new evidence.

---

[129] *The Writings*, p. 124.

[130] Wenzel, S., 'Robert Grosseteste's treatise on Confession, *Deus Est*', p. 218. For references to three MSS employed in this edition and unknown to Thomson, cf. under 'Manuscripts, 1970'.

None of the earliest MSS ascribes the *Deus Est* to Grosseteste, and its inclusion in codices containing authentic works of the bishop is of rather late date;[131] the external evidence of authorship must consequently be regarded as very weak. However, the internal evidence for its authenticity must be judged overwhelming in view of the parallels existing between this work and others of Grosseteste. In addition to the close parallels between the *Deus Est* and the *Templum Domini*, to which Wenzel draws attention, there exist striking resemblances between the *De Cessatione Legalium* and the prologue to the *Deus Est*. Having outlined creation in general in Dionysian terms, Grosseteste endeavours to place man within the whole by using the concept of microcosm and combining it with the other great patristic theme of God's image. In the details both of ideas and of language, the prologue of *Deus Est* is closely akin to Part II of *De Cessatione Legalium*.

Another piece of internal evidence which favours Grosseteste's authorship is a special quality of attention paid by the author of *Deus Est* to the vocation of the clergy. Thus the confessor is instructed to examine priests who come to him in the following terms:

Item quaerendum si causa lucri temporalis ad ordinem accesserit; si in mortali peccato, sive insufficiens in scientia; si ordinis officium fideliter non fuerit executus. (p. 254.)

I conclude that the treatise *Deus Est* is certainly authentic. That the evidence for its authenticity is largely internal is a matter which should not disturb that certitude. Wenzel's edition is critical, being based upon all ten complete MSS of the work.

84. *Constitutiones (Epistola 52)* These *Constitutions*, which Grosseteste issued for his diocese during the earliest years of his episcopacy, were published by Brown (1691) and Luard, *Epistolae*, pp. 154-64. They have been re-edited and studied in the context of English synods in medieval times: Powicke and Cheney, *Councils and Synods with other documents relating to the English Church*, vol. II, pt. 1, A.D. 1205-1313 (Oxford, 1964), pp. 265-78.

---

[131] Ibid., p. 218.

85. *Correctorium Totius Bibliae* This work survives in five MSS, only one of which ascribes it to Grosseteste, and that a late one (*c.*1400). Thomson expressed scepticism as to its genuineness, and (pp. 127-8) accepted it only with hesitation. The authenticity of the *Correctorium* can no longer be maintained; Dales and Gieben have pointed out that on those occasions when it corresponds with the *Prooemium* to the *Hexaemeron* it is so far inferior to the latter work that it would have to be dated to Grosseteste's youth; but it cites John of Garland's *Speculum Ecclesiae* of 1246-49.[132]

87. *Dialogus de Contemptu Mundi* Listed by Thomson as a work of Grosseteste on the evidence of a thirteenth-century MS (pp. 129-30), the so-called *Dialogus* is in fact the first part of Bk. II of the *De Vanitate Mundi* of Hugh of St. Victor.

96. *Moralitates Super Evangelia* Thomson regarded this work highly and placed it firmly within the canon (p. 134). Recently, E. J. Dobson has mounted an impressive attack on both the external and the internal authenticating evidence.[133] Of the two surviving thirteenth-century MSS one (*Lincoln Coll. MS 79*) ascribes the work to Magister Alexander de Ba, and the other (Oxford, *Trinity Coll. MS 50*) to Grosseteste. The ascription to an unknown author should, like the *lectio difficilior*, be given greater weight. While I am not bowled over by Dobson's stylistic arguments against Grosseteste's authorship (Grosseteste wrote in too many styles and genres for such arguments to be straightforward), I can find in the numerous passages which he prints from the *Moralitates* nothing to provide a definite connection with Robert Grosseteste. Although unconvinced by Dobson's other theses (that the *Moralitates* were composed early, *c.*1212, by Master Alexander, Dean of Wells, and were a direct source of the *Ancrene Wisse*), I agree with him that at least the burden of proof has moved to supporters of Grosseteste's authorship.

---

[132] 'The *Prooemium* to Robert Grosseteste's *Hexaemeron*', p. 461 n. 44.

[133] Dobson, *Moralities on the Gospels. A New Source of Ancrene Wisse* (Oxford, 1975). See the predominantly critical review by R. H. Rouse and S. Wenzel in *Speculum* 52 (1977), pp. 648-52.

99. *Parabola Domini Roberti Grosseteste* The *Parabola* forms part of the Lyons dossier and has been edited by Gieben; see no. 107 below.

103. *Templum Domini* Walter of St. Edmund, Abbot of Peterborough and friend of Grosseteste, possessed '*Templum Domini cum arte confessionaria*' of Grosseteste already before his death, which occurred in 1246.[134] Boyle (*Med. Renaiss. Stud.* 8 (1979), p. 46, n. 27) has identified two important sources of the *Templum*, in the confessional books of Richard Wethersett and Robert Flamborough. It may be hoped that further study of the sources will help to fix its date of composition with accuracy.

105. *De Tyrannide (De Principatu Regni)* This brief distinction of true and false exercise of power forms part of the Lyons dossier and has been edited by Gieben; see the following note.

107. Grosseteste at the Papal Curia in 1250 Thomson uncovered and listed the eight documents which tell the story of Grosseteste's visit to Pope Innocent IV in May 1250. An extensive paraphrase and partial rendering of the dossier was published by Pantin.[135] Once again it is Gieben who has placed all students of Grosseteste in his debt by publishing the whole dossier in an edition which fully endorses his reputation as a resourceful editor.[136] He has employed 7 MSS (plus Brown's published text for the *Memoriale*), which among them adequately represent the five groups of sermon-collections, the setting in which the material has been transmitted to us.

109. *Liber Curialis* Thomson's attribution of this poem to Grosseteste (pp. 148-9) depended upon the unique MS known

---

[134] M. R. James, 'Lists of MSS formerly in Peterborough library', in *Supplement to the Bibliographical Society's Transactions* 5 (1926), p. 22.

[135] W. A. Pantin, 'Grosseteste's relations with the Papacy and the Crown', in *Scholar and Bishop*, pp. 178-215. The extracts translated come from docs. 1 and 9 in Thomson's enumeration of the material. In an appendix (pp. 209-15) the author gives a paraphrase (with references to sources and elements of a commentary) covering all nine docs. Extracts from several important letters of Grosseteste are also translated in the course of the article: pp. 184-5, 185-6, 188, 189-190, 197-9.

[136] 'Robert Grosseteste at the papal curia, Lyons 1250. Edition of the documents', in *Coll. Franc.* 41 (1971), 340-93. The text occupies pp. 350-93.

to him as containing it, namely, Oxford, *Trinity College MS
18*, which Thomson dated to before 1235. While preparing an
edition of the work, Gieben turned up two MSS containing
notable fragments of it.[137] He furthermore identified the *Liber
Curialis* with the *Urbanus Magnus* of Daniel of Beccles.[138] The
composite character of the *Urbanus* leaves open the possibility
that the *Liber Curialis* is, after all, an authentic work of
Grosseteste, which was later inserted bodily into the *Urbanus*.
If indeed the poem is authentic, the new MS-evidence would
suggest that Grosseteste must have composed it while he was
a young clerk in the household of the Bishop of Hereford,
before *c*.1198. Only the thorough exploration of the text and
sources of the *Urbanus* can resolve the problem of the authenticity of the *Liber*.

110. *Regulae ad Custodiendum Terras* Miss D. Oschinsky has
argued that the *Regulae* represent a translation of the Anglo-
Norman *Reulles* which she has recently edited; see under no.
120 below.

111. *Stans Puer ad Mensam* This work was first edited from
a single MS in 1868. Thomson lists six MSS (p. 150), of which
two (Oxford, *Bodleian 315*, and *Trinity 18*) contain only
the first seven lines of the poem, called by Thomson 'Recension 1'. Gieben considers that only the longer text is authentic.
He identifies eleven MSS containing the poem in whole or in
part, and offers a critical edition of the text of fifty-six
lines.[139] It was the chief Latin text-book of manners known
in England, and as such obtained wide currency in the grammar schools during the later Middle Ages.

112. *Statuta Familiae* This ordinance for the conduct of
Grosseteste's household has been edited for the first time by
Oschinsky as a companion to the *Reulles* (see below under
no. 120).

120. *Les Reulles de Seint Robert* A glimpse of life in Grosseteste's home is afforded by the reading of the *Statuta Familiae*,

[137] 'Robert Grosseteste and Medieval Courtesy-books', in *Vivarium* 5 (1967),
47–74; cf. p. 69.
[138] J. G. Smyly, ed., *Urbanus Magnus Danielis Becclesiensis* (Dublin, 1939),
lines 875–1546.
[139] 'Robert Grosseteste and Medieval Courtesy-Books', p. 56.

now published for the first time with the Anglo-Norman *Reulles* by Oschinsky.[140] The *Statuta* provided for the regulation of the bishop's *familia* and the domains attaching to the see. Oschinsky has managed to reconstruct the relationship among the three compositions, the *Statuta*, the Anglo-Norman *Reulles*, and the *Latin Rules*. She suggests with a high degree of probability that the *Reulles* represent an amplified version of the *Statuta*; the French was then rendered into Latin, and much later the Latin passed into English in extract form.

Naturally a close relationship obtains among all three works, but the fact that Grosseteste is claimed for all of them as their author unfortunately does not settle the problem of authenticity for each single one. The case of the *Statuta* is uncomplicated: they are Grosseteste's rules for the conduct of his own household and bear his mental signature. The French *Reulles* were sent to the Countess of Lincoln when she was widowed and faced with the management of her large estates. There is a case for regarding Grosseteste as the principal author of the *Reulles* and the immediate author of such of them as are taken more or less directly from the *Statuta*. And indeed the common elements in the two texts manifest a spiritual physiognomy we easily recognize from other sources. The dating of the *Reulles* is unproblematic: the Countess was widowed in 1240, so that her request for assistance must have been made and fulfilled before she took the easy way out of estate management by remarrying in 1242.

E.   Sermons, Letters, *Dicta*

7.   *Exiit Edictum*   The text of this sermon as published by Unger from *Brit. Libr. MS Royal vii.F.2*, is incomplete.[141] Its theme is an important one, namely, the reasons for the Incarnation.[142] The sermon makes allusion to a longer previous treatment of the same subject; this can only be interpreted as a reference to the *De Cessatione*, and so the sermon must have been composed after *c.*1237.

---

[140] Dorothea Oschinsky, *Walter of Henley and other treatises on estate management and accounting* (Oxford, 1971), xxiv+504 pp. For the texts and translation see pp. 387–416; the editor's discussion is found on pp. 191–9. The *Reulles* and the English version of their Latin form have both been printed before; see Thomson, *The Writings*, pp. 151, 159.

[141] See Unger, art. cit., pp. 18–23.

[142] See note on no. 77, pp. 489–90.

16. *Tota Pulchra Es*　The importance of the sermon *Tota Pulchra Es* was recognized many years ago by Longprė, who quoted a brief section of it in a note published in 1933.[143] Gieben prefaces his edition of the sermon with a very scholarly treatment of Grosseteste's teaching on the Immaculate Conception.[144] In the present sermon Grosseteste teaches only the possibility of the Immaculate Conception, whereas in a later one he expounds the belief quite unambiguously.[145]

The date of *c.*1230 suggested by Fr Gieben should be revised in the light of the close parallels which the sermon contains (particularly as regards its concept of *'purgatio'*) with no less than five passages of the *Commentary on the Celestial Hierarchy*. A date around the year 1240 would consequently be more acceptable.

30. *Ecclesia Sancta Celebrat*　This was preached on an Easter Sunday morning to clergy of the Diocese of Lincoln. Grosseteste reflects on the powers of the soul as natural endowments (revealing a strong influence of Aristotle, Avicenna, and Alcher of Clairvaux), before following the way of ecstasy. The close verbal parallels to Grosseteste's own commentary on the *Mystical Theology* suggest that it was composed in 1243 or shortly afterwards.

In editing its text I employed three MSS, one of which was the Prague, *Národni Mus.* XII.E.5.[146] The doctrine and sources of the sermon are explored in the introduction.

33. *Ex Rerum Initiatarum*　In an enlightening introduction, whose footnotes are heavily larded with texts from *inedita* of Grosseteste, Gieben explains the Bishop's attitude to preaching, and offers a few comments upon this long and important sermon on the theme of Redemption.[147]

---

[143] 'Robert Grossetête et l'Immaculée Conception', in *Arch. Franc. Hist.* 26 (1933), 550-1.

[144] 'Robert Grosseteste and the Immaculate Conception . . .', in *Coll. Franc.* 28 (1958), 211-27.

[145] In the sermon *Ex Rerum Initiatarum* (see n. 147).

[146] J. McEvoy, 'Robert Grosseteste's Theory of Human Nature. With the Text of His Conference, *Ecclesia Sancta Celebrat*', in *Rech. Théol. anc. méd.* 47 (1980) pp. 131-187.

[147] 'Robert Grosseteste on Preaching. With the Edition of the Sermon *Ex Rerum Initiatarum* on Redemption', in *Coll. Franc.* 37 (1967), 100-41. For the text of the sermon, pp. 120-41. For the list of MSS containing the sermon, see Thomson, *The Writings*, p. 176, no. 33.

The date of *c.*1240 which Gieben assigns to the sermon takes account of the evident Dionysian influence (which is not confined to the vocabulary of the piece alone, but extends to its thought), and of the knowledge of Greek exhibited (p. 114). Strictly speaking, the year 1240 should be considered rather as a *terminus post quem*, because Grosseteste's explicit defence of the Immaculate Conception contained in this sermon marks a clear advance on the teaching of the *Tota Pulchra Es* of *c.*1240, and so a period of time should be allowed for the development.

Letter CXXVIII This famous letter was written by Grosseteste in protest against the presentation by Pope Innocent of his nephew to a Lincoln canonry. Fr L. Boyle has studied it in depth and corrected Luard's faulty edition from *MS Bodley 42*, to produce 'a working copy', which he prints as an appendix to his study, *Med. Renaiss. Stud.* 8 (1979), 3-51; pp. 40-4 contain the letter.

Letter CXXX Luard edited this letter from *MS Bodley 750* and ascribed it to Grosseteste, assuming that a reference in the *intitulatio (R. miseracione divina lincolniensis episcopus)* was intended to stand for '*Robertus*'. The word 'Grosseteste' had been added in the left-hand margin by a later, but similar, hand. Mantello has re-examined all the relevant evidence and convincingly shown that the letter is almost certainly not the work of Grosseteste but of his second successor as Bishop of Lincoln, Richard Gravesend.[148]

Dicta A paper by E. J. Westermann, published in 1945, examined the numerous parallels which exist between elements of the sermon-collections of Grosseteste and his *Dicta*.[149] Twenty-seven of the latter are in fact sermons; Westermann showed that, of the sixteen sermons existing in the oldest collections and surviving also in the form of *Dicta*, five have received considerable expansion and revision by Grosseteste himself, namely sermons 43, 53, 54, 55, 56 (which appear as

---

[148] F. A. C. Mantello, 'Letter CXXX of Bishop Robert Grosseteste: a problem of attribution', in *Med. Stud.* 36 (1974), 144-59.

[149] 'A comparison of some of the Sermons and the *Dicta* of Robert Grosseteste', in *Med. Human.* 3 (1945), 49-63.

Dicta 21, 72, 50, 2, and 119 respectively);[150] the rest differ little, though in general it can be said that the *Dicta* are more exact in the matter of quotation than are the corresponding sermons, which were intended for a wider audience.

It was not Westermann's intention to edit the full texts of the sermons or *Dicta* discussed, and for the most part he confined himself to citing brief extracts from them. In no case did he give a complete text. His contribution drew attention to the need for a full edition of both works.

Dictum 60 The trinitarian exemplarism which occupied an important place in Grosseteste's theology received full expression in this *Dictum*, '*Omnis creatura speculum est*', which Fr Gieben has printed from three reliable MSS, including the Prague, *Národni* Mus. MS XII. E. 5.[151]

Fr Gieben assigns no date to the *Dictum*, but quotes Grosseteste's remark that the *Dicta* belong to his activity in the schools, which sets the *terminus ante quem* in 1235.

3. EDITIONS PLANNED AND IN PREPARATION

Editions of a number of writings by Grosseteste are planned, although by no means all his most important writings have found an editor. The *Hexaemeron*, edited by Dales and Gieben, is with the printer and will appear as vol. VI in the series *Auctores Britannici Medii Aevi*. The biblical treatise, *De Cessatione Legalium*, is being edited by Professor King (The University of the South, Sewanee, Tennessee). The work is well advanced. The task of editing the *Commentary on the Posterior Analytics* has been taken up by Rossi who hopes to publish the text in 1981. Among other Aristotelean translations, mention has already been made of the *De Caelo* and the commentary of Simplicius, which Bossier intends to publish, and of the *Notulae* of Grosseteste and the translation of the Greek commentators on the *Ethics*, the edition of which Mercken and Reilly are to complete. Dr Holland is working on the Damascene translations. The brief *De Subsistentia Rei*

---

[150] Numbers refer to Thomson's catalogue.

[151] 'Traces of God in nature according to Robert Grosseteste, with the text of the *Dictum* "omnis creatura speculum est"', in *Franc. Stud.* 24 (1964) 144-158. For the text, see pp. 153-8.

which Pelster discovered was edited just before his death by Callus and left with the late Dr Hunt, who did not manage to print it. The important *De Lineis, Angulis et Figuris* is being re-edited by Eastwood. I hope to re-edit the *De Luce* and *De Motu Corporali et Luce*, and the *Expositio in Galatas*. Ruth J. Dean is preparing the edition of the minor Anglo-Norman works.

ADDENDUM

27. *Commentarius in Libros Analyticorum Posteriorum Aristotelis* The reprinted Venice 1514 edition of this work of central importance, which even its editor, Panfilo Monti, acknowledged to be of poor textual quality, has recently been superceded by the appearance of the critical text, an event eagerly awaited for many years by scholars.[152] Dr Rossi has established that Grosseteste depended principally upon the version of the work made from the Greek by James of Venice, but that he had access to two other twelfth-century translations of *Johannes*, and the Arabo-Latin of Gerard of Cremona for parts of the text at least, and that he was in a position to invoke the Greek tradition of paraphrases on the work by such Aristotelean scholars as Alexander of Aphrodisias, Themistius, and Philoponus. Rossi agrees with the generally-accepted view that the work as we have it must have been written between *c*.1220 and 1230 (see our Appendix B). This edition is a landmark in the field. Its quality merits the highest respect.

[152] *Robertus Grosseteste. Commentarius in Posteriorum Analyticorum Libros.* Introduzione e testo critico di Pietro Rossi. Florence 1981, 412 pp. This edition reached me while the present work was in the press. A list of correspondences between the foliation of the Venice 1514 edition and the pagination of this edition, for all the passages referred to in the present work, is given here for the benefit of the reader:

fol.				fol.			
2^d	−	p.	99	20^c	−	p.	240-1
3^b	−		100-1	20^d	−		244
3^{c-d}	−		103	22^{b-c}	−		255-7
4^d	−		111	23^c	−		266
8^b	−		137	25^{a-d}	−		278-86
8^{c-d}	−		139-42	32^c	−		344
10^c	−		157	37^c	−		386-7
11^d	−		171	$39^{d}-40^a$	−		403-5
$16^{d}-17^b$	−		212-6	40^b	−		406

List of the Manuscripts mentioned in Appendix A

MS Harley 3858
MS Harley 5441
MS Royal II. D. xxx
MS Royal VI. E. v
MS Royal VII. F. ii
MS Royal XII. E. 25
MS Lambeth Palace Libr. 523
MS Public Record Office E. 164/2
Longleat House MS
Lyons, MS Bibl. municipale 414
Madrid, MS Bibl. Nacional 3314
Melk, MS Stiftsbibl. 59 (B. 24)
   MS 61 (B. 26)
   MS 427 (H. 46)
Metz, MS Bibl. de la Ville 521
Milan, Bibl. Ambrosiana 35 supp.
   MS E. 71 supp.
Munich, MS Staatsbibl. Clm. 280B
   MS Clm. 4525
   MS Clm. 8827
   MS Clm. 18210
   MS Clm. 18759
Oxford, MS Bodleian Libr. lat. misc. b. 3
   MS Bodleian Libr. Can. gr. 97
   MS Bodleian Libr. e Museo 134
   MS Bodleian Libr. lat. th. c. 17
   MS Bodleian Libr. Bodley 42
         MS Bodley 198
         MS Bodley 310
         MS Bodley 315
         MS Bodley 750
         MS Bodley 857
         MS Digby 104
         MS Digby 190
         MS Digby 220
         MS Rawlinson A. 446
         MS Rawlinson C. 552
   MS Balliol Coll. 99
   MS Balliol Coll. 228
   MS Corpus Christi Coll. 251
   MS Exeter Coll. 28

MS *Magdalen Coll. 109*
MS *Magdalen Coll. 192*
MS *Merton Coll. 14*
MS *Merton Coll. 69*
MS *Merton Coll. 145*
MS *Merton Coll. K. 1. 5. Coxe 295*
MS *Queen's Coll. 312*
MS *St. John's Coll. 93*
MS *St. John's Coll. 190*
MS *Trinity Coll. 18*
Padua, *Bibl. Civica C.M. 187*
Paris, MS *Arsenal 698*
MS *Bibl. nat. gr. 437*
MS *Bibl. nat. 543*
MS *Bibl. nat. lat. 1620*
MS *Mazarine 787*
Prague, MS *Univ. Bibl. N.G. 31*
MS *VIII. F. 3*
MS *XII. F. 21*
MS *Národní Museum XII. E. 5*
Rome, *Bibl. Vaticana Chigi A. V. 129*
MS *Chigi E. VII. 225*
MS *Ottoboni lat. 1870*
MS *lat. 760*
MS *lat. 2171*
MS *Pal. lat. 2088*
Salamanca, MS *Bibl. de la Universidad 2028*
Stockholm, MS *Kungl. Bibl. V. a. 3*
Venice, MS *Bibl. Naz. Marciana Z. lat. 241*
MS *VI. 222*
Vienna, *Dominikanerkloster 192/158*
MS *Nationalbibl. lat. 790*
MS *Nationalbibl. lat. 1619*
MS *Schottenkloster 29*
MS *396*
Wisbech, MS *Town Libr. 5*

Appendix B

The Chronology of Grosseteste's Natural Works

ALTHOUGH Grosseteste's writings on nature and natural philosophy have been more adequately studied, particularly of recent years, than any other part of his output, scholars have reached little agreement on even the essentials of their chronology.[1] The tentative results which I have reached have emerged from the convergence of a number of considerations, not all of which are applicable to each writing. External evidence has a special importance; however, there is not very much of it. The sources employed in a given work constitute a useful criterion for its place within the whole series, as does the degree of exactitude in the understanding of Aristotle's thought and the application of his scientific methodology. Cross-references to other scientific works of Grosseteste are few in number, but important. Finally, there are the inner development of thought and the points of contact between some scientific writings and works which fall into other categories (psychological, theological, and exegetical). When used with care, these are valid and useful aids to the chronological ordering of the natural writings.

EARLY SCIENTIFIC INTERESTS

De Artibus Liberalibus The sources (Augustine's *De Musica* for psychology, Boethius for mathematics) and teaching (the musical harmony of the universe, the utility of astronomy and astrology for medicine and agriculture), indicate that this work comes from that very early period in Grosseteste's career, before 1209, when he taught the arts at Oxford. This is probably the earliest extant work.

---

[1] The only comprehensive attempt to provide a chronology of the writings on natural philosophy has been made by R. C. Dales: 'Robert Grosseteste's scientific works', in *Isis* 52 (1961), 381–402. I am much indebted to it. I have elsewhere set out my own arguments in detail and documented my observations: 'The Chronology of Robert Grosseteste's Writings on Nature and Natural Philosophy', in *Speculum* (forthcoming). Here I have kept references down to a minimum. Baur's edition is referred to unless otherwise specified. The Venice 1514 edition of the *Comm. in Anal. Post.* has been used.

De Generatione Sonorum   Priscian and Isidore are the major influences on the study of language and phonetics. While Avicenna's triple division of the soul (vegative, sensitive, and rational) is invoked, the essay cannot be much later than the *De Artibus*, with which it has much in common, and I am inclined to place its composition before 1209.

ASTRONOMY AND ITS APPLICATIONS:
THE CALENDAR, ASTROLOGY

The interest in astronomy, and of course in astrology, which is a notable feature of the *De Artibus*, seems to have absorbed Grosseteste during the ten or fifteen years following 1209 and never quite to have deserted him, although he became very critical of astrology.

De Sphaera   This work must be placed among the earliest of Grosseteste's scientific treatises. Although he knows the *De Caelo* of Aristotle, he attributes the motion of the heavens to the *anima mundi*. He attempts to explain climatic variations by applied astronomy; there is no sign that he has read the *Meteorologica*. The opposition between the Ptolemaic theory of the precession of the equinoxes and the theory of trepidation of Ibn-Thābīt is clear to him, whereas the far-reaching controversy between the protagonists of astronomy and natural philosophy has not yet appeared above his horizon. The work was probably composed within a few years of 1215. In later writings he will question some of its basic positions and assumptions.

Computus I, Calendarium, Computus Correctorius, Computus Minor   The order in which these works were written is clear, but only the last can be dated with accuracy: '*sed nativitate Domini elapsi sunt 1200 anni et eo scil 44 amplius*', it comments.[2] When we turn back from the work of the ageing bishop to that of the much younger theologian with deep scientific interests, it strikes us that Grosseteste was continually adding to his library and revising his calculations of the Calendar. He began by writing the *Computus* and revising it in his own *Calendarium* and then in his *Correctorius*. The latter was

---

[2] Thomson, *The Writings*, p. 97.

probably written between 1225 and 1230, since the interests in the theological, symbolic and especially physical dimensions of the sun and solar light (light is the unique means by which heavenly influence is exerted upon all sublunary phenomena), affiliate the work to Grosseteste's writings on the metaphysics of light, rather than to the physical treatises which he wrote up until the early 1220s.[3] Grosseteste is already critical of the pagan assumption that the world is eternal and he anticipates the denunciations which will appear with regularity in his writings during the 1230s. Yet his knowledge of Aristotle is faulty, for he clearly fails to realize that the entire conception of planetary motion which he expounds is derived, not from Aristotle but from the reform of Alpetragius, whose *De Motibus Caelestibus* he has studied with attention (p. 217). Only after he has studied the *Physics* and *Metaphysics* will he become aware of the differences. All in all, a date of composition not much later than 1225 is strongly indicated for the *Correctorius*.

De Impressionibus Aeris This treatise marked the high point of Grosseteste's astrological explorations, which went hand in hand with his interest in astronomy and its practical application to the reform of the Calendar (which Grosseteste regarded as a service to the Church and her liturgical worship). He believed that the weather, and hence the harvest, could be predicted years in advance on the basis of astrological indications, and in this treatise he attempted to infer the state of the weather on 15 April 1249, then in the July of that year and the autumn of 1255, from the qualities of the planets. His only authorities were Ptolemy and Theodosius. There is nothing to indicate a date of composition any later than *c.*1220.

De Generatione Stellarum Grosseteste's interest in the heavens extends itself beyond (and perhaps through) astronomy and astrology to the physical theory of the heavens (*De Sphaera*), the nature of the quintessence and of the heavenly bodies, discussed in the present work. His acquaintance with Aristotle (*De Caelo, De Generatione, De Anima, De Animalibus*) is wide but not deep, for he finds nothing implausible about implanting the explicitly alchemical doctrines of sublimation and

---

[3] The *Computus Correctorius* is edited by Steele in *Opera hactenus inedita Fr. Rogeri Baconi*, fasc. vi, pp. 212-67; see p. 217.

humiliation into the framework of the celestial physics of Aristotle and arguing that, although the quintessence itself is incorruptible, the bodies of the stars are elementated (Baur, p. 26.2-11). The great controversies (between physicists and astronomers; concerning the eternity of the world) leave no traces on this work, which was written after the *De Sphaera*, but not much after 1220. It contains a reference to the *De Animalibus* (translated *c.*1217).

Grosseteste would have been at least fifty years of age when he wrote this work, yet the indications are that his Aristotelean adventure, and with that his interest in physical questions, was only beginning. Was he by *c.*1220 starting to make up for lost time in the study of Aristotle? His ignorance of the physical works suggests that he may have studied at Paris (after the dispersal of the Oxford Schools in 1209) in the aftermath of the condemnation of the *libri naturales* at Paris (1210). It also looks as though his study of physical questions was derived from his earlier pursuits, astronomy, astrology, and cosmography.

THE SUBLUNARY WORLD AND METEOROLOGICAL PHENOMENA

De Cometis   This is probably among the earliest attempts made by Grosseteste to explore the causes of a sublunary phenomenon. The complete ignorance it shows with regard to *Meteorologica I*, chs. 6-7 (on the comet) can only mean that that work, which was to become a favourite with Grosseteste, was not yet in his possession at the time of writing. It follows from this that the *De Cometis* antedates the *De Impressionibus Elementorum, De Natura Locorum*, and *De Operationibus Solis*. If the reference in the longer recension to '*cometa qui nuper apparuit*' refers to the appearance of Halley's comet late in 1222, as seems very likely, then the recension must have been written within a year or two. This would square very well with the contents, which show no advance in Aristotelean knowledge by comparison with the *De Generatione Stellarum* and actually need presuppose no other peripatetic reading than that of the *De Caelo*. Grosseteste accepts the principle that nothing new appears in the actual heavens; hence the comet cannot be a star. It is a sublunary phenomenon, of the nature of fire, generated by the virtue of a star or a planet from the portions of fire in the earth and sublimated to the sphere of fire, where it obeys the diurnal motion of the first sphere. When Grosseteste enlarges upon the process of the generation of comets the mixture of astrology and alchemy that is already familiar to us from the *De Generatione Stellarum* recurs.

De Impressionibus Elementorum The developments of thought that are apparent as between the *De Cometis*, on the one hand, and the present work and the treatise on the tides (which probably followed it), on the other, are significant enough to allow us to speak of a new phase of Grosseteste's scientific thinking. Light assumes the place of a central physical phenomenon and some knowledge of geometrical optics is applied to the solution of physical problems, although not as consistently and confidently as will be the case in the latest group of physical treatises. The aim of this opuscule is to explain meteorological phenomena (the warmth of the air, formation of clouds, and precipitation) in terms of their cause, the heat of the sun. Aristotle is actually mentioned only once, but the general stimulus of his meteorological works is to be assumed. The *compositio*, or deduction of the different forms of precipitation (dew, rain, snow, and hail) from varying degrees of heat and cold as their cause (Baur, p. 89.2-35) is mature and methodical, the most worthwhile passage, scientifically speaking, in all the treatises examined up to this point. The little work contains what is probably Grosseteste's first venture into the theory of heat. We may take it to have been written after the *De Cometis* but before Grosseteste had fully developed his metaphysics of light. A date around 1225 would seem to be indicated.

De Fluxu et Refluxu Maris Grosseteste is able to refer to the *Catoptrics* of Euclid and has read the *De Motibus Caelestibus* of Alpetragius (translated 1217-20), though he disagrees with its explanation of the tides. He is writing between 1220 and 1230: the Assisi MS states that the question was determined '*in scolis suis*'; now he gave up his own school when he moved to the friars in 1229-30.

The *De Fluxu* makes the light of the moon the cause of the movement of the tides. All motion in the elements is the effect of a *'species immaterialis'* (Dales, p. 464). Light becomes incorporated into the elements and, by expanding their particles, generates heat. Grosseteste has travelled a long way indeed from the astrological notions that were prevalent in his earlier writings. The date suggested by Dales ('within a year or so of 1227') has a high degree of probability. The author refers to himself as *'disputantes'*; if we take the word at its face value we must conclude that either he brought questions of natural philosophy into his theological lectures, raising them when determining the literal sense of Scripture, or he held concurrent lectures in both domains.

THE ACTION OF LIGHT

We come now to a group of five interlocking writings which represent the crown of Grosseteste's scientific activity and display little evidence of development of thought from one to the other.

De Lineis — De Natura Locorum   This is a single work. Its reading shows how greatly Grosseteste's library has expanded: it includes the *De Animalibus* of Avicenna, the *De Vegetabilibus* or *De Plantis* of Pseudo-Aristotle, Averroës on the *De Anima* and a commentator on the *De Caelo*, who may be Simplicius. Grosseteste's original views on the nature and action of light lie behind the work, which lays down the claim for the rule of geometrical law in nature. The attempt to give physical values to geometrical concepts is quite consistently pursued in the first part of the treatise (*De Lineis*); that is its title to fame. Natural force is strongest when it is propagated in a straight line and is weakened in different ways by reflection and refraction. The figure of the pyramid is of particular interest, as it provides the laws, the climatic effects of which are illustrated in the second half of the treatise (*De Natura Locorum*). The causes of all natural phenomena can be discovered through the power of geometry. Here Grosseteste completes his theory of the tides by invoking his newly-acquired geometrical and optical knowledge. The work places us once more before the conflict of physics with astronomy and looks forward to the *Hexaemeron* (c.1237) in which the same unresolved controversy will be discussed at length. The emphasis on the intricate beauty and wondrousness of the laws of a geometrically-functioning nature is a reliable indication that the metaphysical views expressed in *De Luce* and repeated in subsequent works were formed in Grosseteste's mind at the time he wrote the treatise. He places the rainbow high among these wonders, referring to it in terms which suggest that he felt himself to be already in possession of the key to its explanation.

De Iride, De Colore, De Calore Solis This group of treatises constitutes Grosseteste's last scientific effort to explore the wonders of nature. In the absence of any evidence of development of thought within the group and in the awareness of their common dependence on the *De Lineis — De Natura Locorum*, it is necessary but also sufficient to postulate that they derived from a single inspiration and probably followed each other fairly rapidly, closing the series of opuscules with *éclat* before Grosseteste turned his attention to other interests (mostly philological, theological, and pastoral), and very likely before the close of his academic career at Oxford. An indication that this was probably the case is given by the comparison between the group of late physical treatises and the *De Operationibus Solis*, from which it emerges that the latter work presupposes the *De Lineis — De Natura Locorum* already completed, and that it probably followed this treatise within a short space of time. Similar parallels are found as between the *De Operationibus* and the *De Calore Solis* and *De Colore*. Since the *De Operationibus* was probably written before Grosseteste became bishop and before he had learned Greek, it should be assigned to the years 1230-3. The group of scientific treatises should be given a similar dating.

ARISTOTELEAN DEVELOPMENTS

The order of composition of the scientific treatises and the chronology which has emerged from their investigation serve as a useful guide to the dating of a number of works which grew directly out of Grosseteste's long immersion in the Aristotelean corpus, and also for the two treatises on light.

512   *Appendixes*

Commentarius in Libros Analyticorum Posteriorum There is widespread disagreement concerning the date at which this celebrated commentary was written. Callus placed its composition firmly in the first decade of the century, relying on the testimony of Nicholas Trivet that Grosseteste wrote it while he was a master of arts.[4] How it could at once be so early and yet manage to contain 'explicitly or implicitly all his main theses', he did not feel the need to explain. Crombie tempered his own adhesion to Trivet's statement by the suggestion that Grosseteste may have spread his commentary over several years.[5] There is nothing inherently improbable in this hypothesis, yet it jars against one's strong feeling that, however early some of the material incorporated into and may have been written, the commentary was composed with an unsurpassable uniformity of thought and style and hence without long interruptions.

Dales has argued against Crombie and Callus for a relatively late date of composition, around 1228, basing his arguments upon the doctrinal comparison between the commentary and the scientific writings.[6] I agree on the whole with his reasoning but find that it helps to set the limits of writing at 1220 and 1230, rather than justifies a definite dating of the commentary. I hope that a review of the evidence may narrow these limits somewhat.

Grosseteste based his commentary on the version from the Greek by James of Venice, making reference to '*aliae translationes*'. Despite his complaints about Aristotle's obscurity, it was to Themistius's paraphrase (in the translation of Gerard of Cremona) and not to the Greek text that he resorted for clarification of the difficult passages. Remarks here and there display a general interest in Greek, but nothing that could not have been picked up by someone who read as widely as he did. He wrote his commentary before he began his study of the language, which he did not long after 1230.

The sources of the commentary turn out to be extensive and are representative of Grosseteste's reading during the latter half of the 1220s. The *Physics* is frequently referred to, but, more significantly, he has been inside the covers of the

[4] Callus, in *Scholar and Bishop*, p. 12.
[5] Crombie, *Robert Grosseteste and the Origins of Experimental Science*, p. 47.
[6] Dales, in *Isis* 52 (1961), pp. 395-7.

Metaphysics, which probably brings the writing towards 1230.
He has the *De Animalibus* and probably the *De Partibus Animalium* also. The *Meterologica* has left a deep imprint on the
work and his interest in the rainbow is already awakened.
The *Catoptrics* of Euclid must be counted among his sources,
which again suggests that he is writing in the latter years of
the 1220s. In the discussion of the tides and the flow of the
Nile (fol. 37^d–38^a) he alludes to '*aliae causae coniunctae*',
which looks like a reference to the eight causes of variation in
the height and depth of tides which are enumerated in the
De Fluxu.

The commentary announces (fol. 37^c) that every natural
body has in its composition something of celestial luminosity
and fire; the first incorporation of light is in the subtlest air-
particles. Now the concept of incorporation was not yet present in the *De Impressionibus Elementorum* but was invoked
in the *De Fluxu* and was to be developed in the physical
treatises after 1230 into a theory of the generation of heat.
The commentary does not go so far, which suggests the location of its composition between that of the *De Fluxu* and the
more sophisticated late group of treatises. It is possible that
Grosseteste found no opportunity to introduce the theory of
heat and of the specific properties of geometrical figures (cf.
De Lineis), but it must be admitted that as an expositor he
tended to create such occasions rather than find them.

The study of the sources and scientific content of the commentary point, then, with some consistency to a date of final
composition some time after the awakening of Grosseteste's
scientific interests in the 1220s but before the arrival of
Averroës and the production of his own most mature treatises,
which are marked by the application of the principles of
scientific investigation, including the subalternation of sciences,
which were laid down in the commentary.

Unfortunately, neither of the cross-references in the commentary to other writings of the author can be used here to
argue for a more accurate dating. The first (fol. 5^a) appears
to be a reference to the *De Luce*, but we wish to reserve it for
use in dating the latter work. The second (fol. 12^b) refers to
a certain treatise devoted to the discussion of a geometrical
question raised in the commentary; I have had no more success than other researchers in identifying this work.

The conclusion that the commentary was composed towards 1230 runs counter to Trivet's assertion, '*cum esset magister in artibus, super librum Posteriorum compendiose scripsit*'.[7] I find it wholly inconceivable that the commentary as we have it could have been written before 1209, for there is no evidence that Grosseteste was in possession of the requisite learning until the late 1220s. On the other hand, I am reluctant to discount Trivet's testimony, for he was an annalist seriously interested in the history of his university. I can only conclude that the word '*compendiose*', with its connotations of brevity, refers, not to our full and lengthy commentary, but to an earlier summary which has disappeared.

Commentarius in VIII Libros Physicorum Dales has given a very enlightening discussion of the date of Grosseteste's notes on the *Physics* and has suggested that most of them were jotted down between 1228 and 1232.[8] The following considerations are given by way of supporting and confirming his conclusions.

The very phrasing of Grosseteste's reference to Averroës ('*Unde Averroës, commentator philosophi videlicet Aristotelis*,' Dales's ed., p. 129), suggests that this is a recently-circulated author who still requires to be introduced, in a way which became redundant within a short time of the sudden appearance of his commentaries at Paris, c.1231. Most of these notes were probably written before Grosseteste came upon Averroës (in the seventh book). On the other hand, his thoughts on physical light are fully formed, he no longer professly teaches the existence of the *anima mundi* (it is now only the hypothesis of '*quidam*'), and he is quite clear in Bk. I that Aristotle, Plato, and Melissus taught the eternity of matter and of the world. It would be surprising if any of the notes were written before 1225 at the earliest.

De Finitate Motus et Temporis, De Differentiis Localibus Grosseteste's struggle to understand the *Physics* issued in some more finished writings than the notes. The *De Finitate Motus* opposed the doctrine of the eternity of the world. It quotes from Averroës on the *Physics*, and its verbal parallels with the *Hexaemeron* draw it towards c.1237.

[7] N. Trivet, *Annales*, p. 243.
[8] Dales, in *Med. Human.* 11 (1957), 12-13; also in the introduction to his edition.

Other opuscula likewise take up problem-areas of the *Physics*, handling them with a double interest, expository and exploratory. The first of these, *De Differentiis Localibus*, addresses itself to a number of difficulties concerning the nature and objectivity of place. Notable from our point of view are the interest the author shows in geometry and especially in the properties of the sphere, and the isolated remark that light is the means by which the *substantia separata* moves the heavens (Baur, p. 86.24). The first no doubt represents a connection with the cosmogony of light, the second confirms that the *anima mundi* has yielded place to the peripatetic First Mover. It belongs to the period of Grosseteste's immersion in the *Physics*.

De Motu Supercaelestium, De Motu Corporali et Luce In the first of these treatises Grosseteste tries to get to grips with a much more substantial aspect of the *Physics*, namely, the nature and cause of circular motion. In this largely exegetical essay he shows himself to be in some position to compare the doctrines of the *Physics* and the *Metaphysics* about the movers of the heavens, and he makes more use of the latter work here than anywhere else in his opuscula, something which suggests a date very close to 1230. This is confirmed by a comparison with the *Computus Correctorius*, in which, it will be recalled, Grosseteste hopelessly confused Aristotle's doctrine on the motion of the planets with the revisions made by Alpetragius. The *De Motu Supercaelestium* offers a more cautious exegesis of the genuinely peripatetic theory of motion, noting that, according to *Metaphysics XII*, the movers are numbered according to the number of their bodies and the different motions of these, whence the multiplicity and diversity of heavenly motion over and beyond the perpetual and continuously-uniform motion that derives from the mover of the first sphere. The *De Motu* must have been written after the *Computus Correctorius*, so that a *terminus a quo* not long before 1230 is indicated. It was finished before the *De Lineis*, which refers back to it in the following terms: '*Cum igitur in aliis dictum est de eis quae pertinent ad totum universum et ad partes eius absolute, et de hiis quae ad motum rectum et circularem consequuntur, nunc dicendum est de actione universali*'.[9]

[9] *De Lineis*, ed. Baur, p. 60.

Now all six MS-collections of Grosseteste's physical works contain both the writings here referred to, *De Motu Super-caelestium* (or *De Dispositione Motoris et Moti in Motu Circulari*, as an alternative title calls it), and *De Motu Corporali et Luce*; in three out of the six the latter follows immediately upon the heels of the former. Their combined subject-matter fits the reference in *De Lineis*; no other extant works do so. Both were therefore written before c.1233. They formed part of a plan in the mind of their author. We shall find the key to that plan when examining the *De Luce*.

That Grosseteste wrote these two opuscules as a pair and around the same time is made plain by the first paragraph of the *De Motu Supercaelestium*, which refers in unmistakeable terms to the *De Motu Corporali*: '*Et quia quaesita est dispositio motoris et moti in motu recto, consequenter est quaerendum de motu circulari*'. (Baur, p. 93). The two treatises, *De Motu Corporali* and *De Motu Supercaelestium*, were written in that order, were almost contemporaneous, and took up two problems which arose from the *Physics*, namely, celestial and rectilinear motion.

De Luce   We may now be in a position to apply some relatively well-defined co-ordinates to the dating of the *De Luce*. Solid footing for a reasoned attempt of this kind has hitherto been sadly wanting and scholars have had to make do with *doxa*; datings as far apart as 1215 and 1240 have been advanced.

In his notes on *Physics III,* in the course of an involved argument about the possibility of relative infinities, we find a cross-reference in the following terms: '*Numerus enim infinitus ad alium numerum infinitum potest esse in omni proportione numerali et non-numerali, sicut alibi demonstravimus*'. '*Alibi*' can only be the passage from the *De Luce* from which the following is taken: '*Est autem possibile, ut aggregatio numeri infinita ad congregationem infinitam in omni numerali se habeat proportione et etiam in omni non-numerali*' (Baur, p. 52). In his notes, Grosseteste proceeds to apply this mathematical idea to the infinite replication of the simple essence, matter and form; and this, of course, parallels the doctrine of *De Luce*. No other parallel between the two works is verbally so close as this, but their agreements on the doctrine of light are too many to leave any doubt that the notes depend on and refer to the *De Luce*, which should then be dated with probability to before *c*.1228.

The commentary on the *Posterior Analytics* is anterior to at least some of the notes on the *Physics*.[10] In it Grosseteste affirms that '*alibi exposuimus qualiter linea est ex infinita multiplicatione linee, et quomodo secundum modum alium linea non est ex punctis, nec superficies ex lineis*'.[11] This should, I consider, be understood as referring to a passage of *De Luce* where Grosseteste, treating of the infinite multiplication of matter by a point of light into three-dimensionality, asserts, '*corpora ex superficiebus componi, et superficies ex lineis, et lineas ex punctis*',[12] but maintains at the same time that a magnitude is composed only of real magnitudes. It can be seen that the passage satisfies the cross-reference in both respects. The *De Luce* must therefore be assigned a *terminus ante quem* of not long before 1230.

The *De Luce* is sparing in source-references, but its sources are quite extensive. It is fairly clear that Grosseteste's early faith in Ptolemaic astronomy has been shaken by his study of Alpetragius. The hypothesis sustained by *aliqui* (i.e. Alpetragius) allows that the sphere of fire turns with the heavens, and that the diurnal motion reaches, although weakly, to the seas. Grosseteste admits that the path of comets constitutes evidence for the first assertion (he had supposed as much in his own *De Cometis*), but he is more circumspect with regard to Alpetragius's explanation of the tides; he reserves his judgement. This passage gives rise to a strong presumption that the *De Luce* was composed after the *De Cometis* but before the *De Fluxu*, in which Alpetragius's hypothesis is nearly summarized, criticized, rejected, and replaced. Several years separate the *De Cometis* (of shortly after 1222) from the more mature phase of Grosseteste's thinking which opened with the *De Impressionibus Elementorum* and *De Fluxu*. It is not fanciful to suppose that the differences between them are due to the firm guidance which the light-metaphysics provided for the later scientific explorations. On this argument the *De Luce* is a pivotal work, which marked the end of the vague

[10] The notes on the *Physics* appear to refer back to the *Comm. in Anal. Post.*; see Crombie, op. cit., p. 47.

[11] *Comm. in Anal. Post.* I, 4, fol. 12b.

[12] *De Luce*, ed. Baur, pp. 53–4. Grosseteste understands this to have been the teaching of the ancient Atomists; it is also his own.

astrological-alchemical speculation of the earlier period and laid a solid metaphysical basis for more strictly scientific inquiry. It presupposes a great deal of reading (of theology, physics, and optics) and a profound change in Grosseteste's mind, as compared with the early astronomical and physical works of *c*.1215-23. We cannot place the *De Luce* earlier than *c*.1225.

The little masterpiece has, of course, more sources than Greek and Arabic ones, for it is of profoundly Christian inspiration and was written by a scholar well versed in the theological tradition. The confidence with which the two traditions, the Greek scientific and the Christian theological, are fused into an indissoluble whole argues for its composition in Grosseteste's mature years, when he was already a respected theologian.

Our conclusion is that the *De Luce* can scarcely have been composed more than a year or two before or after 1226. Its place within the sequence of Grosseteste's works is clear and its centrality in his development undeniable. The *De Motu Corporali*, the first direct application of the metaphysics of light, followed it, and was contemporaneous with the *De Motu Supercaelestium*; both of them assumed more knowledge of the *Physics* than is apparent in the *De Luce* and were closely related to their author's struggle to understand that work. The *De Lineis* of after 1230 presented itself as a continuation of the project, and in the *De Operationibus Solis* Grosseteste verified the results achieved, by demonstrating their fertility in exegesis of the literal sense of Scripture. These were his busiest and most fertile years in philosophy.

Suggested dates of composition

Early scientific interests
    De Artibus Liberalibus   before 1209
    De Generatione Sonorum   before 1209

Astronomy and its applications
    De Sphaera   1215-20
    De Impressionibus Aeris   1215-20
    De Generatione Stellarum   1217/20-5
    Computus I   *c.*1215-20
    Calendarium   *c.*1220
    Computus Correctorius   1225-30
    Computus Minor   1244

The sublunary world and meteorological phenomena
    De Cometis   1222-4
    De Impressionibus Elementorum   before *c.*1225
    De Fluxu et Refluxu Maris   *c.*1226-8

The action of light
    De Luce   *c.*1225-8
    De Lineis — De Natura Locorum   *c.*1230-3
    De Iride, De Colore, De Calore Solis   *c.*1230-3

Aristotelean writings
    Commentarius in Libros Analyticorum
        Posteriorum   1228-30
    Commentarius in VIII Libros Physicorum   1228-32
    De Differentiis Localibus   *c.*1230
    De Motu Supercaelestium, De Motu
        Corporali et Luce   *c.*1230
    De Finitate Motus et Temporis   *c.*1237

Bibliography

THE books and articles which appear in this (necessarily selective) bibliographical list are those which have been referred to in footnotes in the work, or from which quotations have been drawn. For a complete bibliography of works relating to Grosseteste, see Gieben, Servus, 'Bibliographia universa Roberti Grosseteste ab an. 1473 ad an. 1969', in *Coll. Franc.* 39 (1969), 362-418, which replaces all previous bibliographical essays. Other helpful bibliographical tools, relating especially to the MS-literature, are the following:

Aristoteles Latinus. Codices descripsit + Georgius Lacombe, in societatem operis adsumptis A. Birkenmajer, M. Dulong, Aet. Franceschini. Pars Prior (Corpus philosophorum medii aevi, Academiarum consociatarum auspiciis et consilio editum. Union Académique Internationale) (Roma, La Libreria dello Stato, 1939), 763 pp.; Pars Posterior, (Cantabrigiae, Typis Academiae, 1955), pp. 759-1388.

Lohr, C. H., 'Aristotle Commentaries. Authors: Robertus—Wilgelmus', in *Traditio* 19 (1973), 93-197; see under 'Robertus Grosseteste', pp. 100-7.

Schneyer, J. B., *Repertorium der lateinischen Sermones des Mittelalters*, BGPM Bd. 42, H.5 (Autoren: R-Schluss[W]) (Münster i. W., 1974); see under 'Robertus Grossatesta', pp. 176-91.

Thomson, S. H., *The Writings of Robert Grosseteste, Bishop of Lincoln 1235-1253* (Cambridge Mass., 1940).

MANUSCRIPTS

Cambridge, *MS Gonville and Caius College 380.*
Firenze, *MS Biblioteca Laurenziana Plut. XIII, dextr. iii.*
London, *MS British Library Royal VI.E.5.*
London, *MS British Library Royal VII.F.2.*
London, *MS British Library Royal VII.D.15.*

Oxford, MS *All Souls College 84.*
Oxford, MS *Bodleian Library lat. th. c.17.*
Oxford, MS *Magdalen College 57.*
Prag, MS *Národní Mus. XII.E.5.*

ANCIENT AND PATRISTIC AUTHORITIES

Aristotle, *The Works of Aristotle Translated into English.*
    Under the editorship of W. D. Ross (Oxford, 1930-52),
    12 vols.
— *De Anima,* ed. Ross (Oxford, 1961).
— *Physics,* ed. Ross (Oxford, 1936).
Aristoteles Latinus (Corpus Philosophorum Medii Aevi; Union
    académique internationale), Oxford-Louvain, 18 vol., in
    progress.
St. Ambrose, *Hexaemeron Libri VI, P.L.* 14, 123-274; *CSEL*
    32 (Vienna, 1897).
— *De Dignitate Conditionis Humanae, P.L.* 17, 1015-18.
St. Augustine, *De Trinitate, P.L.* 42, 819-1098; *Corp. Chr.
    ser. lat.,* vol. L-LA, ed. Mountain, W. J., 1968.
— *Enarrationes in Psalmos, P.L.* 36-37; *Corp. Chr. ser. lat.,*
    XXXVIII (1956); XXXIX (1956); XL (1956).
— *De Genesi ad Litteram, P.L.* 34, 245-486; *CSEL* vol. 27,
    sect. 3, pars 2 (Vienna, 1894).
— *De Genesi Contra Manichaeos, P.L.* 34, 173-220.
— *Retractationum Libri II, P.L.* 32, 583-660; *CSEL* vol.
    36 (Vienna, 1892).
— *De Musica, P.L.* 32, 1081-1194.
— *Soliloquia, P.L.* 32, 869-906.
— *De Quantitate Animae, P.L.* 32, 1035-80.
— *Expositio in Galatas, CSEL* vol. 84 (Vienna, 1971).
— *Expositio in Johannem, P.L.* 35, 1977-2063.
— *Expositio in Romanos, CSEL* vol. 84 (Vienna, 1971).
— *Ad Paulinum de Videndo Deo, P.L.* 33, 596-622.
St. Basil, *Homiliae IX in Hexaemeron, P.G.* 29, 3-208; *Homélies
    sur l'Hexaëmeron.* Texte grec, introd. et trad. de S. Giet
    (Sources chrét. 26) (Paris, 1950).
— *Orationes II de Structura Hominis, P.G.* 30, 10-82.
The Venerable Bede, *Opera de Temporibus,* ed. Jones (Cam-
    bridge Mass., 1943), 416 pp.
Boethius, *De Consolatione Philosophiae, CSEL* vol. 48, pars i

(Vienna-Leipzig, 1906); *Corp Chr. Ser. lat.* 94, ed. Bieler, 1957.

— *Liber de Persona et Duabus Naturis, P.L.* 64, 1337-54.

— *De Musica, P.L.* 63; ed. G. Friedlein (Leipzig, 1867).

Claudianus Mamertus, *De Statu Animae, P.L.* 53, 697-780; *CSEL* vol. 11, (Vienna 1885), 18-197.

St. Gregory, *Homiliae in Evangelia, P.L.* 76, 1075-1312.

— *Moralia in Job, P.L.* 75, 509-1162; *Corp. Chr. ser. lat.* 143-143A, ed. M. Adriaen (1979).

St. Gregory of Nyssa, *De Hominis Opificio, P.G.* 44, 135-258; *La Création de l'homme.* Introd. et trad. de J. Laplace (Sources chrét. 6) (Paris, 1944).

St. Isidore, *Isidori Hispalensis Episcopi Etymologiarum . . . Libri XX,* ed. Lindsay, 2 vols. (Oxford, Scriptorum Classicorum Bibl., 1966).

— *De Natura Rerum, P.L.* 83, 963 ff.; J. Fontaine, *Isidore de Séville, Traité de la nature* (Bordeaux, 1960).

St. John Damascene, *De Fide Orthodoxa, P.G.* 94, 789-1238; in the Latin versions of Burgundio and Cerbanus, ed. E. Buytaert (New York, 1955), 423 pp.

— *Dialectica, P.G.* 94, 525-676.

Nemesius of Emesa, *De Natura Hominis, P.G.* 40, 504-817.

Origen, *Homélies sur la Genèse,* tr. L. Doutreleau (Sources chrét. 7) (Paris, 1943), 262 pp.

Philo Judaeus, *Opera quae supersunt,* ed. K. Cohn, J. Wentland (Berlin, 1896-1930), 7 vols.

Plato, *Timaeus,* tr. Cornford, *Plato's Cosmology* (London, 1937), pp. 376.

— *Corpus Platonicum Medii Aevi. Plato Latinus II.* Phaedo. Interprete Henrico Aristippo. Edidit et praefatione instruxit L. Minio-Paluello, adiuvante H. J. Drossaart Lulofs (London, 1950), XIX-156 pp.

Plotinus, *Plotini Opera,* ed. P. Henry et H. R. Schwyzer. T. 1 (Paris-Bruxelles, 1951); t. 2 1959; t. 3 (Paris-Leyden, 1973) (Museum Lessianum 33-35).

Proclus, *De Fato,* ed. H. Boese (Berlin, 1960) (Latin version of W. of Moerbeke).

Pseudo-Andronicus de Rhodes, 'Peri Pathon'. Édition crit. du texte grec et de la trad. latine médiévale par A. Glibert-Thirry (Corpus latinum commentariorum in Aristotelem graecorum, Suppl. 2) (Leyden, 1977), 360 pp.

Pseudo-Dionysius Areopagita, *Opera, P.G. 3.*
— *La Hiérarchie céleste.* Introd. R. Roques, étude G. Heil,
notes M. de Gandillac (Sources chrét. 58) (Paris, 1958),
218 pp.

MEDIEVAL AUTHORS

Alanus de Insulis, *Distinctiones Dictionum Theologicarum,*
P.L. 210, 685-1012.
Albertus Magnus, *In Sententiarum Libros IV*, ed. Borgnet,
vols. 28-30 (Paris, 1894).
— *Opera Omnia.* T. XIV, pars i, fasc. i: *Super Ethica Com-*
*mentum et Quaestiones.*Tres libros priores primum edidit
W. Kuebel (Münster i. W., 1968), 219 pp.
Alcherius Claravallensis (?) (=Pseudo-Augustinus), *De Spiritu*
et Anima, P.L. 40, 779-832 (Inter opera Augustini).
Alexander Halensis, *Summa Theologica*, 4 vols. (Quaracchi,
1924-48).
Alexander Nequam, *De Naturis Rerum*, ed. T. Wright (Rolls
Series) (London, 1863), 354 pp.
Anselmus Cantuariensis, *Opera Omnia.* Recensuit Fr. Schmitt
O.S.B., 5 vols. (Edinburgh, 1946-61). *Proslogion*, vol. 1,
pp. 97-122; *De Veritate*, vol. 1, 169-200; *Cur Deus*
Homo, vol. 2, 37-133.
Avicenna, *Opera Philosophica* (Venice,1508), (réimp. Louvain,
1961). *Avicenna Latinus. Liber de Anima seu Sextus de*
Naturalibus. Éd. crit. de la traduction latine médiévale
par S. van Riet, vol. 1 (Louvain-Leyden, 1968), 334 pp.;
vol. 2 (1972), 472 pp.
— *De Medicinis Cordialibus. Fragmentum.*, ibid. vol. 1, pp.
187-210.
St. Bonaventure, *Opera Omnia* (Quaracchi 1882-1902), 10 vol.
— *Commentarius in IV Libros Sententiarum*, vols. I-IV.
— *Breviloquium,*
— *Itinerarium Mentis in Deum*, vol. V.
— *Collationes in Hexaemeron*, vol. V.
Dominicus Gundissalinus, *De Anima*, ed. J. T. Muckle, *Med.*
Stud. 2 (1940), 23-103.
Giraldus Cambrensis, *Opera*, ed. J. S. Brewer, J. Dimock and
G. Warner (London, 1861-91), 8 vol., (Rerum Britt.
Medii Aevi Script. 21).

Gulielmus de Conches, *Glosae super Platonem*, ed. É. Jeauneau (Paris, 1965), 358 pp.

Hildegardis de Bingen, *Liber Compositae Medicinae de Aegritudinum Causis, Signis atque Curis*, ed. Pitra, *Analecta Sacra* 8 (1882).

Honorius Augustodunensis, *Elucidarium, P.L.* 172, 1109-76.

Hugo de Sancto Victore, *De Vanitate Mundi*, ed. K. Müller (Kleine Texte für Vorlesungen und Übungen, 123) (Bonn, 1913), 51 pp.

— *De Sacramentis Christianae Fidei, P.L.* 176, 173-618.

— *Expositio in Hierarchiam Coelestem Dionysii Areopagitae*, ibid. 923-1154.

Johannes Duns Scotus, *Opera Omnia*, vol. V (Civitas Vaticana, 1959), 475 pp.

Iohannis Scotti Eriugenae *Periphyseon (De Divisione Naturae)*, lib. 1. ed. I. P. Sheldon-Williams with the collaboration of L. Bieler (Script. Lat. Hib. 7) (Dublin, 1968). Lib. 2 (SLH 9) (Dublin, 1972).

— *Homélie sur le prologue de Jean*, ed. É. Jeauneau (Sources chrét. 151) Paris, 1969), 392 pp.

— *Expositiones in Ierarchiam Coelestem*, ed. J. Barbet, (*Corp. Chr. C.M.* XXXI) (Turnhout, 1975).

Isaac de Stella, *Epistola ad quemdam familiarem suum de Anima, P.L.* 194, 1875-1890.

Liber de Causis, ed. A. Pattin, in *Tijdschrift voor Filosofie* 28 (1966), 195 pp. Publ. separately (Louvain, 1966), 195 pp.

Mattaeus ab Aquasparta, *Quaestiones de Christo* (Bibl. Franc. Med. Aevi I, 2) (Quaracchi, 1914), 226 pp.

Matthaei Parisiensis, Monachi Sancti Albani, Chronica Majora. Ed. H. R. Luard (Rerum Britt. Med. Aevi Script.) (London, 1872-1884), 7 vols.

Historia Anglorum (Rolls Series), ed. F. Madden (London, 1886-9).

Petrus Abaelardus, *Theologia Summi Boni*, ed. H. Ostlender, *BGPM* Bd. 35, H. 2 (Münster i. W., 1939), 118 pp.

— *Theologia Christiana, P.L.* 178, 1113-1330; *Corp. Chr. C.M., Opera Omnia II*, XII, ed. E. M. Buytaert (1969).

Petrus Lombardus, *Libri IV Sententiarum*, 2 vol. (Quaracchi, 1916); *Sententiae in IV Libros Distinctae*, ed. Tert. T. i: *Prolegomena*, L. 1 et 2 (Grottaferrata, 1971).

Fr. R. Bacon Opera Quaedam Hactenus Inedita. Vol. i, containing: I, *Opus Tertium*; II, *Opus Minus*; III *Compendium Studii Philosophiae*, ed. J. S. Brewer (London, 1859), 573 pp.

— *Opus Maius*, ed. J. H. Bridges, 3 vols. (Oxford, 1897).

Suidae Lexicon, ed. A. Adler (Lipsiae, 1931).

Theodoricus Carnotensis, *De Septem Diebus*, ed. N. Häring, *Arch. Hist. Doctr. Litt. M.A.* 22 (1955), 184-200.

S. Thomas de Aquino, *Commentarius in VIII Libros Physicorum*, Leonine ed., vol. 2 (Rome, 1884).

— *Summa Theologiae*, vols. IV-XII of Leonine ed. (Rome, 1888-1906).

— *On Kingship*, to the King of Cyprus, tr. G. B. Phelan, revised by I. T. Eschmann (Toronto, 1949), 119 pp.

Nicholas Trivet, *Annales*, ed. Thomas Hogg (London, 1845).

Thomas de Eccleston, see Little, A. G. (ed.).

MODERN STUDIES

Alessio, Franco, 'Storia e teoria nel pensiero scientifico di Roberto Grossatesta', in *Riv. Crit. Stor. Fil.* 12 (1957), 251-92.

— 'Studi e ricerche su Roberto di Lincoln', in *Riv. Crit. Stor. Fil.* 12 (1957), 231-37.

Allan, D. J., 'Mediaeval versions of Aristotle *De Caelo* and of the commentary of Simplicius', in *Med. Renaiss. Stud.* 2 (1950) 82-120.

Allers, R., 'Microcosmus from Anaximander to Paracelsus', in *Traditio* 2 (1945), 319-407.

Baeumker, Cl., *Der Platonismus im Mittelalter*, BGPM XXV, H. 1-2, pp. 139-79 (Münster i. W., 1927).

— *Witelo, ein Philosoph und Naturforscher des XIII Jahrhunderts*, BGPM III, H. 2 (Münster i. W., 1908), 686 pp.

Barbour, R., 'A Manuscript of Ps.-Dionysius Areopagita copied for Robert Grosseteste', in *Bodl. Libr. Rec.* 6 (1958), 401-16, 5 tab.

Bardenhewer, O., *Die pseudo-aristotelische Schrift über das reine Gute* (Freiburg i. Br., 1882), 330 pp.

Barraclough, G., *Papal Provisions: Aspects of Church History, Constitutional, Legal and Administrative, in the Later Middle Ages* (Oxford, 1935).

Baur, L., 'Das philosophische Lebenswerk des Robert Grosse-teste, Bischofs von Lincoln (m. 1253)', in *Görres-Gesell-schaft zur Pflege der Wissenschaft im katholischen Deutschland*, 3. Vereinsschrift (Cologne, 1910), 58-82.

— *Die philosophischen Werke des Robert Grosseteste, Bischofs von Lincoln* (*BGPM* Bd. IX) (Münster i. W., 1912), 778 pp.

— 'Das Licht in der Naturphilosophie des Robert Grosseteste', in *Abhandlungen aus dem Gebiete der Philosophie und ihrer Geschichte. Eine Festgabe zum 70. Geburtstag Georg Freiherrn von Hertling* (Freiburg i. Br., 1913), pp. 41-55.

— 'Der Einfluss des Robert Grosseteste auf die wissenschaft-liche Richtung des Roger Bacon', in *Roger Bacon. Essays* . . ., ed. Little, A. G., (Oxford, 1914), pp. 33-54.

— *Die Philosophie des Robert Grosseteste, Bischofs von Lincoln (+1253)*, BGPM Bd. XVIII, H. 4-6 (Münster i. W., 1917), 298 pp.

Beaujouan, G., 'La science dans l'Occident médiéval chrétien', in *Histoire générale des sciences*, ed. Taton, R., t. i: La science antique et médiévale (Paris, 1957), pp. 517-82. Engl. transl.: *Ancient and Medieval Science*, ed. R. Taton (London, 1963), pp. 468-532.

Bérubé, C.-Gieben S., 'Guibert de Tournai et Robert Grosse-teste, sources inconnues de la doctrine de l'illumination, suivi de l'édition critique de trois chapitres du *Rudi-mentum Doctrinae* de Guibert de Tournai', in *S. Bona-ventura*, II, pp. 627-54.

Bettoni, E., 'La formazione dell'universo nel pensiero del Grossatesta', in *La filosofia della natura nel Medioevo*, pp. 350-356.

— 'Intorno all'autenticità del *De Anima* attribuito a Roberto Grossatesta', in *Pier Lombardo* 5 (1961), 3-27.

— *Storia della filosofia medioevale* (Milan, 1961), 308 pp.

Birkenmajer, A., 'Robert Grosseteste and Richard Fournival', in *Med. Human.* 5 (1948), 36-41. Repr. in idem, *Etudes d'histoire des sciences et de la philosophie du moyen-âge* (Studia Copernicana 1) (Wroclaw = Breslau 1970), pp. 216-21.

Bolzan, J. E., see Lértora-Mendoza.

528   *Bibliography*

Bougerol, J.-G., 'S. Bonaventure et la hiérarchie dionysienne', in *Arch. Hist. Doctr. Litt. M.A.* 36 (1969), 131-67.

Boyle, L. E., O.P., 'Robert Grosseteste and the Pastoral Care', in *Med. Renaiss. Stud.* N.S. 8 (1979), 3-51.

Brown, Edwardus, see Gratius, Ortwinus.

Bulaeus, C. E., *Historia Universitatis Parisiensis* (Paris, 1666).

Butterfield, H., *The Origins of Modern Science 1300-1800* (London, 1956).

Callus, D. A., O.P., 'Philip the Chancellor and the *De Anima* ascribed to Robert Grosseteste', in *Med. Renaiss. Stud.* 1 (1941), 105-27.

— 'Introduction of Aristotelian Learning to Oxford', in *Proc. Brit. Acad.* 29 (1943), 229-81.

— 'The *Summa Duacensis* and the Pseudo-Grosseteste's *De Anima*', in *Rech. Théol. anc. méd.* 13 (1946), 225-9.

— 'The Date of Grosseteste's Translations and Commentaries on Pseudo-Dionysius and the *Nicomachean Ethics*', in *Rech. Théol. anc. méd.* 14 (1947), 186-210.

— 'The Oxford Career of Robert Grosseteste', in *Oxoniensia* 10 (1945), 42-72.

— 'The *Summa Theologiae* of Robert Grosseteste', in *Studies in Medieval History presented to F. M. Powicke*, pp. 180-208.

— 'The Contribution to the study of the Fathers made by the thirteenth-century Oxford Schools', in *Journ. Eccles. Hist.* 5 (1954), 139-48.

— (ed.) *Robert Grosseteste, Scholar and Bishop* (Oxford, 1955).

— 'Robert Grosseteste as Scholar', in *Scholar and Bishop* (ed. Callus), pp. 1-69.

Cassirer, E., *Individuum und Kosmos in der Philosophie der Renaissance* (Studien der Bibl. Warburg, H. 10) (Leipzig-Berlin, 1927), 458 pp.

Ceccherelli, I., O.F.M., 'Roberto Grossatesta studioso di greco, e una cosiddetta sua introduzione grammaticale allo studio della lingua greca', in *Studi Franc.* 52 (1955), 426-44.

Chambers, R. W., *Thomas More* (London, 1935) (repr. Peregrine Bks, 1963), 413 pp.

Chartularium Universitatis Parisiensis, ed. Denifle and Chatelain, t.i (Paris, 1889), 708 pp.

Cheney, C. R., *English Synodalia of the Thirteenth Century* (Oxford, 1968), 164 pp. (See also under 'Powicke').

Chenu, M.-D., O.P., 'Le dernier avatar de la théologie orientale en Occident au XIIIe siècle', in *Mélanges Pelzer* (Louvain 1947), pp. 159-81.

— *Introduction à l'étude de saint Thomas d'Aquin* (Montreal, 1950) (3me éd., 1974).

— 'Nature ou histoire? Une controverse exégétique sur la création au XIIe siècle', in *Arch. Hist. Doctr. Litt. M.A.* 20 (1953), 25-30.

— *La Théologie au XIIe siècle* (Études de phil. méd. 45) (Paris, 1957).

— *La Théologie comme science au XIIIe siècle* (Bibl. thomiste 33) (3me Paris, 1957), 110 pp.

— 'L'homme, la nature, l'esprit. Un avatar de la philosophie grecque en Occident au XIIIe siècle', in *Arch. Hist. Doctr. Litt. M.A.* (1969), 123-30.

Chevallier, U., see *Dionysiaca. Recueil . . .*

Clagett, M., 'The *Quadratura per Lunulas*. A thirteenth-century fragment of Simplicius' commentary on the *Physics* of Aristotle', in *Essays in Medieval Life and Thought presented in honor of A. P. Evans* (New York, 1955), pp. 99-108.

Collin-Roset, S., 'Le Liber thesauri occulti de Paschalis Romanus', in *Arch. Hist. Doctr. Litt. M.A.* 30 (1963), 111-98.

Colomer, E., 'Individuo y cosmos en Nicolás de Cusa', in *Nicolás de Cusa en el V Centenario de su morte (1464-1964*, vol. i (Madrid, 1967), pp. 67-88.

Congar, Y. M-J., O.P., 'Aspects ecclésiologiques de la querelle entre mendiants et séculiers dans la seconde moitié du XIIIe siècle et le début du XIVe', in *Arch. Hist. Doctr. Litt. M.A.* 28 (1961), 35-151.

— 'Les laïcs et l'ecclésiologie des '*ordines*' chez les théologiens des XIe et XIIe siècles', in *I laici nella 'Societas christiana' dei secoli XI et XII* (Milan, 1968).

— 'Two factors in the sacralisation of Western Society during the Middle Ages', in *Concilium* 5 (1969), 28-35.

Creighton, M., *Historical Lectures and Addresses . . .* ed. Creighton, L. (London, 1903), 346 pp.

Crombie, A. C., 'Grosseteste and scientific method', in *The Month* 191 (1951), 164-74.

— *Robert Grosseteste and the Origins of Experimental Science, 1100-1700* (Oxford, 1953), 369 pp; new ed. (1962), 371 pp.

— 'Grosseteste's Position in the History of Science', in *Scholar and Bishop*, (ed. Callus), pp. 98-120.

— *Augustine to Galilio*, 2nd ed., vol. i: *Science in the Middle Ages, V-XIII Centuries*. Vol. 2: *Science in the Later Middle Ages and Early Modern Times, XIII-XVII Centuries* (London, 1961), 296 and 380 pp.

Crowley, T., O.F.M., *Roger Bacon: The Problem of the Soul in his Philosophical Commentaries* (Louvain-Dublin, 1950), 223 pp.

Dales, R. C., 'Robert Grosseteste's *Commentarius in Octo Libros Physicorum Aristotelis*', in *Med. Human.* 11 (1957), 10-33.

— 'Robert Grosseteste's scientific works', in *Isis* 52 (1961), 381-402.

— 'The authorship of the *Questio de Fluxu et Refluxu Maris* attributed to Robert Grosseteste', in *Speculum* 37 (1962), 582-8.

— 'A note on Robert Grosseteste's *Hexaemeron*', in *Med. Human.* 15 (1963), 69-73.

— 'Robert Grosseteste's treatise *De Finitate Motus et Temporis*', in *Traditio* 19 (1963), 245-66.

— (ed.) *Roberti Grosseteste, Episcopi Lincolniensis, Commentarius in VIII Libros Physicorum Aristotelis*; e fontibus manu scriptis nunc primum in lucem edidit R. C. Dales (Studies and Texts in Medieval Thought) (Boulder, Colorado, 1963), 192 pp.

— 'The authorship of the *Summa in Physica* attributed to Robert Grosseteste', in *Isis* 55 (1964), 70-4.

— 'The text of Robert Grosseteste's *Questio de Fluxu et Refluxu Maris* with an English Translation', in *Isis* 57 (1966), 455-74.

— 'Robert Grosseteste's views on astrology', in *Med. Stud.* 29 (1967), 357-63.

—, and Gieben, S., 'The *Prooemium* to Robert Grosseteste's *Hexaemeron*', in *Speculum* 43 (1968), 451-61.

— 'The influence of Grosseteste's *Hexaemeron* on the *Sentences* commentaries of Richard Fishacre, O.P. and Richard Rufus of Cornwall, O.F.M.', in *Viator*. *Medieval and Renaissance Studies* 2 (1971), 271-300.

— 'A medieval view of human dignity', in *Journal of the History of Ideas* 47 (1977), 557-72.

— 'Adam Marsh, Robert Grosseteste, and the treatise on the tides', in *Speculum* 52 (1977), 900-01.

D'Alverny, M.-T., 'Le cosmos symbolique du XIIe siècle', in *Arch. Hist. Doctr. Litt. M.A.* (1953), 69-81.

Dawson, J. G., 'Necessity and Contingency in the *De Libero Arbitrio* of Grosseteste', in *La filosofia della natura nel Medioevo*, 357-62.

De Bruyne, E., *Études d'esthétique médiévale*. t. 3: Le XIIIe siècle (Bruges, 1946), 400 pp.; see pp. 72-100, 'Grosseteste et l'esthétique mathématique'.

— *L'Esthétique du moyen âge* (Louvain, 1947), 260 pp.

Delhaye, Ph., *Le Microcosmus de Godefroy de S. Victor*, t. 1, *Texte*; t. 2, *Étude théologique* (Lille-Gembloux, 1951), 295 and 324 pp.

De Wulf, M., *Histoire de la philosophie médiévale*. t. 2: *Le treizième siècle*, 6me (Louvain-Paris, 1936), 407 pp.

Dijksterhuis, E. J., *The Mechanization of the World-Picture* (Oxford, 1961).

Dionysiaca. Recueil donnant l'ensemble des traductions latines des ouvrages attribués à Denys de l'Aréopage, et synopse marquant la valeur de citations presque innombrables... (Paris-Bruges, 1937), 2 vols.

Dobson, E. J., '*Moralities on the Gospels*'. *A New Source of 'Ancrene Wisse'* (Oxford, 1975), 182 pp.

Dondaine, H.-F., *Le corpus dionysien de l'Université de Paris au XIIIe siècle* (Rome, 1953), 154 pp.

— 'L'objet et le "*medium*" de la vision béatifique chez les théologiens du XIIIe siècle', in *Rech. Théol. anc. méd.* 19 (1952), 60-130.

Dubabin, J., 'Robert Grosseteste as translator, transmitter and commentator: the *Nichomachean Ethics*', in *Traditio* 28 (1972), 460-77.

Duhem, P., *Le Système du monde. Histoire des doctrines cosmologiques de Platon à Copernic. T. 3: Deuxième partie. L'astronomie latine au moyen âge (suite)* (Paris,

1915); T. 5: *Troisième partie. La crue de l'aristotélisme (suite)* (Paris, 1917); 549 and 596 pp.

Eastwood, B. S., 'Robert Grosseteste's Theory of the Rainbow. A Chapter in the History of Non-experimental Science', in *Arch. internationales de l'histoire des sciences* 19 (1966), 313-32.

—— 'Grosseteste's 'Quantitative' Law of Refraction: A Chapter in the History of Non-experimental Science', in *Journal of the History of Ideas* 28 (1967), 403-14.

—— 'Medieval Empiricism. The Case of Robert Grosseteste's Optics', in *Speculum* 43 (1968), 306-21.

Ebbesen, S., 'Jacobus Venetius on the *Posterior Analytics* and some early thirteenth-century Oxford Masters on the *Elenchi*', in *Cahiers de l'Institut du moyen-âge grec et latin* (Copenhagen, 1977), no. 21, 1-9.

Éliade, M., *Myth and Reality* (World Perspectives 21) (London, n.d.).

Felder, H., *Geschichte der wissenschaftlichen Studien im Franziskanerorden bis um die Mitte des 13. Jahrhunderts* (Freiburg i. Br., 1904), 557 pp.

Fellmann, F., *Scholastik und kosmologische Reform (BGPM, N.F. 6)* (Münster i. W., 1971), 70 pp.

Foster, M. B., 'The Christian doctrine of the creation and the rise of modern natural science', in *Mind* N.S. 43 (1934), 446 ff.

—— 'Christian theology and modern science of nature', in *Mind* N.S. 44 (1935), 439 ff., and 45 (1936), 1 ff.

Franceschini, E., 'Grosseteste's Translation of the Πρόλογος and Σχόλια of Maximus to the Writings of the Pseudo-Dionysius Areopagita', in *Journ. Theol. Stud.* 34 (1933), 355-63.

—— *Roberto Grossatesta, vescovo di Lincoln, e le sue traduzioni latine* (Venice, 1933), 138 pp.

—— 'Intorno ad alcune opere di Roberto Grossatesta vescovo di Lincoln', in *Aevum* 8 (1934), 529-42.

—— 'La revisione Moerbekana della *'Translatio Lincolniensis'* dell'Etica Nicomachea', in *Riv. Fil. Neo-scol.* 30 (1938), 150-62.

—— 'Un inedito di Roberto Grossatesta: la *Quaestio de Accessu et Recessu Maris'*, in *Riv. Fil. Neo-scol.* 44 (1952), 11-21.

— 'Sulla presunta datazione del *De Impressionibus Aeris* di Roberto Grossatesta', in *Riv. Fil. Neo-scol.* 44 (1952), 22-3.

Friedmann, L. M., *Robert Grosseteste and the Jews* (Cambridge Mass., 1934), 34 pp.

Gamba, U., (ed.) *Il commento di Roberto Grossatesta al 'De Mystica Theologia' del Pseudo-Dionigi Areopagita*, Orbis Romana. Bibl. di testi med. 15) (Milan, 1942), 69 pp.

— 'Roberto Grossatesta traduttore e commentatore del *De Mystica Theologia* ' dello Pseudo-Dionigi Areopagita', in *Aevum* 18 (1944), 100-32.

Gatard, A., 'Grosseteste, Robert', in *DTC* t. VI (Paris 1920, col. 1885-87).

Gieben, Servus (de St. Anthonis, O.F.M. Cap.), 'Denys l'Aréopagite. B. Au 13e siècle. 1. Robert Grosseteste, 1168-1253', in *DS* t. III (Paris, 1957), 340-3.

— 'Robert Grosseteste and the Immaculate Conception. With the text of the Sermon, *Tota Pulchra Es*', in *Coll. Franc.* 28 (1958), 211-27.

— 'Le potenze naturali dell'anima umana secondo alcuni testi inediti di Roberto Grossatesta', in *L'Homme et son destin*, pp. 437-43.

— 'Das Abkürzungszeichen Φ des Robert Grosseteste: *Quomodo Philosophia Accipienda Sit a Nobis*', in *Die Metaphysik im Mittelalter*, pp. 522-34.

— 'The pseudo-Bonaventurian work, *Symbolica Theologia*', in *Miscellanea Melchor de Pobladura* (Romae, 1964), pp. 173-95.

— 'Traces of God in Nature according to Robert Grosseteste. With the text of the Dictum, *Omnis Creatura Speculum Est*', in *Franc. Stud.* 24 (1964), 144-58.

— 'Das Licht als Entelechie bei Robert Grosseteste', in *La filosofia della natura nel Medioevo*, 372-78.

— 'Robert Grosseteste on Preaching. With the Edition of the Sermon *Ex Rerum Initiatarum* on Redemption', in *Coll. Franc.* 37 (1967), 100-41.

— 'Robert Grosseteste and Medieval Courtesy-Books', in *Vivarium* 5 (1967), 47-74.

— and Dales, R. C., 'The *prooemium* to Robert Grosseteste's *Hexaemeron*', in *Speculum* 43 (1968), 451-61.

— 'Thomas Gascoigne and Robert Grosseteste: historical and critical notes', in *Vivarium* 8 (1970), 56-67.

— 'Robert Grosseteste at the papal curia, Lyons 1250. Edition of the documents', in *Coll. Franc.* 41 (1971), 340-93.

— see Bérubé, C.

Gilson, É., 'Pourquoi S. Thomas a critiqué S. Augustin', in *Arch. Hist. Doctr. Litt. M.A.* 1 (1926-7), 5-127.

— *History of Christian Philosophy in the Middle Ages* (New York, 1955), 829 pp.

Goldhammer, K., 'Lichtsymbolik in philosophischer Weltanschauung, Mystik und Theosophie vom 15. bis zum 17. Jahrhundert', in *Studium Generale* 11 (1960), 670-82.

Grabmann, M., *Mittelalterliches Geistesleben*, 3 Bde. (Munich, 1926, 1936, 1956).

Gratius, Ortwinus, *Fasciculus rerum expetendarum ac fugiendarum, prout ab O.G. editus est Coloniae, A.D. 1535 . . . ab innumeris mendis repurgatus . . . una cum appendice . . . Scriptorum . . . qui Ecclesiae Romanae errores et abusus detegunt et damnant, necessitatemque Reformationis urgent.* Opera et studio Edwardi Brown (Londini, R. Chiswell, 1690), In-fol., 2 vol.

Gregory, T., 'Note sulla dottrina delle "Teofanie" in Giovanni Scoto Eriugena', in *Studi Med.* ser III, 4 (1963), 75-91.

— *Anima Mundi. La filosofia di Guglielmo di Conches e la scuola di Chartres* (Florence, 1955), 297 pp.

— *Giovanni Scoto Erigena* (Florence, 1963), 84 pp.

Guidubaldi, E., *Dal 'De Luce' di R. Grossatesta all'islamico 'Libro della scala'. Il problema delle fonti arabe una volta accettata la mediazione oxfordiana* (Florence, 1978), 273 pp.

Haskins, C. H., *Studies in the History of Medieval Science*, 2nd. ed. (Cambridge Mass., 1927), 411 pp.

Hedwig, K., 'Literaturbericht: neuere Arbeiten zur mittelalterlichen Lichttheorie', in *Zeitschrift für philosophische Forschung* 33 (1979), 602-14.

— *Sphaera Lucis. Studien zur Intelligibilität des Seienden im Kontext der mittelalterlichen Lichtspekulation (BGPM N.F. 18)* (Münster i. W., 1980), 299 pp. Ch. 5: 'Robert Grosseteste: *Sphaera Lucis*', pp. 119-56.

Henquinet, F., 'Un recueil de questions annoté par s. Bonaventure', in *Arch. Franc. Hist.* 25 (1932), p. 552.

Hill, J. W. F., 'The tomb of Robert Grosseteste with an account of its opening in 1782', in Callus (ed.), *Scholar and Bishop*, pp. 246–50.

Hissette, R., *Enquête sur les 219 articles condamnés à Paris le 7 mars 1277* (Philosophes médiévaux XXII) (Louvain-Paris, 1977), 337 pp.

Hunt, R. W., 'Notable accessions. Manuscripts', in *Bodl. Libr. Rec.* 2 (1941–9), 226–7.

— 'Manuscripts containing the indexing symbols of Robert Grosseteste', in *Bodl. Libr. Rec.* 4 (1952–3), 241–55.

— 'The library of Robert Grosseteste', in *Scholar and Bishop* (ed. Callus), 121–45.

— 'Verses on the life of Grosseteste', in *Med. Human.* N.S. I (1970), 241–51.

Ivánka, E. von, 'Zur Überwindung des neuplatonischen Intellektualismus in der Deutung der Mystik. *Intelligentia* oder *Principalis Affectio*', in *Platonismus in der Philosophie des Mittelalters*, ed. Beierwaltes, pp. 147–60.

St. John Damascene, *Dialectica*. Version of Robert Grosseteste, ed. O. A. Colligan, O.F.M. (Franciscan Institute Publications, Text series 6) (St Bonaventure, N.Y.-Louvain 1953), 63 pp.

Jourdain, C., 'Doutes sur l'authenticité de quelques écrits contre la cour de Rome attribués à Robert Grosse-Tête, évêque de Lincoln', in *Comptes-rendus des séances de l'Académie des Inscriptions et Belles-Lettres*, B. IV (Paris, 1868), 13–29; also publ. separately, (Paris, 1868), 16 pp; repr. in the author's *Excursions historiques et philosophiques à travers le Moyen Âge* (Paris, 1888), pp. 149–71.

Juschkowitsch, A. P., *Geschichte der Mathematik im Mittelalter* (Leipzig, 1964).

Keeler, L. W., S.J., 'The Dependence of Robert Grosseteste's *De Anima* on the *Summa* of Philip the Chancellor', in *New Schol.* 11 (1937), 197–219.

— 'The alleged Revision of Robert Grosseteste's Translation of the *Ethics*', in *Gregorianum* 18 (1937), 410–25.

Klibansky, R., *The Continuity of the Platonic Tradition during the Middle Ages*. Outlines of a *Corpus Platonicum Medii Aevi*, by Raymond Klibansky. With extracts from unpublished commentaries on Plato (London, 1937), 31 pp., 3 pl.

Koch, J., 'Augustinischer und Dionysischer Neuplatonismus und das Mittelalter', in *Kant-Studien* 48 (1956-7), 117-33; repr. in *Platonismus in der Philosophie des Mittelalters*, ed. Beierwaltes, pp. 317-42.

Koyré, A., 'Le vide et l'espace infini au XIVe siècle', in *Arch. Hist. Doctr. Litt. M.A.* 17 (1949), 45-91.

— 'Die Ursprünge der modernen Wissenschaft: ein neuer Deutungsversuch', in *Diogenes* 4 (1957), 421-48.

— *From the Closed World to the Infinite Universe* (Baltimore, 1957), 313 pp.

Kranz, W., *Kosmos* (Bonn, 1958).

Kristeller, P. O., *Eight Philosophers of the Italian Renaissance* (Stanford, 1964).

Kuhn, T. S., *The Copernican Revolution*. Planetary Astronomy in the Development of Western Thought (Cambridge Mass., 1957), 297 pp.

Kurdzialek, M., 'Der Mensch als Abbild des Kosmos', in *Miscell. Med.* 8 (1971), 35-75.

Legrand, J., S.J., *L'Univers et l'homme dans la philosophie de S. Thomas,* 2 vols. (Brussels, 1946).

Lértora Mendoza, C. A., 'Los comentarios de Santo Tomas y de Roberto Grossatesta a la Física de Aristóteles', in *Sapientia* 25 (1970), 179-208; 257-88.

— 'La *Summa Physicorum* y la filosofía natural de Grosseteste', in *Sapientia* 26 (1971), 199-216.

— and Bolzán, J. E., 'La *Summa Physicorum* atribuida a Roberto Grosseteste', in *Sapientia* 26 (1971), 21-74 (the text of the *Summa* is printed with a Spanish transl. on pp. 24-73).

— *Roberto Grosseteste, Suma de los ocho libros de la Física de Aristóteles (Summa Physicorum).* Texto latino, traducción y notas de J. E. Bolzán y C. Lértora Mendoza (Col. 'Los Fundamentales') (Buenos Aires, 1972), 150 pp.

Lewis, C. S., *The Discarded Image* (Cambridge, 1964), 232 pp.

Little, A. G., *The Grey Friars in Oxford. Part I: A History of the Convent. Part II: Biographical Notices of the the Friars.* Together with Appendices of Original Documents (Oxford Hist. Society XX) (Oxford 1891-2), 369 pp.

— *Studies in English Franciscan History* (Ford Lectures, 1916) (Manchester-London, 1917), 248 pp.

— 'Roger Bacon. Annual Lecture on a Master Mind', *Proc. Brit. Acad.* XIV; publ. separately by Oxford U.P., n.d.

— 'The Franciscan School at Oxford in the Thirteenth Century', in *Arch. Franc. Hist.* 19 (1926), 803-74.

— *Franciscan Papers, Lists and Documents* (Publ. of the Univ. of Manchester. Hist. Series 81) (Manchester, 1943), 262 pp.

— (ed.) *Tractatus Fr. Thomae vulgo dicti de Eccleston de adventu Fratrum Minorum in Angliam* ... (Coll. d'Etudes et de Documents VII) (Paris, 1909), 227 pp.

— (ed.) *Fratris Thomae vulgo dicti de Eccleston tractatus de adventu Fratrum Minorum in Angliam; denuo ed. A. G. Little* (Manchester, 1951), 115 pp.

Loewe, R., 'The Mediaeval Christian Hebraists of England. The *Superscriptio Lincolniensis*', in *Hebrew Union College Annual* (Cincinnati) 28 (1957), 205-52.

Longpré, E., O.F.M. Cap., 'Robert Grossetête et l'Immaculée Conception', in *Arch. Franc. Hist.* 26 (1933), 550-1.

Luard, H. R., (ed.) *Roberti Grosseteste Episcopi quondam Lincolniensis Epistolae* . . . (Rerum Brit. med. aevi Script.) (London, 1861), 467 pp. (reprinted New York, 1965).

Lynch, L. E., 'The doctrine of divine ideas and illumination in Robert Grosseteste, Bishop of Lincoln', in *Med. Stud.* 3 (1941), 163-73.

Maier, A., *Zwei Grundprobleme der scholastischen Naturphilosophie* (2nd ed., Rome, 1951), 318 pp.

— *Metaphysische Hintergründe der spätscholastischen Naturphilosophie* (Rome, 1955), 405 pp.

— *Zwischen Philosophie und Mechanik.* Studien zur Naturphilosophie der Spätscholastik (Rome, 1958), 385 pp.

— *Ausgehendes Mittelalter. Gesammelte Aufsätze* ... (Rome, 1964), 508 pp.

Major, K., 'The *Familia* of Robert Grosseteste', in *Scholar and Bishop* (ed. Callus), 216-41.

Mansion, A., 'Quelques travaux récents sur les versions latines des *Éthiques* et d'autres ouvrages d'Aristote', in *Rev. Néoscol. Phil.* 39 (1936), 78-94.

— 'La version médiévale de l'*Éthique à Nicomaque.* La *translatio Lincolniensis* et la controverse autour de la révision attribuée à Guillaume de Moerbeke., in *Rev. Néoscol. Phil.* 41 (1938), 401-27.

Mantello, F. A. C., 'Letter CXXX of Bishop Robert Grosseteste: a problem of attribution', in *Med. Stud.* 36(1974), 144-59.

Mazzeo, J. A., 'Light metaphysics, Dante's *Convivio* and the letter to Can Grande Della Scala', in *Traditio* 14 (1958), 191-229.

McEvoy, J., 'The sun as *res* and *signum*: Grosseteste's Commentary on Ecclesiasticus ch. 43, vv. 1-5', in *Rech. Théol. anc. méd.* 41 (1974), 38-91.

—— 'Microcosm and macrocosm in the writings of St. Bonaventure', in *S. Bonaventura*, vol. II (Rome, 1974), pp. 309-43.

—— 'Robert Grosseteste and the reunion of the Church', in *Coll. Franc.* 45 (1975), 39-84.

—— 'Medieval cosmology and modern science', in *Philosophy and Totality*, ed. J. McEvoy, (Belfast, 1977), pp. 91-110.

—— 'La connaissance intellectuelle selon Robert Grosseteste', in *La Revue philosophique de Louvain* 75 (1977), 5-48.

—— 'The metaphysics of light in the Middle Ages', in *Philosophical Studies* (Dublin) 26 (1979), 124-43.

—— 'The correspondence between Grosseteste's quotations from the Latin translation of the *Physics* and the Greek text', in *Bulletin de philosophie médiévale* 21 (1979), 52-62.

—— 'The absolute predestination of Christ in the theology of Robert Grosseteste', in *Mélanges Bascour* (Leuven, 1980), pp. 212-30 (see under 'Collective Works').

—— 'Robert Grosseteste's Theory of Human Nature. With the Text of His Conference, *Ecclesia Sancta Celebrat*', in *Rech. Théol. anc. méd.* 47 (1980), 131-87.

—— 'Language, tongue and thought in the writings of Robert Grosseteste', in *Sprache und Erkenntnis im Mittelalter. Miscellanea Mediaevalia* 13/2 (Berlin, 1981), pp. 585-92.

—— 'The influence of Aristotle at Oxford 1200-1250', in *World Congress on Aristotle* (Thessalonica 1978) (forthcoming).

—— 'Der Brief des Robert Grosseteste an Magister Adam Rufus (Adam von Oxford, OFM): ein Datierungsversuch', in *Franziskanische Studien* 63 (1981).

—— 'Questions of authenticity and chronology concerning works of Grosseteste edited 1940-1980', in *Bulletin de*

philosophie médiévale, part 1: vol. 23 (1981); part 2: vol. 24 (1982).
— Art. 'Robert Grosseteste', in *Theologische Realenzyklopädie*, hrsgeg. von G. Krause und G. Müller, Band 13 (Berlin-New York, 1984).
— 'The Chronology of Robert Grosseteste's Writings on Nature and Natural Philosophy', in *Speculum* (forthcoming).
Mercken, H. P. F., *Aristoteles over de menselijke volkomenheid.* Boeken I en II van de Nicomachische Etiek met de commentaren van Eustratius en een Anonymus in de Latijnse vertaling van Grosseteste (Verhandeling van de Koninkl. Vlaamse Acad. voor Wetenschappen, Letteren en Schone Kunsten . . . J.G. XXVI, 53) (Brussel, 1964), 72+210 pp.
— *The Greek Commentaries on the Nicomachean Ethics of Aristotle in the Latin translation of Robert Grosseteste* (Corpus lat. Commentariorum in Aristotelem graecorum VI, 1), vol. I: Eustratius on bk I and the Anon. Scholia on bks II, III and IV (Leyden, 1973), 134*+371 pp.
Miano, V., 'La teoria della conoscenza in Roberto Grossatesta (n. c. 1170, m. 1253)', in *Giorn. Metaf.* 9 (1954), 60–88.
Michaud-Quantin, P., 'La notion de loi naturelle chez Robert Grosseteste', in *Actes du XIᵉ Congrès Intern. de Philosophie*, vol. 12 (Amsterdam, 1953), pp. 166–70.
Minio-Paluello, L., 'Note sull'Aristotele latino medievale I–III', in *Riv. Fil. Neo-scol.* 42 (1950), 222–37.
Mohl, R., *The Three Estates in Medieval and Renaissance Literature* (New York, 1933), 425 pp.
Moorman, J. R. H., *Church Life in England in the Thirteenth Century* (Cambridge, 1945), 444 pp. (repr. with corrections, 1955).
Moraux, P., '*Quinta essentia*', in Pauly-Wissowa, *Real-Enzyklopädie der Klass. Altertumswissenschaft*, Bd. 24 (Stuttgart, 1963), cols. 1171–1263.
Morgan, M. M., 'The Excommunication of Grosseteste in 1243', in *Engl. Hist. Rev.* 57 (1942), 244–50.
Muckle, J. T., 'The *Hexameron* of Robert Grosseteste. The first twelve chapters of Part Seven', in *Med. Stud.* 6 (1944), 151–74.
— 'Robert Grosseteste's use of Greek sources in his *Hexameron*', in *Med. Human* 3 (1945), 33–48.

—— 'Did Robert Grosseteste attribute the *Hexameron* of St
Venerable Bede to St Jerome?', in *Med. Stud.* 13 (1951),
242-4.

Murray, J., *Le 'Château d'Amour' de Robert Grosseteste,
évêque de Lincoln* (Paris, 1918), 183 pp.

Nolan, E. and Hirsh, S. A., *The Greek Grammar of Roger
Bacon* (Cambridge, 1902).

Oschinsky, D., *Walter of Henley and other treatises on estate
management and accounting* (Oxford, 1971), 504 pp.

Palma, R. J., 'Grosseteste's ordering of *scientia*', in *New
Schol.* 50 (1976), 447-63.

Pantin, W. A., 'Grosseteste's relations with the Papacy and
the Crown', in *Scholar and Bishop* (ed. Callus), pp. 178-
215.

Parent, J. M., *La doctrine de la création dans l'école de Chartres* (Paris, 1938).

Pelster, F., 'Zwei unbekannte philosophische Traktate des
Robert Grosseteste', in *Scholastik* 1 (1926), 572-3.

Pelzer, A., 'Les versions latines des ouvrages de morale con-
servés sous le nom d'Aristote en usage au XIIIe siècle',
in *Rev. Néoscol. Phil.* 23 (1921) 316-41, 378-412; repr.
in Pelzer, *Études* . . . (Louvain, 1964), pp. 120-87.

Phelan, G. B., 'An unedited text of Robert Grosseteste on the
subject-matter of theology', in *Rev. Néoscol. Phil.* 36
(1934), (= *Hommage à M. De Wulf*), pp. 172-9.

Pouillon, H., O.S.B., 'La Beauté, propriété transcendentale,
chez les Scholastiques (1220-1270)', in *Arch. Hist.
Doctr. Litt. M.A.* 15 (1946), 263-329.

—— 'Grosseteste's Contribution to the History of Philosophy',
in *Proc. Am. Cath. Phil. Assoc.* 27 (1953), 142-4.

Powicke, F. M., 'Robert Grosseteste and the *Nicomachean
Ethics*', from *Proc. Brit. Acad.* (London, 1930), 22 pp.

—— 'Robert Grosseteste, Bishop of Lincoln', in *Bull. J. Ryl.
Libr.* 35 (1953), 482-507.

—— 'Introduction', in *Scholar and Bishop* (ed. Callus), pp.
XIII-XXV.

Powicke, F. M. and Cheney, C. R., *Councils and Synods with
other Documents relating to the English Church.* vol. II,
part 1. A.D. 1205-1313 (Oxford, 1964).

Ratzinger, J., *Die Geschichtstheologie des hl. Bonaventura*
(Munich, 1957), 166 pp.

Reichl, K., *Tractatus de Grammatica. Ein fälschlich Robert Grosseteste zugeschriebene spekulative Grammatik. Edition und Kommentar* (Veröffentlichungen des Grabmann Instituts, N.F. 28) (Munich, 1976), 224 pp.

Robert Grosseteste on Light (De Luce). Trans. from the Latin, with an introduction by C. C. Riedl (Med. Phil. Texts in Trans., 1) (Milwaukee, 1942), 17 pp.

— *On Light.* Trs. C. F. Terrell, Paideuma 2, no. 3 (1973), 455-62.

— *Aristoteles Latinus*, vol. XXIV, fasc. tertius: *Ethica Nicomachea, translatio Roberti Grosseteste Lincolniensis sive 'Liber Ethicorum'. A. Recensio pura*, ed. R. A. Gauthier (Leyden-Bruxelles, 1972), pp. 141-370.

— *Habes accuratissime lector Aristotelis posteriorum opus ac eius luculentissimum interpretem linconiensem burleumque . . .* (Venetiis, per Gregoriorum de Gregorijs, 1514), 51 fols. (Nachdruck Frankfurt/Mainz, 1966).

— *Commentarius in Posteriorum Analyticorum Libros.* Introduzione e testo critico di Pietro Rossi (Unione Academica Nazionale, *Corpus Philosophorum Medii Aevi*, Testi E Studi II), Florence 1981, 412 pp.

Roques, R., 'La notion de Hiérarchie selon le Pseudo-Denys', in *Arch. Hist. Doctr. Litt. M.A.* (1949), 183-222; (1951), 5-44.

— *L'Univers dionysien. Structure hiérarchique du monde selon le Pseudo-Denys* (Paris, 1954).

Rossi, P., 'Per l'edizione del *Commentarius in Posteriorum Analyticorum Libros* di Roberto Grossatesta', in *Riv. Fil. Neo-scol.* 67 (1975), 489-515.

— 'Tracce della versione latino di un commento greco ai *Secondi Analitici* nel *Commentarius in Posteriorum Analyticorum Libros* di Roberto Grossatesta', in *Riv. Fil. Neo-scol.* 70 (1978), 433-9.

Ruello, F., 'Étude du terme Ἀγαθοδότις dans quelques commentaires médiévaux des *Noms divins*', in *Rech. Théol. anc. méd.* 24 (1957), 225-66; 25 (1958) 5-25.

— 'La *Divinorum Nominum Reseratio* selon Robert Grosseteste et Albert le Grand', in *Arch. Hist. Doctr. Litt. M.A.* 26 (1959), 99-197.

— *Les 'noms divins' et leurs 'raisons' selon saint Albert le Grand commentateur du 'De divinis nominibus'* (Bibl.

thomiste 35) (Paris, 1963), 242 pp.; pp. 155-73, 'La raison de nom divin selon Robert Grosseteste'.

Russell, J. C., 'Hereford and Arabic Science in England about 1175-1200', in *Isis* 18 (1932), 14-25.

—— 'The preferments and '*Adiutores*' of Robert Grosseteste', in *Harv. Theol. Rev.* 26 (1933), 161-72.

—— *Dictionary of Writers of Thirteenth Century England.* (Special suppl no. 3 to Bull. of the Inst. of Hist. Research) (London-New York-Toronto, 1936), 210 pp.

—— 'Richard of Bardney's Account of Robert Grosseteste's Early and Middle Life', in *Med. Human* 2 (1944), 45-54.

—— 'Phases of Grosseteste's Intellectual Life', in *Harv. Theol. Rev.* 43 (1950), 93-116.

—— 'Some Notes upon the Career of Robert Grosseteste', in *Harv. Theol. Rev.* 48 (1955), 199-211.

Saccaro, G. B., 'Il Grossatesta e la luce', in *Medioevo* 2 (1976), 21-75.

Schaefer, A., 'The position and function of man in the created world according to St Bonaventure', in *Franc. Stud.* 21 (1961), 233-382; see under same title, Cath. Univ. Studies in Theology 154 (Washington D.C., 1965), 67 pp.

Shipperges, H., 'Einflüsse arabischer Medizin auf die Mikrokosmosliteratur des 12. Jahrhunderts', in *Miscell. Med.* 1 (1962), 129-53.

Serene, E. M., 'Robert Grosseteste on induction and demonstrative science', in *Synthesis* 40 (1979), 97-115.

Sharp, D. E., *Franciscan Philosophy at Oxford in the Thirteenth Century* (Brit. Society of Franciscan Stud. 16) (London, 1930), 419 pp; (reprinted Farnborough, 1966).

Smalley, B., *The Study of the Bible in the Middle Ages*, 2nd ed. (Oxford, 1952), 406 pp.

—— 'The Biblical Scholar', in *Scholar and Bishop* (ed. Callus), pp. 70-97.

Smith, A. L., *Church and State in the Middle Ages* (Ford lectures 1905) (Oxford, 1913).

Smyly, J. G., (ed.), *Urbanus Magnus Danielis Becclesiensis* (Dublin, 1939), 102 pp.

Stadter, E., 'Die Seele als *minor mundus* und als *regnum.* Ein Beitrag zur Psychologie der mittleren Franziskanerschule', in *Miscell. Med.* 5 (1968), 56-72.

Stevenson, F. S., *Robert Grosseteste, Bishop of Lincoln. A*

Contribution to the Religious, Political and Intellectual History of the Thirteenth Century (London, 1899), 348 pp; (reprinted Dubuque, Iowa, 1962).

Stinissen, W., (ed.) *Aristoteles over de vriendschap*. Boeken VIII en IX van de Nicomachische Ethiek met de commentaren van Aspasius en Michaël in de Latijnse vertaling van Grosseteste (Verhandelingen van de Koninkl. Vlaamse Acad. voor Wetenschappen, Letteren en Schone Kunsten van België, Kl. der Lett. XXV, No. 45) (Brussels, 1963), 183 pp.

Strawley, J. H., *Robert Grosseteste, Bishop of Lincoln 1235-1253* (Lincoln Minster pamphlets, 7) (Lincoln, 1953), 36 pp.

— 'Grosseteste's Administration of the Diocese of Lincoln', in *Scholar and Bishop* (ed. Callus), 146-77.

Tavard, G., 'Die Engel', *Handbuch der Dogmengeschichte*, Bd. II, Fasc. 2b, hrsg. von Schmaus, Grillmeier und Scheffczyk (Freiburg-Basel-Wien, 1968), 96 pp.

Taylor, E. G., 'Robert Grosseteste as an observer', in *Isis* 55 (1964), p. 342.

Theiler, W., 'Die Sprache des Geistes in der Antike', in *Forschungen zum Neuplatonismus* (Quellen und Studien zur Geschichte der Phil. X) (Berlin, 1966), pp. 81-7.

Théry, G., 'L'entrée du Ps.-Denys en occident', in *Mélanges Mandonnet*, t. II (Paris, 1930), pp. 23-30.

Thomson, S. H., 'The *De Anima* of Robert Grosseteste', in *New Schol.* 7 (1933), 201-21.

— 'A Note on Grosseteste's work of Translation', in *Journ. Theol. Stud.* 34 (1933), 48-52.

— 'The text of Grosseteste's *De Cometis*', in *Isis* 19 (1933), 19-25.

— 'The *Summa in VIII libros Physicorum* of Grosseteste', in *Isis* 22 (1934), 12-18.

— 'The *Notule* of Robert Grosseteste on the *Nicomachean Ethics*', from *Proc. Brit. Acad.* (London, 1933), 195-218; printed separately, London, 1934, 26 pp.

— 'Grosseteste's Topical Concordance of the Bible and the Fathers', in *Speculum* 9 (1934), 139-44.

— *The Writings of Robert Grosseteste, Bishop of Lincoln, 1235-53* (Cambridge Mass., 1940), 302 pp.

— 'Two Early Portraits of Robert Grosseteste', in *Med. Human.* 8 (1954), 20-21.

— 'Grosseteste's concordantial signs', in *Med. Human.* 9 (1955), 39-53, 6 tab.

— 'Grosseteste *Quaestio de Calore, De Cometis,* and *De Operacionibus Solis*', in *Med. Human.* 11 (1957), 34-43.

— 'An unnoticed autograph of Grosseteste', in *Med. Human.* 14 (1962), 55-60, 1 tab.

Thorndike, L., *A History of Magic and Experimental Science during the first thirteen centuries of our era,* vol. II (New York, 1923), 1036 pp.

— 'Notes upon some medieval astronomical, astrological and mathematical MSS at Florence, Milan, Bologna, and Venice', in *Isis* 50 (1959), 33-50.

Thunberg, L., *Microcosm and Mediator. The Theological Anthropology of Maximus the Confessor* (Lund, 1965), 500 pp.

Tierney, B., 'Grosseteste and the Theory of Papal Sovereignty', in *Journ. Eccles. Hist.* 6 (1955), 1-17.

— 'Limits to obedience in the Thirteenth Century: the case of Robert Grosseteste', in *Contraception, Authority and Dissent,* ed. C. E. Curran (New York, 1969), pp. 76-100.

Tognolo, A., 'Il *De Artibus Liberalibus* di Roberto Grossatesta', in *Arts libéraux et philosophie au moyen-âge,* pp. 593-7.

Troilo, S., 'Due traduttori dell'*Etica Nicomachea:* Roberto di Lincoln e Leonardo Bruni', in *Atti del Reale Istituto Veneto di scienze, lettere ed arti,* an. acad. 1931-32, t. 91, 275-305.

Tropia, L., 'La versione latina medievale del *Peri Pathon* dello Pseudo-Andronico', in *Aevum* 26 (1952), 97-112.

Turbayne, C. M., 'Grosseteste and an Ancient Optical Principle', in *Isis* 50 (1959), 467-72, 6 fig.

Unger, D. J., O.F.M. Cap., 'Robert Grosseteste, Bishop of Lincoln (1235-1253), on the Reasons for the Incarnation', in *Franc. Stud.* 16 (1956), 1-36.

Van Steenberghen, F., *La Philosophie au XIIIᵉ siècle* (Philosophes médiévaux IX) (Louvain-Paris, 1966), 594 pp.

Vaughn, R., *Matthew Paris* (Cambridge, 1958), 288 pp., 21 pl.

Völker, W., *Kontemplation und Ekstase bei Pseudo-Dionysius Areopagita* (Wiesbaden, 1958), 263 pp.

Weisweiler, H., 'Die Dionysiuskommentare *In Caelestem Hierarchiam* des Skotus Eriugena und Hugos von St-Viktor', in *Rech. Théol. anc. méd.* 19 (1952), 26-47.

Wenzel, S., 'Robert Grosseteste's treatise on Confession, *Deus Est*', in *Franc. Stud.* 30 (1970), 218-93.

Westermann, E. J., 'A Comparison of some of the Sermons and the *Dicta* of Robert Grosseteste', in *Med. Human.* 3 (1945), 49-68.

Wharton, H., *Anglia Sacra, sive Collectio Historiarum antiquitus scriptarum de Archiepiscopis et Episcopis Angliae* . . . Pars secunda (Londini, Impendii Richardi Chiswel, 1691), 706 pp.

Wiersma, W., 'Die aristotelische Lehre vom *Pneuma*', in *Mnemosyne*, 3. F., 11 (1942), 102-7.

Zedler, B. H., 'Comment on Dom Pouillon's paper: Robert Grosseteste and the unity of man', in *Proc. Am. Cath. Phil. Ass.* 27 (1953), pp. 144-55.

Zellinger, E., *Cusanus-Kondordanz* (Munich, 1960), 331 pp.

Zöckler, O., 'Grosseteste, Robert', in *Realencyklopädie für protestantische Theologie und Kirche*, VII (Leipzig, 1899), 193-9.

FESTSCHRIFTEN, CONGRESSES, AND COLLECTIVE WORKS

Actes du XIe Congrès International de Philosphie. Bruxelles, 20-26 août 1953. *Vol. XII: Histoire de la philosophie: méthodologie, antiquité et Moyen-Âge* (Amsterdam, 1953).

Arts libéraux et philosophie au moyen-âge. Actes du Quatrième Congrès international de philosophie médiévale. Université de Montréal, Montréal, Canada, 27 aout-2 sept 1967 (Montréal-Paris, 1969).

Die Metaphysik im Mittelalter. Ihr Ursprung und ihre Bedeutung. Vorträge des II. internationalen Kongresses für mittelalterliche Philosophie, Köln, 31. August bis 6. September 1961 (*Miscellanea Mediaevalia*, 2) (Berlin, 1963).

Essays in Medieval Life and Thought, presented in honor of Austin Patterson Evans. Edited by John H. Mundy, Richard W. Emery, Benjamin N. Nelson (New York, 1955), 258 pp.

La filosofia della natura nel Medioevo. Atti del Terzo Congresso Internazionale di filosofia medioevale. Passo della Mendola (Trento), 31 agusto-5 settembre 1964. (Milan, 1966).

Le Soleil à la Renaissance. Colloque International, Université Libre de Bruxelles (Brussels, 1965).

L'homme et son destin d'après les penseurs du moyen-âge. Actes du premier congrès international de philosophie médiévale, Louvain-Bruxelles, 28 aout–4 septembre 1958 (Louvain-Paris, 1960).

Mélanges Mandonnet. Études d'histoire littéraire et doctrinale du moyen-âge (Bibliothèque thomiste 13, 14), 2 t. (Paris, 1930).

Mélanges A. Pelzer. Études d'histoire littéraire et doctrinale de la scolastique médiévale (Univ. de Louvain. Recueil de travaux d'histoire et de philologie. Sér. 3, 26) (Louvain, 1947), 662 pp.

Miscellanea Melchor de Pobladura. Studia Franciscana Historica P. Melchiori a Pobladura dedicata . . . ed. Isidorus a Villapadierna O.F.M. Cap. (Bibl. Seraphico-Capuccina 23, 24), 2 vols. (Rome, 1964), 487, 557 pp.

Philosophy and Totality. Lectures delivered under the auspices of the Department of Scholastic Philosophy. Edited by J. McEvoy (Belfast, 1977), 147 pp.

Platonismus in der Philosophie des Mittelalters. Hrsg. von Werner Beierwaltes (Darmstadt, 1969) (Wege der Forschung Bd. CXCVII), 534 pp.

Roger Bacon. Essays contributed by various writers on the occasion of the commemoration of the seventh centenary of his birth, collected and edited by A. G. Little (Oxford, 1914), 426 pp.

Sanctus Bonaventura 1274-1974. Volumen Commemorativum . . . *cura et studio Commissionis Internationalis Bonaventurianae*, 5 vols. (Rome, 1974).

'Sapientiae Doctrina'. Mélanges de théologie et de littérature médiévales offerts à Dom Hildebrand Bascour O.S.B. (Recherches de Théologie ancienne et médiévale, n. spécial 1) (Leuven, 1980).

Sprache und Erkenntnis im Mittelalter. VI. Internationaler Kongress für Mittelalterliche Philosophie, Bonn 28 August–3 September 1977 (*Miscellanea Mediaevalia* 13, 1/2) (Berlin, 1981).

Studies in Medieval History presented to Frederick Maurice Powicke. Edited by R. W. Hunt, W. A. Pantin, R. W. Southern (Oxford, 1948), 504 pp.

World Congress on Aristotle. Thessalonica, August 1978. 4 vols.
(Athens, 1981-84).

UNPUBLISHED THESES

Bossier, Ferdinand (Prof., Antwerp), Edition and critical study
of Simplicius's Commentary on the *De Caelo* in the Latin
versions of Grosseteste and Moerbeke (D. Phil., Leuven,
1975, in Dutch) (to be publ. in the series, Corp. lat.
Comm. in Aristotelem graec., Leuven).

Creek, M. I., *The Sources and Influence of Robert Grosseteste's 'Le Château d'Amour'* (Yale Univ. Ph. D., 1941);
see Diss. Abstr. 31 A (1970-1), 2339-40.

Eastwood, B. S., *The Geometrical Optics of Robert Grosseteste* (Ph. D., Wisconsin Univ., 1964); Univ. Microfilms
no. 64-10, 228.

Servus, A St.-Anthonis (Gieben), *De Metaphysica Lucis apud
Robertum Grosseteste* (Ph. D., Rome, Gregorian Univ.,
1953), 206+98 pp.

Holland, M., *An Edition of Three Unpublished Translations
by Robert Grosseteste of Three Short Works of John of
Damascus* (Ph. D., Harvard Univ., 1980).

King, Ed. B. (Prof., The Univ. of the South, Tenn.), *Robert
Grosseteste and the Pastoral Office* (Ph. D., Duke Univ.,
1969); Univ. microf. 70-2157.

Lee, A. M. *Robert Grosseteste's 'De Cessatione Legalium'.* A
crit. ed. from the extant MSS (Ph. D., Univ. of Colorado,
1942), 198 pp.

McEvoy, J., *Robert Grosseteste on the Celestial Hierarchy of
Pseudo-Dionysius.* An ed. and transl. of his commentary,
chs. 10-15 (M.A., Queen's Univ., Belfast, 1967), 124+
240 pp.

— *Man and Cosmos in the Philosophy of Robert Grosseteste*
(D. Phil., Louvain, 1974), 2 vols., 554 pp.

McQuade, J. S., *Robert Grosseteste's Commentary on the
'Celestial Hierarchy' of Pseudo-Dionysius the Areopagite:*
an ed., trans. and introd. of his text and commentary
(Ph. D., Queen's Univ., Belfast, 1961), 141+22+482 pp.

Palma, R., *Grosseteste's Concept of Truth* (Ph. D., Hope Coll.
Holland, Michigan, 1973).

Indexes

Index of Ancient and Medieval Authors

Index of Names: Modern Authors

Systematic Index